Smart Innovation, Systems and Technologies

Volume 167

The Smart Innovation, Systems and Technologies book series encompasses the topics of knowledge, intelligence, innovation and sustainability. The aim of the series is to make available a platform for the publication of books on all aspects of single and multi-disciplinary research on these themes in order to make the latest results available in a readily-accessible form. Volumes on interdisciplinary research combining two or more of these areas is particularly sought.

The series covers systems and paradigms that employ knowledge and intelligence in a broad sense. Its scope is systems having embedded knowledge and intelligence, which may be applied to the solution of world problems in industry, the environment and the community. It also focusses on the knowledge-transfer methodologies and innovation strategies employed to make this happen effectively. The combination of intelligent systems tools and a broad range of applications introduces a need for a synergy of disciplines from science, technology, business and the humanities. The series will include conference proceedings, edited collections, monographs, handbooks, reference books, and other relevant types of book in areas of science and technology where smart systems and technologies can offer innovative solutions.

High quality content is an essential feature for all book proposals accepted for the series. It is expected that editors of all accepted volumes will ensure that contributions are subjected to an appropriate level of reviewing process and adhere to KES quality principles.

**** Indexing: The books of this series are submitted to ISI Proceedings, EI-Compendex, SCOPUS, Google Scholar and Springerlink ****

More information about this series at http://www.springer.com/series/8767

Álvaro Rocha · José Luís Reis ·
Marc K. Peter · Zorica Bogdanović
Editors

Marketing and Smart Technologies

Proceedings of ICMarkTech 2019

Springer

Editors
Álvaro Rocha
DEI/FCTUC
University of Coimbra
Coimbra, Portugal

Marc K. Peter
FHNW School of Business
FHNW University of Applied Sciences
and Arts
Olten, Switzerland

José Luís Reis
University Institute of Maia—ISMAI
Maia, Portugal

Zorica Bogdanović
Faculty of Organizational Sciences
University of Belgrade
Belgrade, Serbia

ISSN 2190-3018 ISSN 2190-3026 (electronic)
Smart Innovation, Systems and Technologies
ISBN 978-981-15-1563-7 ISBN 978-981-15-1564-4 (eBook)
https://doi.org/10.1007/978-981-15-1564-4

This Springer imprint is published by the registered company Springer Nature Singapore Pte Ltd.
The registered company address is: 152 Beach Road, #21-01/04 Gateway East, Singapore 189721, Singapore

Preface

This book is composed of the papers written in English and accepted for presentation and discussion at the 2019 International Conference on Marketing and Technologies (ICMarkTech'19). This conference had the support of the University Institute of Maia and Polytechnic Institute of Maia, and AISTI (Iberian Association for Information Systems and Technologies). It will take place at Maia, Porto, Portugal, during November 27–29, 2019.

The 2019 International Conference on Marketing and Technologies (ICMarkTech'19) is an international forum for researchers and professionals to present and discuss the latest innovations, trends, results, experiences, and concerns in the various fields of marketing and technologies related to it.

The Program Committee of ICMarkTech'19 was composed of a multidisciplinary group of 152 experts and those who are intimately concerned with marketing and technologies. They have had the responsibility for evaluating, in a 'double-blind review' process, the papers received for each of the main themes proposed for the conference: (A) artificial intelligence applied in marketing; (B) virtual and augmented reality in marketing; (C) business intelligence databases and marketing; (D) data mining and big data—marketing data science; (E) Web marketing, e-commerce, and v-commerce; (F) social media and networking; (G) omnichannel and marketing communication; (H) marketing, geomarketing, and IoT; (I) marketing automation and marketing inbound; (J) machine learning applied to marketing; (K) customer data management and CRM; and (L) neuromarketing technologies.

ICMarkTech'19 received 104 contributions from 19 countries around the world. The papers accepted for presentation and discussion at the conference will be published by Springer (this book) and by AISTI and will be submitted for indexing by ISI, Ei Compendex, Scopus, DBLP, and/or Google Scholar, among others.

We acknowledge all those who have contributed to the staging of ICMarkTech'19 (authors, committees, and sponsors). We deeply appreciate their involvement and support that was crucial for the success of ICMarkTech'19.

Maia, Porto, Portugal
November 2019

Álvaro Rocha
José Luís Reis
Marc K. Peter
Zorica Bogdanović

Contents

About the Editors

Álvaro Rocha holds the title of an Honorary Professor, D.Sc. in Information Science, Ph.D. in Information Systems and Technologies, M.Sc. in Information Management, and B.C.S. in Computer Science. He is a Professor of Information Systems at University of Coimbra, researcher at CISUC (Centre for Informatics and Systems of the University of Coimbra), and a collaborator researcher at LIACC (Laboratory of Artificial Intelligence and Computer Science) and at CINTESIS (Center for Research in Health Technologies and Information Systems). He is also President of AISTI (Iberian Association for Information Systems and Technologies), Chair of IEEE Portugal Section Systems, Man, and Cybernetics Society Chapter, Editor-in-Chief of JISEM (Journal of Information Systems Engineering & Management), and Editor-in-Chief of RISTI (Iberian Journal of Information Systems and Technologies). Moreover, he has acted as Vice-Chair of Experts in the Horizon 2020 of the European Commission, as Expert in the Ministry of Education, University and Research of the Government of Italy, and as Expert in the Ministry of Finance of the Government of Latvia.

José Luís Reis holds a Ph.D. in Technologies and Information Systems from the University of Minho and a Professor with the title of specialist in Management and Administration by IPAM – Porto. He is a Professor at ISMAI, IPAM, and as a Guest Professor at the Catholic Porto Business School. It carries out activities in the areas of training and information systems and technologies in various organizations, coordinating various national and international projects in the areas of information management, applied marketing, and strategic regional planning. He is the author of scientific papers and articles in the fields of information systems modeling, multimedia, gamification, and data mining. He is the author and co-author of several books, namely "Personalization in Marketing – Technologies and Information Systems, "Marketing in agri-food – fundamentals and case studies" and "Gamification Model for SMEs.

Marc K. Peter holds doctorate from CSU Sydney, an Executive M.B.A. from UAS Bern/Babson College/PKU Beijing, and a Master of Marketing from the University of Basel. He is a Fellow of both the British Computer Society and the Chartered Institute of Marketing. At the FHNW School of Business, he is building and leading the Competence Center for Digital Transformation in Olten, Switzerland. His research and teaching focuses are digital transformation, digital marketing, and cybersecurity. Before joining FHNW, he held management positions at Swiss bank PostFinance, eBay International, E*TRADE Australia, and LexisNexis in the areas of strategy, technology, sales, and marketing in Europe and Asia-Pacific.

Zorica Bogdanović holds Ph.D. and is an Associate Professor at Faculty of Organizational Sciences, University of Belgrade, Serbia. She teaches subjects in the areas of e-business and e-business technologies on B.Sc., M.Sc., and Ph.D. studies. Her professional and scientific interests include e-business Internet marketing, Internet technologies, and Internet of things. Results of her researches have been published in many well-known international journals and conference proceedings. She is member of IEEE and secretary of IEEE Computer Chapter CO 16. She is in Chair of seminar of IEEE Computer Chapter CO 16. She is in Chair of the summer school "E-business technologies" at Faculty of Organizational Sciences since 2014. Since 2016, she is in Chair of Center for Internet of things, and since 2017, she is the Head of Department of e-business at Faculty of Organizational Sciences.

Chapter 1
The Digital Advertising Conceptual Flow: A Literature Review

Valerio Stallone[1,2(✉)]

[1] Zürich University of Applied Sciences, Stadthausstrasse 14, 8401 Winterthur, Switzerland
valerio.stallone@zhaw.ch
[2] Maastricht University, 6200 Maastricht, MD, The Netherlands

Abstract. This paper presents a systematic review of the literature of 22 research articles, which used a visualization of the digital advertising conceptual flow with a focus on programmatic advertising. This review aims to address a gap in the literature by identifying a standardization of digital advertising conceptual flow visualization and explanation. Achieving our aims will bring together fragmented literature to serve as a foundation for research into digital advertising conceptual flows and other related research. In our literature review, we have found discrepancies, which imply different understandings of how the advertising exchange between a publisher and an advertiser works. Our work should minimize the risk of wrong conclusions in literature as well as in practice. We therefore recommend a visualization in the field of digital advertising, which should be challenged in future research.

1.1 Introduction

Visiting a website or opening a free app often results as a web user often results in a certain amount of frustration: advertisements popping up can disturb the user's experience. Advertising intrusiveness is a cognitive process in which users perceive ads in a media vehicle to be distracting [1]. The perception of advertising intrusiveness therefore tends to evoke, in the web user, negative emotional reactions, such as annoyance [2]. In a web context, this means avoidance of the site [3]. The digital advertising conceptual flow ("DACF") describes the steps starting with the user's visit of a website or of an app and ending with an interaction of the very same user with the advertising element itself. This flow contains steps in the connection via an exchange platform ("ad exchange") between a website or an app ("publisher") and a company, which wants to advertise a product or a service ("advertiser"). A setting of intermediaries takes part in auctions to decide, which advertisements to show. In this paper, we focus on the steps in the conceptual flow and on how researchers have tried to model the flow in their work. We have considered literature, which discussed at least two steps of the DACF and dated 2015 or younger. Due to this approach, we are able to suggest a visualization of the DACF, which researchers should rely upon for their future studies. Our work starts with an introduction to the historical background and moves forward to the

Á. Rocha et al. (eds.), *Marketing and Smart Technologies*, Smart Innovation, Systems and Technologies 167, https://doi.org/10.1007/978-981-15-1564-4_1

methodologies scholars have worked with when using the DACF in literature. In the last chapter, we suggest a standardization of the DACF visualization for scientific literature as well as for work in nonacademic fields.

1.2 Historical Background

The DACF explains how a website's visitor gets to see an advertising element (e.g., a banner or a video) and how the information and data flow via the intermediaries. The flow starts with the user, with human eyes, which should generate the interaction that is the main resource elements on both ends of the flow, as well as the intermediaries, depend upon: attention [2]. This stream has grown in the last decade to an extent with an increasing amount of companies trying to be a part of it [4]. The simple bilateral relationship between publisher and advertiser evolved into an aggregation of subsequential connections. First descriptions date beginning of the millennium [5] and did not change much. To find visualizations, we have to rely on work created toward the end of last decade. The DACF contains the first sparkle, the user [6], the ad market [7], the data logging [8], and the delivery of [9], as well as the redirection to the destination URL [8].

1.3 Research Method

This review investigates the digital advertising conceptual flow from a multidisciplinary perspective. We selected peer-reviewed articles from four bibliographic databases (IEEE Xplore, Sage journals online, Science Direct, and Google Scholar). We considered peer-reviewed journals and conference proceedings based on their knowledge validity and their highest impact on the research field. In these very journals, the author considered only full papers. To ensure the relevance of the articles in this study, we used a combination of 11 keywords. Each of the selected articles is required to contain one of the following combinations of keywords at least in the abstract:

1. *digital advertising* AND *conceptual flow*,
2. *digital advertising* AND *ecosystem flow*,
3. *display advertising* AND *conceptual flow*,
4. *display advertising* AND *process*,
5. *display advertising* AND *ecosystem flow*,
6. *programmatic advertising* AND *conceptual flow*,
7. *programmatic advertising* AND *process*,
8. *programmatic advertising* AND *ecosystem flow*,
9. *real-time bidding* AND *conceptual flow*,
10. *real-time bidding* AND *process*, and
11. *real-time bidding* AND *ecosystem flow*.

The search for articles took place between May and June 2019 and was limited to English-speaking journals between 2013 and 2019. The rationale is to capture the latest change in investigating the digital advertising ecosystem.

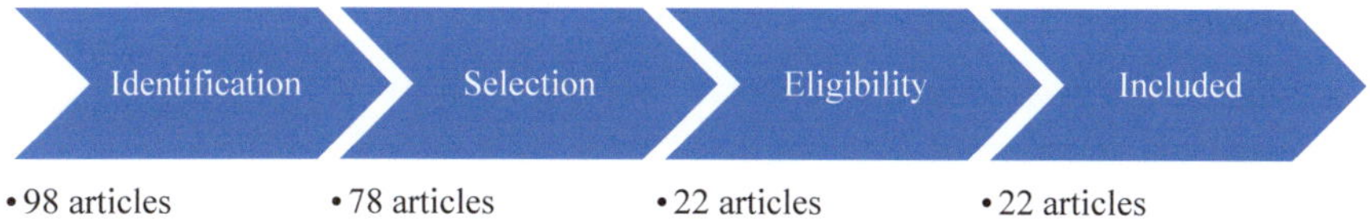

Fig. 1.1. The graph of the overview of the reviewed papers

As shown in Fig. 1.1, the identification phase generated 98 articles, which were further limited to those actually mentioning the digital advertising conceptual flow as a whole. The author selected 78 articles. Out of these articles, only 22 visualized the digital advertising conceptual flow, which the author fully included in the literature review of this paper.

1.4 Descriptive Results

1.4.1 Year of Publication

Table 1.1 and Fig. 1.2 show that the highest number of publications between in 2017, which slightly declined in 2018. The trend is evident: researchers obviously make use of a visualization of the DACF in their analyses, whereas the need is growing more and more.

Table 1.1. The overview of the reviewed papers

Year	2014	2015	2016	2017	2018
Amount	2 [10, 11]	1 [12]	4 [8, 9, 13, 14]	8 [4, 6, 7, 15–19]	7 [20–26]

This academic development reflects itself in the industry: programmatic advertising has turned into a globally 200-billion-dollars per annum business [24] and has grown, since 2014, at a 40% year-on-year pace [27].

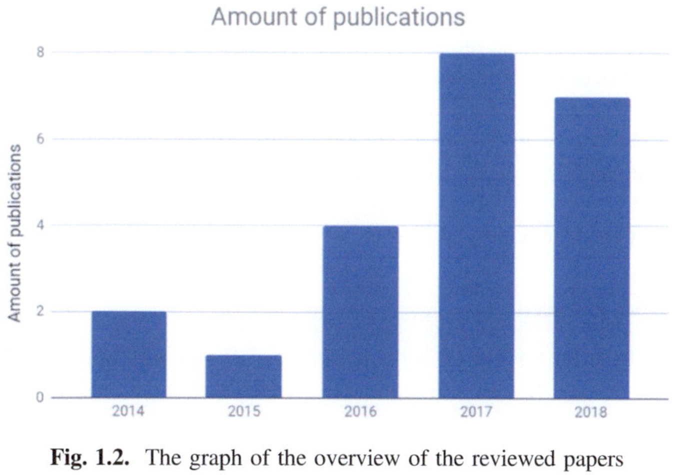

Fig. 1.2. The graph of the overview of the reviewed papers

1.4.2 Publication Typology and Objectives

To categorize the review of the typology and objective of the reviewed paper, the author used the structure of Yuan et al. [10]. In their survey, platform-level papers made almost 80% of all the reviewed papers. In this literature review, it is around 50%.

Table 1.2. Typology and objective of the reviewed papers

Level	Objectives
Platform level	Market specifications [8, 10, 19, 20, 24] Bidding algorithm [11, 15, 25] Mechanism design [4, 12, 14, 18] Market security [6, 9, 23]
Individual level	Behavioral analysis [7, 13, 26] Inventory pricing [17, 22] Market information structure [16, 21]

Table 1.2 also shows that literature, containing the visualization of the conceptual flow, mostly focused on market specifications—containing information about the digital advertising ecosystem in general—and on the mechanism design, which means describing how the elements of the digital advertising ecosystem communicate with each other.

1.4.3 Theoretical Framework

The DACF contains the following interaction categorizations: user interaction, ad market interaction, data logging, and destination information. The user initiates the flow when, via browser, requesting the HTML to be loaded, but she also interacts further on, when interacting with the advertising element (e.g., a click on a banner, playing or pausing a video). The complexity of the ad market depends on the amount of intermediaries in consideration. For the purpose of this paper, we will work with the following intermediaries: an ad exchange, a demand, a supply-side platform ("DSP" and "SSP"), and an ad network [15]. In the ad market steps, publisher and advertiser meet to exchange the slot for money based on a second price auction [23]. Data logging describes the process of tracking the relevant interactions and delivering the information to the elements, who are involved. The destination information is necessary to be able to send the user, in case of a click on the advertising element, to the URL of the advertiser. The analysis of the literature—journals, book chapters, conference proceedings, and theses—controls for the presence of the five categories presented as well as the depth of the interactions within those categories.

1.4.4 Content Differences

Descriptions of the DACF vary from very simple versions with only 4 [17] to 13 steps [12]. Research reports the time component differently: from the advertising request due to the HTML request of a user on supply side to the delivery of the ad, their opinions differ. Latter go from less than 100 ms [9, 16, 21], to around 100 ms [5, 12, 14, 20], to longer than 100 ms [4].

Descriptions in literature did not see the need to mention all the steps involved in the DACF. Some of them even ignore the entire interaction categories we described in this chapter. It leads to ignoring the user interactions (3 out of 22), to not mentioning the data logging (11 out of 22) and to not listing the destination information steps (15 out of 22). Six out of 22 descriptions do not mention steps within the ad market (e.g., bids and auctions). There is no description containing all of the interaction categories—a few contain at least three categories [4, 6–8, 14, 16, 20].

1.5 Results

The author conducted a review of DACF descriptions in literature to find a wide assortment of different versions. In this analysis, the author focused on digital advertising literature mentioning at least two key players or elements. The author did not find any description containing all categories presented in Chap. 4. In Fig. 1.3, the author created a DACF visualization. We based this "minimum viable version" on the literature review, similar to what had been done by Stallone and Klaas [28] for the visualization of the whole ecosystem.

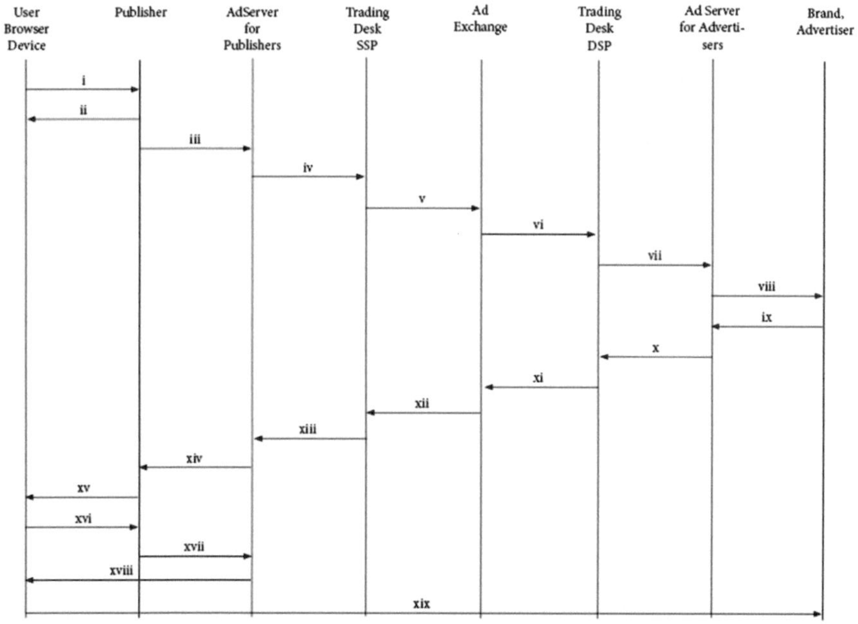

Fig. 1.3. DACF according to the authors

The DACF starts with a user sending out a web request to a publisher's website (i), who again sends the HTML as well as ad frames within the HTML back to the user (ii). The publisher sends the ad request to fill the ad frame sent to the user to the ad server (iii). The ad server prepares itself to the auction and sends the requirements given by the ad frame and the publisher (e.g., size, format) to the SSP (iv). The SSP collects the different auction requests from the different publishers and forwards them to the ad exchange (v), where the latter confronts the auction requests with the auction bids (xi). Before it can do so, the ad exchange informs the DSP about the potential auction and sends the auction requirements (vi), where the latter checks for the advertising material fitting them (vii). The ad server controls for budget restrictions of the advertiser (viii), which decides, whether to participate in the auction (ix) or not. In case the advertiser agrees, the ad server sends the destination URL information to the DSP (x), which places the bid (xi), as already stated before. The ad exchange decides, which bid and therefore which advertiser has won the auction and delivers this information to the SSP (xii). The SSP forwards the destination URL to the ad server (xiii)—this is where the parties involved log the impression. The advertising element is sent to the publisher (xiv) and then to the ad frame on the user's browser (xv).

In case the user interacts with advertising element with a click (xvi), the ad server requests the destination URL (xvii)—this is where the parties involved log the click. The user's browser receives the destination URL (xviii), which then redirects to the new website (xix).

1.6 Conclusion

This paper aimed at giving an overview of the visualization and the use of the DACF in literature. The heterogeneity in the visualization itself is surprising: the fact that most of the papers did not represent the flow in its completeness reflects the lack of transparency of the ecosystem and its intermediaries. The rising trend in analyzing the field of programmatic advertising and of DACF shows the increasing interest of academia, which should reflect in a better understanding of the ecosystem and its functions. This paper presents a summarized version of the DACF, which scholars can use as a starting point for their analyses: by doing so and referring to it, they should be able to make sure to show all needed intermediaries involved in the DACF itself.

1.7 Limitations and Future Research Direction

1.7.1 Limitations

The focus of this review was on the digital advertising conceptual flow visualization and not on the visualization of the digital advertising ecosystem as done elsewise [26]. Furthermore, this review looked at literature coming from different fields of academia (e.g., marketing, information technology, game theory).

1.7.2 Further Research

In the opinion of the authors, future research should focus on the following questions: does the industry accept our DACF in Fig. 1.2? If not, which elements are gone missing? For future research, we therefore suggest carrying an in-depth description and qualitative analysis of the various descriptions displayed in scientific and nonscientific visualizations based on this paper and move the discussion to the public eye. As mentioned in a recent paper, "there needs to be a significant awareness campaign to support transformation toward a healthier online advertising ecosystem" [24].

References

1. Li, H., Edwards, S.M., Lee, J.-H.: Measuring the intrusiveness of advertisements: scale development and validation. J. Advert. **31**(2), 37–47 (2002). https://doi.org/10.1080/00913367.2002.10673665
2. Redondo, I., Aznar, G., Krammer, V.: An effective defense against intrusive web advertising. In: 2008 Sixth Annual Conference on Privacy, Security and Trust, pp 3–14. IEEE (2018). https://doi.org/10.1109/PST.2008.10
3. McCoy, S., Everard, A., Polak, P., Galletta, D.F.: The effects of online advertising. Commun. ACM **50**(3), 84–88 (2007). https://doi.org/10.1145/1226736.1226740; Muthukrishnan, S.: Ad Exchanges: research Issues, pp. 1–12. Springer, Berlin, Heidelberg (2009). https://doi.org/10.1007/978-3-642-10841-9_1
4. Sayedi, A.: Real-time bidding in online display advertising. SSRN Electron. J. (2017). https://doi.org/10.2139/ssrn.2916875
5. Goldman, D.: I always feel like someone is watching me: a technological solution for online privacy. Hastings Commun. Entertain. Law J. **28**(3) (2006). https://repository.uchastings.edu/hastings_comm_ent_law_journal/vol28/iss3/1. Accessed 28 May 2019
6. Kumar, J.: Timeout analysis, troubleshooting and notification in real time bidding advertising system with implementation. Comput. Sci. Eng. **7**(3), 67–78 (2017). https://doi.org/10.5923/j.computer.20170703.01
7. Högström, J., Wallin, E.: Programmatic advertising: a quantitative study of consumer attitudes towards highly frequent online ads (2017). http://hh.diva-portal.org/smash/record.jsf?pid=diva2%3A1106048&dswid=5593. Accessed 28 May 2019
8. Chen, G., et al.: In-depth survey of digital advertising technologies. IEEE Commun. Surv. Tutor. **18**(3), 2124–2148 (2016). https://doi.org/10.1109/COMST.2016.2519912
9. Dwyer, C., Kanguri, A.: Gone in 200 Milliseconds: the challenge of Blocking Malvertising. Student and Faculty Research Days (2016). https://digitalcommons.pace.edu/sfresearchday/3 . Accessed 28 May 2019
10. Yuan, Y., et al.: A survey on real time bidding advertising. In: Proceedings of 2014 IEEE International Conference on Service Operations and Logistics, and Informatics, pp. 418–423.. IEEE (2014). https://doi.org/10.1109/soli.2014.6960761
11. Zhang, C., Zhang, E.: Optimized bidding algorithm of real time bidding in online ads auction. In: 2014 International Conference on Management Science & Engineering 21th Annual Conference Proceedings, pp. 33–42. IEEE (2014). https://doi.org/10.1109/icmse.2014.6930205
12. Yuan, S.: Supply side optimisation in online display advertising. Doctoral thesis, UCL (University College London). UCL (University College London) (2015). http://discovery.ucl.ac.uk/1463229/. Accessed 28 May 2019

13. Yamaguchi, Y., et al.: Web advertising recommender system based on estimating users' latent interests. In: Proceedings of the 18th International Conference on Information Integration and Web-based Applications and Services—iiWAS '16, pp. 42–49. ACM Press, New York, USA (2016). https://doi.org/10.1145/3011141.3011180
14. Zhou, B., Shariat, S.: Finding needle in a million metrics: anomaly detection in a large-scale computational advertising platform (2016). http://arxiv.org/abs/1602.07057, Accessed 28 May 2019
15. Chen, Y., et al.: Measuring network reputation in the ad-bidding process, pp. 388–409. Springer, Cham. https://doi.org/10.1007/978-3-319-60876-1_18
16. Li, J., et al.: Research on information structure of programmatic advertising markets. IFAC-PapersOnLine (Elsevier) **50**(1), 13587–13592 (2017). https://doi.org/10.1016/J.IFACOL.2017.08.2377
17. Papadopoulos, P., et al.: If you are not paying for it, you are the product: how much do advertisers pay to reach you? (2017). https://doi.org/10.1145/3131365.3131397
18. Wang, J., Zhang, W., Yuan, S.: Display advertising with real-time bidding (RTB) and behavioural targeting. Foundat. Trends® Inf. Retr. (Now Publishers, Inc.,) **11**(4–5), 267–435 (2017). https://doi.org/10.1561/1500000049
19. Zhu, X., et al.: Ad ecosystems and key components, pp. 7–18. Springer, Cham (2017). https://doi.org/10.1007/978-3-319-56793-8_2
20. Alaimo, C., Kallinikos, J.: Objects, metrics and practices: an inquiry into the programmatic advertising ecosystem, pp. 110–123. Springer, Cham (2018). https://doi.org/10.1007/978-3-030-04091-8_9
21. Arab, S.A., et al.: Two-layer recommendation-based real time bidding (RTB). In: 2018 IEEE 27th International Conference on Enabling Technologies: Infrastructure for Collaborative Enterprises (WETICE), pp. 91–94. IEEE (2018). https://doi.org/10.1109/wetice.2018.00024
22. Li, J., Ni, X., Yuan, Y.: The reserve price of ad impressions in multi-channel real-time bidding markets. IEEE Trans. Comput. Soc. Syst. **5**(2), 583–592 (2018). https://doi.org/10.1109/TCSS.2018.2831234
23. Niu, C., et al.: ERA: towards privacy preservation and verifiability for online ad exchanges (2018). https://doi.org/10.1016/j.jnca.2017.08.012
24. Parssinen, M., et al.: Is blockchain ready to revolutionize online advertising? IEEE Access **6**, 54884–54899 (2018). https://doi.org/10.1109/ACCESS.2018.2872694
25. Ren, K., et al.: Bidding machine: learning to bid for directly optimizing profits in display advertising. IEEE Trans. Knowl. Data Eng. **30**(4), 645–659 (2018). https://doi.org/10.1109/TKDE.2017.2775228
26. Zhang, Y.: Mathematical approaches to real time bidding strategy (RTB) for advertising campaigns for demand side platforms (2018). https://www.semanticscholar.org/paper/Mathematical-Approaches-to-Real-Time-Bidding-(-RTB-Zhang/0a56be9385783d2ca57406083bccc4078144d4ce. Accessed 28 May 2019
27. Fisher, L.: US programmatic ad spending forecast 2019—eMarketer trends, forecasts and statistics. emarketer.com (2019). https://www.emarketer.com/content/us-programmatic-ad-spending-forecast-2019. Accessed 29 May 2019
28. Stallone, V., Klaas, M.: The Digital advertising ecosystem visualization—literature review (2019). http://www.iadisportal.org/digital-library/the-digital-advertising-ecosystem-visualization-literature-review. Accessed 29 May 2019

Chapter 2
Featured Snippets Results in Google Web Search: An Exploratory Study

Artur Strzelecki(✉) and Paulina Rutecka

Department of Informatics, University of Economics in Katowice, Katowice 40-287, Poland
{artur.strzelecki,paulina.rutecka}@ue.katowice.pl

Abstract. In this paper, authors analyzed 163412 keywords and results with featured snippets collected from localized Polish Google search engine. A methodology for retrieving data from Google search engine was proposed in terms of obtaining necessary data to study featured snippets. It was observed that almost half of featured snippets (48%) is taken from result on first ranking position. Furthermore, some correlations between prepositions and the most often appearing content words in keywords was discovered. Results show that featured snippets are often taken from trustworthy websites like Wikipedia and are mainly presented in form of a paragraph. Paragraph can be read by Google Assistant or Home Assistant with voice search. We conclude our findings with discussion and research limitations.

2.1 Introduction

Google is now the most important Internet search engine. Google engineers are constantly developing the search algorithm to make it the most consistent with mobile standards and voice search adapted.

The mobile-first idea assumes that the first step of the web design process is making a mobile view. The website must meet the requirements of the mobile user and give him or her the best experience. UX of websites is first designed for mobile devices and affects the desktop version (progressive enhancement). This idea arose from the fact that it is easier to add elements to desktop version than to wrap it down to the mobile version. This tendency is also correlated with the growing number of mobile users. The website has to load in a short time and does not need any further actions like taps or user decisions. In effect, websites are designed in a one-page layout that does not require any action except scrolling the screen.

The statistics show that mobile users are the largest group of web users. In 2016 it was the first time when mobile traffic was bigger than desktop traffic and amounted to 51,3% (according to statcounter.com data). According to data from Global Digital 2019 report made by We Are Social in Hootsuite in January 2019 [1], there was 5.11 billion unique mobile users, which means an increase of 100 million (2%) compared to 2018. In 2019 mobile devices with the result of 51.6% are on the first place among the devices through which users browse the network.

Á. Rocha et al. (eds.), *Marketing and Smart Technologies*, Smart Innovation, Systems and Technologies 167, https://doi.org/10.1007/978-981-15-1564-4_2

The first snippets were introduced by Google in 2012. Their goal was to show the most important information on the Search Engine Result Page (SERP). In addition to the snippets, which have been popular for several years like a knowledge graph or multimedia carousel, direct answer is becoming an increasingly common snippet.

The direct answer snippet also called the featured snippet was introduced by Google in 2016 (in Poland on April 2016). It appears as a succinct response in the form of a paragraph, list or table. The direct answer gives a highly quick response to the user query. This kind of snippet is also suitable for voice search. Voice search is becoming more popular by introducing it not only on Google Home devices but also as Google's voice assistant on mobile phones. The assistant initially debuted in May 2016 (in Poland on January 2019). Direct answer snippet is placed on the top of the search results page. It is also designed to be read by Google's Voice Assistant.

Motivation behind this study is to explore the new featured snippet that appears in web search results. Authors want to gain knowledge what factors on a keyword level and on a website level cause displaying of this snippet. The goal of our study is to analyze 163 412 keywords along with search quantity data and direct answers' source to answer the question about the correlation between types of keywords and direct answers, as well as the correlation between keywords and the answer given.

The paper is organized as follows. Section 2 contains a review of the relevant literature on snippets. Section 3 includes the methodology for data retrieval and processing, while Sect. 4 presents the data and quantitative results. In Sect. 5, the authors highlight the contribution of the research, discuss its limitations, and finally, draw conclusions about the results and propose possible future research avenues.

2.2 Literature Review

Search engines display different snippets in SERPs. Snippets can be divided into five categories [2]: regular snippets (1), rich snippets (2), Google News (3), entity types (4), and featured snippets (5).

The first category is regular snippet displayed for typical, organic results. Since the beginning of search engines, regular snippets were displayed as two lines of description placed below the title and URL of the result [3]. For last few years, it has been observed that some tests of changing length of regular snippets either on desktop or on mobile version have been done [4, 5]. Research interest for regular snippets is about their function of being enough informative for users [6]. Recent research was conducted on different age groups to determine how regular snippets are perceived by younger or older search engine users [7].

The second category are rich snippets that come from structured data dictionary schema.org [8]. Google, Microsoft (Bing), Yahoo, and Yandex have founded schema.org and are recognizing structured data included in RDFa, Microdata, or JSON [9]. Rich snippets created on structured data are displayed along with regular snippets [10]. Search engines show results with expanded data about products availability, price and condition, recipes, reviews, jobs, music, video, and others, included in schema.org. Rich snippets are considered as important variable, especially when examining bottom-ranked results [11].

The third category are snippets displayed from Google News. These snippets are part of specialized news service and are created automatically [12]. By news publishers, they are considered in different ways. Recently in Germany and Spain Google news was restricted, cause displaying snippets of news releases violates copyrights of news publishers [13]. Some propositions are made to solve this possible violation, e.g., a plan for ancillary copyright is proposed, by creating original snippets [14].

The fourth category is entity type snippet. Entity types in Google are known as Knowledge Graph introduced in 2012 year and in Bing are known as Satori, launched in the same year [15]. These entities are constructed object and concepts, including people, movies, places, events, books, arts, science, etc. These entity databases are considered by users as an important part of search results [16].

The last category are featured snippets. It is the latest improvements of SERP. The search engine retrieves pieces of information from web pages and displays it in a box, above organic results together with a source URL. Google programmatically determines that a page contains a likely answer to the user's question and presents the result as a featured snippet. The other known name for this snippet is a direct answer or answer box. Direct answer supposed to deliver answers for keywords, without need to visit the result presented in search engine [17]. This snippet is displayed in four different forms like paragraph [18], table [19], and ordered or unordered list [20].

In early stage of featured snippets deployment, one research examined the possible factors of presenting piece of information from particular URL in featured snippets [20]. Featured snippets was named as Google Answer Box and noticed that some prerequisites need to be met if a website could be included in the featured snippet, like high ranking, multiple keyword inclusion in website's content, different locations for keyword like headings, title, URL, paragraph, image ALT, links, and structured content in the form of ordered or unordered list. Search engines create objects displayed among search results, and also they delist results because of the variety of reasons [21].

2.3 Research Method

Dataset was collected using Senuto (https://www.senuto.com/). Senuto is an online tool that retrieves data on websites visibility from Google search engine. Visibility in search engine in its basic structure is a set of three elements: keyword, ranking, and result [22]. On a daily basis Senuto extracts data from Google based on the keywords list and saves results along with the position. Senuto has a database of 20 million keywords. Each keyword is at least once in a month entered to Polish localized Google search engine and a list of top 50 results is returned.

A dataset from Senuto was acquired in July 2019. Except this basic structure, authors asked Senuto owners to modify tool, to be able extracting more data from SERP. The goal of this research is to examine featured snippets, thus extended structure for extracting results was prepared. In extended visibility structure, along with basic set, additional elements are collected, and dataset contains the following elements: keyword, number of monthly searches, date of last extraction, parameters, featured domain, featured main domain, featured position, featured type, and featured URL.

Dataset with extended visibility structured contains 163 412 examined keywords that were resulting in SERP with featured snippet. Keyword (also called as query) is the term entered to search engine interface. Keyword length is from one term up to 25 terms in case of Google. Other search engines have other maximum number of terms included in one query. Number of monthly searches is number of searches for this keyword in Google polish localized version. Data is imported through API from Google Ads. Date of last extraction is date for retrieving data.

Parameters contains internal list of snippets displayed for particular keyword. Parameters inform that for this particular keyword except featured snippet, in close proximity of featured snippet also can be displayed as site links (1), image thumbs (2), Google Ads (3), map (4), brand (5), wiki on right side (6), city (7), news (8), and name (9).

Site links are additional URL results displayed only for domain on first position in organic results. Image thumbs show a row with images as a piece of result from image search. Google Ads snippet present sponsored search results from Google Ads platform. Map shows a piece of map connected with the query. Brand means that for search keyword, some information can be taken from other services offered by search engine like Google My Business. Top info is displayed above the featured snippet. Wiki on right contains piece of information taken from Wikipedia.

Featured domain is an exact part of URL between protocol (HTTP of HTTPS) and first slash. Featured main domain is normalized domain name. It does not contain any prefixes before the name, such a subdomain. Featured position is ranking position of results displaying the same URL that is displayed in featured snippets. Featured snippets are taken from results 1 to 10, only from SERPs first page. Featured type defines three dimensions of displayed featured snippet: paragraph, list, and table. Featured URL is the final URL of the result that piece of information is displayed from.

2.4 Data and Results

This section presents the data description and summary of results from data.

2.4.1 Data Description

Domains, most often appearing as a source of direct answer and their position in search results, as well as links between direct answer and other snippets and the type of direct answer were analyzed.

Table 2.1. Type of featured snippet

Featured type	Frequency	Percentage
Paragraph	114465	70.05
List	46509	28.46
Table	2438	1.49

The answer type: Direct answers appear in three basic forms: as an article, a list, and a (Table 2.1). The paragraph understood as the text response to the query appeared most often—114298 times, which is 69.94% of the total direct answers. The second in order of frequency is the list form, which appeared in 46509 results, and that is 28.46% of the analyzed phrases. The least frequent direct answer appearing is the table which constitutes only 1.49% of results (2438 occurrences). The unclassified response type appeared 167 times, representing 0.10% of the results.

Table 2.2. Ranking position for featured snippet

Position	Frequency	Percentage
0	485	0.30
1	79867	48.87
2	30618	18.74
3	20878	12.78
4	14469	8.85
5	9582	5.86
6	2860	1.75
7	1909	1.17
8	1319	0.81
9	860	0.53
10	554	0.34

The position of the answer's source domain: A factor, which had also been analyzed, was the position of the website which was the source from which the answer quoted as the direct answer was downloaded. The vast majority of the fragments used in direct answer came from the first 10 search results domains.

Most often are cited domains from the first search position—79868 results were cited from these domains (48.88%). The results from the second position were cited 30618 times (18.74%), from the third position—20878 times (12.78%). Other items are shown in Table 2.2. Domains that were below the first 10 search results were cited one time (domain from positions 14, 15, 25, 32, 33, 41, 42) or three times (domain from item 11). For 485 records, a result of 0 appears, which should be classified as a data read error.

Table 2.3. Other snippets displayed along with featured snippet

Params	Frequency	Percentage
Image thumbs	102934	62.99
Site links	41348	25.30
Brand	24214	14.82
Wiki	18675	11.43
Ads	3148	1.93
Name	2850	1.74
Map	1807	1.11
City	1062	0.65
News	107	0.07

Occurrence with other snippets: Direct answers also occur along with other types of snippets as shown in Table 2.3. They appear most often in the company of an image —102934 times (62.99%) or site links—41348 times (25.30%). Less frequently, along with featured snippets, brand—24214 times (14.82%), wiki (knowledge graph)—18675 times (11.43%). The other snippets shown in the Table 2.3, that appear together with the direct answer are marginal. These are Google Ads (1.93%), name (1.74%), map (1.11%), city (0.65%), and news (0.07%). A direct answer alone with no other snippets appears 26813 times, which is 16.41% of the analyzed results. Brands are keywords, which contain a term with a brand name and are recognize based on site links, e.g., brand keyword is "refrigerator samsung".

Source domains: Among the emerging domains cited by Google in direct answer the most-appearing is wikipedia.org—Internet encyclopedia (48242 times), medonet.pl —health category (2352 times), fandom.com—entertainment (1259 times), kwestias-maku.com—cooking (1033 times), apteline.pl—online pharmacy (941 times), sciaga.pl —education (913 times), goodreads.com—books (847 times), quora.com—Q&A service (835 times), chillizet.pl—radio information service (833 times), and mp.pl—health (762 times).

2.5 Results

The analysis of the possessed data allowed to divide the keywords into three main sections due to the syntax of the key phrase. First one contains keywords started with pronouns or containing pronouns characteristic for the syntax of the interrogative sentence. The second section contains keywords beginning with or containing an attribute such as cost, weight, and capacity. The third section of keywords contains phrases beginning with or containing an adverb's or adjective's superlative form.

Table 2.4. TOP pronouns in keywords

Pronouns	Frequency of keywords beginning with	Frequency of keywords containing
co	3029	4604
jak	2724	1328
ile	843	670
kiedy	674	365
czy	106	295
kim	182	229
dlaczego	488	220
kim	379	131
dlaczego	187	123
jakie	220	90
jaka	148	87
kim	64	30
czy	3	16
dlaczego	79	14

Keywords syntax—pronouns: From the keywords, there were separated pronouns as follows: what, how, when, how much/how many, who, where, why, with what, whom, and which. Among the analyzed 163 412 keywords, 9131 starts with one of the words mentioned above (Table 2.4), which is 5.59% of the keywords analyzed in general. 8192 analyzed key phrases contain the pronoun, but it does not appear as the first word in the phrase. This is 5.01% of the entire analyzed group of keywords.

Attributes of keywords content: Queries often refer to attributes that define a feature. From the analyzed keywords, these keywords were distinguished, which contain the attribute. 89 phrases which are the query about the object's, product's, country's, or person's feature have been separated. Table 2.5 shows the 10 most often used keywords and 10 attributes that appear in a keyword, but not on the first position.

Table 2.5. TOP 10 attributes

Keywords beginning with attribute	Frequency	Keywords contain attribute	Frequency
objawy	1143	objawy	3164
przepis	665	definicja	2550
rodzaje	664	cena	1403
definicja	356	przepis	1140
wymiary	307	wymagania	906
cena	273	znaczenie	841
wymagania	271	wymiary	547
temperatura	202	rodzaje	420
znaczenie	160	kalorie	411

89 separated attribute definitions appear as the first word in key phrase for 10173 phrases, which account for 6.23% of the analyzed keywords. In the next positions, there are up to 18050 key phrases, which account for 11.05% of the keywords analyzed.

Table 2.6. TOP 10 superlatives

Keywords beginning with superlative	Frequency	Keywords contain superlative	Frequency
najładniejsze	210	najpiękniejsze	43
najmłodsza	190	największi	38
najdroższe	121	najładniejsze	20
najpiękniejsze	117	najdroższa	18
najwięksi	104	najlepsza	18
najlepsza	69	najwięcej	16
najdroższa	60	najmłodsza	14
najdłuższy	60	najważniejsze	14
najdłuższa	59	najlepiej	14

Keywords with superlative prefix (Table 2.6): In the analyzed keywords, 202 keywords with the superlative prefix "naj" (suggesting the best or the worst) were extracted. 2401 key phrases started with the words "the best or the worst", which is 1.47% of the words analyzed in general. 378 contained these phrases on a different than the first position of the phrase (0.23% of the analyzed keywords). The same words appear in the most popular list, regardless of whether it is at the beginning or elsewhere in the key phrase. Definitely more often the words the best and the worst appear at the beginning of the key phrase formation. The analysis at the level of individual URLs indicates that the majority of the phrases that result in the display of featured snippet refers to queries for culinary, medicine, and IT.

2.6 Conclusion and Discussion

In this paper, we presented an analysis of the dataset that causes appearance of featured snippets (direct answer or Google answer) in the search engine. The findings of our study indicate that the Google search engine is being developed in the direction of displaying the query response from the search results page. Google does not discriminate blue links, but makes the valuable site stand out. Because there is no automatic inclusion and no mark that can be added to website for being taken into consideration for featured snippets displaying, our findings on a keyword level help to prepare better content on the website, to be considered for this snippet.

The analysis of data shows that the appearance of a direct answer is closely related to the question form of the phrase (the occurrence of pronouns) or the occurrence of words in the phrase specifying an attribute that has a specific value, considered as the answer to the query, e.g., product price. Frequently appearing phrases are also queries containing adjectives in superlative form, e.g., the highest peak, the largest city. All these queries have unambiguous and undeniable value.

Google uses as a source from which direct answer is cited only pages with high ranking in search engines which meet the criteria of Google's ranking factors. Meeting these criteria defines the site as containing content of a high value, popular among users. This means that the website enjoys the trust of both the search engine and the users themselves. This makes it very likely that the information in the direct answer is correct. The same websites, such as Wikipedia, are often cited.

The most common types of direct answers are responses in the form of a paragraph. This kind of answer is the most legible and at the same time the most convenient to be read by voice search systems. It potentially allows the computer for a voice response using a speech synthesizer. Results in the form of a list or table appear less frequently. The form of a list usually appears in cookery recipes (ingredients for preparing a dish) or in medicine, when symptoms of a disease are mentioned. The table appears most frequently for inquiries regarding prices or other values related to financial products (such as taxes, loans, insurance). Featured snippets often occur along with other types of snippets, of which the most common are images that draw the user's attention by highlighting themselves in the search results list, as well as links to the pages from which the data used for the answer come. By correlation with links to direct pages, they are called zero position in search results, and SEO specialists are analyzing how one can meet enough ranking factors to make the page cite the direct answer. This item is very often clicked, and the page in this position gains the position of an expert.

The limitation of our research was the fact of having a set of data concerning 163 412 keywords only in Polish. All data concern the Google search engine, which is dominant in Poland. However, we realize that results containing featured snippets types are observed in other language versions of Google as well. Other limitation is data dataset is relatively small when comparing it to volume of searches made daily by search engine users.

Future research will be conducted to investigate the factors affecting the display of results from specific websites in the snippets area. Another direction of future studies is to analyze content of featured snippet by extracting paragraphs, tables, and lists from SERPs and study them. However, this direction demands to extend even more, already extended dataset for studying visibility in search engines.

References

1. Global Digital Report 2019—We Are Social, https://wearesocial.com/global-digital-report-2019
2. Strzelecki, A., Rutecka, P.: The snippets taxonomy in web search engines. In: Pańkowska, M., Sandkuhl, K. (eds.) Perspectives in Business Informatics Research. pp. 177–188. Springer, Cham (2019). https://doi.org/10.1007/978-3-030-31143-8_13
3. Lewandowski, D.: The retrieval effectiveness of web search engines: considering results descriptions. J. Doc. **64**, 915–937 (2008). https://doi.org/10.1108/00220410810912451
4. Sachse, J.: The influence of snippet length on user behavior in mobile web search: an experimental eye-tracking study. Aslib J. Inf. Manag. **71**, 325–343 (2019). https://doi.org/10.1108/AJIM-07-2018-0182
5. Kim, J., Thomas, P., Sankaranarayana, R., Gedeon, T., Yoon, H.-J.: What snippet size is needed in mobile web search?. In: Proceedings of the 2017 Conference on Conference Human Information Interaction and Retrieval—CHIIR '17. pp. 97–106. ACM, New York, NY (2017). https://doi.org/10.1145/3020165.3020173
6. Kattenbeck, M., Elsweiler, D.: Understanding credibility judgements for web search snippets. Aslib J. Inf. Manag. **71**, 368–391 (2019). https://doi.org/10.1108/AJIM-07-2018-0181
7. Bilal, D., Huang, L.-M.: Readability and word complexity of SERPs snippets and web pages on children's search queries. Aslib J. Inf. Manag. **71**, 241–259 (2019). https://doi.org/10.1108/AJIM-05-2018-0124
8. Hop, W., Lachner, S., Frasincar, F., De Virgilio, R.: Automatic web page annotation with Google rich snippets. In: Meersman, R., Dillon, T., Herrero, P. (eds.) Lecture Notes in Computer Science. pp. 957–974. Springer, Berlin, Heidelberg (2010). https://doi.org/10.1007/978-3-642-16949-6_21
9. Steiner, T., Hausenblas, M., Troncy, R.: How Google is using linked data today and vision for tomorrow. In: Auer, S., Decker, S., Hauswirth, M. (eds.) Proceedings of Linked Data in the Future Internet at the Future Internet Assembly (FIA 2010), pp. 1–10, Ghent (2010)
10. Khalili, A., Auer, S.: WYSIWYM authoring of structured content based on schema.org. In: Lecture Notes in Computer Science, pp. 425–438. Springer, Berlin, Heidelberg (2013). https://doi.org/10.1007/978-3-642-41154-0_32

11. Marcos, M.-C., Gavin, F., Arapakis, I.: Effect of snippets on user experience in web search. In: Proceedings of the XVI International Conference on Human Computer Interaction—Interacción ’15, pp. 47. ACM, New York, NY (2015). https://doi.org/10.1145/2829875.2829916
12. Das, A.S., Datar, M., Garg, A., Rajaram, S.: Google news personalization: scalable online collaborative filtering. In: Proceedings of the 16th International Conference on World Wide Web—WWW ’07, pp. 271–280. ACM, New York, NY (2007). https://doi.org/10.1145/1242572.1242610
13. Calzada, J., Gil, R.: What do news aggregators do? evidence from Google news in Spain and Germany. Universitat de Barcelona, Barcelona (2016). https://doi.org/10.2139/ssrn.2837553
14. Potthast, M., Chen, W., Hagen, M.: A plan for ancillary copyright: original snippets. In: Albakour, D., Corney, D., Gonzalo, J., Martinez, M., Poblete, B., Vlachos, A. (eds.) Proceedings of the Second International Workshop on Recent Trends in News Information Retrieval at ECIR, pp. 3–5., Grenoble, France (2018)
15. Uyar, A., Aliyu, F.M.: Evaluating search features of Google knowledge graph and bing satori. Online Inf. Rev. **39**, 197–213 (2015). https://doi.org/10.1108/OIR-10-2014-0257
16. Juel Vang, K.: Ethics of Google’s knowledge graph: some considerations. J. Inf. Commun. Ethics Soc. **11**, 245–260 (2013). https://doi.org/10.1108/JICES-08-2013-0028
17. Miklosik, A.: Search engine marketing strategies: Google answer box-related search visibility factors. In: Carvalho, L., Isaías, P. (eds.) Handbook of Research on Entrepreneurship and Marketing for Global Reach in the Digital Economy, pp. 463–485. IGI Global, Hershey, PA (2019). https://doi.org/10.4018/978-1-5225-6307-5.ch020
18. Zhao, Y., Zhang, J., Xia, X., Le, T.: Evaluation of Google question-answering quality. Libr. Hi Tech. **37**, 312–328 (2019). https://doi.org/10.1108/LHT-10-2017-0218
19. Hancock, B., Lee, H., Yu, C.: Generating titles for web tables. In: WWW ’19: Proceedings of the 2019 World Wide Web Conference, pp. 638–647. ACM, New York, NY (2019). https://doi.org/10.1145/3308558.3313399
20. Miklošík, A., Hlavatý, I., Daňo, F., Červenka, P.: Google answer box keyword-related analysis a case study. Eur. J. Sci. Theol. **12**, 185–194 (2016)
21. Strzelecki, A.: Website removal from search engines due to copyright violation. Aslib J. Inf. Manag. **71**, 54–71 (2019). https://doi.org/10.1108/AJIM-05-2018-0108
22. Strzelecki, A.: Google web and image search visibility data for online store. Data **4**, 125 (2019). https://doi.org/10.3390/data4030125

Chapter 3
Willingness to Pay a Premium Price for Streaming Services: The Role of Trust in Services

Luísa Augusto[1(✉)], Sara Santos[2], and Pedro Espírito Santo[3]

[1] Centro de Estudos Em Educação, Tecnologias E Saúde, Instituto Politécnico de Viseu, Campus Politécnico, 3504-510 Viseu, Portugal
laugusto@esev.ipv.pt

[2] Instituto Politécnico de Viseu, Campus Politécnico, 3504-510 Viseu, Portugal
ssantos@esev.ipv.pt

[3] Instituto Politécnico de Coimbra, ESTGOH, Rua Gen. Santos Costa, 3400-124 Oliveira do Hospital, Portugal
pedro.santo@estgoh.ipc.pt

Abstract. Online business has seen a strong growth, namely, the on-demand digital streaming service, which means that the brands have a constant concern in the provision of an increasingly high-quality technology service to meet the needs of consumers. Closely associated with this quality is the convenience of the service and the sophistication of it, the trust, and the willingness to pay a premium price. Thus, this work aims to understand if there is availability to pay a premium price for this type of service. In this way, we constructed a questionnaire that was self-administered by 156 users of Netflix branded streaming services. Our study highlights the mediating role of trust in the effects that the sophistication and convenience of service have on the willingness to pay a premium price. In this way, this study brings contributions to the academy and to the business world.

3.1 Introduction

Internet-based business development has been accelerating in recent years, and in this context, many innovations have emerged, notably television content is now accessible via mobile devices connected to the Internet [29]. One of the services that has seen its popularity grow is the on-demand digital streaming service where the best-known international brands in this type of service offering are Netflix, YouTube, and Hulu. This popularity has grown due to the technology and improvement that these services have been targeting over the years [24].

Convenience of service emerges as an important factor in decision-making, customer engagement, and satisfaction, and it is critical for companies to focus their efforts on promoting convenience of service [25]. Another relevant concept associated with perceived quality of service and brand personality is that of service sophistication. These concepts have aroused the interest of the academy in various fields and scientific

Á. Rocha et al. (eds.), *Marketing and Smart Technologies*, Smart Innovation, Systems and Technologies 167, https://doi.org/10.1007/978-981-15-1564-4_3

areas. In this study, we intend to understand the role of service convenience and service sophistication in providing a quality service in the case of on-demand digital streaming.

3.1.1 Contextualization Netflix

The Netflix brand was founded in 1997 and today offers its subscribers the ability to watch movies and television series via streaming. Present in over 190 countries, and with over 130 million subscribers, it is a successful example of streaming [4]. Their strategy differs across markets, and many subscribers prefer programming in their local language. However, there are still many potential subscribers, accustomed to free content, still hesitate to pay for streaming services [4]. Netflix, with its global reach, has more subscribers worldwide than all other pure combined streaming services [4].

3.2 Literature Review

3.2.1 Convenience of Service

The concept of convenience of service has been in huge demand from consumers [11], as there is a direct, significant, and positive relationship between convenience of service and value perceived by consumers [13], being a strong motivating agent for consumer engagement [25]. Convenience is a nonmonetary investment [5] and concerns two variables: temporal and economic, corresponding to the time and effort that consumers use to purchase goods or services. It may be influenced by a sociological and a philosophical dimension, that is, the concept of convenience is dependent upon social class, race, values, and personal and cultural norms [32]. Farquhar and Rowley [11] add a demographic dimension that may condition the orientation of convenience, namely, age, occupation, working hours, education, and lifestyles.

Seiders et al. [26] understand the concept of convenience from five dimensions that served as the basis for the construction and validation of the SERVCON scale, aiming to measure the convenience of the perceived service. Thus, this scale integrates items such as convenience decision, convenience access, convenience benefit, transaction convenience, convenience post-benefit, satisfaction, shopping fun, product category involvement, behavioral intentions, repurchase visits, and repurchase of spending.

3.2.2 Service Sophistication

Sophistication is one of the dimensions of brand personality [1], being defined as "the degree of elegance and style in a brand's personality trait" [19], as glamorous, pretentious aspects, charming and romantic [1], elegance and charm [23] good-looking, glamorous, elegant, confident, persistent, and leader. Barnes and Corbitt [3] state that consumers have service sophistication needs and that service sophistication can promote good consumer relations with a particular brand and can affect the perception of brand quality [23]. Lin [19] argues that sophistication has a positive influence on the affective loyalty and action of consumers to a brand and interferes with service quality [15].

Wentzel [31] adapts Aaker's scale and measures the concept of sophistication from the following indicators: not upper-class/upper-class, not good-looking/good-looking, not charming/charming, and not glamorous/glamorous. This scale is also the guiding basis for the study of sophistication as a dimension of brand personality in the present investigation.

3.2.3 Brand Trust

Brand trust has been cited as a central concept in long-term relationships. To build customer trust, brands give them the information they need [8] and trust is important throughout the buying process [22]. Chaudhuri and Holbrook [6] define trust as "the average consumer willingness to rely on a brand's ability to perform its stated function." Thus, trust is one of the main antecedents of loyalty [6, 8], which translates into a willingness for consumers to co-create information and become more loyal [6]. Faced with situations of uncertainty, information asymmetry, or fear of opportunism, trust helps reduce uncertainty or lack of information [8] and profoundly influences "interpersonal and intergroup behavior" [14]. Both trust and commitment are extremely important variables in successful business and bring sustainable competitive advantages [7]. Studies by Thomson [28] show that strong brand links are predictive of satisfying relationships, with trust and commitment from consumers. According to this literature review, we propose the following research hypotheses:

H1: Convenience has positive effects on trust in services.
H2: Sophistication has positive effects on service trust.

3.2.4 The Importance of Paying a Premium Price

Marketing has evolved toward the development of a relational mindset that involves the characteristics of a product or brand [10]. Brand actions foster a sense of loyalty that is more profitable, as consumers are more willing to pay more for one brand than another. Willingness to pay a premium price is defined as "the amount a customer is willing to pay for their preferred brand over comparable/ smaller brands of the same size/ package quantity" [21]. Studies by Anselmsson et al. [2] show that quality is a significant determinant of premium pricing, but the addition of other image dimensions doubles predictability and understanding of premium pricing. The strongest determinants are the social image, exclusivity, and the country of origin, and further studies on this subject are needed. Thus, our study will test the following research hypotheses:

H3: Convenience of service positively affects willingness to pay premium price.
H4: The sophistication of the service has positive effects on the willingness to pay a premium price.

Dewar [9] considers that a brand can charge a premium price for simplifying the buying process and reducing consumer risk. Consumers having an ongoing relationship with brands are less price sensitive [27]. On the other hand, studies show that there is a positive relationship between perceived quality and premium price with impact on behavior [21], preference and purchase intention [30]. In this sense, our study aims to test the following research hypotheses:

H5: Trust in services positively affects customers' willingness to pay a premium price for the service.

By deducting the research hypotheses, we constructed the following conceptual model (Fig. 3.1).

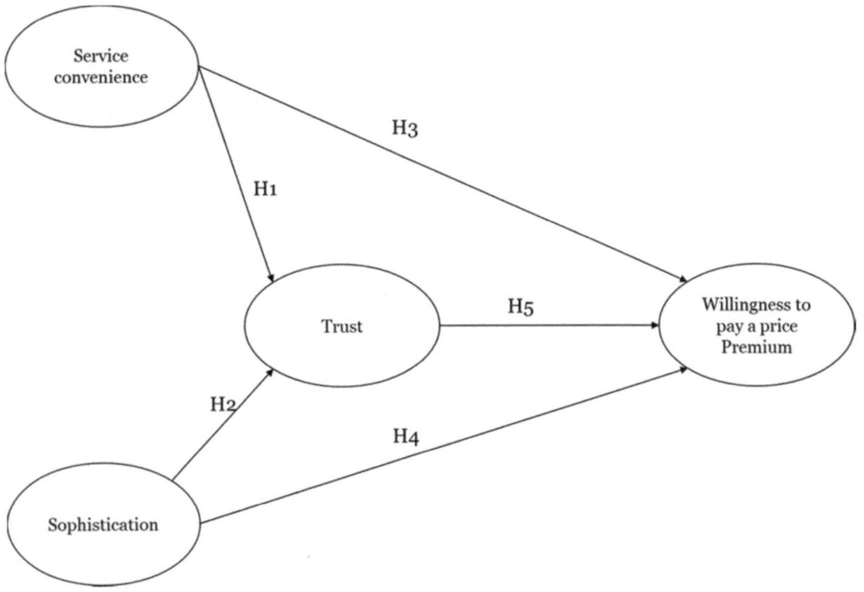

Fig. 3.1. Conceptual model

3.3 Methodology

This work was developed through an exploratory causal study, whose data collection instrument was an online self-administered questionnaire by users of streaming services. The sampling technique chosen was convenience sampling, as we wanted to collect data from individuals who were users of streaming services available in Portugal. To do so, we chose users of Netflix brand services, as this brand is internationally recognized and has sophisticated and robust services. Thus, data were obtained from a valid sample of 159 individuals in 2019, between January and March, after conducting a pretest through 14 individuals.

The sample consists of equal proportions of males (50.0%) and females (50.0%) whose ages are between 20 and 29 years in most cases (53.5%). Most of these individuals have higher education qualifications and their income is in most cases below €3000. However, a large proportion of individuals (68.8%) have used the streaming service more than 10 times. The operationalization of the questionnaire was developed through multi-item, reflexive constructs that were adapted from other authors and measured using five-point Likert scales, ranging from strongly disagree (1) to strongly agree (5). Thus, the convenience of service was adapted from Seiders et al. [24]. For sophistication, the Wentzel [30] measurement scale was used. Trust was measured through the items proposed by Thomson [28]. Willingness to pay a premium price was developed through the study of Anselmsson et al. [2]. The items are presented in Table 3.1.

From the obtained data, it was verified the existence of multicollinearity (VIF coefficient—variance inflation factor) whose values should be less than 5 as indicated in the literature [20]. From this analysis, due to the violation of the high VIF values, the items SCO3 and TRT3 were removed from the analysis, as shown in Table 3.1.

Table 3.1. Items used in the questionnaire

Description	Mean	Standard deviation	Loadings
Service convenience (SCO)			
This service is fast	4.15	0.818	0.944**
This service is easily accessible	4.25	0.780	0.928**
This service has high availability.	4.01	0.854	Deleted
Service sophistication (SPH)			
I consider the Netflix service to be high class	3.88	0.863	0.895**
I find the Netflix service attractive	4.03	0.922	0.893**
I consider the Netflix service to be global	3.92	0.964	0.742**
Trust (TRT)			
I trust in the Netflix service	3.97	0.946	0.935**
Whenever I need to know I can count on the Netflix service	3.69	1.040	0.924**
I can count on the Netflix service	3.76	0.980	Deleted
Willingness to pay a premium price (WPP)			
I am available to pay more for Netflix services compared to other similar services	2.64	1.249	0.872**
Price paid for Netflix services is appropriate	3.44	1.055	0.863**
If the price of Netflix services doubles, I will continue to be customer	1.94	1.195	0.727**

**$p < 0.01$

3.4 Results

Following the previous analysis of the data, it was decided to estimate the model through the method of structural equation analysis by partial least squares (PLS). PLS analysis is divided into two steps [17]. The first step to perform is the analysis of the measurement model whose objective is to evaluate the ability of the items under study to measure their concepts. This evaluation of the measurement model is made in terms of reliability and validity [20]. Construct validity is a measure of analysis of the validity of the measurement model. In this measure, from the data presented in Table 3.1, it is verified that the items under analysis have standardized coefficients between 0.727 and 0.944 and its statistical significance exists ($p < 0.01$). Thus, it is considered that the data obtained attribute the validity of the factors under analysis. For the convergent validity analysis, the extracted mean variance and the composite reliability were analyzed. The analysis of the average variance extracted (AVE) is considered acceptable since each construct has AVE > 0.5 (between 0.678 until 0.876) as shown in Table 3.2. The indicated values suggest that the factors also have internal consistency, since Cronbach's α values (between 0.772 and 0.859) are greater than 0.7 in all cases [16]. In addition, the factors under study have composite reliability (CR) greater than 0.7 (between 0.862 until 0.934). Then, the verification of discriminant validity is also analyzed and our study supports the existence of bivariate correlations inferior to the square root of the extracted mean variance (AVE) construct (on the diagonal of Table 3.2) which confirms the criterion of Fornell and Larcker [12] to affirm that the concepts under study have discriminant validity.

Table 3.2. Correlations between constructs and validity of the measurement model

	α Cronbach	CR	AVE	SCO	SPH	TRT	WPP
SCO	0.859	0.934	0.876	**0.936**	0.818	0.751	0.504
SPH	0.801	0.883	0.716	0.681*	**0.846**	**0.855**	0.682
TRT	0.843	0.927	0.864	0.643*	0.797*	**0.930**	**0.733**
WPP	0.772	0.862	0.678	0.457*	0.569*	0.625*	**0.823**

*p < 0.05; **p < 0.01, *Note* The diagonal elements (bold) are the square root values of the extracted mean variance. The values below the diagonal are the correlation coefficients. The elements above the diagonal are the HTMT ratio values of the correlations of Henseler et al. [18]

The second step of the model analysis through partial least squares structural equations (PLS) is the structural model test. For this purpose, the bootstrap resampling technique with 5000 subsamples was used to have stability in the obtained results. To test the hypotheses under study, we used the structural equations with partial least squares estimates using SMART PLS v3.2.8 software. This analysis allowed to evaluate the relationships under study as well as the indirect effects present in the model (Fig. 3.2).

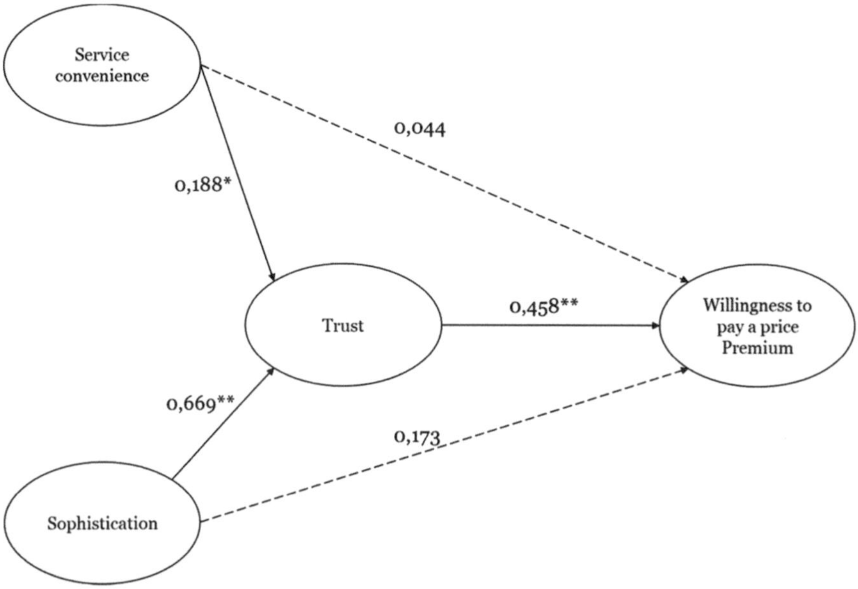

Fig. 3.2. Structural model tested, *p < 0.05; **p < 0.01

Although our results show no direct effect on service sophistication and service convenience in willingness to pay a premium price, from Table 3.3, we conclude that service convenience affects trust in service ($\beta_{SCO \rightarrow TRT} = 0.188$; $p < 0.05$). Similarly, our results corroborate that the sophistication of the streaming service affects trust in the streaming service ($\beta_{SPH \rightarrow TRT} = 0.669$; $p < 0.01$). The results provide evidence to assert that trust has direct effects on willingness to pay more for a service ($\beta_{TRT \rightarrow WPP} = 0.458$; $p < 0.01$).

Table 3.3. Hypothesis test

Hypothesis	B	T value	CI
H1: SCO → TRT	0.188*	2.095	[0.018; 0.366]
H2: SPH → TRT	0.669**	8.708	[0.508; 0.811]
H3: SCO → WPP	0.044	0.442	[−0.152; 0.238]
H4: SPH → WPP	0.173	1.336	[−0.083; 0.426]
H5: TRT → WPP	0.458**	4.733	[0.258; 0.641]

β—Standardized path coefficients; CI—confidence interval 95%

Our work also sought to evaluate the indirect effects present in the research model that we show in Table 3.4.

Table 3.4. Summary of indirect effects

Path	B	p value	Confidence interval 95%
SCO → TRT → WPP	0.086*	0.040	[0.01..0.17]
SPH → TRT → WPP	0.307**	0.000	[0.15..0.47]

β—Structural coefficient; *p < 0.05; **p < 0.01

The indirect effects present in the model are significant, as both sophistication influences the willingness to pay a premium price ($\beta_{SPH \rightarrow TRT \rightarrow WPP} = 0.307$; $p < 0.01$) as convenience of service influences the willingness to pay a premium price for streaming services ($\beta_{SCO \rightarrow TRT \rightarrow WPP} = 0.086$; $p < 0.05$).

3.5 Conclusions

The adoption of streaming services has grown, and, in this context, our study makes important contributions to the management of these services. Regarding theoretical contributions, our study shows that the effects of the variables under study on willingness to pay more for these services are supported by this study. In this sense, trust assumes a mediating role in the effects of service sophistication and service convenience. Our study also points out that the willingness of individuals to pay a premium price has as its central variable the trust that, in the absence of this, the influence of the sophistication and convenience of the streaming service ceases to exist. As practical inputs, service managers in general, and streaming service managers, should focus their actions on increasing individuals' trust in order to increase their willingness to pay more for the services they enjoy. To increase trust the streaming providers should be showed content according to each customer's profile. It is also suggested that a customer support system should exist and provide testimonials from satisfied customers.

Although our study shows important conclusions, the fact that the sample consists of young people with higher qualifications is a limitation. In addition, it was requested that only respondents who were users of Netflix streaming services should participate in our study, which may be a limitation. Thus, we suggest studies that diversify the sample presented here and investigate other variables that determine the willingness to pay a premium price that have not been studied in this paper.

References

1. Aaker, J.L.: Dimensions of brand personality. J. Mark. Res. **34**(3), 347–356 (1997)
2. Anselmsson, J., Vestman Bondesson, N., Johansson, U.: Brand image and customers' willingness to pay a price premium for food brands. J. Prod. Brand Manag. **23**(2), 90–102 (2014)
3. Barnes, S.J., Corbitt, B.J.: Mobile banking: concept and potential. IJMC **1**(3), 273–288 (2003)

4. Brennan, L.: How Netfix Expanded to 190 Countries in 7 Years. Harvard Business Review. https://www.hbsp.harvard.edu/product/H04LEY-PDF-ENG?itemFindingMethod=Catalog. Last accessed 14 July 2019
5. Chang, M.Y., Chen, K., Pang, C., Chen, C.M., Yen, D.C.: A study on the effects of service convenience and service quality on maintenance revisit intentions. Comput. Stand. Interfaces **35**(2), 187–194 (2013)
6. Chaudhuri, A., Holbrook, M.B.: The chain of effects from brand trust and brand affect to brand performance: the role of brand loyalty. J. Mark. **65**, 81–93 (2001)
7. Chen, Y.H., Lin, T.P., Yen, D.C.: How to facilitate inter-organizational knowledge sharing: the impact of trust. Inf. Manag. **51**(5), 568–578 (2014)
8. Chiu, C.M., Huang, H.Y., Yen, C.H.: Antecedents of online trust in online auctions. Electron. Commer. Res. Appl. **9**, 148–159 (2010)
9. Dewar, N.: What are brands good for? MIT Sloan Manag. Rev. **46**(1), 86–94 (2004)
10. Dwivedi, Y.K., Kapoor, K.K., Chen, H.: Social media marketing and advertising. Mark. Rev. **15**(3), 289–309 (2015)
11. Farquhar, J.D., Rowley, J.: Convenience: a services perspective. Mark. Theory **9**(4), 425–438 (2009)
12. Fornell, C., Larcker, D.F.: Evaluating structural equation models with unobservable variables and measurement error. J. Mark. Res. **18**(1), 39–50 (1981)
13. García-Fernández, J., Gálvez-Ruíz, P., Fernández-Gavira, J., Vélez-Colón, L., Pitts, B., Bernal-García, A.: The effects of service convenience and perceived quality on perceived value, satisfaction and loyalty in low-cost fitness centers. Sport Manag. Rev. **21**(3), 250–262 (2018)
14. Golembiewski, R.T., McConkie, M.: The centrality of interpersonal trust in group processes. In: Cooper, C.L. (ed.) Theories of Group Process. John Wiley, New York (1975)
15. Gorla, N., Somers, T.M., Wong, B.: Organizational impact of system quality, information quality, and service quality. J. Strateg. Inf. Syst. **19**(3), 207–228 (2010)
16. Hair, J., Black, W., Babin, B., Anderson, R.: Multivariate Data Analysis: global Edition, 7th edn. Pearson Higher Education, New York (2010)
17. Hair, J., Hult, G., Ringle, C., Sarstedt, M.: A Primer on Partial Least Squares Structural Equation Modeling. SAGE Publications, PLS-SEM) (2016)
18. Henseler, J., Ringle, C.M., Sarstedt, M.: A new criterion for assessing discriminant validity in variance-based structural equation modeling. J. Acad. Mark. Sci. **43**(1), 115–135 (2015)
19. Lin, L.Y.: The relationship of consumer personality trait, brand personality and brand loyalty: an empirical study of toys and video games buyers. J. Prod. Brand Manag. **19**(1), 4–17 (2010)
20. Maroco, J.: Análise Estatística com o SPSS. 6ª Edição. Pêro Pinheiro: Report Number (2014)
21. Netemeyer, R.G., Krishnan, B., Pullig, C., Wang, G., Yaci, M., Dean, D., Ricks, J., Ferdinand, W.: Developing and validating measures of facets of cus-tomer-based brand equity. J. Bus. Res. **57**(2), 209–224 (2004)
22. Powers, T., Advincula, D., Austin, M.S., Graiko, S., Snyder, J.: Digital and social media in the purchase decision process: a special report from the advertising research foundation. J. Advert. Res. **52**(4), 479–489 (2012)
23. Ramaseshan, B., Tsao, H.Y.: Moderating effects of the brand concept on the relationship between brand personality and perceived quality. J. Brand Manag. **14**(6), 458–466 (2007)
24. Ramasubramanian, S., Murphy, C.J.: Experimental studies of media stereotyping effects. In: Webster, M., Jr., Sell, J. (eds.), Laboratory Experiments in the Social Sciences, pp. 385–402. Elsevier Academic Press, San Diego, CA, US (2014)
25. Roy, S.K., Shekhar, V., Lassar, W.M., Chen, T.: Customer engagement behaviors: the role of service convenience, fairness and quality. J. Retail. Consum. Serv. **44**, 293–304 (2018)

26. Seiders, K., Voss, G., Godfrey, A., Grewal, D.: SERVCON: development and validation of a multidimensional service convenience scale. J. Acad. Mark. Sci. **35**(1), 144–156 (2007)
27. Thomson, M., MacInnis, D.J., Park, C.W.: The ties that bind: measuring the strength of consumers' emotional attachments to brands. J. Consum. Psychol. **15**(1), 77–91 (2005)
28. Thomson, M.: Human brands: investigating antecedents to consumers' strong attachments to celebrities. J. Mark. **70**(3), 104–119 (2006)
29. Vaala, S.E., Jordan, A.B.: Chapter 15—Children's learning in a mobile media environment: policies, practices, and possibilities. In: Blumberg, F.C., Brooks, P.J. (eds.), Cognitive Development in Digital Contexts, pp. 305–324. Academic Press, San Diego (2017)
30. Washburn, J.H., Plank, R.E.: Measuring brand equity: an evaluation of a consumer-based brand equity scale. J. Mark. Theory Pract. **10**(1), 46–62 (2002)
31. Wentzel, D.: The effect of employee behavior on brand personality impressions and brand attitudes. J. Acad. Mark. Sci. **37**(3), 359–374 (2009)
32. Yale, L., Venkatesh, A.: Toward the construct of convenience in consumer research. ACR North American Advances (1986). http://acrwebsite.org/volumes/5956/volumes/v13/NA-13. Last Accessed 6 July 2019

Chapter 4
Websites as Spaces of Confluence: Narratives and the Pursuit of Legitimacy

Fernando Pinto Santos(✉) and Mafalda Nogueira

IPAM, Porto, Portugal
fernando.santos@ipam.pt, mafalda.nogueira@ipam.pt

Abstract. Narratives are important means through which entrepreneurs can present rationales for their activities and influence assessments of their company's plausibility. The purpose of this research is to understand how can websites be engaged in the construction of entrepreneurial marketing narratives. To this end, we have followed a company operating in the fashion industry for over a year. As the initial fieldwork evolved, websites emerged as critical in the articulation of the company's narrative. Consequently, this research was driven by the goal to develop an understanding on how websites can be employed in the construction of narratives. The results allowed developing two main contributions. First, we suggest that websites can gain a central role in the construction and dissemination of narratives, enabling to make them tangible and persistent over time. Narratives in websites can become reference points for companies, influencing internal stakeholders and feeding further accounts. Second, we identify and explore five qualities of narratives in websites that influence how these narratives can be constructed and disseminated: (1) tangibility; (2) persistence; (3) ubiquity; (4) multimodality; and (5) hyper-intertextuality.

4.1 Introduction

Early stages of entrepreneurial activities are usually characterized by the need to attract resources required for growth, and prior research has highlighted the role of identity narratives as a mechanism to influence positive assessments of a company's plausibility [1–3]. Although different semiotic resources can be employed in narratives, as video, sound, photos, and a combination of these [4, 5] like it happens in websites and some social media, research has not addressed how different communication channels and distinct semiotic modes might influence the entrepreneurial narrative construction and dissemination [6, 7]. Accordingly, we have decided to pursue the following research question: how can websites be engaged in the construction of entrepreneurial marketing narratives?

Á. Rocha et al. (eds.), *Marketing and Smart Technologies*, Smart Innovation, Systems and Technologies 167, https://doi.org/10.1007/978-981-15-1564-4_4

4.2 Related Literature

Following the linguistic turn in social science, there has been a growing interest of scholars in the discursive analysis of organizations. Different areas have been explored with an interest in narratives such as identity, marketing, strategy, and entrepreneurship, among others (for a review see [8]). It is the temporal dimension of narratives that distinguishes these from other kinds of discourse [8, 9]. We will use the term narrative in an encompassing way, drawing from literature that employs both the concepts of story and narratives, and considering these interchangeably. Aldrich and Fiol [10, p. 652] have early addressed entrepreneurial narratives, suggesting that entrepreneurs may use narratives to "make a case that their ventures are compatible with more widely established sets of activities," thus creating legitimacy. Subsequently, Lounsbury and Glynn [2] proposed the concept of "cultural entrepreneurship" to address the role of narratives in the construction of a new organization's identity that might facilitate legitimacy attribution by different stakeholders and access to acquisition of resources.

The way a narrative is communicated bears implications for the narrative itself [6, 7, 11]. However, the range of types of communications that have been studied in entrepreneurial domain has been extremely limited [7]. On the one hand, there has been attention to oral accounts [10]. On the other hand, there is also attention to written accounts of narratives [3]. Fischer and Reuber [7] argue that entrepreneurial firms can articulate their narratives not only in conventional means of communication (such as prospectuses or personal interactions) but also in social media, as Twitter. In overall terms, a focus on textual forms is consonant with the broader contemporary narrative research focus on written texts and oral communication [8].

Kallinikos et al. [12] have synthesized four key features of digital artifacts: first, these are editable and, at least in principle, are always possible to modify or update. The editability of the content does not imply interference on the logical structure that governs the object. Second, digital artifacts are interactive, offering different embedded options that allow users to explore information. Third, digital artifacts are open to be approached and modified by means of other digital objects, being reprogrammable in their underlying structure and functioning (and thus expandable and reusable, see [13]. Openness is thus seen differently from editability. Finally, digital artifacts are easily distributed and can spread over information infrastructures and the Internet. Websites exemplify well the editability, interactivity, openness, and distributedness of digital artifacts as proposed by Kallinikos et al. [12]. Although digital technologies are becoming increasingly ubiquitous in organizations, our understanding of the consequences and opportunities created by the organizational pervasiveness of this informatization continues to be well underdeveloped [14, 15].

Kress [16, p. 79] defines mode as "a socially shaped and culturally given semiotic resource for making meaning. Image, writing, layout, music, gesture, speech, moving image, soundtrack and 3D objects are examples of modes used in representation and communication." In the entrepreneurial domain, Navis and Glynn [6] have highlighted that different communication modalities affect the communication of narratives, echoing a wider recognition that if "modes differ in their affordances, it is evident that they configure the world differently, often profoundly…" [16, p. 79]. Navis and Glynn [6] suggested further studies to explore different modalities in narratives but this path of inquiry remains largely underexplored. This study addresses this challenge.

4.3 Methods

This article builds on a case study, within the qualitative and interpretative research tradition [17, 18]. The company was selected due to its information richness and potential to learn about the phenomena under investigation, according with the purposes of this study [19]. The case of Lovia was considered as relevant to study the phenomenon being researched and a good opportunity to learn [20]. According to the case study research logic, a variety of qualitative sources of data were employed [21]. Full access was granted to the organization and this allowed developing interviews with the entrepreneurs as well as interviews with others, like interns. The collection of empirical material started in January 2015 and ended in March 2016. Sources of data include transcripts of 12 semi-structured interviews with the entrepreneurs and other persons in the company; notes from numerous informal talks and from observations made in visits to the company; copies of internal documents, press releases, and the content of the website and social media of Lovia. News in the media about the company and public presentations of the entrepreneurs were also analyzed. The collection, analysis and interpretation of empirical material took place simultaneously throughout the process of research, in an iterative way. A concern with triangulation of different sources was always present in the analysis of data. The complete set of empirical material was analyzed, in different moments in time, in an open and holistic manner, according with the purpose of this research. Interpretive content analysis guided the process of analysis and allowed to thematize this material [22].

4.4 Results

Lovia was founded in the summer of 2014 by three Finnish entrepreneurs. The first products became available in the end of January of 2015 and in May the company's online shop was launched. Lovia designs and manufactures women clothes and bags. Clothes are produced in its atelier in Helsinki and bags are manufactured in the nearby Estonia. Sustainability is a central concern of the entrepreneurs. Products are made using only recycled and organic materials and this is highly emphasized in the communication of the company. Lovia's products are targeted at high-end consumers, with a special focus on international markets.

Lovia's communication revolves around a narrative that connects the origins of the company with what it is doing nowadays and how it is doing it. This narrative is the dominant one in the organization and its hegemonic nature is promoted by the entrepreneurs. Although we recognize that there are other narratives circulating in the company, our focus was set on the dominant official narrative of the organization.

Essentially, Lovia's narrative presents the identity of the company as based on three central themes: sustainability, luxury, and transparency in its production processes. The following excerpt from a company's presentation document illustrate these themes, describing the beginning of the narrative and establishing a temporal sequence with the present and the intended future, as well as introducing the entrepreneurs as central characters in the narrative, and producers as secondary characters:

> Do you have any idea who has made your luxury bag? We didn't, and that is why we started Lovia. Lovia makes traceable high-end design from excess materials and wild nature. All products have been designed based on the materials. All Lovia products are traceable to the material sources, production methods and locations. With our DNA concept we can introduce all these and the people and their faces along the production chain. The product DNA also works as a marketing tool (…).
>
> For us, elegance goes beyond beautiful surface, and this is the elegance we are to reveal. The core idea of Lovia is to make elegant accessories and clothes with elegant values - transparency, sustainability and sharing (…).

Lovia's core team consists of three founders who are professionals from complementary fields. Lovia's designer Outi Korpilaakso has years of work experience on design and production in a large commercial company as well as in a smaller fashion start-up. Tiina Kosonen is the sales expert of the team and has specialized on fashion and lifestyle business. A former journalist Anniina Mustalahti takes care of the DNA concept and PR.

This language-based account is intertwined with photos in the web page, and the visual dimension assumes a dominant presence. In the website the "DNA" section provides information about the production process: "As transparency is one of the core values of Lovia, the product DNA was created to provide you with information about our production chain. Scroll down to learn more about the materials and their suppliers, production and makers of our products." The photos and the detailed descriptions of the people involved in the production and of the spaces where Lovia's manufactures its products, presented in the website, are regarded by the entrepreneurs as a manifestation of the genuineness of their narrative. As one of the founders told us: "we can actually prove the sustainability as we have the DNA concept." So, in this entrepreneurs' perspective, the information in the website proves that Lovia is developing its operation in a sustainable way, by opposition to the competition that "don't actually prove what they're doing differently compared to others…" (Tiina).

The website seems to enable an ontological accomplishment: the narrative of Lovia is real because one can check the photos and the information of the producers, of real people. As the fieldwork evolved, we were able to increasingly realize that the website assumes a critical role in the company's activities. The website is clearly a relevant communication channel for Lovia and its narrative, but it assumes several functions beyond the simple conveyance of information between the organization and its external audiences. The website communicates to different stakeholders (internal and external), crossing over organizational boundaries, binding the entrepreneurs to their commitments, and enabling their own reflection based in the exposure assumed:

> I think it's more important that they [company's values] are on the website than in our papers… Because I think it's very important for our customer… in our point of view… and with our ideology… to tell them, so I think it's actually more important to inform people very clearly and very widely… It also makes us responsible for, realizing those values and not only, saying that's something that we would like to do… (Anniina)

I think it's good place to check our own, values again and again and again, because it shows OK, this is what we are making, you can actually see it, everybody can see it. This is what we made. Should we change something or, we should, whatever we are doing, the new things we are doing? …We just check that we are doing things according to our vision. (Outi)

Moreover, the information in the website serves as reference point for the entrepreneurs, supporting further communication and decisions:

…it kind of helps that we have the DNA concept, that through we check the new materials or the manufacturers or suppliers through this DNA concept. And, then we can see right away if they're suitable or not, because of course we don't want to put anything this kind of unclear stuff or use this kind of manufacturer or anything. So, it's very good way of keeping the standard… (Outi).

Furthermore, and related to this stabilization, the narrative in the website endures through time, as it is recognized by the entrepreneurs:

I think it's really good that we have this because our concept, we have this DNA concept, so it's our core message for everywhere, we can always refer that OK we have this, kind of library for all the things we are making. So I think if we would only, talk about it in social media, the message will always disappear because the feed is full and, bla bla bla… So now we have the core place that we, the information and it's all there. (Anniina)

In overall terms, the entrepreneurs recognize that websites play a critical role in contemporary business practices, as illustrated in their own behavior when an organization contacts Lovia: "… if people contact us, of course we Google it. And then we are right away like… OK, what is this?" This view on the importance of websites influenced the efforts made in developing Lovia's own website, as explained by the entrepreneurs:

Especially when we decided that we don't want to have a, proper store, from the beginning, the website had to be, the channel to showcase the company, in just a second. And if, the new customer, a possible customer goes on our site, she or he will see immediately… what Lovia is about.

In overall terms, Lovia's website gradually assumed a central importance in the company's activities. The website is the main store of the organization, presenting and making available all its products, as well as it is the principal channel through which the narrative is articulated. Through the different modes, the narrative of the company is made quite vivid and secondary narratives that complement the main one also are highlighted. As analyzed, the website serves not only as a communication channel to external audiences, as it also became a reference point for the entrepreneurs themselves. Thus, the website became an essential artifact in the company's activities, making tangible an ideational level of intent and making this enduring through time.

4.5 Discussion

In the case of Lovia, we found that the overall logic of the company's website content was organized around a dominant discourse in the form of narrative. Thus, the narrative establishes a coherence between different elements of the website. Lovia's narrative presents the temporal evolution of the organization, its characters, connections with broader issues and an overall sense of purpose. Lovia's website anchors the narrative and provides tangibility to it. Although digital, websites are tangible [23] and persistent [24]. And these qualities have been essential to the company's activities. Entrepreneurs can rely on the persistence of the narrative in the website and use it as a template for further accounts of the organization's narrative. Moreover, Lovia's entrepreneurs use the narrative in the website as a means to their own sensemaking, being able to reflect on their company's purposes and activities, while considering how to progress and imagine the future [23]. Thus, Lovia website became a "self-enhancing reference point" [25] for the entrepreneurs; influencing them through time.

Our findings challenge perspectives on websites as ever-changing [26]. Although previous works have characterized digital artifacts, like websites, as continually in the making [12] and as presenting a lack of the stability characterizes traditional objects [27], still these can be regarded as tangible and persistent [24, 28]. Importantly, the qualities of Lovia's website became a central aspect of the legitimacy claims of the entrepreneurs. The website provides an ontological existence to the organization's activities, showcasing the products and their makers, as well as the entrepreneurs, in their physical form (through their photos) and their ideas, though the explanation of their motivations. The transparency of the production processes, in particular, is vividly portrayed in the website with photos and videos of the producers and their working environments. These multimodal qualities enhance the vividness of Lovia's account and by consequence also the plausibility of the company's legitimacy claims.

As it became evident in Lovia's narrative construction, websites enable different features. First, and as stated, narratives in websites can become tangible and persistent. Narratives can also become multimodal as we have seen, and photos, graphical elements, and videos, for example, allow crafting a vivid account. Having the narrative presented in different modes allows the web page visitors to read the narrative according to their modal preferences. Moreover, and related to this openness in reading the narrative, in the website we have hyperlinks (what [29], has called hyper-intertextuality) that connect different parts of the website and that readers can follow at their will. Finally, websites make narratives ubiquitous, and this feature impacts the way one can relate with the narrative: we know that the narrative is tangible, persistent, and always available in the website and thus we can access it and use it whenever we want, through a computer or mobile phone, for example. From blending this study's insights with existing theory, we have developed the (Table 4.1).

Table 4.1. Qualities of narratives in website

Qualities of narratives in websites	Lovia's case	Literature
Tangibility	Lovia's narrative aggregated different ideational elements and the website enabled to make these elements available together and tangible. Websites are a medium where the narrative can be weaved together and reified	Materiality of digital artifacts [28]
Persistence	The website provides the principal channel through which Lovia's narrative has been enduring. The narrative in the website became its official account and a reference point for further oral and written accounts. Stakeholders can rely on the narrative in the website: it persists through time	Affordances of social media: persistence [24]
Ubiquity	Lovia's narrative in the website is always accessible, at any moment and from almost any place. There are no geographical and time constraints to access it. Thus, websites imply a recontextualization of interpretation	Affordances of social media: ubiquity [16]
Multimodality	Lovia's narrative in the website is constructed with different modes. The ensemble of these modes, with a particular dominance of the visual one, makes the narrative vivid. The multimodal qualities of narratives in websites can facilitate the reader's engagement. Also, multimodality allows the reader of the narrative to freely choose the modes he/she prefers to explore	Multimodality [3, 16, 30, 31]
Hyper-intertextuality	The narrative in the website of Lovia is presented with hyperlinks that connect different elements of the narrative. Website visitors have a high degree of freedom to explore the narrative according to their interests	Hyper-intertextuality of corporate blogs [29]

4.6 Conclusions

Websites can serve as a confluence space of different discourses of companies and assume multiple simultaneous roles: marketing, corporate communication, and online sales, among others. In this multi-purposive space, different functions can become entangled and divides between the content blurred. This study suggests that websites might assume a central role in the articulation of narratives, impacting how entrepreneurs present and sustain their legitimacy claims. First, websites assume an ontological function, demonstrating that the new organizations do exist. Websites were found to be especially important to make the entrepreneurs' narrative tangible and persistent through time. These qualities make websites suitable to become a reference point for the stakeholders, a stable platform from where entrepreneurs can develop further accounts of narratives and reflect on the future. Furthermore, websites make narratives ubiquitous: the narrative is tangible, persistent, and always available online. Finally, the multimodal qualities of websites enable to present marketing narratives in a vivid way, which endows the account with aesthetic properties and makes them potentially more engaging.

References

1. Aldrich, H., Fiol, M.: Fools rush in? The institutional context of industry creation. Acad. Manag. Rev. **19**(4), 645–670 (1994)
2. Lounsbury, M., Glynn, M.A.: Cultural entrepre-neurship: stories, legitimacy, and the acquisition of resources. Strat. Manag. J. **22**, 545–564 (2001)
3. Martens, M.L., Jennings, J.E., Jennings, P.D.: Do the stories they tell get them the money they need? the role of entrepreneurial narratives in resource acquisition. Acad. Manag. J. **50** (5), 1107–1132 (2007)
4. Meyer, R., Höllerer, M., Jancsary, D., van Leeuwen, T.: The visual dimension in organizing, organization, and organization research: core ideas, current developments, and promising avenues. Acad. Manag. Ann. **7**(1), 489–555 (2013)
5. Kress, G., Van Leeuwen, T.: Reading Images. Routledge, London (1996)
6. Navis, C., Glynn, M.A.: Legitimate distinctiveness and the entrepreneurial identity: influence on investor judgments of new venture plausibility. Acad. Manag. Rev. **36**(3), 479–499 (2011)
7. Fischer, E., Reuber, A.R.: Online entrepreneurial communication: mitigating uncertainty and increasing differentiation via Twitter. J. Bus. Ventur. **29**(4), 565–583 (2014)
8. Vaara, E., Sonenshein, S., Boje, D.: Narratives as sources of stability and change in organizations: approaches and directions for future research. Acad. Manag. Ann. 1–71 (2016)
9. Chung, H.H.: Narrative Leadership: exploring the Concept of Time in Leader Storytelling. Seattle Pacific University (2018)
10. O'Connor, E.: Storied business: typology, intertextuality, and traffic in entrepreneurial narrative. J. Bus. Commun. **39**(1), 36–54 (2002)
11. Clarke, J.S., Holt, R.: Images of entrepreneurship: using drawing to explore entrepreneurial experience. J. Bus. Ventur. Insights **11** (2019)
12. Kallinikos, J., Aaltonen, A., Marton, A.: The ambivalent ontology of digital artifacts. MIS Q. **37**(2), 357–370 (2013)
13. Kallinikos, J., Mariátegui, J.C.: Video as digital object: production and distribution of video content in the internet media ecosystem. Inf. Soc. **27**(5), 281–294 (2011)
14. Kallinikos, J.: On the computational rendition of reality. Organization **16**(2), 183–202 (2009)
15. Yoo, Y.: Computing in everyday life: a call for research on experiential computing. MIS Q. **34**(2), 213–231 (2010)
16. Kress, G.: Multimodality: a Social Semiotic Approach to Contemporary Communication. Routledge, Abingdon (2010)
17. Stake, R.E.: The Art of Case Study Research. Sage (1995)
18. Simons, H.: Case Study Research in Practice. Sage publications, London (2015)

19. Patton, M.Q.: Qualitative Research and Evaluation Methods, 4th edn. Sage Publications, Inc., Thousand Oaks (2015)
20. Stake, R.E.: The Art of Case Study Research. Sage (2009)
21. Piekkari, R., Welch, C., Paavilainen, E.: The case study as disciplinary convention: evidence from international business journals. Organ. Res. Methods **12**(3), 567–589 (2009)
22. Miles, M.B., Huberman, A.M.: Qualitative Data Analysis: a Source Book of New Methods. Sage Publications, Newbury Park (1984)
23. Gartner, W.B.: Entrepreneurial narrative and a science of the imagination. J. Bus. Ventur. **22** (5), 613–627 (2007)
24. Treem, J.W., Leonardi, P.M.: Social media use in organizations: exploring the affordances of visibility, editability, persistence, and association. Commun. Yearb. **36**, 143–189 (2012)
25. Coupland, C., Brown, A.D.: Constructing organizational identities on the web: a case study of Royal Dutch/Shell. J. Manag. Stud. **41**(8), 1325–1347 (2004)
26. Brown, A.D.: A narrative approach to collective identities. J. Manag. Stud. **43**(4), 731–753 (2006)
27. Ekbia, H.R.: Digital artifacts as quasi-objects: qualification, mediation, and materiality. J. Am. Soc. Inf. Sci. Technol. **60**(12), 2554–2566 (2009)
28. Leonardi, P.: Materiality, sociomateriality, and socio-technical systems: what do these terms mean? How are they different? Do we need them?. In: Leonardi, P.M., Nardi, B.A., Kallinikos, J. (eds.), Materiality and Organizing: Social Interaction in a Technological World, pp. 25–48. Oxford University Press, Oxford, UK (2012)
29. Barros, M.: Tools of legitimacy: the case of the petrobras corporate blog. Organ. Stud. **35**(8), 1211–1230 (2014)
30. Kress, G.R., Van Leeuwen, T.: Multimodal Discourse: the Modes and Media of Contemporary Communication. Arnold, London (2001)
31. Jones, R.H., Chik, A., Hafner, C.A. (eds.): Discourse and Digital Practices: doing Discourse Analysis in the Digital Age. Routledge, Abingdon (2015)

Chapter 5
User Behavior: The Case of Instagram

Teresa Guarda[1,2](✉), Isabel Lopes[2,3], José Avelino Victor[4,5], and Encarnación González Vázquez[6]

[1] Universidad Estatal Península de Santa Elena–UPSE, La Libertad, Ecuador
tguarda@gmail.com
[2] Algoritmi Centre, Minho University, Guimarães, Portugal
isalopes@ipb.pt
[3] UNIAG (Applied Management Research Unit), Polytechnic Institute of Bragança, Bragança, Portugal
[4] Instituto Politécnico Da Maia, Maia, Portugal
javemor@ismai.com
[5] Instituto Universitário Da Maia, Maia, Portugal
[6] Universidad de Vigo, Vigo, Spain
egzlez@uvigo.es

Abstract. Digital marketing is the promotion of products or brands through digital media, being one of the main ways that companies have to communicate with the public, directly, personalized, and at the right time. The big trend in digital marketing is social media marketing. This tool, available to everyone today, enables companies to communicate with a large number of consumers, enabling them to create, enhance, and optimize customer relationships, with the aim of promoting their products or brands using digital media. This finding raises the following question: Digital marketing is the promotion of products or brands through digital media? Thus, the main objective of this study was to understand the behavior of Instagram users, i.e., whether respondents actually use this social network, for what purpose they exploit it, and whether or not they are influenced by it. If they are influenced, which behaviors they adopt. The results are discussed in the light of the literature, and future work is identified to mitigate this recent phenomenon to which some companies are still suspicious.

5.1 Introduction

The emergence of the Internet is a challenge for classical economic analysis, as it has a huge impact on people's habits and also on their living standards all over the world [1]. Several studies have focused in recent years on the effect of Internet use and, in particular, the impact of social networks on consumer behavior.

The aim of this study is to analyze in a student community, in a young population, to what extent they are influenced by Instagram, and if they are influenced which behaviors they adopt.

Á. Rocha et al. (eds.), *Marketing and Smart Technologies*, Smart Innovation, Systems and Technologies 167, https://doi.org/10.1007/978-981-15-1564-4_5

Nowadays, social media content has been used by various brands for competing with the competitors, promoting products and offers, and maintaining a reputation among the stakeholders [2, 3].

Social networks as a facilitating medium that allows communication with a large number of consumers, achieving greater proximity between companies and their target audience are constantly evolving, and social networks have quickly become part of the corporate communication strategy [4].

On the other hand, with the reach of the Internet, there is more competition between companies seeking the attention of the same customers, so it is necessary to retain some aspects such as authenticity and credibility able to increase the probability of success [4].

This success is verified through the influence that is created on consumers, which depends a lot on their behavior. Consumer behavior can be understood as consumer behavior in finding, buying, using, evaluating, and disposing of products and services that they expect to meet their needs. More than understanding what drives consumers to buy it is necessary to understand what drives them to these attitudes and preference for a particular brand [5]. Among consumers there is also the ability to influence, for example, opinions and decisions about brands, products, and services [6].

It is in this context that the present work fits in, which structurally after this introduction, proceeds with the review of the literature about social networks and the marketing in these networks. Then, in Sect. 5.3, we present the social network under study, as well as its emergence, evolution, and purpose. In Sect. 5.4, we discuss how the survey was conducted, focusing on the population covered and its structure. Finally, the results and conclusions are presented in the light of the research data, the limitations of this study are identified, and future work is proposed.

5.2 Social Networks

Social networks are virtual spaces where groups of people or companies relate through sending messages and sharing content, among others.

Currently, there are different social networks, each with a purpose and a specific audience; we can highlight the following: Facebook, YouTube, WhatsApp, Instagram, Twitter, Pinterest, Skype, LinkedIn, Snapchat, and Messenger.

Social networking, provide a means of communication for users by connecting people, some of which might not ever meet face-to-face [7]. Millions of users select these networks for sharing their information and experiences. Most social networks are open environments that anyone can freely enter to generate contents and share them [8].

Social networks are already part of the daily lives of consumers and companies, and a significant part of the interaction between consumers and businesses is made through these platforms. Brands are increasingly integrating these tools into their marketing strategies [9]. Companies of all sizes find on the Internet the possibility of communicating with customers, suppliers, and consumers in an agile and unlimited way.

The emergence of social networks has changed the way brands interact with the market, providing advantages in bringing them closer together. Because Instagram (the social network under study in this paper) is a powerful digital platform with visual content, it gives brands the opportunity to motivate customers with its functionality [10].

5.3 Instagram

Social media is unarguably one of the wealthiest sources of information. The opinions shared on social platform have an immense influence toward the brands equity. Social media has flourished with platforms like Facebook, Twitter, and Snapchat. However, over the past decade, Instagram, one of the most famous photo posting social media, has dominated the youth's attention with its unique feature of being the first-ever photo-sharing application [11].

Thus, among the various existing social networks, the choice for Instagram is due to the fact that it is a recent social network that has achieved a large membership in a short time. Second, Instagram is the social media that has the most power of influence due to its aesthetics, sharing, and instantaneity. The first factor popularizing Instagram is aesthetics, complemented by its image filters. Also, it is possible capture images and sharing them instantly [12].

5.3.1 Origin and Evolution

Instagram appeared on October 6, 2010, to the public, by Mike Krieger and Kevin Systrom. Instagram came from the simplification of another application called Burbn, also developed by the same Instagram engineers. Burbn was an application for iPhones that worked through geolocation, allowing you to find locations, book future visits, earn points with visits, capture photos.

Due to the difficulty of using the Burbn application, the two engineers decided to focus on one feature only: sharing photos from their cell phone, which they thought was identical to an instant telegram, was renamed Ins-tagram ("Insta" comes from "Instant Camera", which means instant camera, in free translation "Gram" was taken from "telegram". Initially, the application was only available for electronic equipment equipped with Apple's iOS operating system. Only from April 2012, it was made available for devices with Android operating system. A week after the availability of the Instagram App for Android devices, Facebook creator Mark Zuckerberg announced the purchase of the company that developed the app.

Instagram can be used for a variety of purposes, such as social, marketing, branding. The most usual is to be used socially without any purpose, just entertainment. On Instagram you put content like photos or videos, this content will be seen by our followers if our account is private or by everyone who has access to our account (in this case, the account is public). With this, a simple Instagram user sees what people do (travel, parties, daily live) wear or eat, which can serve as inspiration, motivation, or influence.

In Instagram, the user signs up, starts by creating a profile. With this register, it is possible to take photos, insert special filters, making the image have artistic and differentiated traces. Instagram has filters that soften the lighting, brighten the image, increase the contrast, change the color saturation, and generate black-and-white images, among many other effects.

Thus, Instagram has a set of tools that can be used for different purposes: photo/video publishing, Instagram Web, Instagram Direct, Newsfeed, activity pages, hashtag and location, and Instagram stories.

5.4 Research Methodology

In order to understand and identify the pattern of behavior of Instagram users, it was considered appropriate to apply the Survey technique, as it could provide a clear, direct and objective answer to the questions presented to respondents. Moreover, since the aim was to broadly characterize the population, consisting of students from five schools in a higher education institution, it was considered that this number made it impossible or inappropriate to use other alternative research techniques.

The survey stands out as the social research technique most suited to large-scale studies. Apparently easy and simple to use, relatively fast, standardized and routinized, favoring comparable data, generalizable, and amenable to quantitative analysis.

Another factor that contributed to the choice of the survey was the fact that the study was conducted in a higher education institution, within the scope of the Marketing II course of management, whose respondents were basically students, thus the applicability of a online survey, to understand the behavior of users on Instagram, has become quite appropriate. For the elaboration and collection of data, we used Google forms.

5.4.1 Population

The study had as its scope a higher education institution, which consists of five schools, having a student population of about 7000 students.

In order to draw representative conclusions about a population from a sample, it is necessary to take into account the sample and the sampling error, which is the difference between a sample result and the true population result. Normally the set value is 5% and the sample required for our study of 384 responses. However, only 284 survey responses could be collected, so in this case, the sampling error is 5.82% and the confidence level used is 95%.

5.4.2 Structure

The structure of the Survey resulted from the literature review on social networks in general and Instagram in particular. The survey questions, the individual response, and confidential nature were organized into three groups.

The first section/group consisted of a brief description of the scope of the study and an indication that participation in the study was completely voluntary, the confidentiality of the data collected was ensured and it was reported that it was a one-time questionnaire, exclusively academic.

The second group aimed to obtain a brief characterization of the respondent, namely, gender, age, residence, and academic qualifications.

The third group focused on the main object of the study, the Instagram app. It consisted of 12 questions, among them, if you use social networks in your daily life, if you use Instagram, how much time per day, what publications you do, what you research, among others.

An attempt was made to make a survey simple to answer and not too long to get as many answers as possible.

5.5 Results

Many studies in literature focus on different forms of social media including blogs and conclude that these media outlets gain immense traction and have a subsequent social influence on the information consumers [13, 14, 15].

This study tried to verify this reality, gathering the necessary data, proceeding to a simple and bivariate descriptive analysis of the data and hypothesis tests.

5.5.1 Simple Descriptive Data Analysis

Looking at the data obtained in the survey, we found that out of 284 responses, 70.4% corresponded to females and 29.6% to males.

Regarding the age group, the histogram analysis (see Fig. 5.1) shows that the answers are mostly centered between 20 years and 25 years. This is due to the fact that the surveys were conducted at the academic level.

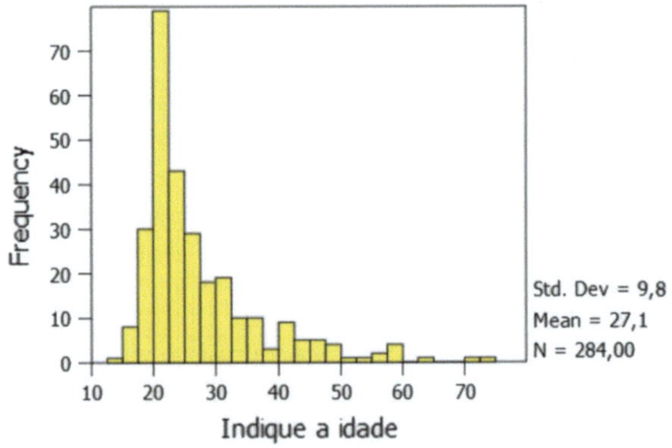

Fig. 5.1. Age range histogram

It is noteworthy that we are facing a quantitative variable, that is, we were able to determine several parameters, such as the mode, median, and mean that belong to the location measurements and the standard deviation and variance that belong to the variation (see Fig. 5.2).

N	*Valid*	284
	Missing	0
Mean		27,14
Mode		22,00
Std Dev		9,85
Variance		97,00
Minimum		14,00
Maximum		74,00
	50 (Median)	24

Fig. 5.2. Age range parameters

Regarding Instagram, asked whether the account is private or public, 74.5% of the surveyed population has a private account and 25.5% has a public account, meaning that most people are stuck with their privacy.

Regarding the question of how much time you spend on Instagram daily, it was found that people are on Instagram between 30 min at 1:30 pm a day.

Regarding the publications you make on Instagram, there were 236 answers to this question, 85.6% publish photos, 67.4% stories, 16.1% videos, and lastly 5.9% direct (see Fig. 5.3).

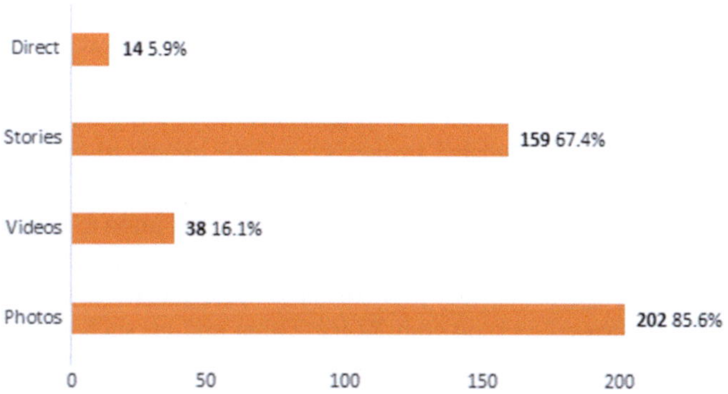

Fig. 5.3. Instagram posts

Asked what they do when they see a post that for some reason piqued their interest, 66.2% of people like it. Only 18.1% of people keep the post in their collection. The other options also have some adherence, but it is not so usual, such as commenting, identifying someone to view the post, sharing the post in their stories among others.

Regarding the question of whether you are an influenced person, the answers can be seen in Fig. 5.4. Where 199 (65%) people are influenced by the Instagram pages, only 35 (15%) people answered that they do not consider themselves influenced by the pages that follow. This proves that Instagram is a social network that influences people in their daily lives.

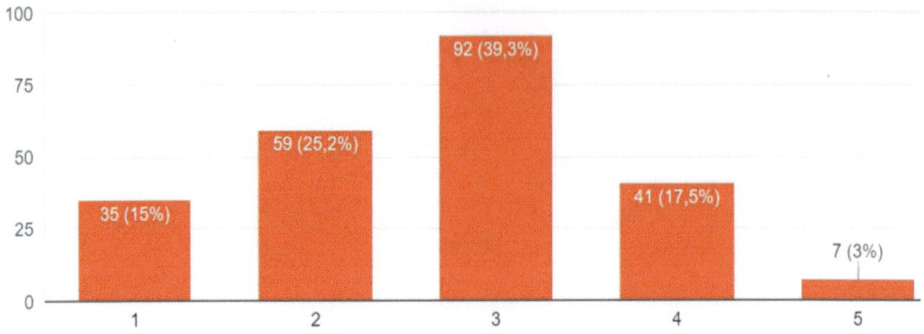

Fig. 5.4. Instagram influence—Do you consider yourself a person influenced by the pages you follow? 1—Never, 2—Rarely; 3—Sometimes; 4—Often; 5—Always

When asked what purpose Instagram uses, most respondents only use Instagram socially, about 83.8% of the population. A minimum percentage of 5% uses as a form of marketing tool. About 11.3% use Instagram for another purpose.

Regarding the type of pages you like to follow, it was found that the most followed according to our sample are pages of "Fashion" and "Blogs" with a high adherence, with 25 and 20%. These values are also influenced by the fact that we have more females responding than males. The pages of "Memes" also have great adherence with 23%.

When asked which types of posts have the greatest power of inspiration/influence, it was found that the variable that most influences or inspires people is the posts with 38.1% motivational phrases. Fashion-related posts also have a big influence on people, about 18.8%. The other posts related to meals, recipes, brands or exercises have relatively the same influence, ranging from 9.4 to 11.7%.

To the question of what more Instagram search, of the known options, the most searched is Humor with 42.2% and right after that we have Style\Beauty with 41.4%, this value may be related to the fact that we have more female answers. Next up is Travel with 37.6% followed by food with 29.1%. Other options such as animals, dance, TV and movies, architecture, sports, shopping, and car have roughly the same grip. The ones we have the least research on are decoration and music with 0.4% each.

5.5.2 Bivariate Data Analysis

The relationship between the use of social networks and the use of the Instagram social network can be observed in Fig. 5.5, where we can see that of the people who use social networks, 239 people or 87.23% of our sample uses the Instagram social network, only 35 people, about 13% do not use this social network. We can also conclude that 10 people do not use any social networks. With this study, we can conclude that social networks are part of most people's daily lives.

Using social networks * using social network Instagram

Using social networks	Using social networks Instagram		Total
	Yes	No	
Yes	239.00	35.00	274.00
	87.23%	12.77%	100.00%
	100.00%	77.78%	96.48%
	84.15%	12.32%	96.48%
No	.00	10.00	10.00
	.00%	100.00%	100.00%
	.00%	22.22%	3.52%
	.00%	3.52%	3.52%
Total	239.00	239.00	284.00
	84.15%	84.15%	100.00%
	100.00%	100.00%	100.00%
	84.15%	84.15%	100.00%

Fig. 5.5. Using social networks and using Instagram

The relationship between sex and the use of Instagram can be seen in Fig. 5.6, where we can verify that of the 84 male respondents 64 (76%) use the Instagram social network, and only 20 (24%) do not use Instagram. For females, 200 people answered the survey. Of these people, 175 (88%) use Instagram, and 25 (12%) do not use Instagram. Although we find that females use Instagram more than males, we cannot conclude that this is overall, as our sample had less than half male responses.

Gender * using social networks Instagram

Gender	Using social networks Instagram		Total
	Yes	No	
Male	64.00	20.00	84.00
	76.19%	23.81%	100.00%
	26.78%	44.445	29.58%
	22.54%	7.04%	29.58%
Female	175.00	25.00	200.00
	87.50%	12.50%	100.00%
	73.22%	55.56%	70.42%
	61.62%	8.80%	70.42%
Total	239.00	45.00	284.00
	84.15%	15.85%	100.00%
	100.00%	100.00%	100.00%
	84.15%	15.85%	100.00%

Fig. 5.6. Relationship between gender and use of Instagram

5.5.3 Statistical Inference

The Chi-square test is based on data grouped into mutually exclusive and exhaustive classes in the form of counts. This allows us to verify whether or not two variables are independent (Fig. 5.7).

The use of social networks versus gender:

H_0 Using Instagram is gender independent.

H_1 Using Instagram is gender dependent.

ET (α)	p-value	Conclusions
5%	0.02	p-value=0.017 <0.05 so we reject H_0

Fig. 5.7. The use of social networks versus gender

Regarding the variables use of social networks and gender, we obtained a p-value of 0.021 which is less than 0.05, so we rejected H_0, i.e., the variables are not independent, so there is a relationship between them (Fig. 5.8).

What kind of publications do you make versus what kind of pages do you like to follow:

H_0 The posts you make are independent of the pages you like to follow.

H_1 The posts you make depend on the pages you like to follow.

ET (α)	p-value	Conclusions
5%	0.149	p-value =0.149>0.05 so we don't rejects H_0

Fig. 5.8. What kind of publications do you make versus what kind of pages do you like to follow

Regarding the variables under study, we obtained a p-value of 0.149, which is greater than 0.05, so we did not reject the null hypothesis, thus concluding that the publications people make do not depend on the type of pages they like to follow.

5.6 Conclusion

This study allowed us to conclude that with regard to gender of respondents, mode corresponds to female gender, as it represents about 70.4% while male gender represents only 29.6%.

Regarding the influence of Instagram on people, we can conclude that 92 respondents, i.e., 39.3%, are influenced by Instagram "sometimes". In a sample of 234 respondents, only 35 people, or 15%, are not influenced by the pages that they follow. So we can conclude that Instagram influences people in their daily lives.

On average, out of 200 responses, Instagram usage time is approximately 1 h: 35 m per day on Instagram.

Thus, according to the study, we were able to conclude that Instagram is part of people's daily lives, and due to its differentiating feature achieved a great adhesion in a short time. It is also the social networks that have the most power of influence thanks to their aesthetics, sharing, and instantaneity.

The results obtained from this study may help those using this tool to identify important factors that may be considered to eventually formulate business strategies to achieve customer satisfaction.

As future work, we consider it relevant to analyze user behavior on two different social networks and see the possible variations and reasons for them. In this way, a limitation of this research work that only focused on one tool is found Instagram.

References

1. Nedra, B.A., Hadhri, W., Mezrani, M.: Determinants of customers' intentions to use hedonic networks: the case of Instagram. J.Retail. Consum. Serv. **46**, 21–32 (2019). Elsevier
2. Brennan, R., Croft, R.: The use of social media in B2B marketing and branding: An exploratory study. J. Cust. Behav. **11**(2), 101–115 (2012)
3. Chen, C.P.: Exploring Personal Branding on YouTube. J. Internet Commer. **12**(4), 332–347 (2013)
4. Platon, O.E.: An Analysis of the Social Media Presence of the Brands, Global Economic Observer, "Nicolae Titulescu" University of Bucharest, Faculty of Economic Sciences; Institute for World Economy of the Romanian Academy **2**(1), 192–201 (2014)
5. Shiffman, L.G., Kanuk, L.L.: Comportamento do consumidor. 9.ed. Rio de Janeiro: LTC (2009)
6. Chen, Y., Wang, Q., Xie, J.: Online Social Interactions: a natural experiment on word of mouth versus observational learning. J. Mark. Res. **XLVIII**, 238–254 (2011)
7. Gerhart, N., Koohikamali, M.: Social network migration and anonymity expectations: What anonymous social network apps offer. Comput. Human Behav. **95,** 101–113 (2019). Elsevier
8. De Meo, P., Ferrara, E., Rosaci, D., Sarné, G.: Trust and compactness in social network groups. IEEE Trans. Cybern. **45**(2), 21–32 (2015)
9. Santos, R.: A influência do Instagram na atitude do consumidor: o caso da Levi Strauss and Co. Universidade Europeia, Mestrado em Marketing Digital (2016)
10. Pereira, R.: As motivações e o envolvimento do consumidor com as marcas do setor vestuário e acessórios-A influência do Instagram. Dissertação de Mestrado em Marketing Digital, ISCAP, Instituto Politécnico do Porto (2018)

11. Desai, S. Han, M.: Social media content analytics beyond the text: a case study of university branding in Instagram. In: Proceedings of the 2019 ACM Southeast Conference, pp. 94–101 (2019)
12. Pires, G.: Fotografia através de dispositivos móveis: estudo de caso sobre o Instagram. Revista da Graduação **6**(1) (2013)
13. Berthon, P.R., Pitt, L.F., Plangger, K., Shapiro, D.: Marketing meets Web 2.0, social media, and creative consumers: Implications for international marketing strategy. Bus. Horiz. **55**(3), 261–271 (2012)
14. McCormick, K.: Celebrity endorsements: Influence of a product-endorser match on Millennials attitudes and purchase intentions. J. Retail. Consum. Serv. **32**(1), 39–45 (2016)
15. Prentice, C., Han, X.Y., Hua, L.-L., Hu, L: The influence of identity-driven customer engagement on purchase intention. J. Retail. Consum. Serv. **47**, 339–347 (2019)

Chapter 6
The Influence of YouTubers in Consumer Behavior

Isabel Lopes[1,2], Teresa Guarda[2,3](✉), José Avelino Victor[4,5], and Encarnación González Vázquez[6]

[1] UNIAG (Applied Management Research Unit), Polytechnic Institute of Bragança, Bragança, Portugal
isalopes@ipb.pt
[2] Algoritmi Centre, Minho University, Guimarães, Portugal
tguarda@gmail.com
[3] Universidad Estatal Península de Santa Elena–UPSE, La Libertad, Ecuador
[4] Instituto Politécnico Da Maia, Maia, Portugal
javemor@ismai.com
[5] Instituto Universitário Da Maia, Maia, Portugal
[6] Universidad de Vigo, Vigo, Spain
egzlez@uvigo.es

Abstract. The massive use of the Internet, the development of new information and communication technologies and their consequent appropriation by individuals, transformed the way of acting, being and communicating in today's society. A growing number of companies that invest in audiovisual content, as part of the outreach strategy of its products/brands. It's a valuable, elaborate tool that can help a variety of marketing strategies reach new audiences, create more compelling, and far-reaching content. In this sense, this research was carried out in order to verify the behavior of consumers regarding this tool, in order to understand the influence of YouTubers on consumer behavior, essentially at the time of purchase decision. We opted for a quantitative method, where questionnaire surveys were applied to 428 people, mostly young people attending higher education. The results are discussed in light of the literature and future work is identified to mitigate this recent phenomenon.

6.1 Introduction

Digital Marketing is the promotion of products or brands through digital media. It is considered one of the main ways that companies have to communicate with the public directly, personalized, and at the right time. The big trend in digital marketing is social media marketing. This tool available to everyone today enables businesses to communicate with a large number of consumers, enabling them to create, enhance, and optimize customer relationships with the goal of promoting their products or brands using digital media.

Á. Rocha et al. (eds.), *Marketing and Smart Technologies*, Smart Innovation, Systems and Technologies 167, https://doi.org/10.1007/978-981-15-1564-4_6

Today social networks play an important role in the social and personal lives of most individuals. In addition to allowing you to communicate with friends, they allow users to connect with people around the world with whom they share common interests [1]. The development of the internet has enabled new platforms to emerge, on which anyone can produce and share content. One such platform is YouTube, where user-produced videos are increasingly popular. Digital influencers like YouTubers are more credible sources than opinion leaders and traditional advertising. It then becomes increasingly important to realize the influence that YouTubers have on their audience.

This work, structurally after this introduction, continues with a review of the literature on social networks, and marketing on these networks. Then, in Sect. 6.3, we present the social network under study, as well as YouTubers as influencers and YouTube advertising, as well as the intention of buying followers. In Sect. 6.4, we cover how the survey was conducted, focusing on the target population and its structure. Finally, the results and conclusions are presented in the light of the research data, the limitations of this study are identified and future work is proposed.

6.2 Social Networks

Social networking, provide a means of communication for users by connecting people, some of which might not ever meet face-to-face [2]. Millions of users select these networks for sharing their information and experiences. Most social networks are open environments that anyone can freely enter to generate content and share them [3].

Social networks are already an integral part of consumers' lives and increasingly important for brands to interact with their audience. O'Brien [–4] says that with social networks the consumer has become increasingly demanding, changing the dynamics of their relationships with companies, which have gone from being unilateral to being a two-way social bond. These networks allow for greater segmentation and interaction between brands and consumers.

Currently, there are different social networks, each with a purpose and a specific audience, we can highlight the following: Facebook, YouTube, WhatsApp, Instagram, Twitter, Pinterest, Skype, LinkedIn, Snapchat, and Messenger.

6.3 YouTube

The development of the Internet has also changed the way consumers search for and buy products and the way they interact with each other and with brands. People are more informed, increasingly being less influenced by traditional advertising. It is increasingly difficult to capture consumer attention. For this reason, more personalized communication is required, tailored to each consumer and with a specific interest for him. This led companies to use YouTube as a communication resource for their products and brand. By using social networks, companies can increase their exposure, captivate their website traffic, and develop greater loyalty to their followers [5].

YouTube is the most widely used platform for audiovisual content today. In addition, videos represent the most shared content on social networks. It is precisely for this double trend that companies should use this platform to promote their products/brands.

Since a lot of user-generated content is posted on YouTube every day, and some of it is product related, YouTube has become one of the platforms where consumers seek product information before making purchase decisions and these contents significantly affect purchase intent [6].

6.3.1 YouTubers as Influencers

YouTubers are the people who produce videos for YouTube. They are considered influencers and opinion leaders capable of developing an opinion and directing it to their followers, promoting greater consumer interest than the advertising brand itself [7].

With the phenomenon of social networking and user-created content, many social media users have eventually garnered a large audience and had become opinion leaders themselves. With the democratization of information and communication, this opinion leadership has expanded its reach, leading initially unknown people to make the first leap and become the new influencers [8].

6.3.2 YouTube Advertising and Purchase Intent

The influence of online recommendation on the purchase decision process has been widely studied, and several studies indicate that recommendation has more persuasive power than advertising and promotions. The strategic importance of the recommendation remains one of the most effective ways to message, attract, and retain customers [9]. The advertising that appears on YouTube can affect brand awareness as well as consumers' purchase intent.

Virtual interactivity in social networks is essential for building a positive brand image, branding, and a positive environment to ensure favorable brand assessments in the virtual world [10]. Arantes [11] also highlights the growth of digital video ads, the predominant advertising format on YouTube, both in investment and in profit generated. A key reason for this growth is the ability to effectively exploit emotions that distinguish video advertising from other advertising formats [1].

Purchase intent is one of the most relevant measures to understand if a brand communication effort was relevant. The purchase decision of customers is a complex process that is generally related to consumer behavior, perceptions, and attitudes [12].

6.4 Methodology

In order to understand and identify the influence of YouTubers on consumer behavior, it was considered appropriate to apply the Inquiry technique, as it provides a clear, direct, and objective answer to the questions presented to respondents. In addition, since the aim was to broadly characterize the population consisting of students from five schools in a higher education institution, it was considered that this number made it impossible or inappropriate to use other alternative research techniques.

The survey stands out as the social research technique most suited to large-scale studies. Seemingly easy and simple to apply, relatively fast, standardized and routinized, giving priority to comparable, generalizable and quantitative analysis data.

Respondents were primarily students, so the applicability of an online survey to understand user behavior towards YouTubers became quite appropriate. Data collection took place between April 16 and May 8, 2019. For the elaboration and collection of data, the Google forms were used, for the analysis and treatment of data, using the software SPSS and Microsoft Excel.

6.4.1 Population

The study had as its scope a higher education institution, which consists of five schools. Having a student population of about 7000 students.

In order to draw representative conclusions about a population from a sample, it is necessary to take into account the sample and the sampling error, which is the difference between a sample result and the true population result. Normally the set value is 5% and the sample required for our study of 384 responses. However, as 417 responses were collected, the sampling error was readjusted to 4.8% and the confidence level used was 95%.

6.4.2 Survey Structure

The structure of the Survey resulted from the literature review on social networks in general and YouTube in particular. The survey questions, which were individually answered and confidential, were organized into three sections.

The first section concerns the identification of respondents' personal data, such as age, gender, and others. The second one is the use of YouTube and the preferences within it. The third section aims to identify purchase intent by watching YouTubers videos. The questions "What kind of video content do you usually watch on YouTube?" and "Witch Portuguese YouTubers do you normally watch?" generated several variables, as each multiple-choice generated one variable.

An attempt was made to make a survey that was simple to answer and not too long to get as many answers as possible.

6.5 Results

Many studies in literature focus on different forms of social media including blogs and conclude that these media outlets gain immense traction and have subsequent social influence on the information consumers [13, 14, 15].

This study tried to verify this reality, gathering the necessary data, proceeding with simple descriptive analysis and statistical inference.

A total of 439 questionnaires were collected, and only 428 were considered to have correct and valid answers since the others were poorly answered. Of the sample, 417 claims to use the social network YouTube. The study has 38 variables, 7 of which are ordinal.

6.5.1 Descriptive Analysis

Regarding the gender of respondents, the sample is mostly made up of females, corresponding to about 76% of the total respondents, corresponding to 324 responses. In turn, males represent 24% of the sample, representing 104 responses.

Most respondents are between 17 and 25 years old. The mode is 19 years old and corresponds to 82 of the respondents

From the sample, it is possible to verify that most of the interviewed individuals live in the North of Portugal, with 315 answers corresponding to 74%; in the Center with 87 responses (20%) and a minority of the South and the islands, out of 26 responses representing the 6% sample.

Survey participants using YouTube, as shown in Fig. 6.1, were 428 individuals who volunteered to participate in the study, 417 (97%) confirmed using the YouTube social network and, therefore, only 11 did not.

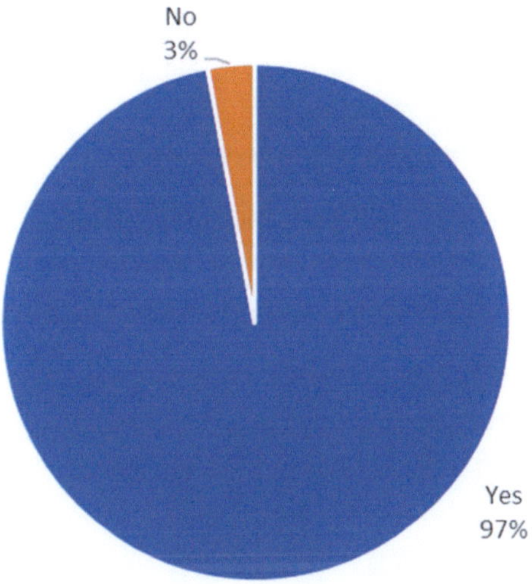

Fig. 6.1. YouTube social network users

Regarding the time devoted to the YouTube platform, it appears that the vast majority of respondents, about 302, access YouTube daily (72%). About 77 respondents go to YouTube more than three times a week (19%), 26 respondents go once a week (6%) and lastly, a minority of 12 respondents go to the weekends only (3%).

Asked about the most viewed videos, the vast majority (about 283 respondents) use YouTube primarily for listening to music. Next, the most-watched type of videos are Fashion and Beauty (197 people) and Comedy (184), see Fig. 6.2.

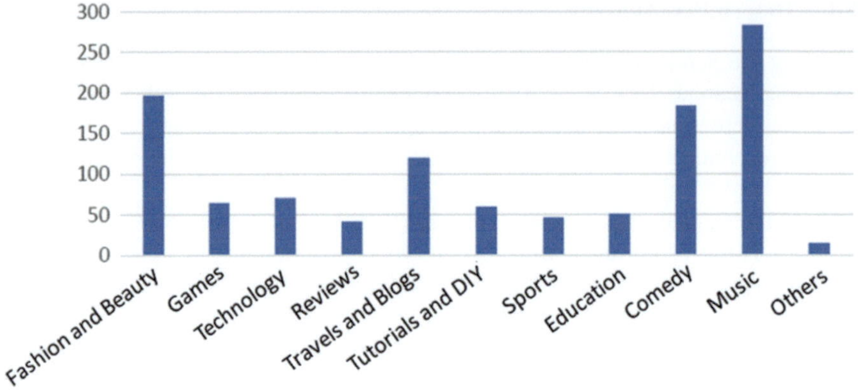

Fig. 6.2. Most viewed videos by type

Although most respondents use the YouTube platform, when asked whether "YouTube videos lead to changes in my personal attitudes and changes in information", the most answered was "Sometimes".

In the question "Watching YouTubers videos makes me want to buy the advertised products", most participants answered "Sometimes". Regarding the question "Brands advertised on YouTube videos have a positive influence on my buying decision", most answered "Sometimes", although very close to "Rarely" and "Never" (see Fig. 6.3).

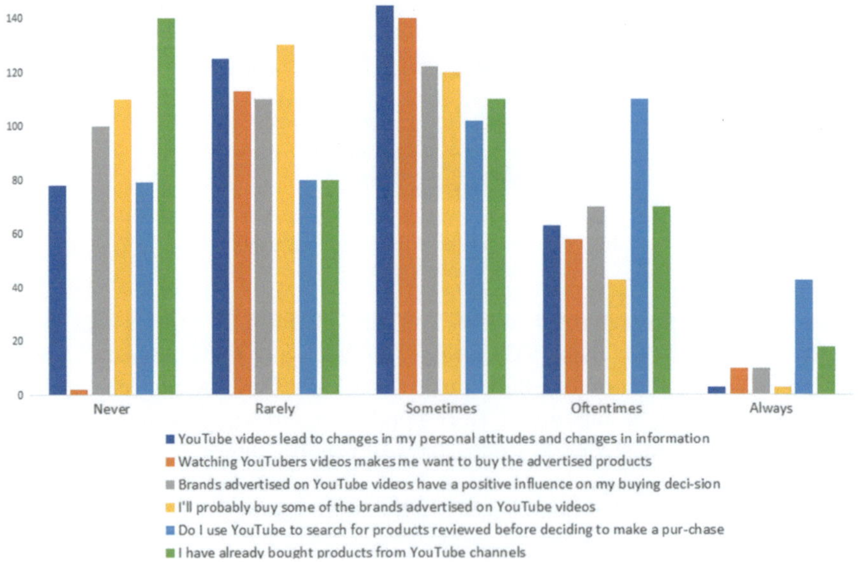

Fig. 6.3. The influence of YouTubers on consumer behavior

When asked about "I'll probably buy some of the brands advertised on YouTube videos", most respondents said "Rarely" and "Sometimes". However, to the question "Do I use YouTube to search for products reviewed before deciding to make a purchase", most of the answers were "Oftentimes" and "Sometimes". Relating the influence of YouTubers to consumers, "I have already bought products from YouTube channels" most answered "Never" and "Sometimes".

6.5.2 Statistical Inference

Statistical inference aims to draw conclusions about a population through a sample of that same population.

5.2.1 6.Gender Randomness Test

Randomness tests assume that observations are classified into two mutually exclusive categories. In the present study, the gender randomness test was performed in order to verify whether or not the sample is random in relation to this variable (Table 5.1).

H_0 Sample is random with respect to gender.

H_1 Sample is not random about gender.

Table 6.1. Gender randomness test

	p-value	α
Gender randomness test	0.839	0.05

For a significance level of 5%, we obtained a *p-value* of 0.839, which is greater than 0.05. As such, we do not reject H_0, so there is not enough statistical evidence to state that the sample is not random about gender.

5.2.2 6.Chi-Square Independence Test

This test allows to analyze the independence between two variables and can be expressed on any scale that is grouped into mutually exclusive classes. Table 6.2 summarizes all the Chi-Square independence tests performed.

Table 6.2. Chi-square independence tests

Test	Hypotheses	*p-value*	Conclusion
1st	H_0: The purchase of products indicated in the videos is independent of net income. H_1: The purchase of products indicated in the videos is dependent on net income.	0.831	We do not reject H_0
2nd	H_0: Frequency of access to YouTube is independent of the current situation of the individual.	0.273	We do not reject H_0

(*continued*)

Table 6.2. (*continued*)

Test	Hypotheses	*p-value*	Conclusion
	H_1: Frequency of access to YouTube is dependent on the individual's current situation.		
3rd	H_0: Frequency of access to YouTube is independent of the individual's current situation. H_1: Frequency of access to YouTube is dependent on the individual's current situation.	0.000	We reject H_0
4th	H_0: The influence of watched videos on the purchase decision is gender independent. H_1: The influence of watched videos on the purchase decision is gender dependent.	0.004	We reject H_0
5th	H_0: Frequency of access to YouTube is independent of the current situation of the individual. H_1: Frequency of access to YouTube is dependent on the individual's current situation.	0.010	We reject H_0

1st Test–Buying products featured in videos versus net income

H_0 The purchase of products indicated in the videos is independent of net income.
H_1 The purchase of products indicated in the videos is dependent on net income.

$$p - values = 0.831 \tag{6.1}$$

Since we get a *p-value* = 0.831, which is greater than 0.05, we do not reject H_0. There is insufficient statistical evidence to claim that the purchase of products depends on net income. We thus conclude that the variables are independent.

2nd test—YouTube access versus current status

H_0 Frequency of access to YouTube is independent of the current situation of the individual.
H_1 The frequency of access to YouTube is dependent on the current situation of the individual.

$$p - values - = 0.275 > 0.05 \tag{6.2}$$

Since *p-value* = 0.275, greater than 0.05 is obtained, we do not reject H_0, which means that there is not enough statistical evidence to state that the variables are dependent.

3rd test—Desire to buy the products advertised in the video's versus gender

H_0 The desire to buy the products advertised in the videos is gender-independent.
H_1 The desire to buy the products advertised in the videos is gender-dependent.

$$p - values = 0.001 < 0.05 \tag{6.3}$$

Since we get a *p-value* very close to zero, less than 0.05, we reject H_0. There is a relationship between the two variables as the desire to buy products is gender-dependent.

4th test—Influence of videos on purchase decision versus gender

H_0 The influence of videos on the purchase decision is gender-independent.

H_1 The influence of videos on the purchase decision is gender-dependent.

$$p - values = 0.004 < 0.05 \tag{6.4}$$

Since we get a *p-value* very close to zero, we reject H_0. There is a relationship between the two variables since the influence of videos on the purchase decision is gender-dependent.

5th test—Desire to buy the advertised products versus region

H_0 The desire to buy the advertised products is independent of the region.

H_1 The desire to buy the advertised products is dependent on the region.

$$p - values = 0.010 < 0.05 \tag{6.5}$$

Since we get a *p-value* = 0.010, less than 0.05, we reject H0. Therefore, the desire to buy the advertised products is dependent on the region.

6.6 Conclusion

The study aimed to understand the influence of YouTubers on consumer behavior, essentially at the time of purchase decision. One of the main conclusions we can draw is that consumers are increasingly aware and not let themselves be influenced so easily. It can be seen that the higher the income of the individual, the less he is influenced to buy. In other words, the individuals who have already been influenced and who have bought the most products indicated in the videos are those who have no income. As your income increases, it can be seen that people are not so influenced by YouTube videos.

Respondents are mostly female students between the ages of 17 and 25. About 72% of respondents go to YouTube every day, and about 283 respondents use YouTube primarily to listen to music.

The results obtained from this study may help those using this tool to identify important factors that may be considered to eventually formulate business strategies to achieve customer satisfaction.

As future work, we consider it relevant to analyze user behavior in two different social networks and to see possible variations and their motives. In this way, a limitation of this research work that focused only on one tool YouTube is overcome.

References

1. Wnent, S.: Product Placement on Instagram—the Sponsored and Fabricated EWOM: Comparing Product Placement on Instagram and Banner Advertisement of a Weight Loss Product. Marketing Communications, University of Twente, Netherlands, Master specialization (2016)
2. Gerhart, N., Koohikamali, M.: Social network migration and anonymity expectations: what anonymous social network apps offer. Comput. Human Behav. **95**, 101–113 (2019). Elsevier
3. De Meo, P., Ferrara, E., Rosaci, D., Sarné, G.: Trust and compactness in social network groups. IEEE Trans. Cybern. **45**(2), 21–32 (2015)
4. O'Brien, C.: The emergence of the social media empowered consumer. Irish Market. Rev. **21** (2), 32–40 (2011)
5. Phua, J., Jin, S.V., Kim, J.: Gratifications of using Facebook, Twitter, Instagram, or Snapchat to follow brands: the moderating effect of social comparison, trust, tie strength, and network homophily on brand identification, brand engagement, brand commitment, and membership intention. Telemat. Inform. **34**(1) (2017)
6. Yüksel, H.F.: Factors affecting purchase intention in YouTube videos. J. Knowl. Econ. Knowl. Manag. **11**(2) (2016)
7. Hall, A., Towers, N.: Understanding how Millennial shoppers decide what to buy Digitally connected unseen journeys. Int. J. Retail Distrib. Manag. **45**(5) (2017)
8. Segarra-Saavedra, J., Hidalgo-Marí, T.: Influencers, moda feminina e Instagram: el poder de la influencia en la era 2.0. Revista Mediterránea de Comunicación/Mediterranean J. Commun. **9**(1) (2018)
9. Teixeira, N.R.: O papel da recomendação online no processo de tomada de decisão de compra. Tese de Mestrado em Marketing, ISCTE-IUL (2010)
10. Barreda, A., Okumus, F., Nusair, K., Bilgihan, A.: The mediating effect of virtual interactivity in travel-related online social network websites. Int. J. Hosp. Tourism Adm. **17** (2) (2016)
11. Arantes, M., Figueiredo, F., Almeida, J.M.: Understanding video-ad consumption on YouTube: a measurement study on user behavior, popularity, and content properties. In: Proceedings of the 8th ACM Conference on Web Science (2016)
12. Mirabi, V., Akbariyeh, H., Tahmasebifard, H.: A study of factors affecting on customers purchase intention. case study: the agencies of bono brand tile in Tehran. J. Multidiscip. Eng. Sci. Technol. **2**(1) (2015)
13. Berthon, P.R., Pitt, L.F., Plangger, K., Shapiro, D.: Marketing meets Web 2.0, social media, and creative consumers: implications for international marketing strategy. Bus. Horiz. **55**(3), 261–271 (2012)
14. McCormick, K.: Celebrity endorsements: influence of a product-endorser match on Millennials attitudes and purchase intentions. J. Retail. Consum. Serv. **32**(1), 39–45 (2016)
15. Prentice, C., Han, X.Y., Hua, L.-L., Hu, L: The influence of identity-driven customer engagement on purchase intention. J. Retail. Consum. Serv. **47**, 339–347 (2019)

Chapter 7
Service Loyalty in Retail Banking: An Empirical Study

Mathias Mujinga(✉)

School of Computing, University of South Africa, Johannesburg, South Africa
mujinm@unisa.ac.za

Abstract. The main objective of the paper is to report on the service loyalty of retail banking customers in South Africa. The findings are based on an empirical study that used SERVLOYAL measurement scale and obtained 520 complete and valid responses. The results show that service loyalty in retail banking in South Africa does not depend on gender, age or income of customers. However, the bank plays a crucial role in how customers perceive service loyalty.

7.1 Introduction

Retail banking institutions operate in a competitive environment as customers can shift from one service provider to the other. Hence, to satisfy customers and maintain loyalty banks offer a number of initiatives such as cost-effective transactional bundles and loyalty programmes. Quality of service is one aspect that banks also strive to improve on for customer satisfaction. Service providers need to deliver high-quality service to achieve customer satisfaction and realise customer retention [1, 2]. Especially given the all-encompassing nature of customer satisfaction in all aspects of an organisation's activities from profitability and sustainability. This is critical in the service industry, banking, in particular, were value-added services form a large part of the business activities.

The paper investigates service loyalty in retail banking industry using the SERVLOYAL measurement scale [3] based on customers' perceptions. The paper is organised as follows: customer satisfaction and service loyalty are discussed first followed by a discussion of SERVLOYAL scale and methodology. This followed by descriptive statistics and inferential statistical analysis. Lastly the conclusion is provided.

7.2 Customer Satisfaction

Organisations that sell a product or provide a service rely on customer satisfaction for long-term viability. Customer satisfaction is critical as it ensures customers come back to acquire the products or services. This leads to developing customer loyalty and retention [4]. Hence, to be successful and ensure further growth the business needs to strive to a higher level of customer satisfaction by fulfilling the customer's needs.

Á. Rocha et al. (eds.), *Marketing and Smart Technologies*, Smart Innovation, Systems and Technologies 167, https://doi.org/10.1007/978-981-15-1564-4_7

There are a number of aspects that leads to customer satisfaction. When a customer's goals are met, and the interaction is pleasant, they then become satisfied. This is vitally important in building customer loyalty. Customer satisfaction is important as it plays a vital role in establishing customer loyalty [4]. Satisfaction is attained when the customer can achieve the goal they set out without any limitations and they are pleased with their purchase outcome [5]. When satisfaction is refined by the customer it helps form part of their customer experience, this then starts to develop affective connotations [4].

Customer satisfaction is a key factor in customer retention [6]. It is the customer's attitude formed subsequent to purchasing a product or using a service. It can be viewed as a manifestation of a customer feeling content with the product or service. However, there are a number of factors that influence customer satisfaction, which includes price, product quality and service quality, as well as other contextual and personal aspects have been named [7].

According to Hallowell [8], customer satisfaction leads to increased customer loyalty that in turn increases the bank's profitability. Ngo and Nguyen [9] also established between service quality, customer satisfaction, and customer loyalty in retail banking. Additionally, other research has found that customer satisfaction is a baseline measurement of excellence that can be applied to any organisation [10]. Hoffmann and Birnbrich [11] states that satisfaction is accomplished not only matching the customer expectations but surpassing them to promote customer's intention to remain with the present service provider.

7.3 Service Loyalty

Customer loyalty is vital for any business organisation. Loyalty leads to customer retention, which enables the organisation to realise business goals [12]. Customer loyalty plays a role in leveraging an organisation's competitive advantage [13]. South African organisations and banks in particular offer loyalty programmes as a means to attain their loyalty. These programmes are often in partnerships with other organisations for rand value cash backs or loyalty points that are converted into cash.

Additionally, researchers have suggested that satisfaction contributes to customer loyalty [14, 15]. Satisfaction is also described as a key element of brand loyalty, as well as an imperative fluctuating component of the definition of loyalty. Other researchers have suggested that satisfaction is an emotive antecedent of brand loyalty [7]. There have been recurring evidence in consumer marketing research proving that high customer satisfaction leads to behavioural and repurchase intentions, retention of customers and customer loyalty [7, 16].

Customer commitment to the service provider is a key factor of customer loyalty in the service industries [17]. It has also been proved integral to relationship quality as it promotes successful customer service provider relationships. When considering this association between customer commitment and loyalty, researchers have implied that customer commitment may possibly lead to customer word of mouth communication, which is a central feature of attitudinal loyalty [7]. Therefore, it is necessary for service providers to strategically align their focus and emphasis on improving total customer satisfaction, instead of focusing exclusively on improving service quality [18].

7.4 SERVLOYAL

Loyalty can be defined as a standard of continuous sponsorship/funding, with the customers' inclination in terms of their preferences and objectives [3]. Service loyalty depends on a number of factors that include customer service satisfaction [3, 19] and the quality of service delivered by the service provider [3]. Sudhahar et al. [3] developed a service loyalty (SERVLOYAL) measurement scale that is conceptualised into seven dimensions. These include behavioural, attitudinal, cognitive, conative, affective, trust and commitment. A brief discussion on each is given in the next section.

Dick and Basu [20] conceptualised customer loyalty as an interaction of attitude and behaviour such that the behaviour (loyalty) is determined by the strength of the relationship between relative attitude and repeat patronage. The behavioural loyalty measures include attributes such as brand allegiance, price elasticity, share of category (number of times a brand is purchased in a given period) and price until switching [21]. Behavioural dimension measures the customer's behavioural intent to engage with the service provider going forward. The attitudinal dimensions of loyalty includes attributes such as word of mouth, complaining behaviour and purchase intentions [22]. Attitudinal dimension attempt to determine the customer's attitude based on slight changes in services delivered. The cognitive loyalty component includes attributes like preference to the service organisation, the belief that the service organisation provides best offer and suiting customer needs [23]. Cognitive dimension relates to the customer's perceptions in relation to other banks and how the customer recognises with the bank. Conative behaviour strives to predict some specific customer behaviour. Affective dimension measures the customer's affection towards the service provider. Other major components of customer loyalty are trust and commitment that measure how reliable the service provider is in addressing customer issues. Lastly, action loyalty deals with how likely the customer is inclined to engage in activities that advertise the brand.

7.5 Methodology

The study conducted a quantitative survey to collect empirical data using the SERVLOYAL measurement scales adopted from the literature and previous studies. Retail banking users with one of the five major retail banks in South Africa were eligible respondents for the study. The SERVLOYAL scale [3] consists of eight constructs with a total of 38 items. An online survey tool was used to administer the survey.

The survey consists of a number of moderating variables for group comparisons. These include gender, age, income and bank. These factors provide an insight when used for detailed group comparisons analysis within a moderating variable. A 5-point Likert scale was used to measure customers' perception from strongly disagree to strongly agree for each of the 38 statements.

7.6 Data Analysis

Data analysis is presented in three sections; frequencies, descriptive statistics and inferential analysis.

7.6.1 Descriptive Statistics

The descriptive statistics of the collected data is shown in Table 7.1. There were 520 responses with male and female respondents 44% and 56%, respectively. The age groups of respondents consist of five groups and six income groups.

Table 7.1. Frequencies (N = 520)

Variable	Category	Frequency	Percentage (%)	Cumulative percentage (%)
Gender	Male	227	44	44
	Female	293	56	100
Age	Younger than 20 years	15	3	3
	20-29 years	212	41	44
	30-39 years	154	30	74
	40-49 years	53	10	84
	50 years or older	86	16	100
Income	Less than R10 000	119	23	23
	R10 000 – R19 999	102	20	43
	R20 000 – R29 999	103	20	63
	R30 000 – R39 999	74	14	77
	R40 000 – R49 999	53	10	87
	R50 000 or more	69	13	100
Bank	FNB	189	36.3	36.3
	S. Bank	86	16.5	52.9
	ABSA	89	17.1	70.0
	Nedbank	46	8.8	78.8
	Capitec	103	19.8	98.7
	Investec	7	1.3	100.0

The majority of respondents came from the five major retail banks with only 1.3% using Investec, a private bank.

7.6.2 Instrument Reliability Test

Instrument reliability was performed using IBM SPSS software package to determine the reliability of the SERVLOYAL scale. Out of the total of 38 statements that constitute the eight dimensions only two were discarded as they did not contribute to the construct reliability, each for *affective* and *commitment* dimensions. After which all eight dimensions met the Cronbach's Alpha's threshold of 0.7 as alluded by Pallant [24]. Table 7.2 shows the Cronbach's Alphas for each of the dimensions and the whole set of 36 statements with the lowest alpha of 0.830. The table also shows the mean scores and standard deviation for each dimension.

Table 7.2. Instrument reliability

Item	No of items	Cronbach's Alpha	Mean	Std. Dev
Behavioural	4	0.898	4.0659	0.85348
Attitudinal	4	0.848	3.4130	0.92206
Cognitive	8	0.941	3.6281	0.94213
Conative	5	0.830	3.5412	0.81716
Affective	5	0.728	3.5785	0.71816
Trust	4	0.873	3.5207	0.87126
Commitment	3	0.870	3.6442	0.91966
Action loyalty	3	0.926	3.7987	0.97895
SERVLOYAL	36	0.976	3.6376	0.78194

7.6.3 Inferential Statistical Analysis

The study tests the null hypothesis that SERVLOYAL mean scores are equal for all moderating factors variables, hence service loyalty does not depend on any group demographics. Accepting the alternative hypothesis suggests that service loyalty is dependent on customer profiles and banks need to approach each group of customers with group-specific strategies for customer retention. Independent samples tests were conducted on gender, which is the only factors that meet the dichotomous criteria. While one-way ANOVA analysis was performed on all other variables to test the difference in means scores.

Independent Samples T-Test: Gender. Independent samples *t*-tests for SERVLOYAL mean scores was performed for each dimension based on gender groups. The *Levene's test for equality of variances* was performed and none of the dimensions had any significant differences in mean scores at 95% confidence level. Therefore, the study accepts the null hypothesis that there is no significant difference between the means and concluded that the variance in mean scores for males and females are equal.

When considering the *equal variances assumed*, the results show that the difference in means for all dimensions are statistically insignificant as well, as *p* values are all above the threshold of 0.05% level.

One-way ANOVA Analysis. SERVLOYAL mean scores for all eight dimensions was used to conduct the ANOVA analysis based on three remaining moderating factors; age, income and bank. According to Hair et al. [25], one-way ANOVA analysis compares two or more independent groups in order to determine whether there is statistical evidence that the associated population means are significantly different. Table 7.3 shows the ANOVA *F* statistic and significance levels. The results show that only the bank has significant differences in mean scores at 0.05% level. This suggests service loyalty is not dependent on age or income of customers. However, it significantly depends on the customer's bank. This is also supported by a significant higher *F* value of 13.612 that implies the difference could not be attributed to chance

Table 7.3. One-way ANOVA: SERVLOYAL mean scores

Variable	df	F	Sig
Age	4	1.194	0.312
Income	5	0.311	0.906
Bank	5	13.612	0.000

This means banks are offering services that lead customers to associate with the brand. To avoid prejudice the study sidesteps multiple comparisons based on the bank names as the sample is considered non-representative.

7.7 Conclusion

Customer satisfaction is a major determinant of customer loyalty and retention and banks need to determine factors that improve satisfaction. SERVLOYAL scale is one of the measurement scales used to investigate service loyalty in the service industry. The main objective of this study was to measure service loyalty in retail banking using SERVLOYAL scale. The overall null hypothesis of the study states that mean scores of SERVLOYAL scale are equal for all investigated moderating factors. The findings of inferential statistical analyses were used to test the null hypothesis

The independent samples t-tests based on gender groups indicate that service loyalty is not dependent on gender as mean scores are not statistically different based on gender at 95% confidence level. The study accepts the null hypothesis. One-way ANOVA analysis based on age, income and bank were conducted. The results show that mean scores are statistically insignificant for multi-groups within age and income variables, which also suggests service loyalty has nothing to do with age or income of customers. However, the bank which the customer use has an effect on how loyal customers are to the bank. This implies that the bank is offering different services that makes customers loyal to the service provider. Hence, the study accepts the alternative hypothesis that the mean difference in SERVLOYAL scores is statistically significant for different banks. This suggests that banks can improve their competitive advantage by concentrating on brand loyalty to retain customers who are loyal based on the bank's offerings.

References

1. Jamal, A., Naser, K.: Customer satisfaction and retail banking: an assessment of some of the key antecedents of customer satisfaction in retail banking. Int. J. Bank Market. **20**(4), 146–160 (2002)
2. Rahman, M.S., Khan, A.H., Haque, M.M.: A conceptual study on the relationship between service quality towards customer satisfaction: Servqual and Gronroos's service quality model perspective. Asian Soc. Sci. **8**(13), 1227–1232 (2012)
3. Sudhahar, J.C., Israel, D., Britto, A.P., Selvam, M.: Service loyalty measurement scale: a reliability assessment. Am. J. Appl. Sci. **3**(4), 1814–1818 (2006)
4. Oliver, R.L.: Whence consumer loyalty? J. Market. **63**(1999), 33–44 (1999). https://doi.org/10.2307/1252099
5. Paul, J., Mittal, A., Srivastav, G.: Impact of service quality on customer satisfaction in private and public sector banks. Int. J. Bank Market. **34**(5), 606–622 (2016). https://doi.org/10.1108/ijbm-03-2015-0030
6. Kottler, P., Keller, K.L.: Marketing Management, 15th edn. Pearson Pretice Hall, New Jersey (2016)
7. Rauyruen, P., Miller, K.E.: Relationship quality as a predictor of B2B customer loyalty. J. Bus. Res. **60**(1), 21–31 (2007). https://doi.org/10.1016/j.jbusres.2005.11.006
8. Hallowell, R.: The relationships of customer satisfaction, customer loyalty, and profitability: an empirical study. Int. J. Serv. Ind. Manag. **7**(4), 27–42 (1996). https://doi.org/10.1108/09564239610129931
9. Ngo, V.M., Nguyen, H.H.: The relationship between service quality, customer satisfaction and customer loyalty: An investigation in Vietnamese retail banking sector. J. Compet. **8**(2), 103–116 (2016). https://doi.org/10.7441/dokbat.2016.43
10. Mihelis, G., Grigoroudis, E., Siskos, Y., Politis, Y., Malandrakis, Y.: Customer satisfaction measurement in the private bank sector. Eur. J. Oper. Res. **130**(2), 347–360 (2001). https://doi.org/10.1016/s0377-2217(00)00036-9
11. Hoffmann, A.O.I., Birnbrich, C.: The impact of fraud prevention on bank-customer relationships: An empirical investigation in retail banking. Int. J. Bank Market. **30**(5), 390–407 (2012)
12. Singh, H.: The importance of customer satisfaction in relation to customer loyalty and retention. Acad. Market. Sci. **60,** 193–225 (2006)
13. Bharadwaj, S.G., Varadarajan, P.R., Fahy, J.: Sustainable competitive advantage in service industries: a conceptual model and research propositions. J. Market. **57**(4), 83–99 (1993). https://doi.org/10.1177/002224299305700407
14. Kaura, V., Prasad, C.S.D., Sharma, S.: Service quality, service convenience, price and fairness, customer loyalty, and the mediating role of customer satisfaction. Int. J. Bank Market. **33**(4), 404–422 (2015). https://doi.org/10.1108/ijbm-04-2014-0048
15. Amin, M.: Internet banking service quality and its implication on e-customer satisfaction and e-customer loyalty. Int. J. Bank Market. **34**(3), 280–306 (2016)
16. Al-Hawari, M.A.: How the personality of retail bank customers interferes with the relationship between service quality and loyalty. Int. J. Bank Market. **33**(1), 41–57 (2015). https://doi.org/10.1108/ijbm-09-2013-0096
17. Makanyeza, C., Chikazhe, L.: Mediators of the relationship between service quality and customer loyalty: Evidence from the banking sector in Zimbabwe. Int. J. Bank Market. **35**(3), 540–556 (2017). https://doi.org/10.1108/ijbm-11-2016-0164
18. Cronin Jr., J.J., Taylor, S.A.: Measuring service quality: a reexamination and extension. J. Market. **56**(3), 55–68 (1992). https://doi.org/10.2307/1252296

19. Luarn, P., Lin, H.-H.: A customer loyalty model for e-service context. J. Electron. Commer. Res. **4**(4), 156–167 (2003). https://doi.org/10.4324/9781315162829-2
20. Dick, A.S., Basu, K.: Customer loyalty: toward an integrated conceptual framework. J. Acad. Mark. Sci. **22**(2), 99–113 (1994). https://doi.org/10.1177/0092070394222001
21. Rundle-Thiele, S., Mackay, M.M.: Assessing the performance of brand loyalty measures. J. Serv. Mark. **15**(7), 529–546 (2001). https://doi.org/10.1108/eum0000000006210
22. De Ruyter, K., Wetzels, M., Bloemer, J.: On the relationship between perceived service quality, service loyalty and switching costs. Int. J. Serv. Ind. Manag. **9**(5), 436–453 (1998). https://doi.org/10.1108/09564239810238848
23. Harris, L.C., Goode, M.M.H.: The four levels of loyalty and the pivotal role of trust: a study of online service dynamics. J. Retail. **80**(2), 139–158 (2004). https://doi.org/10.1016/j.jretai.2004.04.002
24. Pallant, J.: SPSS Survival Manual: A Step by Step Guide to Data Analysis using IBM SPSS, 5th edn. McGraw-Hill Education, Berkshire (2013)
25. Hair Jr., J.F., Hult, G.T.M., Ringle, C., Sarstedt, M.: A Primer on Partial Least Squares Structural Equation Modeling (PLS-SEM), 2nd edn. Sage Publications, London (2016)

Chapter 8
The Use of Virtual Reality in Marketing: Exploring the Need for Technology and Language Adaptation to Create High Quality Immersive Experiences

Markus Rach[1(✉)] and Russell Scott[2]

[1] University of Applied Sciences Northwestern Switzerland, Riggenbachstrasse 16, 4600 Olten, Switzerland
markus.rach@fhnw.ch
[2] Cognetic Technologies, 186 Cook Ave, Hillarys, Perth, Australia
russ@cogneticmedia.com

Abstract. This paper seeks to provide an overview of the current practitioners and scientific knowledge with regard to the use of Virtual Reality in marketing. A literature review was employed to conduct a systematic analysis of 50 sources to comprehensively capture the state of research and implementation of Virtual Reality. Findings reveal, that the majority of articles focus on either area of application or the effectiveness of Virtual Reality versus other media technologies. This paper critiques the current range of VR application and performance measures by proposing an alternate perspective, provided by narrative theory, as derived from the animation and filmmaking industry.

8.1 Introduction

Immersive technologies, such as Virtual Reality (VR), have experienced unprecedented growth in the application in recent years [1]. The VR retail market is expected to reach USD 41.5 Billion by 2020, displaying double-digit annual growth rates, on the back of large investments in VR technology. Facebook's 2014 acquisition of Oculus for USD 2 Billion serves as an example [2]. Scholarly interest in VR has likewise increased, which is evident by the growth in published papers [3].

Despite the intensified interest in the topic, little research and application evidence has been conducted to understand how qualitative Virtual Reality experiences are formed.

We, therefore, aim to systematically address this research gap by answering the following questions through a practitioner-led discussion and literature review.

1 What are the Virtual Reality use cases being researched in marketing?
2 How is the success or failure of Virtual Reality applications being determined?

Á. Rocha et al. (eds.), *Marketing and Smart Technologies*, Smart Innovation, Systems and Technologies 167, https://doi.org/10.1007/978-981-15-1564-4_8

8.2 Virtual Reality

Virtual Reality has a number of general definitions both in literature and in colloquial use; however, these definitions share common factors: computer simulated 3D environments and a level of user interaction [4, 5, 6]. Others [7] have more recently defined Virtual Reality to be an artificial environment, which is experienced through sensory stimuli (as sights and sounds) provided by a computer and in which one's actions partially determine what happens in the environment.

The term Virtual Reality can encompass immersive and non-immersive experiences. Non-immersive VR is displayed via a computer screen, while traditional media such as a keyboard and mouse are used for interaction. Immersive VR is achieved when users are completely encompassed by the virtual environment. Technologies such as the Cave Automatic Virtual Environment (CAVE) [8] or the consumer Head Mounted Display (HMD) products, such as Oculus and HTC Vive, can be used to create such immersive VR experiences [1]. Immersive experiences can be further defined by physical and mental immersion [6, 9].

Mental immersion refers to being deeply engaged with the VR experience, which by most definitions is a goal of content creators. Physical immersion, on the other hand, refers to the body entering into a medium, thus reacting to a technology's synthetic stimuli on the body′s senses [6, 10]. Sensory feedback is vital to the creation of a state of physical immersion. VR systems provide sensory feedback by the means of visual feedback (e.g. through a display), haptic feedback (e.g. through touch enabled gloves or a controlling device) and sound or motion feedback.

This paper will examine the research and analysis solely on immersive VR experiences.

8.3 Method

Figure 8.1 presents the methodology applied for this paper.

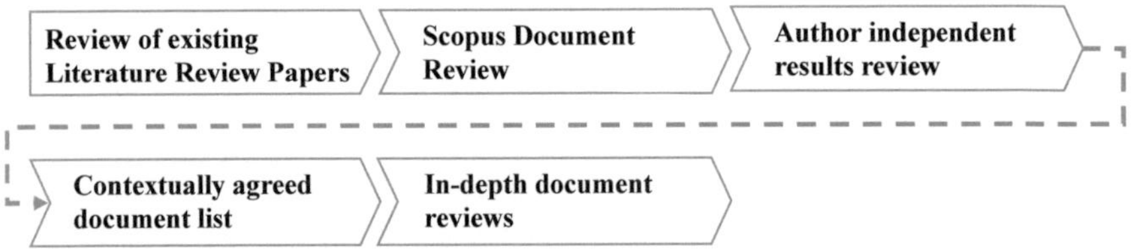

Fig. 8.1. Schematic visualization of the literature review process

8.3.1 Literature Review

An existing research paper [3] analyzed 150 VR and marketing related articles from 115 journals through a text mining approach, which proved effective. Similarly, the authors of this paper decided to rerun a new text-based literature review to assure an unbiased research focus.

Elsevier Scopus was used to identify relevant academic literature with regards to the application of Virtual Reality in Marketing terms. Since 800 documents were identified, inclusion criteria contained documents from 2010 to 2019; reducing search results to 426 documents. These 426 results were reviewed by the authors to independently identify the most relevant documents. Exclusion criteria included papers extending their research or scope to Augmented Reality, Mixed Reality, addressing VR from a technical point of view and literature addressing VR from a non-marketing point of view.

As a final step, these independent lists of documents were combined to the final list of documents in a mutual discussion of the contextual relevancy of each independently identified source.

Each of the documents in the final list was then successively analyzed for its applied performance criteria, that is, how it determined the success or failure of Virtual Reality applications for marketing purposes.

Sources include academic papers, books, and conference proceedings to reflect the latest thoughts and findings on the topic.

8.4 Results and Discussion

8.4.1 Areas of Application

Based on the definition of sighted papers [3] and to extend on this work, findings from this review were grouped in four dominant areas of application: “Virtual Setting”, “Interaction”, “Experiential Marketing”, and “Manufacturing & new product development”.

“Virtual Setting” is found in research such as in [11], where the effectiveness of Virtual Reality to simulate a retail store environment was considered. [12] analyzed the fit of Virtual Reality for the consumption of sporting events while [13] considered applications in high-involvement purchasing decisions.

References [14, 15, 16, 17] researched, amongst others, areas of application within the tourism sector. These papers fall into the topic category “Interaction”.

The category of “Experiential Marketing” proved very interesting due to the novelty of applications and thus the evident growing application potential of VR. Examples found for this category include the works of [18], with their focus on creating memorable customer experiences by the means of VR applications, or [19], who researched the effect of VR on intent and thus decision-making factors.

The last category, “Manufacturing & new product development” is probably the most narrowly defined category [3]. It entails findings such as [20] assessing the impacts of VR on B2B marketing and [21], who demonstrated the effect of VR on customers’ automotive design experiences.

It can be said, that the areas of application are diverse yet seem to somewhat relate to the categories applied from [3]. Many studies did not specify a single application category [22], while others [23] related their work to multiple categories or a much broader spectrum of application.

8.4.2 How Successful Virtual Reality Is Determined

The majority of Marketing research considers success factors via a media comparison (Table 8.1), where the effectiveness of scenarios presented in VR is compared to the effectiveness of similar scenarios presented via Non-VR media. [24], for example, compared images versus videos versus Virtual Reality videos for apartment evaluation, while [11] contrasted the effects of VR versus desktop computing on a shopping experience. Some research conducted studies on the immersion effect of VR on consumer behavior, often in single VR applications [25, 26, 27, 28, 29]. The majority of studies found that VR has a positive potential for marketing.

Table 8.1. Summary of marketing success measures

# of papers	Citation example	VR success measure
11	(Chandra Sekhar et al. 2018)	Immersion
6	(Cacho-Elizondo et al. 2018)	Mixed
3	(Laurell et al. 2019)	NA
7	(Barnes 2016)	Presence
1	(Cheng and Wang 2011)	Recall
22	(Guttentag 2010)	Media comparison

8.4.3 Presence

One key factor to a successful VR experience is described in broader literature as "presence", a condition where users report a sensation of being in the virtual world [3]. Presence is regarded as a vital component to elicit an emotional response from within a virtual environment [30, 31].

Presence is a subjective measure and as such there are varied models on how to consider the phenomena. According to [32], there are two main pillars to presence: context and immersion. Others, like [33] suggest three factors are vital: spatial presence, involvement, and realness. [34] consider presence a property of normal everyday experience, thus while the sensory input is affected by technology, our end processing remains the same. In this model, the successful state of presence is achieved through suppression of the mismatch between the predicted and the actual interoceptive state [28].

[35] argued that the sense of presence is highly individual yet can be approximated by the user's propensity to respond to virtually generated sensory data as if they were real [36]. The sense of presence exists on a scale, where differences in the quality of the sensory input and the subsequent cognitive processing will vary the perceived degree of presence [37].

8.4.4 Factors Affecting Presence

The presentation technology used to present the VR experience can influence the level of presence. Higher frame rates in the HMDs are positively correlated with reported presence with a minimum framerate of 15 Hz, as being found in two separate studies [38. 36] consider the literature surrounding sound in VR and report various studies that suggest sound has a large effect on presence. Perhaps as interestingly, they connect this with literature examining the importance of audio on a person's real-world "presence", describing an "inverse presence" effect when a person suffers sudden and lasting hearing loss.

Body Engagement and haptic feedback are other areas that can affect engagement [39]. Slater et al. [40] examined the effects walking has on presence, reporting that "real walking" provided the greatest sense of presence when compared to walking in place or moving via mouse.

Diemer et al. [28] examined current literature and found that generally, research has shown that more sophisticated technology results in an increased level of presence. This fits in with models presented by [34], where better technology reduces the perceived mismatch between the expected and the simulated worlds.

Perhaps a surprising addendum to the consideration of technology's role in presence: visual realism has not been shown with certainty to contribute to the effect. One study with a driving simulation context found that visual quality had a clear impact on presence [41], while others, such as that conducted by [42], found no relationship.

Diemer et al. [28] consider an addition to [34] model where presence "is the result of a cognitive judgment from the immersiveness, the interactivity provided by the VR system and the emotional arousal from the perceived content of the VR scenario". Put more simply, emotion evoked by the content has a role to play in presence. This relationship is also suggested by research from [37]. Baños et al. [43] showed that technological factors had a greater effect on presence when the user was experiencing emotionally neutral content than it did when the user was exposed to emotionally raised content.

8.4.5 Divergences in Analysis

If the above is accepted, that emotion plays a key role in creating presence, consideration of the narrative rules of VR is a vital aspect of analysis [43]. To properly analyze the effectiveness of a VR project in a marketing context, one must consider how well the emotional message itself was communicated to be able to draw any effective conclusions.

A crude example to highlight this point: a 2D feature film generally adheres to what is known as a three-act structure: beginning, middle, and end [44]. If a 2D film is created with only one act, poor response from an audience can more likely be identified as a problem with the act structure, rather than the medium itself being unsuitable for the task. In other words, the construction of content itself has affected the outcome negatively. Likewise, if a research project uses VR content that has been created in a way that conflicts with emerging best practices for VR, this content will likely have a negative impact on the viewer's experience or presence and will, therefore, influence any findings in the research.

Research examining how to generate an emotional response from content is encompassed by "narrative theory". Notably, the initial keyword search involving "marketing", "Immersion" "Virtual Reality", and "presence" provided no papers involving narrative theory. In addition, analysis of papers within the original keyword search found no consideration for the role content itself played in influencing results.

Expanding the keyword search to examine papers that consider narrative theory in VR revealed a number of content-based impactors on degrees of presence.

This is a key finding in this review; by not considering the impact of content in an analysis of VR efficacy, the reliability of a conclusion is at the very least diminished and at very worst invalidated.

8.4.6 Narrative Factors Affecting Presence

Language is the system of conventions by means of which humans achieve communication. “A language is a social fact, a kind of social contract. It exists not in an individual, but in a community” [45]. Language, in this case, is not limited to human speech [46], it is both a social product of the faculty of speech and a collection of conventions that have been adopted by a social body to allow individuals to exercise that faculty.

VR will, like all media, require its own language to achieve effective communication, but as a new medium, this language is both emerging and evolving. Dooley [47] analyzed three Australian VR film case studies and concluded that while there were emerging approaches, there are as yet no fully defined rules for VR narrative grammar. That being said, [48] tested 50 participants with various VR videos and found a number of impactors that have direct negative consequences for presence.

For example, the starting position of the viewer (being seated or standing) can have a dramatic impact on the viewer’s comfort, with 88% of test subjects preferring the standing position. Moreover, being in a seated position reduces a user’s willingness to look around the VR environment.

An orientation phase, a moment for the viewer to orient themselves within the virtual world, has been shown to be necessary for a positive experience. This phase is possible as long as 40 s [48] for those new to VR.

The editing style for transitioning between scenes can also impact a viewer’s experience. The user′s choice to look in one direction will impact their ability to understand the narrative if the intent was for them to look in a different direction. Jessica Brillhart posited a new editing style dubbed “probabilistic experiential editing” [48], describing the notion that there is a chance the user won’t follow the storyline due to their attention being on an unintended point of interest (POI). Probabilistic editing accounts for POIs within a scene and the chance a viewer may be looking at them at key points in time.

Gödde et al. [48] noted that when presented a video with highly diverged POIs, key informational elements were missed by the majority of the viewers and only 20% of viewers could answer four basic questions regarding the plot of the video just watched, with 8% being unable to answer any questions at all. This is a clear example of the dramatic impact of poorly constructed content can have on the user’s experience.

8.4.7 The Need for a Paradigm Shift

It is apparent that two streams of research have emerged. One stream is driven by a functional focus on application areas such as Marketing. This stream focuses heavily on the effective measures of Virtual Reality against other media or technologies. This stream largely ignores the effects narrative components have on the construction of presence and immersion in Virtual Reality.

The other stream of research is of creative nature and stems from the entertainment, animation, and filmmaking industry. This stream of research focuses heavily on the narratives of Virtual Reality and is summed up effectively by [49]: "Virtual Reality requires its very own syntax".

The success of Virtual Reality applications, particularly within a business setting, cannot be solely determined in isolation by either technical or creative standards but requires a combined consideration of both streams of research. We, therefore, propose the following model as a success determinant for further VR research (Fig. 8.2).

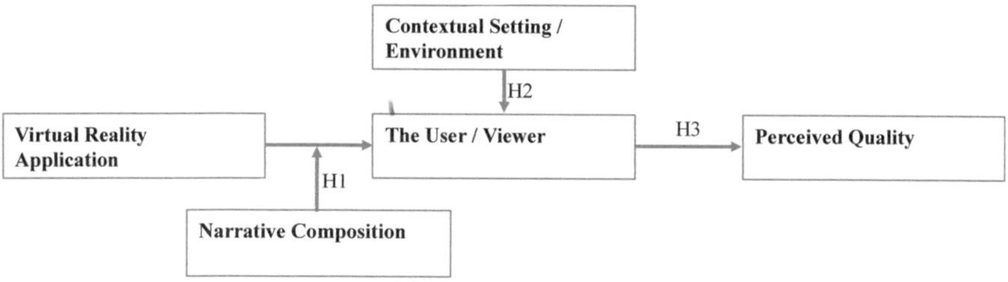

Fig. 8.2. Proposed conceptual model for further virtual reality research

The core concept of the above model is to incorporate the narrative composition of a VR application as a moderating variable and the contextual setting as a mediating variable, therefore, explain the user's or viewer's reaction to the VR experience.

8.5 Conclusions and Implications

The aim of this paper was to consider how VR is being used in marketing and how the success of those projects is being determined. The finding that divergent streams of research exist reflects somewhat the early stage that VR development is at. Including content considerations in future research will not guarantee effective emotional engagement from the audience; it will, however, serve as a minimum standards framework to analyze effectiveness.

8.5.1 Limitations

Despite all efforts undertaken, this paper has several limitations that need consideration when applying the paper's findings.

First, since articles, conference proceedings and book sources were selected on the basis of a multistep process, involving both scholarly key word searches and human sorting; there may be literature that was missed.

Second, this study focuses exclusively on VR technologies and thus excludes work from AR or Mixed Reality studies.

Third, although all possible efforts have been undertaken to provide a robust literature review of identified documents, interpretation errors cannot be excluded due to the mix of the terminology used by both the practitioners and scholars, often causing divergent perspectives.

8.5.2 Managerial Implications

This paper offers insights and adaptable learnings for marketing managers, marketing agencies and creative professionals. Marketing managers are encouraged to question pure media translation efforts from 2D media to Virtual Reality; not in the sense of semantics but in the syntax needed to create successful VR experiences. This requires managers to gain a deeper understanding of their target audience, the setting of application and also their objective of use. Marketing agencies are encouraged to question customer requests and propose a narrative adaption of VR experiences for the benefit of both the user and the client.

With more companies turning to digital solutions and aiming to increase their use of Virtual Reality applications [50], the role and importance of VR in marketing will continue to increase. For this, it is imperative that practitioners do not solely focus on substituting current media or technology with VR, but also focus on creating high-quality VR experiences to extrapolate the positive effect VR can have in identified and future areas of application.

8.5.3 Future Research

Based on the papers analyzed and the practitioner discussion following, we propose two avenues for further research.

First, scholars need to develop a deeper understanding of a user's predisposition towards technology. Specifically, as digital natives emerge as the dominant generation, it is vital for scholarly work to research generation impacts, alongside education levels and prior technology exposure.

Second, narrative theories cited remain dislocated from functional research in business or marketing sense. Confirming the level of impact these concepts have in functional settings is vital in both a business to consumer and a business to business context.

References

1. Suh, A., Prophet, J.: The state of immersive technology research: a literature analysis. Comput. Human Behav. (86), 77–90 (2018)
2. Kilic, K., Saygi, M.G., Sezer, S.O.: Exact and heuristic methods for personalized display advertising in virtual reality platforms. J. Ind. Manag. Optim. **15**(2), 833–854 (2019)

3. Loureiro, S.M.C., Guerreiro, J., Eloy, S., Langaro, D., Panchapakesan, P.: Understanding the use of virtual reality in marketing: a text mining-based review. J. Bus. Res. **100**, 514–530 (2019)
4. Steuer, J.: Defining virtual reality: Dimensions determining telepresence. J. Commun. **42**(4), 73–93 (1992)
5. Yoh, M.S.: The reality of virtual reality. In: Proceedings Seventh International Conference on Virtual Systems and Multimedia, pp. 666–674. IEEE (2001)
6. Sherman, W.R., Craig, A.B.: Understanding Virtual Reality: Interface, Application, and Design. Morgan Kaufmann (2018)
7. Ni, D., Nee, A.Y.C., Ong, S.K., Li, H., Zhu, C., Song, A.: Point cloud augmented virtual reality environment with haptic constraints for teleoperation. Trans. Instit. Meas. Control, **40** (15), 4091–4104 (2018)
8. Cruz-Neira, C., Sandin, T.A., DeFanti R.V.: Surround screen projection-based virtual reality: the design and implementation of the cave. In: Proceedings of SIGGRAPH 1993, pp 135–142 (1993)
9. Kwon, J.H., Powell, J., Chalmers, A.: how level of realism influences anxiety in virtual reality environments for a job interview. Int. J. Human-Comput. Stud. **71**(10), 978–987 (2013)
10. Tham, J., Duin, A.H., Gee, L., Ernst, N., Abdelqader, B., McGrath, M.: Understanding virtual reality: presence, embodiment, and professional practice. IEEE Trans. Prof. Commun. **61**(2), 178–195 (2018)
11. Schnack, A., Wright, M.J., Holdershaw, J.L.: Immersive virtual reality technology in a three-dimensional virtual simulated store: investigating telepresence and usability. Food Res. Int. **117**, 40–49 (2019)
12. Rynarzewska, A.I.: Virtual reality: a new channel in sport consumption. J. Res. Interact. Market. **12**(4), 472–488 (2018)
13. Sihi, D.: Home sweet virtual home: the use of virtual and augmented reality technologies in high involvement purchase decisions. J. Res. Interact. Market. **12**(4), 398–417 (2018)
14. Guttentag, D.A.: Virtual reality: applications and implications for tourism. Tourism Manag. **31**(5), 637–651 (2010)
15. Huang, Y.C., Backman, K.F., Backman, S.J., Chang, L.L.: Exploring the implications of virtual reality technology in tourism marketing: An integrated research framework. Int. J. Tourism Res. **18**(2), 116–128 (2016)
16. Tussyadiah, I.P., Wang, D., Jung, T.H., Tom Dieck, M.C.: Virtual reality, presence, and attitude change: Empirical evidence from tourism. Tourism Manag. **66**, 140–154 (2018)
17. Marasco, A., Buonincontri, P., van Niekerk, M., Orlowski, M., Okumus, F.: Exploring the role of next-generation virtual technologies in destination marketing. J. Destination Market. Manag. **9**, 138–148 (2018)
18. Cacho-Elizondo, S., Álvarez, J.D.L., Garcia, V.E.: Exploring the adoption of augmented and virtual reality in the design of customer experiences: proposal of a conceptual framework. J. Market. Trends **5**(2), 1961–7798 (2018)
19. Bogicevic, V., Seo, S., Kandampully, J.A., Liu, S.Q., Rudd, N.A.: Virtual reality presence as a preamble of tourism experience: the role of mental imagery. Tourism Manag. **74**, 55–64 (2019)
20. Boyd, D.E., Koles, B.: Virtual reality and its impact on B2B marketing: a value-in-use perspective. J. Bus. Res. **100**, 590–598 (2019)
21. Kato, T., Tsuda, K.: December. does virtual reality influence customer behavior in car design?. In: ISPIM Innovation Symposium, pp. 1–15. The International Society for Professional Innovation Management (ISPIM) (2018)

22. Saren, M., Harwood, T., Ward, J., Venkatesh, A.: Marketing beyond the frontier? Researching the new marketing landscape of virtual worlds (2013)
23. Singh, H., Singh, C., Majumdar, R.: Virtual reality as a marketing tool. In: Emerging Trends in Expert Applications and Security, pp. 445–450. Springer, Singapore (2019)
24. Grudzewski, F., Awdziej, M., Mazurek, G., Piotrowska, K.: Virtual reality in marketing communication–the impact on the message, technology and offer perception–empirical study. Econ. Bus. Rev. **4**(3), 36–50 (2018)
25. Biocca, F.: Communication within virtual reality: Creating a space for research. J. Commun. (1992)
26. Bates, R., Vickers, S., Istance, H.O.: Gaze interaction with virtual on-line communities: levelling the playing field for disabled users. Univers. Access Inform. Soc. **9**(3), 261–272 (2010)
27. Eisenbeiss, M., Blechschmidt, B., Backhaus, K. Freund, P.A.: The (real) world is not enough: motivational drivers and user behavior in virtual worlds. J. Interact. Market. **26**(1), 4–20 (2012)
28. Diemer, J., Alpers, G.W., Peperkorn, H.M., Shiban, Y., Mühlberger, A.: The impact of perception and presence on emotional reactions: a review of research in virtual reality. Frontiers Psychol. **6**, 26 (2015)
29. Willems, K., Brengman, M., Van Kerrebroeck, H.:. The impact of representation media on customer engagement in tourism marketing among millennials. Eur. J. Market. (2019)
30. Parsons, T.D., Rizzo, A.: Affective outcomes of virtual reality exposure therapy for anxiety and specific phobias: a meta-analysis. J. Behav. Ther. Exp. Psychiatry **39**, 250–261 (2008)
31. Price, M., Mehta, N., Tone, E.B., Anderson, P.L.: Does engagement with exposure yield better outcomes? Components of presence as a predictor of treatment response for virtual reality exposure therapy for social phobia. J. Anxiety Disord. **25**, 763–770 (2011)
32. Slater, M., Wilbur, S.: A Framework for Immersive Virtual Environments (FIVE): Speculations on the role of presence in virtual environments. Presence Teleoperators Virtual Environ. **6**(6), 603–616 (1997)
33. Schubert, T., Friedmann, F., Regenbrecht, H.: The experience of presence: factor analytic insights. Presence: Teleoperators Virtual Environ. **10**(3), 266–281 (2001)
34. Seth, A.K., Suzuki, K., Critchley, H.D.: An interoceptive predictive coding model of conscious presence. Frontiers Psychol. **2**, 395 (2011)
35. Milgram, P., Takemura, H., Utsumi, A., Kishino, F.: Augmented reality: a class of displays on the reality-virtuality continuum. systems research. In: Das, H. (ed.) Telemanipulator and Telepresence Technologies. Telemanipulator and Tele- Presence Technologies,vol. 2351, pp. 282–292. Photonics for Industrial Applications, Boston (1994)
36. Sanchez-Vives, M.V., Slater, M.: From presence to consciousness through virtual reality. Nat. Rev. Neurosci. **6**(4), 332 (2005)
37. Lorenz, M., Brade, J., Diamond, L., Sjolie, D., Busch, M., Tscheligi, M., Kilasmant, P., Heyde, C., Hammer, N.: Presence and User Experience in a Virtual Environment under the Influence of Ethanol: An Explorative Study (2018)
38. Barfield, W., Hendrix, C.: The Effect of Update Rate on the Sense of Presence (1995)
39. Slater, M., Usoh, M., Steed, A.: Depth of presence in virtual environments. Presence: Teleoperators Virtual Environ. **3**(2), 130–144 (1994)
40. Slater, M., Usoh, M., Steed, A.: Taking steps: the influence of a walking technique on presence in virtual reality. ACM Trans. Comput.-Human Interact. (TOCHI), **2**(3), 201–219 (1995)
41. Welch, R.B., Blackmon, T.T., Liu, A., Mellers, B.A., Stark, L.W.: The effects of pictorial realism, delay of visual feedback, and observer interactivity on the subjective sense of presence. Presence: Teleoperators Virtual Environ. **5**(3), 263–273 (1996)

42. Zimmons, P., Panter, A.: The influence of rendering quality on presence and task performance in a virtual environment. In: IEEE Virtual Reality, 2003. Proceedings, pp. 293–294. IEEE (2003)
43. Baños, R.M., Botella, C., Alcañiz, M., Liaño, V., Guerrero, B., Rey, B.: Immersion and emotion: their impact on the sense of presence. Cyberpsychology Behav. **7**(6), 734–741 (2004)
44. Higgins, S.: Suspenseful situations: melodramatic narrative and the contemporary action film. Cinema J. 74–96 (2008)
45. Bauer, L.: Linguistics Student's Handbook. Edinburgh University Press (2007)
46. Saussure, F.D.: Course in general linguistics (W. Baskin, Trans.). Philosophical Library, New York (1959)
47. Dooley, K.: Storytelling with virtual reality in 360°: a new screen grammar. Stud. Australasian Cinema **11**(3), 161–171 (2017)
48. Gödde, M., Gabler, F., Siegmund, D., Braun, A.: Cinematic Narration in VR–Rethinking Film Conventions for 360°. In: International Conference on Virtual, Augmented and Mixed Reality, pp. 184–201 (2018)
49. Barricelli, B., Gadia, D., Rizzi, A., Marini, D.: Semiotics of –virtual reality as a communication process. Behav. Inform. Technol. **35**, 879–896 (2016)
50. Boyd, D., Koles, B.: An introduction to the special issue "virtual reality in marketing": definition, theory and practice. J. Bus. Res. (2019)

Chapter 9
Social Media Usage Among Wine Tourism DMOs

Filipa Jorge[1,2(✉)], Mário Sérgio Teixeira[1,2], Carlos Fonseca[1,2], Ricardo Jorge Correia[2,4], and Ramiro Gonçalves[1,3]

[1] Universidade de Trás-Os-Montes e Alto Douro, UTAD, Vila Real, Portugal
filipajorge@utad.pt
[2] Centro de Estudos Transdisciplinares para o Desenvolvimento, CETRAD, Vila Real, Portugal
[3] Instituto de Engenharia de Sistemas e Computadores, Tecnologia e Ciência, INESC TEC, Porto, Portugal
[4] Instituto Politécnico de Bragança, IPB, Bragança, Portugal

Abstract. Social media is an important tool for tourism destination promotion. The usage and the contents published on these platforms have an important role in visitors' decision-making process. Information and communication technologies are changing DMOs' markets and communication paradigm, since they allow the interaction between these organizations, visitors, and stakeholders. Therefore, social media are increasing their relevance on DMOs' marketing strategies. This purpose of this study is to analyze comparatively social media platforms' usage by six wine tourism DMOs. The results were provided by the analysis of secondary data available on these platforms and DMOs posts on the four most popular social media platforms to tourism industry: Facebook, Instagram, Twitter, and YouTube. Results indicate that DMOs use their official accounts on these platforms on different ways. This study also reveals that Facebook is more interactive than other platforms, and YouTube achieves less engagement levels.

9.1 Introduction

Tourism sector is facing a technological revolution. The number of tourists booking online travels, accommodation, and other tourism products and services is increasing and even for those who don't buy online, at least they use Internet to search information about destination [1, 2]. Social media platforms allow to share information through the use of texts, images [3], but also short duration microfilms for tourism destinations promotion [4]. These types of contents may be important to potential visitors who are planning their travel, but also for visitors who already are in the destination and these information may improve their tourism experience [3].

Destination Marketing Organizations (DMOs) have as mission the representation of their destinations at the marketplace in order to attract more tourists [2]. According to

Á. Rocha et al. (eds.), *Marketing and Smart Technologies*, Smart Innovation, Systems and Technologies 167, https://doi.org/10.1007/978-981-15-1564-4_9

[5], DMOs have several responsibilities concerning the development of the destination. The roles more mentioned for these organizations are the marketing strategies formulation, which includes promotion activities, and the communication between tourism stakeholders from the destination. To DMOs activities, social media platforms may act as a powerful and low-cost way to overcome this challenge [6]. For [2], the use of technologies by DMOS can be an opportunity and also another challenge.

Even though there is a growing interest in social media usage in tourism industry and the existence of empirical evidences that support the strategical importance of these platforms to achieve a better competitiveness of this industry, there are few exploratory researches that are focused on the social media platforms usage by DMOs as was indicated by [6] and [7]. Although this research is also motivated by the future investigation recommendation of [8], which is to analyze the social media use by DMOs of emerging destinations.

This research has the purpose to analyze comparatively the usage of four social media platforms (Facebook, Instagram, Twitter, and YouTube) by six DMOs that promote small wine tourism destinations. The present study aims to contribute to scientific discussion that are concerned about the use of different social media platforms to tourism destinations' promotion. Besides, it is also expected to yield some good practices to destination marketers who are interested in social media marketing.

The present article is divided into five sections. In this section, the social media thematic on tourism was introduced. In the following section, relevant literature toward this mentioned theme will be exposed. The third section will describe the methodology used to accomplish this study objective. In the fourth section, the results will be presented. Lastly, in the fifth section, the principal conclusions and contributions of this research, and the limitations and suggestions for future studies will be enunciated.

9.2 Literature Review

Social media marketing should be part of a more holistic strategy. According to [9], social media marketing has as objectives: improving sales, increase brand awareness, improve brand image and brand awareness, reduce marketing costs, and promote user interactivity on social media platforms. Social media platforms also provide an opportunity for organizations to try to understand how their consumers view their actions and organization.

Tourism is a very competitive industry, and destinations compete to attract more visitors. Internet has changed the way that potential visitors access information, the way they plan and book tourism products, and allows them to share contents about the destination [10]. Previous researches have revealed that digital channels have a significant impact in tourism sector since tourists use them to have information about the destinations [11]. For that reason, Internet presents itself as a growing promotional channel for tourism industry because it is cheaper and more efficient than traditional marketing tools [6].

Tourism industry is very intensive in information [12]. This explains that tourists' decision-making process is highly influenced by the information expressed on social media comments or reviews made by other consumers [13, 14]. Therefore, social media plays a vital role in tourists' decision-making [12]. The use of social media platforms by tourists is changing this sector and the way how tourists research and find information, because these platforms provide the collaborative production of information about tourism destinations and organizations [15].

Social media can be extremely useful for tourism stakeholders, since they represent an information source, which allow them to understand tourists' preferences and how to respond [16].

Therefore, DMOs should engage with their tourists, which means that DMOs should encourage them to share images, videos, and other contents about the destination in order to influence other potential tourists [17].

Some studies indicate that DMOS are not using the total potential of social media, since most of DMOs are only using these platforms as promotional tool and not using them to mitigate potential customer services problems or as a co-creation place, which will allow them to gain competitiveness [6, 8, 18].

9.3 Methodology

The previous research was focused on the social media use by national DMOs. Nevertheless, this research will be dedicated to emerging wine tourism destinations. In order to have similar destinations, it was selected from Forbes (2018) list of the ten most beautiful wine tourism destinations: Douro Valley (Portugal), Franschhoek (South Africa), Saint Emilion (France), Valle de Guadalupe (Mexico), Barossa Valley (Australia), Piedmont (Italy), Napa Valley, (USA), Hawke's Bay (New Zealand), Moselle Valley (Germany), and La Rioja (Spain) [19].

From these ten wine tourism destinations, we selected six to use as sample. The information about the DMOs responsible for these six wine tourism destinations are expressed in Table 9.1. Almost all DMOs' websites are focused on a wine tourism destination, with exception of Douro Valley. This DMO is responsible for the promotion of this wine tourism destination but also for a more extensive region denominated Porto and North of Portugal. Besides, two of these six wine tourism destinations, explicitly Douro and Saint Emilion are classified as world heritage by UNESCO [20].

Data collection was done in October 2018 on social media official pages of these six DMOs in four social media platforms: Facebook, Instagram, YouTube, and Twitter. These social media platforms were chosen for two reasons: these social media were considered important for tourism sector [12], [17], and all the selected DMOs have an official page in these social media, which allows a better comparison. Having into consideration that social media platforms have several publications by day, this research limits the period of observations to 3 months (July to September 2018).

9.4 Results

Results are presented in tables below and they are discussed in order to afford a comparative analysis.

This research follows the recommendation of [8] to describe the audience of each DMO for each social media platform. According to these authors, the social media

Table 9.1. Destinations, DMOs, and social media platforms used as sample

Wine Tourism Destinations	DMOs	Facebook page	Instagram username	Twitter username	YouTube page
Douro Valley	Turismo do Porto e Norte de Portugal	facebook.com/TurismoPortoNortePortugal	@turismo_portoenorte	@PortoeNortePT	youtube.com/channel/UCdBiRW6kEaFKplpd4JXW5IA
La Rioja	La Rioja Turismo	facebook.com/turismodelarioja	@lariojaturismo	@lariojaturismo	youtube.com/user/Tierraconombredevino
Saint Emilion	Saint Emilion Tourisme	facebook.com/saintemiliontourisme/	@mysaintemilion	@saint_emilion	youtube.com/channel/UCZrUIiNm4xg_g1tranZY1_w
Napa Valley	Visit Napa Valley	facebook.com/VisitNapaValley	@visitnapavalley	@VisitNapaValley	youtube.com/user/VisitNapaValley
Franschhoek	Franschhoek Wine Valley	facebook.com/Franschhoekwinevalley	@franschhoek_sa	@Franschhoek_SA	youtube.com/channel/UCcy6piUBEN8tpiYjmLreFng
Barossa Valley	Tourism Barossa	facebook.com/BarossaWines/	@barossawines	@BarossaWineAu	youtube.com/user/barossadirt

platform audience can be measured by the number of followers on Twitter and Instagram, the number of likes on the Facebook page, and by the total number of views on YouTube. This information is summarized in Table 9.2.

Table 9.2. Social media audience for each DMO

	Facebook page likes	Instagram followers	Twitter followers	YouTube views
Douro Valley	151,644	5,325	5,336	54,262
La Rioja	32,917	8,146	47,000	2,123,721
Saint Emilion	11,981	4,195	2,256	25,936
Napa Valley	147,274	76,500	28,700	1,791,685
Franschhoek	18,930	21,900	9,760	31,235
Barossa Valley	6,751	15,500	18,800	27,846
Total	369,497	131,566	111,852	4,054,685

Considering the total audience of the six DMOs, YouTube is the social media platform that has more audience, followed by Facebook, Instagram and, lastly, Twitter. However, it is important to have into consideration that one user may view a video several times, but users cannot follow or like a same page twice. As positive indicators related with the audience size: La Rioja DMO has the highest number of total views in YouTube, and more followers on Twitter; Douro Valley DMO has the most likes on their Facebook official page; Napa Valley DMO has the highest number of followers on Instagram. In the negative side, Saint Emilion is the DMO with less audience in Instagram, YouTube, and Twitter; in Facebook, Barossa Valley DMO has lesser audience.

In Table 9.3, the date on which each DMO has created its official account in the four social media platforms is reported. This information is important to identify the early adopters for all the four platforms.

Table 9.3. Membership dates of the DMOs for each social media platform

	Facebook	Instagram	Twitter	YouTube
Douro Valley	Jan./2010	Mar./2015	Dec./11	Apr./2012
La Rioja	Sept./2010	Apr./2013	Dec./09	Oct./2008
Saint Emilion	Aug./2011	Sept./2013	Feb./10	Feb./2013
Napa Valley	Jan./2010	Aug./2012	Mar./10	Apr./2012
Franschhoek	Nov./09	May/2013	Oct./09	Feb./2010
Barossa Valley	2008	Jul./2012	Apr./10	Aug./2010

Barossa Valley DMO emerges as early adopter on Facebook and Instagram. La Rioja DMO was the first to have an official account on YouTube. On Twitter, Franschhoek DMO was the earliest to develop an account on this social media platform. Saint Emilion was the last DMO to join Facebook and YouTube and Douro Valley DMO was the last to create an official account on Instagram and Twitter.

In the table below (Table 9.4) are presented the total number of posts that each DMO has published in Instagram and Twitter, and the total number of videos uploaded in YouTube on their official account. In this table, no data concerning to Facebook are expressed, since this social media platform does not make this information available to the users who view the profiles, unlike with the other three social media platforms (Instagram, YouTube, and Twitter).

Table 9.4. Total number of posts for each DMO

	Instagram	Twitter	YouTube	Total number of posts for each DMO
Douro Valley	292	15,400	74	15,766
La Rioja	355	16,800	107	17,262
Saint Emilion	696	2,679	52	3,427
Napa Valley	1,711	18,500	23	20,234
Franschhoek	1,210	23,700	5	24,915
Barossa Valley	1,494	8,570	84	10,148
Total number of posts by social media platform	5,758	85,649	345	91,752

Twitter is the social media platform where DMOs post more, followed by Instagram and, last of all, YouTube. One of the reasons for YouTube to appear as the social media with the least posts is because this platform is orientated toward uploading videos, and this type of content consumes more time and money than others types of contents used on the others platforms like Twitter or Instagram. Besides, different DMOs lead total posts for each social media platform. In Instagram, Napa Valley is the DMO with the highest number of posts. In Twitter, it is Franschhoek DMO who posts more. In YouTube, it is La Rioja DMO who uploads more videos.

In the following table (Table 9.5), the data collected about DMOs' activity on the four social media platforms is presented. As indicators of this activity, the number of posts made in the 3 months selected and the interactivity generated by these posts were used. The number of posts made during the 3 months was considered every post made or shared by the DMOs. Besides, a number of interactions from these posts, i.e., every likes, comments, shares, and views (in the case of videos) were considered. For social media platform, Twitter, the data was not collected for the number of interactions.

Similarly to the total number of posts, also in the period analyzed, YouTube was the social media platform less used to make publications. In fact, only La Rioja and Saint Emilion DMOs have published videos in this platform during these 3 months. Furthermore, during this period, La Rioja DMO was the one that made more posts on Facebook, YouTube, and Twitter. Napa Valley DMO was the one who posted more on

Table 9.5. Social media activity

	Facebook		Instagram		Twitter	YouTube	
	Posts in 3 months	Interactions	Posts in 3 months	Interactions	Posts in 3 months	Posts in 3 months	Interactions
Douro Valley	12	15,827	0	0	8	0	0
La Rioja	291	294,309	28	17,479	>732	11	340,738
Saint Emilion	60	18,373	57	11,270	128	4	622
Napa Valley	146	8,076,312	119	106,168	363	0	0
Franschhoek	284	211,679	96	27,629	557	0	0
Barossa Valley	75	9,466	88	14,767	0	0	0

Instagram. It is important to note that although Napa Valley is not the one who makes the most posts, this DMO is the one who achieved more interactivity in Facebook and Instagram. This may indicate that this DMO strategy focuses more on the quality than the quantity of their posts and seems to have success because it has more user interaction than other DMOs with more posts.

9.5 Final Considerations

Results indicate that wine tourism DMOs are using social media platforms to promote their destinations. With regard to Facebook data, there are several facts that justify to be highlighted. The first DMO to join this social network was Barossa Valley, but is the one that has the least followers. Besides, Napa Valley DMO has less followers than Douro Valley DMO, but have converted more of these followers into interactions. This best result may be explained by the fact that the Napa Valley DMO performs a larger number of posts compared to Douro Valley DMO.

In Instagram, all DMOs have lower number of followers than in the previous social media platform, what can be explained by this social network be more recent. Napa Valley DMO has the highest number of followers, more posts in the periods analyzed, as well as more interactions. On opposing, Douro Valley DMO was the least active in Instagram, since it did not perform any post in the 3 months analyzed and published fewer posts when we look all posts.

In YouTube, the first DMO to create an official page on this social network was La Rioja, and is also the one that has published more videos in total and in the period analyzed and who has more visualizations. In the opposite, Saint Emilion DMO is the last to join the YouTube. However, this last DMO and La Rioja DMO were the only ones that uploaded videos during the analyzed period. Although, Napa Valley DMO has far more views than most DMOs with more videos, which may indicate that despite publishing fewer videos, they are often viewed.

Concerning to Twitter, the first DMOs to create a page on this social media were Franschhoek and La Rioja. The last DMO has the better performance, which means that has more publications in the total and in the analyzed period and has a higher number of followers. Further, during the analyzed period, Barossa Valley DMO did not post any publication, and, similarly, Douro Valley DMO only performed eight posts. This seems to indicate a disinvestment in this particular social media by these DMOs. On other hand, Saint Emilion DMO has a lower total number of posts and a lower number of followers than these two DMOs, but performed more posts in the analyzed period, which may indicate that they are investing more on this social media platform.

This research contributes to the discussion about the social media usage by DMOs, with the innovation of being applied to wine tourism destinations. For these organizations, the results obtained leave some information to allow them to improve their effective communication using the most relevant social media platforms associated to tourism promotion.

The present research has several limitations that should be outpaced in future research. The first limitation is the sample used, since it should have a larger size, what could be suggested for future research. Second, this study utilizes only secondary data available on the social media platforms. Hence, future studies may use primary data, such as interviews with DMOs' marketers may be interesting to have a better understanding of social media marketing strategies that can explain the data collected. Third, the present research only analyzes the social media platforms. Thus, it is proposed that future research analyzes other digital marketing tools, such as websites, that will allow gaining a better understanding of DMOs' marketing strategies.

Acknowledgements. This research was funded by European Regional Development Fund (FEDER) through NORTE 2020, operation number NORTE-01–0145-FEDER-000014.

References

1. Castañeda, J.A., Frías, D.M., Rodríguez, M.A.: Antecedents of internet acceptance and use as an information source by tourists. Online Inf. Rev. **33**(3), 548–567 (2009)
2. Li, S.C.H., Robinson, P., Oriade, A.: Destination marketing: the use of technology since the millennium. J. Destin. Mark. Manag. **6**(2), 95–102 (2017)
3. Roque, V., Raposo, R.: Social media as a communication and marketing tool in tourism: an analysis of online activities from international key player DMO. Anatolia **27**(1), 58–70 (2016)
4. Shao, J., Li, X., Morrison, A.M., Wu, B.: Social media micro-film marketing by Chinese destinations: the case of Shaoxing. Tour. Manag. **54**, 439–451 (2016)
5. Wang, Y.C.: Destination marketing and management: scope, definition and structures. In: Wang, Y., Pizam, A. (eds.) Destination Marketing and Management: Theories and Applications, pp. 1–20. CABI, Wallingford
6. Uşaklı, A., Koç, B., Sönmez, S.: How 'social' are destinations? Examining European DMO social media usage. J. Destin. Mark. Manag. **6**(2), 136–149 (2017)
7. Leung, D., Law, R., van Hoof, H., Buhalis, D.: Social media in tourism and hospitality: a literature review. J. Travel Tour. Mark. **30**(1–2), 3–22 (2013)
8. Uşaklı, A., Koç, B., Sönmez, S.: Social media usage among top European DMOs. In: Kozak, N., Kozak, M. (eds.) Tourist Destination Management. Tourism, Hospitality & Event Management, pp. 1–14. Springer, Cham (2019)
9. Felix, R., Rauschnabel, P.A., Hinsch, C.: Elements of strategic social media marketing: a holistic framework. J. Bus. Res. **70**, 118–126 (2017)
10. Mariani, M.M., Di Felice, M., Mura, M.: Facebook as a destination marketing tool: evidence from Italian regional destination management organizations. Tour. Manag. **54**, 321–343 (2016)
11. Buhalis, D., Law, R.: Progress in information technology and tourism management: 20 years on and 10 years after the Internet-the state of eTourism research. Tour. Manag. **29**(4), 609–623 (2008)
12. Hays, S., Page, S.J., Buhalis, D.: Social media as a destination marketing tool: its use by national tourism organisations. Curr. Issues Tour. **16**(3), 211–239 (2013)
13. Filieri, R., Alguezaui, S., McLeay, F.: Why do travelers trust TripAdvisor? Antecedents of trust towards consumer-generated media and its influence on recommendation adoption and word of mouth. Tour. Manag. **51**, 174–185 (2015)
14. Reza Jalilvand, M., Samiei, N.: The impact of electronic word of mouth on a tourism destination choice. Internet Res. **22**(5), 591–612 (2012)
15. Zeng, B., Gerritsen, R.: What do we know about social media in tourism? A review. Tour. Manag. Perspect. **10**, 27–36 (2014)

16. Dellarocas, C.: The digitization of word of mouth: promise and challenges of online feedback mechanisms. Manag. Sci. **49**(10), 1407–1424 (2003)
17. Molinillo, S., Liébana-Cabanillas, F., Anaya-Sánchez, R., Buhalis, D.: DMO online platforms: Image and intention to visit. Tour. Manag. **65**, 116–130 (2018)
18. Lalicic, L., Huertas, A., Moreno, A., Jabreel, M.: Emotional brand communication on Facebook and Twitter : are DMOs successful ? J. Destin. Mark. Manag. 1–11 (2019)
19. Forbes: 10 beautiful travel destinations for wine lovers. *Forbes* (2018)
20. UNESCO: World Heritage List. [Online] Available: https://whc.unesco.org/en/list/ (2018)

Chapter 10
Tactical Approaches to Disclose Influencers' Advertising Partners

Ana Filipa Couto[1] and Pedro Quelhas de Brito[2(✉)]

[1] Faculdade de Economia, Universidade Do Porto, Porto, Portugal
filiparibeiro5@gmail.com
[2] LIAAD/INESC-Tec e Faculdade de Economia, Universidade Do Porto, Porto, Portugal
pbrito@fep.up.pt

Abstract. This study analyzes the decision-making process of digital influencers pertaining to the promotion of monetary partnerships. We undertook fifteen interviews with Instagram influencers from different lifestyle orientations and business profiles. The content analysis allowed us to identify four determinant dimensions of decision process shaping the influencers relationship with their followers. Specifically, once the brand marketers give incentives to the influencers, they decide what, when, and how to communicate their followers the motivation behind the initiative to show/promote that brand. Such tactical decision process to disclose the nature of the partnership depends not only on the type of compensation but also on the influencer's criteria. We look at the justifications, arguments to frame such decision/behavior. The practical, ethical, and theoretical implications will be discussed.

Keywords: Digital influencers · Tactical · Disclosure · Advertising

10.1 Introduction

Social media is "one of the latest effective communication tools" (p. 2) [1]. The increasing use of applications known as social networks or social media has added a powerful communication channel to the market. Consumers now have a greater power to influence other consumers as social media allows them to communicate with a vast network of consumers using the same platform. Not only do consumers have this power, but brands have also identified this potential of social media. Brands are increasingly devoting more of their marketing resources to this domain as it offers an opportunity to reach a wide audience [2].

As Campbell and Marks point out, brands quickly realized that sales and promotional messages are not usually the kind of messages consumers want to see, and have taken a more subtle approach, indicating this approach as the beginning of native advertising [2].

Á. Rocha et al. (eds.), *Marketing and Smart Technologies*, Smart Innovation, Systems and Technologies 167, https://doi.org/10.1007/978-981-15-1564-4_10

10.2 Literature Review

Native advertising, according to Nieto, regards the publication of content promoted on a platform, respecting the format and style of the editorial content of that platform. Native ads are so cohesive with the page content, similar in design, and so consistent with the platform behavior that the viewer simply feels they belong [3]. The lack of regulation of digital advertising has allowed brands and influencers to take advantage of legal gaps and disregard transparency of the content they sponsor [4].

Concerning the risks of this form of advertising, Taylor advocates that "in the absence of stricter regulation (…) I see that [native advertising as the new marketing trend] as very negative for the marketing field as it raises the spectrum of intentional deception designed to deceive consumers" (p. 207) [5]. However, Nieto argues that similarly to native advertising, contents are usually accompanied by captions identifying them as advertising content, which are used for ethical reasons and also to enhance the brand image near consumers, since the clear indication that the brand is allocating resources to share content in a respectful, non-intrusive way gives the users a positive memory of the brand [4].

10.3 Influencer Marketing

Influencer marketing is so-called due to engaging digital influencers. Brands aim to encourage influencers to showcase and support their products through their social networking activities [6]. The correct identification of the appropriate digital influencer for the brand or product is essential to this marketing strategy: the influencer personal brand characteristics must be common to those of the brand/product [7]. This type of marketing shows the importance of word-of-mouth in decision-making for advertisers and consumers [8]. The influencer pay is directly related to their number of followers. Micro-influencers are compensated with free products or up to $150, whereas a macro-influencer with over 1 million followers may receive up to $15,000 per post [9].

Hwang and Zhang [10] clarify that compared to traditional celebrities, digital celebrities are "typical" people who have become famous due to content sharing on blogs or social networks. The term microcelebrity was already referred in previous studies to describe digital celebrities; the authors further stated that a microcelebrity was "a mind-set and a set of practices that courts attention through insights into its practitioners' private lives, and a sense of realness that renders their narratives, their branding, both accessible and intimate" [11] (p. 202). Instafamous is the term used for digital influencers who became famous through their Instagram profiles [1]. Also, Freberg et al. [12] define social media influencer as someone who shapes consumer attitudes through tweets, blogs, or any other communication format on social networks.

The definitions of the above terms translate the evolution of the study of these individuals and direct us to a broader definition that would include all aspects included in the presented concepts. Dhanesh and Duthler [7] (p. 3) did this work, defining the digital influencer as "a person who, through personal branding, builds and maintains relationships with multiple followers on social media, and has the ability to inform, entertain, and potentially influence followers' thoughts, attitudes, and behaviors". These content creators accumulate a solid following base, allowing their followers some insight into their personal lives, experiences, and opinions [6]. Their social influence is recognized according to their number of followers. The larger the number, the more social influence is perceived [13]. The most distinguishing feature of digital influencers is the relationship they build and nurture between their personal brand and their followers [7], achieving a dynamic relationship with the followers as they are technically accessible in a variety of social networking platforms, which facilitates a strong level of parasocial interaction between the source and the followers [14].

10.4 Electric Word-of-Mouth

As already mentioned, the influencers communicate with their audience/followers through online publications on the various social media platforms. The electronic word of mouth (eWoM) [15] is the dissemination of online reviews, arguments, and recommendations related to personal experiences with specific products or service providers, with the purpose of generating persuasive effects on the target consumers.

Thoumrungroje [16] states that eWoM generates a network of consumers who are connected to each other, while Van Doorn et al. [17] adds that likes, shares, and comments are regarded as manifestations of ad engagement by users who join eWoM. The eWoM has stronger consumer decision-making effects than traditional advertising techniques [18]. De Veirman et al. [6] report that a similar message is perceived as more authentic and reliable when communicated by a consumer in place of an advertising professional.

10.5 Self-branding

Khamis et al. [11] reveal a new era of self-branding, with social networks allowing any user to create an image sharing through the content they post, share, and like. This self-created image may allow them to create a fan base from their followers to earn social media fame, allowing ordinary users to achieve online fame.

Marwick underlines the importance of implementing the necessary practices to obtain this type of fame and consequent financial gains, indicating that digital influencers as "niche celebrities" get a livelihood through their audience, even though they are unknown to most of the population and lack the attention of the popular media. This author reinforces the importance of microcelebrity practices, increasingly linked with the ability to sell one's image as a consumer of products, and with the notion, already widespread by the abovementioned authors, that those practices are within the reach of any social network user [19].

10.6 Digital Advertising Regulation

In a study conducted through various interviews with US marketing agency professionals, Childers et al. [8] identified that digital influencers sign a contract that addresses what the influencer may or may not say and how to disclose that the post is sponsored. The need to track the shared posts in real time by the influencers they work with arises from the possibility that by mistake they do not include information about paid content. This tracking will allow editing the post in order to include the missing information [8].

Several institutions issued guides to take into account when practicing social media advertising. The guidelines "Endorsement Guides: Truth-in-advertising and Disclosure", issued by the USA Federal Trade Commission and the "Guidelines for Digital Influencers" by the International Consumer Protection and Enforcement Network both provide up-to-date guidance for influencers and marketers [7]. However, although these guides are public, marketers as a whole are still in the process of discovering this new approach to social media. In fact, no agency has yet implemented a complete process to incorporate influencer marketing in general advertising strategy [8].

In Portugal, there is no known legislation directly directed to this new reality of advertising on social networks, carried out by digital influencers. Still, taking into consideration Articles 9 and 10 of the Portuguese Advertising Code, the legislation demands a transparent behavior: "It is forbidden the use of subliminal images or other dissimulating means that exploit the possibility of conveying advertising without the knowledge by the recipients of the advertising nature of the message" (Art. 9, Nr.1) and also "Advertising must respect the truth, not distorting the facts" (Art. 10, Nr.1) and "The statements regarding the origin, nature, composition, properties, and conditions for the acquisition of goods or services advertised must be accurate and subject to proof at all times before the competent authorities" (Art. 10, Nr.2) [20].

10.7 Effects of Paid Content Disclosure

According to the Persuasion Knowledge Model (PKM), "when individuals perceive a message to have persuasive intent, their prior persuasion knowledge will be activated and they will process the message differently than if they were unaware of the persuasive intent of the message" (p. 4) [7]. Studies on native advertising have shown that the disclosure of sponsored content may trigger advertising recognition or knowledge of conceptual persuasion. This knowledge of conceptual persuasion (recognizing the content as advertising) leads to knowledge of attitudinal persuasion (critical attitude) generating trust/distrust on the marketer [21].

However, Krouwer et al. [22] tested the hypothesis of acknowledgment of disclosure to increase knowledge of attitudinal persuasion and concluded that this was not the case. Dhanesh and Duthler [7] determined that if influencers disclose that advertising is paid, they could significantly increase the level of trust and satisfaction. They also revealed that ad recognition does not influence the relationship between followers and influencers. Therefore, indicating that if a post is perceived as an ad/advertising this will not affect the relationship between follower and influence. Also, followers are more likely to buy the products/services or share the message of influencers.

10.8 Disclosure Decision Process

Campbell and Marks advocate that planning a coherent native advertising strategy is critical to the success of brands with an online presence and underline that brands sharing misleading content will unlikely achieve long-lasting relationships with the consumer (considering non-disclosure of sponsored content). The disclosure of content as paid advertising positively affects the followers' perception of honesty and transparency, and subsequently credibility of the influencer [7]. That said, why do we find advertising in the social media that chooses not to be disclaimed as advertising? In this study, we tried to understand the decision process of advertising disclosure on Instagram, through interviews with Instagram influencers working in Portugal.

10.9 Method

This study was based on primary data, collected by the researcher specifically to perform this study [23]. This is a qualitative study that will allow understanding how individuals "act and reconstruct the system of social representations in their individual practices" (p. 30), focusing on obtaining information about what the respondents think about the research subject, and also about their way of acting regarding that subject, through in-depth interview technique [24]. An interview guide was used with the topics to address, in order to conduct semi-structured interviews, considering the requirement to answer a core set of questions while keeping the flexibility to include new elements, unknown at the beginning of the study.

10.9.1 Data Collection

The use of this type of interview will facilitate analysis, quantification, and comparison [24]. According to Cunningham and Haley, this approach also allows exploring the context, considering the participants as experts, and the results based on the experiences of the participants and not on assumptions from the researcher [25].

An essential part of this form of data collection is the preparation of the various relevant aspects when conducting the interview. The preparation of the interview guide in order to create an environment in which the interviewee feels comfortable with the dialog and thus is more likely to confide [24]. Before the interviews, the profiles and pages of the interviewees were analyzed in order to extract as much information as possible and so that the interviewer could be more informed about the interview. The anonymity of all respondents was guaranteed, with identification merely according to the group in which they were included as per the segmentation performed according to the content shared.

10.9.2 Data Collection

The participants fulfill the following criteria: acting as an influencer in Portugal and having over 10,000 followers [26], on Instagram.

Within the universe of 103 selected and contacted influencers, only those interviewed were counted as "yes", and positive responses that were not carried out due to interview format or unavailability of the parties, negative answers or lack of response were counted as "no". Table 10.1 shows a positive response rate of 14.56% from the influencers contacted via private message on Instagram and email, as applicable.

Table 10.1. Analysis of the relationship between contacts and response to an interview

	Absolute N	Relative N (%)
Total	103	100
Yes	15	14.6
No	88	85.4

In order to facilitate the content analysis of the 15 influencers interviewed in the scope of this research, they were divided into three groups according to the content shared in their profiles: Healthy Lifestyle (HL), with 5 influencers on this group, all within the age range between 25 and 35 years old; Lifestyle, Beauty and Fashion (LBM), with 9 influencers, 3 of which aged 18–24 years old and the remainder 25–35 years old; and Savings (P), consisting of one influencer aged between 35 and 45 years old. See de Segmentation chart of interviewed influencers in Table 10.2. All the interviewed were female.

Table 10.2. Segmentation chart of interviewed influencers

Ref	Age	Area of the country	Nr of followers (m)	Group
HL1	25–35	Lisbon	13.1	*Healthy lifestyle*
HL2	25–35	Lisbon	14.2	*Healthy lifestyle*
HL3	25–35	Lisbon	17.3	*Healthy lifestyle*
HL4	25–35	Metropolitan Porto	12.2	*Healthy lifestyle*
HL5	25–35	Aveiro	20.4	*Healthy lifestyle*
LBM1	18–24	Metropolitan Porto	25.9	LBM
LBM2	18–24	Porto	14.8	LBM
LBM3	18–24	Metropolitan Porto	32.9	LBM
LBM4	25–35	Porto	26.8	LBM
LBM5	25–35	Porto	22.5	LBM
LBM6	25–35	Porto	15.6	LBM
LBM7	25–35	Porto	14.4	LBM
LBM8	25–35	Porto	163	LBM
LBM9	18–24	Porto	11.9	LBM
P1	35–45	Algarve	34.8	Savings

10.9.3 Data Analysis

The interviews were followed by the analysis of the content obtained. Bardin (2010) describes the following stages in content analysis: (a) organization of the analysis; (b) coding (transformation of data into units, coding according to precise rules); (c) categorization (decomposing the text into smaller units/classes by differentiation and regrouping according to previously established criteria); (d) treatment of results, inference, and interpretation of results [27].

The analysis of the interviews through content analysis was organized according to the major topics identified and the segmentation of the interviewees. This sort of analysis allowed two types of links: the identification of the intervening dimensions in the disclosure decision process (overall analysis) and the identification of the arguments used by the influencers in those decisions (individual analysis).

10.10 Results and Discussion

Four dimensions were identified in the advertising disclosure decision process: influencer behavior regarding advertising, influencer's attitude towards advertising, brand behavior and forms of remuneration. The characterization of the identified dimensions is shown in Table 10.3.

Table 10.3. Characterization of the identified dimensions

Influencer behavior towards advertising	Explicitly declares
	Explicitly declares only when the brand requests
	Does not declare explicitly/declares implicitly
	Does not declare
Brand behavior	Requests brand identification + specific hashtags + advertising hashtags (#ad or #pub)
	Requests brand identification + specific *hashtags*
	Requests only brand identification
	Does not give any kind of instructions
Forms of remuneration	Receives a percentage of sales through the coupon the brand delivers to the influencer (credit in online brand store)
	Receives products/services in exchange for disclosure
	Receive products/services and cash compensation
	Receive in cash
Influencer attitude towards advertising	Any divulgation (paid or not) is advertising
	If divulgation is paid, it is divulgation
	If only receives products, it's not advertising
	If receives products + coupon through which receives a percentage of the sales with the coupons (credit on online brand store), it is not advertising

It is possible to conclude from the analysis of the interviews that the determining factor for explicit disclosure of paid advertising/partnership is brand behavior, i.e., the instructions given for communication. If the brand demands the express communication of partnership, the influencers will comply and declare. However, when the brand does not issue specific instructions, it is up to the influencer to decide whether or not to disclose. See in Fig. 10.1 The relationship between identified dimensions.

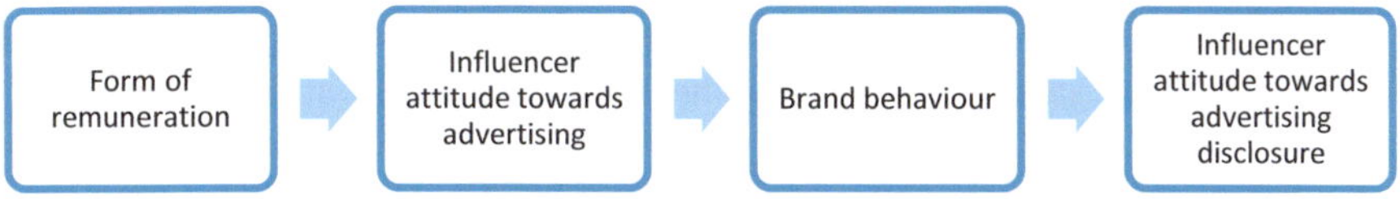

Fig. 10.1. Relationship between identified dimensions

Concerning the arguments used by influencers to justify their decision to expressly disclose whether the content they share is advertising, in the group of interviewed influencers we found professionals who always declare when it is a paid partnership/advertising, as two of the influencers from the Healthy Lifestyle group: "(…) I always use identifiers (…) I try to make everything very genuine, but I think people realize from the identifications that advertising is often indicated as "pub"." [HL4] and "I immediately say it is advertising (…)" [HL1].

On the other hand, as already mentioned, some interviewees do not always declare their posts as paid advertising/partnership. One of the most common justifications among influencers is the request or absence of indication by the brand to share content. One of the interviewed influencers stated that "I usually place a hashtag, because sometimes we identify the brand and it's not advertising, sometimes the brands send us products and we like them and we share, (…) if a brand has sent a product and does not ask for anything in return, and I like the product, I share, but it is not advertising. If they send me the product and ask me for something in return, then it would be advertising; but if they send it because they are simply sending it, we (influencers) are not obliged [to share]." [LMB2].

Most of the interviewed individuals argue that publicizing without a regular payment (in cash) is not paid advertising or paid service. They do not consider receiving products for free thus taken as a gift is an effective payment—"I don't currently receive money for the photos I posted. So, I never make that sort of partnership with brands; that don't pay me. So, I've never done those publications with hashtags [advertising], I haven't reached that level yet." [HL3].

According to the literature, "The possibility of deception poses challenges for sponsoring organizations and their SMIs in terms of fair disclosure, openness, and transparency practices required by law." (p. 4) [7]. The analysis of our interviews corroborates this argument. One of the influencers commented—"(…) I always try to introduce it in a natural way; some people don't even realize what is advertising and what is let's say, organic." [LMB5]—and—"(…) videos of the looks of the day help me a lot to introduce the products I receive in a very natural way, because I don't make that video of "oh, I received this", because (…) in advertising that doesn't work; people are already tired of seeing what we receive and they think "oh, she received it but she won't use it", so, I really like to show that "ok, I will use it, I am using it", and always I try to introduce things in a very natural way." [LMB5].

Moreover, regarding the recognition of content as advertising, influencers indicate that they share in the same post products that are offered and products that are bought by them, without distinction. One of the influencers of the LBM group stated: "(…) I have looks with clothes I buy, as I have looks with little known pieces of clothing, as I have looks in which I can combine things I receive with things I buy." [LBM4]. When questioned whether it was complicated to identify the products offered and those which she simply use and shows, one of the influencers said: "(…) but one of my goals is always not to make such distinction clear-cut on my followers' minds. Of course, some are noticed, for instance, if I am a brand ambassador it is obvious that I am working with them. But it is always my goal that my followers never realize where the stuff came from, but simply that I have that, some products I have acquired it, and others I like it." [LBM6].

In addition, the influencer states that she does not announce it is a paid partnership due to the potential loss of credibility in sharing the product: "(…) if the brand gives you the product people may think you are just doing it as a job, they don't understand if you really like it or not" [LMB6]. According to Campbell and Marks, good native advertising is not based on secrecy or deception, because the disclosure of paid partnership will make the platform user realize the brand is allocating resources to offer content and does so in a respectful and non-intrusive manner, the user will remember this brand positively [4].

On the other hand, some influencers believe that followers trust the influencer and for that reason, even if they don't realize it is a sponsored post, they trust the product and the opinion conveyed: "I think they have that perception [of when a posting is paid advertising/partnership], but I also think that if they trust that person, they will believe they are doing the advertising they really believe in." [LMB1]. The most recent literature indicates that the disclosure of paid partnership posts positively influences the followers' perceptions of honesty, transparency, and thus, the credibility of the influencer [7].

In the study by Dhanesh and Duthler there is a noticeable evolution in awareness of these issues, with indication of the issuance of a guide entitled "Truth-in-advertising and Disclosure" by the Federal Trade Commission of the United States and the "International Consumer Protection and Enforcement Network's Guidelines for Digital Influencers", both dedicated to providing up-to-date guidance for digital influencers and marketers. This may potentially lead organizations to comply with these guidelines, as indicated in October 2017 by the changes in Facebook policy regarding transparency in advertising [7].

The interviews revealed the experiences lived by influencers with brands that already considered this transparency in their strategy, and other brands that do not "(…) there are brands that really require that. There are brands that you have to advertise a service, (…) they always require that you inform in stories or publication that there is a partnership. Through hashtags or whatever, you are required to say it is a partnership. There are these situations, and there are other situations that no, they don't tell you anything." [LBM3].

10.11 Conclusions

This study contributes to the perspective about advertising of digital influencers on Instagram in Portugal. We can understand that brands have control over the transparency of the communication made in their favor.

This study also showed that when the brand does not issue explicit instructions on the disclosure of the content shared by digital influencers as advertising, the influencers don't assume them as advertising because: (I) they don't consider they are being paid when they only receive products/services; (II) they don't consider they are being paid when they receive products/services and a percentage of sales made using their coupon, in the form of credit offered by the brand; (III) if they receive products from a brand, but the brand doesn't ask them to share, they don't consider they are advertising; (IV) their followers may discredit their opinion if they know it is advertising; and (V) they try to make shared content look more natural than traditional advertising.

Although the literature advises against secrecy surrounding the disclosure of advertising due to harmful consequences, as mentioned throughout this paper, and highlights the benefits for brands and digital influencers from disclosure and underlying transparency, there is a challenge to be faced that can only be overcome by dissemination of information about the best practices of native advertising near digital influencers in Portugal. The dissemination of this information could be enhanced by a follow-up of the legislation regarding the new reality of online marketing. It is expected that the evolution of the awareness concerning the best practices of native advertising will progressively make brands more assertive regarding the express disclosure of published content as advertising.

The main limitation of this work is that it was thoroughly based on interviews with Digital Influencers who agreed to participate in the study. It will only reflect the context in which they operate: the reality of Instagram presence in Portugal.

10.11.1 Academic Contribution

The study allowed identifying the four dimensions involved in the decision-making process concerning the disclosure of paid advertising. It was also identified which of these dimensions is the determining factor for influencers to actually disclose content as paid advertising: brand behavior. The brand influences this decision by defining through briefing that the influencer should place hashtags such as #ad, #pub or #sponsored in the post. Although the literature largely contradicts less transparent behaviors on these issues, the present study reveals arguments used by influencers to justify deviant behavior.

10.11.2 Contribution to Marketers and Digital Influencers

The accomplishment of this work allowed knowing the opinion and the way the interviewed influencers act. In the light of the studied topic, the conclusions of this study call for awareness of issues such as the need for clear information on what may constitute a form of payment/remuneration (for instance, several influencers consider that receiving brand products does not constitute remuneration) and the transparency of paid shared content. Marketers should work closely with influencers to ensure that they receive enough information on these issues so that the brand image and the personal brand of the influencer are not hindered by the possibility of being perceived by the public as misleading or lacking transparency.

10.11.3 Limitations and Future Research

This work was heavily supported by interviews with digital influencers who agreed to participate in the study, and will only reflect the context in which they operate. Due to the difficulty in getting answers from more influencers, the group of participants has a large percentage of influencers from the north of the country and dedicated to beauty and fashion. The difficulty in obtaining answers from the influencers impacted the group in yet another way: one of the identified segments consisted only of one element, which affects representativeness.

It was possible to identify a trend among influencers to shift from sharing content on a specific topic such as beauty or fitness or connected topics, such as fitness and healthy eating, to sharing content on topics that may attract more followers (such as fashion, beauty, fitness, healthy eating, and travels), which makes clear segmentation more difficult.

Further research on this issue may overcome these issues by strategically including influencers from various parts of the country; and clearly dedicated to one particular topic, although they may also share some content on other topics. In order to address the gender issue (all influencers interviewed were female), it would also be useful to take into account the topics they deal with, such as technology and sports, which have more male influencers (such as the influencers Bernardo Almeida and Nuno Agonia).

References

1. Djafarova, E., Trofimenko, O.: 'Instafamous'—credibility and self-presentation of micro-celebrities on social media. Inf. Commun. Soc. 1–15. https://doi.org/10.1080/1369118x.2018.1438491 (2018)
2. Campbell, C., Marks, L.J.: Good native advertising isn't a secret. Bus. Horiz. **58**(6), 599–606 (2015). https://doi.org/10.1016/j.bushor.2015.06.003
3. Interactive Advertising Bureau: IAB—Native Advertising Playbook. https://www.iab.com/wp-content/uploads/2019/05/IAB-Native-Advertising-Playbook-2_0_Final.pdf (2013)
4. Nieto, B.G.: The influencer: a key tool in the digital context of false advertising. Methaodos-Revista De Ciencias Sociales **6**(1), 149–156 (2018). https://doi.org/10.17502/m.rcs.v6i1.212
5. Taylor, C.R.: Native advertising: the black sheep of the marketing family. Int. J. Advert. **36** (2), 207–209 (2017). https://doi.org/10.1080/02650487.2017.1285389
6. De Veirman, M., Cauberghe, V., Hudders, L.: Marketing through Instagram influencers: the impact of number of followers and product divergence on brand attitude. Int. J. Advert. **36** (5), 798–828 (2017). https://doi.org/10.1080/02650487.2017.1348035
7. Dhanesh, G.S., Duthler, G.: Relationship management through social media influencers: Effects of followers' awareness of paid endorsement. Pub. Relat. Rev. **45**(3). https://doi.org/10.1016/j.pubrev.2019.03.002 (2019)

8. Childers, C.C., Lemon, L.L., Hoy, M.G.: #Sponsored#Ad: agency perspective on influencer marketing campaigns. J. Curr. Issues Res. Advert. (2018). https://doi.org/10.1080/10641734.2018.1521113
9. Crain, A.: What happens when you reach a million instagram followers—the coveted 'M' can bring influencer status and lucrative marketing deals. Wall Str. J. https://www.wsj.com/articles/now-you-too-can-get-1-million-instagram-followers-1515599740 (2018)
10. Hwang, K.J., Zhang, Q.: Influence of parasocial relationship between digital celebrities and their followers on followers' purchase and electronic word-of-mouth intentions, and persuasion knowledge. Comput. Hum. Behav. **87**, 155–173 (2018). https://doi.org/10.1016/j.chb.2018.05.029
11. Khamis, S., Ang, L., Welling, R.: Self-branding, 'micro-celebrity' and the rise of social media influencers. Celebr. Stud. **8**(2), 191–208 (2017). https://doi.org/10.1080/19392397.2016.1218292
12. Freberg, K., Graham, K., McGaughey, K., Freberg, L.A.: Who are the social media influencers? A study of public perceptions of personality. Pub. Relat. Rev. **37**(1), 90–92 (2011). https://doi.org/10.1016/j.pubrev.2010.11.001
13. Jin, S.A.A., Phua, J.: Following celebrities' tweets about brands: the impact of Twitter-based electronic word-of-mouth on consumers source credibility perception, buying intention, and social identification with celebrities. J. Advert. **43**(2), 181–195 (2014). https://doi.org/10.1080/00913367.2013.827606
14. Jin, S.V.: "Celebrity 2.0 and beyond!" effects of Facebook profile sources on social networking advertising. Comput. Hum. Behav. **79**, 154–168 (2018). https://doi.org/10.1016/j.chb.2017.10.033
15. Shin, D., Song, J.H., Biswas, A.: Electronic word-of-mouth (eWOM) generation in new media platforms: the role of regulatory focus and collective dissonance. Mark. Lett. **25**(2), 153–165 (2014). https://doi.org/10.1007/s11002-013-9248-z262
16. Thoumrungroje, A.: The influence of social media intensity and EWOM on conspicuous consumption. In: 2nd International Conference on Strategic Innovative Marketing, vol. 148, pp. 7–15. https://doi.org/10.1016/j.sbspro.2014.07.009 (2014)
17. van Doorn, J., Lemon, K.N., Mittal, V., Nass, S., Pick, D., Pirner, P., Verhoef, P.C.: Customer engagement behavior: theoretical foundations and research directions. J. Serv. Res. **13**(3), 253–266 (2010). https://doi.org/10.1177/1094670510375599
18. Goldsmith, R.E., Clark, R.A.: An analysis of factors affecting fashion opinion leadership and fashion opinion seeking. J. Fash. Mark. Manag. **12**(3), 308–322 (2008). https://doi.org/10.1108/13612020810889272
19. Marwick, A.E.: Instafame: luxury selfies in the attention economy. Pub. Cult. **27**(1), 137–160 (2015). https://doi.org/10.1215/08992363-2798379
20. Código da Publicidade Português. (1990)
21. Boerman, S.C., van Reijmersdal, E.A., Neijens, P.C.: Sponsorship disclosure: effects of duration on persuasion knowledge and brand responses. J.Commun. **62**(6), 1047–1064 (2012). https://doi.org/10.1111/j.1460-2466.2012.01677.x
22. Krouwer, S., Poels, K., Paulussen, S.: To disguise or to disclose? The influence of disclosure recognition and brand presence on readers' responses toward native advertisements in Online news media. J. Interact. Advert. **17**(2), 124–137 (2017). https://doi.org/10.1080/15252019.2017.1381579
23. Malhotra, N.K.: Pesquisa de Marketing-: Uma Orientação Aplicada: Bookman Editora (2001)
24. Aires, L.: Paradigma Qualitativo e Práticas de Investigação Educacional. Universidade Aberta (2015)

25. Cunningham, A., Haley, E.: A look inside the world of advertising-free publishing: a case study of Ms. Magazine. J. Curr. Issues Res. Advert. **22**(2), 17–30. https://doi.org/10.1080/10641734.2000.10505105 (2000)
26. Gottbrecht, L.: The Three Types of Influencers All Marketers Should Know. https://www.mavrck.co/the-three-types-of-influencers-all-marketers-should-know-infographic/ (2016)
27. Bardin, L.: Análise de Conteúdo. Lisboa: Edições70 (2010)

Chapter 11
The Characteristics of Digital Influencers and Their Ethically Questionable Attitudes

Sara Silva[1] and Pedro Quelhas de Brito[2(✉)]

[1] Faculdade de Economia da, Universidade do Porto, Porto, Portugal
silvasaradaniela@gmail.com
[2] LIAAD/INESC-Tec e Faculdade de Economia da Universidade do Porto, Porto, Portugal
pbrito@fep.up.pt

Abstract. The aim of the research is to analyze how social network users define and characterize digital influencers' task (DI) and how they perceive possible ethically questionable attitudes. We undertook five focus group involving 34 DI's followers aged 18-29 years old. Six major dimensions qualifying and justifying DI underlying intention and behavior were identified (Profile, Interaction, Trust, Recommendations, Ethics, and Bought Products). The nature of the unethical DI's behavior were described as well as their consequences.

11.1 Introduction

The world is constantly changing as businesses only endure and become successful if they can adapt. Social networks have become an important part of people's daily lives, and after brands decided to use these digital platforms by creating a profile on social networks, they adopted them, incorporating them in their marketing strategies after perceiving the influence a user can have on their followers. Brands quickly realized that the fastest way to gain space on a social network from an influencer would be to contact influencers and reward them monetarily or non-monetarily if they conveyed positive information to their followers.

Social network users have access to a set of publications from users with a large number of followers, who use the content they post to display products of a particular brand, such as clothing brands. This content hence is liable to advertise a product or brand. This sort of publication is considered a new form of marketing, the influencer marketing, which is increasingly used by brands. The objective is that the conclusions from this study may allow understanding how brands can use these digital influencers in a positive way. In addition, it is important for digital influencers to understand how to proceed when representing a brand.

Every day new digital influencers are born, accepting any proposal by any brand. As a result, they end up advertising for brands and products without quality. Sometimes they do not recognize the lack of quality simply because they did not try the product, i.e., they just share the feedback the brand requested. That is why brands should create

Á. Rocha et al. (eds.), *Marketing and Smart Technologies*, Smart Innovation, Systems and Technologies 167, https://doi.org/10.1007/978-981-15-1564-4_11

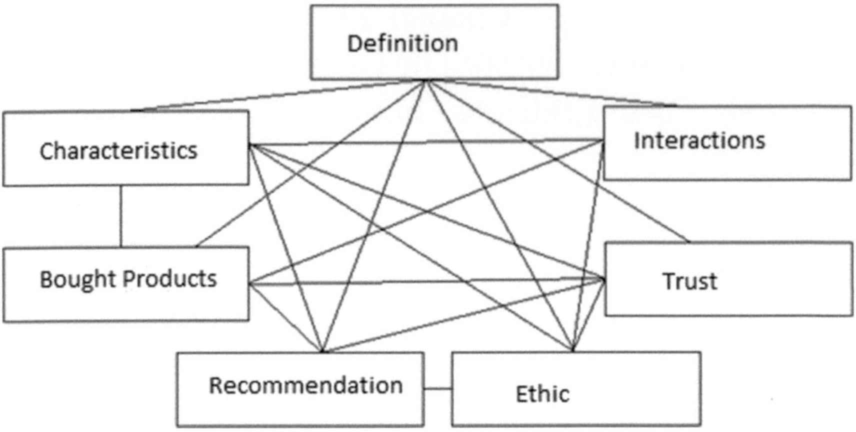

Fig. 11.1. Representation of the relationships among the different dimensions

a set of criteria to select the digital influencer that will represent their brand and products.

Therefore, this research analyzes the influence of certain users of social networks on the purchase decision process of a particular audience. However, the target of the investigation is to understand what happens when these same influencers are associated with ethically questionable attitudes, although unintentional or unconscious. This negative part then becomes our main research question. When an unethical behavior by a digital influencer is made public, does the audience lose confidence in the influencer? Do such attitudes have any impact on influencer marketing?

11.2 Literature Review

11.2.1 Digital Influencers

Based on the literature we may define Digital influencers as content generators with a large number of followers on the social media and are able to influence others. Therefore, they use many of their publications to advertise a particular brand or product to manipulate and persuade the purchase decision of those watching the publication [1–6].

These digital influencers are employed by brands to promote products and create electronic word-of-mouth. They are rewarded by offering them those same products or unique experiences [7]. This electronic word-of-mouth (e-wom) consists of a positive or negative opinion about a brand or product conveyed by a consumer or potential consumer to an audience, in this case in an online context via social networks such as Instagram [8].

The goal of digital influencers is to capture the attention of their followers and other Instagram users [9] creating strong self-branding that can be used by brands and advertisers in a positive way for both sides [10].

Self-branding consists of creating a personal brand, developing a "distinctive public image" that stands out among other users. Many self-branding celebrities associate their names to brands in a profitable way, making their followers associate the person's success and positive traits with the brand [11].

Most digital influencers are opinion leaders, meaning "someone whose opinion is highly respected and used by their followers as an aid in decision-making in various situations, such as what kind of clothing to wear, where to repair household items, how to discipline children…" [12]. When these opinion leaders are used to advertise a brand or product through their social networks, the result is often better than using traditional advertising, such as billboards at bus stops or shopping malls [13].

This occurred because they are considered as experienced individuals with no commercial benefit, and therefore their opinion is always more reliable than brand advertising campaigns aimed at promoting products [14].

11.2.2 Influencer Marketing

Brands began to realize that the real opinions and recommendations that certain social network users conveyed really influenced consumers. Therefore, they became more active on social networks to communicate faster and more easily with consumers. In addition, social media as brand communication channel began reaching more prominence by being present in the "promotion" section of the business's marketing mix [2, 15]. Influencer marketing thus stemmed from harnessing the influence of such individuals for their own products [16].

Influencer marketing is "the art and science of engaging people who are influential online to share brand messaging with their audiences in the form of sponsored content" [17].

This form of marketing is increasingly used since it is "fast, targeting a specific audience and cheaper than traditional advertising" [18] and is more efficient than the latter concerning the influence on consumer buying decisions [3]. Moreover, studies show that individuals who are exposed to product recommendation are more likely to purchase it than individuals who are not exposed to product recommendation, and online influence is greater than face-to-face influence [8].

For companies to enjoy the full benefits of influencer marketing, the selection of digital influencers requires caution [2]. This is because as soon as brands begin to use influencer marketing and rely on an influencer to promote their name, product, or service, they must be prepared to lose control of the advertising process, handling it to the influencer [19].

The knowledge of the influencer on a subject is important to influence those who view their contents. However, according to previous studies, there are other factors that contribute to this influence and in turn to select a good influencer. These factors may be the number of followers, the ratio between the number of followers and the number of profiles the influencer follows [3], the number of times an influencer content is shared by other users, among others. Numbers are important to define the best influencer for a campaign. However, engagement is also important, as well as the relationship between the influencer and his followers, and other factors that allow assessing the quality of the influencer [24–26]. This is the most important and most difficult step for brands [16]. "More important than winning the audience is to win the right audience so that the return is greater" [13]. If the digital influencer is not chosen correctly it can bring unfavorable results to the brand. For this reason, many brands discredit influencer marketing because they think it does not bring positive results when the problem is often simply in the decision process of choosing the influencer. A cheaper choice may become more expensive [16].

11.2.3 Ethics in Influencer Marketing

Although, in recent years many companies have created and even increased the budget for this marketing strategy [29, 2], if it is not done correctly and ethically it can harm both the influencer and the brand. Initially, it is important to highlight that ethics "is the code of moral principles and values that governs the behaviors of a person or group with respect to what is right or wrong" [30].

Hence, while these users are viewed as reliable, some of them use fake likes and followers through paid websites and apps to reach greater fame among the audience. Thus, although achieving greater reputation, brands may lose money by using an influencer who actually has no influence and therefore cannot reach the intended audience [7].

Nowadays, most influencers advertise many products, often competing products, even, and they are starting to lose credibility [31]. Therefore, sometimes, even if the content is not sponsored and sincere, viewers perceive it as something sponsored or encouraged by a brand.

It is already public and quite usual to talk about this marketing strategy that created the new profession of digital influencer. So, when a social media user views a photo of a celebrity or person with many followers talking about a product, it is normal to assume it is brand-sponsored content, regardless of whether it is sponsored or not [32].

As a result, when content shared by an influencer is perceived as paid for by the brand, the information conveyed may be received with some doubt from the followers because they do not know if that information is sincere or not, decreasing the credibility of the influencer [28, 33]. In addition, it may not lead to the purchase goal because they are unsure of product quality and do not want to waste money. Thus, sponsored content does not usually lead to a positive attitude but rather to a neutral, sometimes even negative attitude from the follower [27, 28, 32].

Influencers often receive brand products for review. In this case, it is important that influencers really experience the product and convey all the characteristics and true information about it whether positive or negative. In most cases, followers try to ascertain if the product has been experienced, to increase the credibility of the influencer and the certainty that the information received is true and proven by the influencer they trust [27, 34]. Still, when shared content is sponsored by a brand, it can never be said to be e-wom, since even if it is a sincere opinion, if sponsored it becomes commercial information [28].

On the other hand, influencers often admit the shared content is sponsored by a brand. But many prefer to hide this information, either at their own discretion or because the brand requires it, so that followers are more likely to buy the product, and thus brands perceive these influencers as the most influential [27]. While it may have initial benefits for the influencer, concealment of this information becomes serious because it is "essentially an attempt to mislead consumers" which is unethical and illegal in some countries can severely harm both the influencer and the brand [28].

It is therefore very important for the influencer to always use a clear message. If it is publicity he should admit so, and if not, he should communicate his opinion which is not encouraged by a brand. Thus, while it may contribute negatively to the consumer's buying decision, the consumer perceives the influencer and the brand as true and correct [18, 28].

This concealment of information is often due to the existing pressure from the company. When the brand rewards the influencer monetarily or offering products, it can make the influencer feel obliged to point out the positive features of a product even when there are none. This pressure may be purposeful, intended by the brand to ensure the influencer will say exactly what the brand wants [28, 32]. However, the brand should encourage the influencer to communicate with the followers that the content is sponsored when the content is sponsored [35].

Another important issue in influencer marketing is that companies do not always understand how they should act towards influencer. As a result, many brands ship products to a large number of influencers leading to "aggressive sales and advertising campaigns with no framing, market analysis or consumer studies", which in some cases fills the consumer feed with a large amount of content from different influencers advertising the same product, with similar noncreative content and the text required by the brand, identical for everyone [16].

This causes what Monteiro [16] defines as pseudo-influencers who are "people with some notoriety in social networks who decide to compromise with this type of aggressive sales actions in exchange for remuneration". These individuals who call themselves influencers end up conveying the information required by the brand, often without considering whether it is true or not [16].

If, as mentioned above, brands should exercise caution in choosing influencers, influencers should also exercise caution in choosing the brands and products they will represent. This is because many pseudo-influencers only agree to represent a product because they will receive something for free or because they will be rewarded monetarily. What happens is that usually, these products do not meet their values or lifestyle. In addition to being unethical toward their followers, this type of content may harm the influencer who becomes viewed as untrustworthy, with negative impact on the number of followers and likes and potentially damaging the brand reputation [16].

When all brands chose carefully the influencers and the use of this marketing strategy, pseudo-influencers will begin to disappear. In this, influencer marketing universe would, therefore, remain only those who do their jobs in a correct and ethical manner. This way, brands can have more confidence when they transfer the process to the hands of influencers and followers may again trust the opinions and recommendations from social networks [16].

11.3 Methodology

The main objective of this research is to understand if the less ethical aspect of digital influencers is perceived by the followers, thus harming their influencer marketing and in turn harming the brands that hire them to advertise their products.

Hence, the research question (RS) is: Are ethically questionable behaviors of digital influencers perceived by their followers?

In order to answer the research question, we will conduct an exploratory research that will allow us to explore the problem at hand and draw relevant conclusions for influence marketing [36].

Initially, primary data were collected on purpose to research the problem. Such data allowed for a qualitative survey based on a smaller sample than in the case of a quantitative survey and nonstatistical [36].

The research was conducted with five focus groups to stimulate rich conversation, each gathering a small group of individuals in a relaxed environment. To be considered valid, the groups selected included between 6 and 10 participants, and each one lasted approximately one hour. These focus groups were recorded and transcribed into a digital document. At the beginning of each focus group, each participant answered an individual questionnaire mostly intended to understand some demographic characteristics, the activity they usually perform in social networks and their opinion about digital influencers and influence marketing. These questionnaires were conducted anonymously so that respondents could give their opinion truly, not feeling judged by what others might think. Focus groups allow the participants to listen to the information conveyed by other participants, thus encouraging to share opinions on a particular topic more easily [36].

11.4 Data Analysis and Results

11.4.1 Sample Characterization

In the 5 focus groups conducted, it was possible to collect data from 34 participants aged between 18 and 29. 53% of the participants were female and 47% male. The network most used by respondents was Instagram. Of the 34 respondents, 29 follow digital influencers on social networks and only 5 show no interest in this matter. In addition, about 53% of respondents admit that digital influencers drive them to buy new products.

11.4.2 Content Analysis

Based on the answers of the respondents, obtained during the focus groups, it was possible to create a model to visualize the common aspects between the various focus groups, so that as a result we may identify the various dimensions that allow answering the research question.

After analyzing the common points, it was possible to define and characterize each of the dimensions that allow justifying and drawing conclusions about the research question. Thus, seven dimensions were found as can be seen in Table 11.1.

Table 11.1. Identification and characterization of dimensions

Dimension	Characteristics
Digital influencer	Persuasion; high number of followers; arises naturally or intentionally; conveys information; advertises products; rewarded
Characteristics	Honesty; communication; attractiveness; empathy; charisma; confidence; consistency; creativity; innovation
Interaction	Likes; giveaways; quick interactions
Bought products	Technology; beauty products; others
Trust	Empathy; area of interest; honesty; coherence; experience with the product; consistency; opinion leader; time
Recommendation	None; save for the future; confirm through third party; brand/product research Paid; not paid
Ethics	Lying or occultation of information; incorrect attitudes; false advertising

With defined and characterized dimensions it is necessary to understand how they are related. This interconnection between the various dimensions is shown in the following figure (Fig. 11.1).

11.4.3 Comparison of Results with Research Question

After the content analysis performed on the collected data, it can be concluded that the research question defined results in a positive answer.

This research question aims to understand if social network users can interpret ethically questionable actions taken by digital influencers. Thus, the research question was confirmed by the ethics dimension in which the participants of the various focus groups refer to different actions by digital influencers perceived by the respondents as less correct and unethical.

According to the respondents, these behaviors by digital influencers which are ethically questionable can be divided into three categories: lying or withholding information, incorrect attitudes, and false advertising. In the first category, the influencer conveys wrong information to followers or does not convey important information, which in the case of influencer marketing may interfere with the purchase decision process. As an example of this first category is a case from a focus group in which the participant argued that when influencers say that "… they bought the product X in the Y store, when the product brand offered it" it is a way of lying to their followers. The second category is associated with incorrect and unethical attitudes by digital influencers. An example from the focus group was the recent case of "an influencer who said she was vegan and was seen eating meat."

These attitudes could quickly become highly polemical and reach out to other social media users besides his followers. Lastly, false advertising is more common among pseudo-influencers. To illustrate this untrue advertising, the focus group highlighted the case of an actress who gave an interview to a radio station "where she was asked if she had ever advertised a product she never used. She said yes, Becel butter."—Or "a channel talking about movies and the critics spoke well of a movie that no one liked, so the fans complained a lot".

Therefore, digital influencers must be very careful concerning unethical situations, so that their credit in the digital world is not harmed. Brands should also choose influencers carefully so that they do not damage their image and reputation.

11.5 Conclusions

11.5.1 Overview

This research allowed the identification of a set of characteristics that are expected in digital influencers, according to a sample of social network users. Moreover, it was possible to understand that social network users are able to perceive some attitudes from digital influencers as ethically questionable, and even identify some of those situations through actual examples.

We identified that the fact that brands pay social media users to advertise their products has become so common that one of the characteristics respondents use to define a digital influencer is that they are paid. In addition, honesty is the characteristic followers value, the most in an influencer. This suggests that if influencers are insincere about the information they share, it may cost them the trust of their followers, damaging their reputation. Along with honesty, characteristics such as communication, attractiveness, empathy, charisma, trustworthiness, consistency, creativity, and innovation are important for an individual to become a digital influencer.

Most followers who participated in focus groups admitted they have purchased products based on recommendations from digital influencers. It is important for the influencer to demonstrate the product experience over time so that followers can really see how the product behaves. However, the decision to buy a product is not something that happens immediately and unreasonably.

Another conclusion from the study is that trust in the influencer is very important to increase the purchase intention. This confidence is mainly based on the fact that the influencer has the ability to convey positive and also negative opinions about the products they try. Negative feedback becomes very important for followers to perceive the user as a trusted influencer. In addition, empathy, honesty, consistency, opinion leadership, the area in which the digital influencer operates, the experience he or she demonstrates with the product, and how long an individual follows him also contribute toward the confidence deposited by the follower in the influencer.

On the other hand, the unethical attitudes of the influencer may negatively influence purchase decisions, as they undermine the confidence of the follower in the influencer.

Thus, influencers should have correct attitude and convey truthful information so that followers do not feel misled and persuaded to buy a product or purchase a service because of the opinion of a digital influencer.

Answering the questions posed in the introduction, the public will, in fact, lose confidence in the influencer due to unethical attitudes. For this reason, those attitudes end up having a negative impact on influencer marketing. Finally, brands may continue to invest in this marketing strategy but always with the utmost precaution in selecting the influencer, in order to achieve the proposed goal. Otherwise, brands may lose customers and damage their reputation in the market.

11.5.2 Contribution to Management

This research allows marketers and brands to understand how social network users perceive digital influencers. A greater understanding of the point of view of these users makes it possible to use influence marketing more efficiently. In addition, making a good selection of the digital influencer may safeguard their own profession and contribute toward its future, preventing it from being marginalized and viewed as untrue. This way, brands can understand what are the characteristics and attitudes an influencer should have and based on that make a selection of the influencer that best fits the mission and purpose of the brand. This allows the brand to incorporate the digital strand into the marketing strategy, achieving good sales and brand recognition. By selecting the best influencer, the brand may reach maximum results with minimum effort.

For influencers, this research is also important. For those who are not in the digital field, they are able to understand the importance of the presence in this media and help them understand this domain. For digital influencers, this research is useful as they can understand what characteristics they need to develop and what attitudes they should have in order for their followers to trust the information they convey and subsequently buy the products they advertise. The greater the trust and engagement between the follower and the influencer, the better the results the latter can achieve in influence marketing, and subsequently the greater the value perceived by brands.

11.5.3 Research Limitations

This research was conducted with five focus groups totaling 34 participants. In the future, it would be important to increase the number of participants in order to collect more data. In addition, the participants originated from the same geographical group in the north of the country and included young adults. Thus, it would be relevant to obtain data from individuals from other geographic areas of the country, and older participants, with different purchasing powers and other interests, since these may have different perceptions concerning digital influencers, whose influence may be different.

Although it was possible to canvass the final consumers' opinions and perceptions, it would be important to canvass the point of view of influencers themselves. Hence, future studies may incorporate the views of digital influencers regarding their ethical and unethical attitudes and their concerns. On the one hand, to please their followers so they will not stop following them, but be as sincere as possible in the conveyed information.

In addition, since the present research is based on qualitative methods, it would be relevant for future research to apply quantitative methods so that the dimensions found may be analyzed based on a combination of the two methods.

Acknowledgements. This work is financed by National Funds through the Portuguese funding agency, FCT - Fundação para a Ciência e a Tecnologia within project: UID/EEA/50014/2019.

References

1. Hund, E.: Measured Beauty: Exploring the aesthetics of Instagram's fashion influencers. In: Paper presented at the Proceedings of the 8th International Conference on Social Media & Society (2017)
2. Gross, J., Wangenheim, F.V.: The big four of influencer marketing. A Typology of Influencers. Marketing Review St. Gallen, 2, 30–38 (2018)
3. De Veirman, M., Cauberghe, V., Hudders, L.: Marketing through Instagram influencers: the impact of number of followers and product divergence on brand attitude. Int. J. Advertising **36**(5), 798–828 (2017)
4. Abidin, C.: Visibility labour: engaging with Influencers' fashion brands and# OOTD advertorial campaigns on Instagram. Media Int. Australia **161**(1), 86–100 (2016)
5. Jin, S.V., Muqaddam, A., Ryu, E.: Instafamous and social media influencer marketing. Market. Intell. Planning (2019)
6. Lou, C., Yuan, S.: Influencer marketing: how message value and credibility affect consumer trust of branded content on social media. J. Interact. Advertising, 1–16 (2019)
7. Sen, I., Aggarwal, A., Mian, S., Singh, S., Kumaraguru, P., Datta, A.: Worth its weight in likes: towards detecting fake likes on instagram. Paper presented at the Proceedings of the 10th ACM Conference on Web Science (2018)
8. Hennig-Thurau, T., Gwinner, K.P., Walsh, G., Gremler, D.D.: Electronic word-of-mouth via consumer-opinion platforms: what motivates consumers to articulate themselves on the internet? J. Int. Market. **18**(1), 38–52 (2019)
9. Lee, D.H.: Digital cameras, personal photography and the reconfiguration of spatial experiences. In: Understanding Creative Users of ICTs, pp. 44–53. Routledge (2013)
10. Hearn, A., Schoenhoff, S.: From Celebrity to Influencer: London Wiley (2016)
11. Khamis, S., Ang, L., Welling, R.: Self-branding, 'micro-celebrity' and the rise of social media influencers. Celebrity Stud. **8**(2), 191–208 (2017)
12. Cosmas, S.C., Sheth, J.N.: Identification of opinion leaders across cultures: an assessment for use in the diffusion of innovations and ideas. J. Int. Bus. Stud. **11**(1), 66–72 (1980)
13. Costa, I., Alturas, B.: Líderes de Opinião Digital portugueses, e o seu impacto, na promoção de produtos, serviços e eventos nas redes sociais. Paper presented at the 13th Iberian Conference on Information Systems and Technologies (CISTI'2018) (2018)
14. Brown, J., Broderick, A.J., Lee, N.: Word of mouth communication within online communities: conceptualizing the online social network. J. Interact. Market. **21**, 2–20 (2007). https://doi.org/10.1002/dir.20082
15. Kotler, P., Armstrong, G.: Principles of marketing: Pearson education (2010)
16. Monteiro, M.: Os Pseudo Influenciadores Digitais. Retrieved from https://lsdmlondon.com/pt/academico-pt/os-pseudo-influenciadores-digitais/ (2018)
17. Sammis, K., Lincoln, C., Pomponi, S.: Influencer Marketing for Dummies. Wiley (2015)

18. Evans, N.J., Phua, J., Lim, J., Jun, H.: Disclosing instagram influencer advertising: the effects of disclosure language on advertising recognition, attitudes, and behavioral intent. J. Interact. Advertis. **17**(2), 138–149 (2017). https://doi.org/10.1080/15252019.2017.1366885
19. Cao, P., Meister, S., Klante, O.: How social media influence apparel purchasing behavior. Market. Rev. St. Gallen **31**(6), 77–86 (2014)
20. Iyengar, R., Van den Bulte, C., Valente, T.W.: Opinion leadership and social contagion in new product diffusion. Market. Sci. **30**(2), 195–212 (2011)
21. Brown, D., Fiorella, S.: Influence marketing: How to create, manage, and measure brand influencers in social media marketing: Que Publishing (2013)
22. Keller, E., Berry, J.: The influentials: One American in ten tells the other nine how to vote, where to eat, and what to buy: Simon and Schuster (2003)
23. Weimann, G.: The influentials: People who influence people: SUNY Press (1994)
24. Basille, D.: Social Media Influencers are not Traditional Influencers. Retrieved from https://www.briansolis.com/2009/11/social-media-influencers-are-not-traditional-influencers/ (2009)
25. Straley, B.: How To: Target Social Media Influencers to Boost Traffic and Sales. Retrieved from https://mashable.com/2010/04/15/social-media-influencers/?europe=true (2010)
26. Freberg, K., Graham, K., McGaughey, K., Freberg, L.A.: Who are the social media influencers? A study of public perceptions of personality. Public Relations Rev. **37**(1), 90–92 (2011)
27. Stephen, A., Bart, Y., Du Plessis, C., Goncalves, D.: Does paying for online product reviews pay off?. The effects of monetary incentives on content creators and consumers, ACR North American Advances (2012)
28. Tsao, W.C., Mau, T.-C.: Ethics in social media marketing: How should sponsorship information be disclosed in online product reviews? ASLIB J. Informat. Manage. (2019)
29. Bevilacqua, J.: Why Brands Need to Utilize Influencer Marketing in 2018. Retrieved from https://stjoseph.com/insight/influencer-marketing-2018-infographic/ (2018)
30. Daft, R.L.: Management, 9th edn. Cengage Learning, Mason, USA (2010)
31. Tripp, C., Jensen, T.D., Carlson, L.: The effects of multiple product endorsements by celebrities on consumers' attitudes and intentions. J. Consum. Res. **20**(4), 535–547 (1994)
32. Silvera, D.H., Austad, B.: Factors predicting the effectiveness of celebrity endorsement advertisements. Eur. J. Mark. **38**(11/12), 1509–1526 (2004)
33. Tsao, W.-C., Hsieh, M.-T., Shih, L.-W., Lin, T.M.: Compliance with eWOM: the influence of hotel reviews on booking intention from the perspective of consumer conformity. Int. J. Hospitality Manage. **46**, 99–111 (2015)
34. Oberhofer, M., Füller, J., Hofmann, V.: Tryvertising-what makes consumers share product innovations with others? ACR North Am. Advanc. (2014)
35. Tong, Y., Wang, X., Teo, H.H.: Understanding the intention of information contribution to online feedback systems from social exchange and motivation crowding perspectives. Paper presented at the 2007 40th Annual Hawaii International Conference on System Sciences (HICSS'07) (2007)
36. Malhotra, N.K.: Pesquisa de Marketing: Uma Orientação Aplicada (2006)

Chapter 12
The Use of Influencers in Social Media Marketing

Mariana Oliveira[1], Renata Barbosa[1], and Alexandre Sousa[2(✉)]

[1] ISMAI, University Institute of Maia, Maia, Portugal
{A033452, rtv}@ismai.pt
[2] ISMAI, Research unit CITEI, University Institute of Maia, Maia, Portugal
avs@ismai.pt

Abstract. In conventional marketing, companies invest in television, radio ads, billboards, etc. and are forced to create a standard campaign for the general public. In Social Media Marketing, companies develop techniques and methods to locate and reach their target audience creating a precise differentiation of customers, determining values, needs, desires, etc. In this context, the Influencers represent a recent kind of social media product placement and major/minor celebrity recommendation. By analyzing Social Media Marketing, we can obtain a way to categorize and evaluate the Influencers, considering numeric data, such as number of views, followers/subscribers, likes, and engagement rate, but also more subjective data such as personality.

12.1 Introduction

The objective of this paper is to analyze the rise and impact of social media influencers, a recent and important development in online marketing. The remaining of this paper is structured as follows. Section 2 compares digital and conventional marketing and discusses the impact of social networks and social media influencers. Section 3 presents the key factors governing the influencers impact. Section 4 presents some of the findings of a study on Portuguese influencers. Section 5 has the conclusion and suggestions for future work.

12.2 Marketing and Social Networks

Marketing changes the sale of a product from a mere sale into something with value, concept, and purpose. Through marketing, sales are transformed into acquisition of values and services that represent a statement about life choices and quality of life [1]. Marketing is governed by the process of winning and retaining customers. It can also be said that marketing is a business function that generates value for the customer while at the same time giving the company a stable competitive advantage by managing the variables present in marketing: product, price, communication, and distribution. Kotler describes marketing as a social process, where individuals acquire what they need and

Á. Rocha et al. (eds.), *Marketing and Smart Technologies*, Smart Innovation, Systems and Technologies 167, https://doi.org/10.1007/978-981-15-1564-4_12

want through the offerings, creations, and negotiations of valuable products and services [2].

The process of producing sales and services is no longer enough to achieve business goals. With marketing the customer is determinant in the sales, having decision power. Companies recognizing this change in the decision paradigm took the initiative to develop market analysis and interaction techniques to enable them to communicate the qualities and benefits of their products through "vehicles," expanding and diversifying distribution channels, and allowing products to be tailored to customer needs [3].

12.2.1 Digital Versus Conventional Marketing

Conventional marketing uses traditional means to promote campaigns. Those campaigns/ads are launched in media such as radio, TV, newspapers, and magazines. The approach is direct, using advertisements placed on television breaks and pauses between songs on the radio or available space in the print media. But one of the permanent features of Conventional marketing is its cost, requiring massive and complex productions to carry out campaigns and advertisements. Beyond its high cost, obtaining statistical data that correlates sales with ads is complex to achieve. Customer-to-customer questionnaires are required to determine which purchases were actually triggered by that advertisement, thus determining its effectiveness, which makes the analysis of results time-consuming, and requires conducting research with customers on the street or via telephone. Conventional Marketing also covers billboards, leaflet distribution, physical store merchandising and telemarketing as promotional tools. Conventional marketing uses a message that might be felt intrusive by the public, trying to develop the customer/public interest for the product even when the potential customer is not looking for that product.

Marketing generally focuses on three factors: customer, customer satisfaction, and the customer's dazzle with the outcome. Digital marketing uses these three features in an improved manner. The changes made by digital marketing are based on the fact that the relationship between companies and customers has changed dramatically with the popularization of the Internet [4]. And with the Internet constantly evolving, digital marketing kept up with that evolution. The ingredients that, to date, had the most impact on digital marketing were the vulgarization of the use of social networks and the emergence of the ability to perform data analytics in a big data context.

12.2.2 The Impact of Social Networks

Social networks are based on a community that connects people, companies, institutions, etc. The ability to disseminate and expand information was the key element that propelled social networks' success on the Internet age [5].

Over the years and to the present day, social networks have evolved into a worldwide social phenomenon, making such progress that today they represent work tools, social interaction, and information dissemination. All these changes and metamorphoses made social networks significantly impact society, the consumer, and also the market itself. In this way, social networks have expanded the horizons for consumers/users. The consumer has become decisive for the implementation of new trends, products, services, etc. [6]. This opens the door to a new online trading market. New techniques to drive the market, companies, and the consumer themselves were eventually created to adapt to this new universe described as social media marketing [7].

With the constant use of the Internet and social networks, digital marketing has for a long time stopped being considered just another web page or a simple banner at the top of the page. Digital marketing is necessarily the new trend and will continue to be so for some time so companies need to adapt to it, particularly in the context of the restrictions placed by the EU Regulation 2016/679 of 27 April 2016 on the protection of individuals with regard to the processing of personal data and on the free movement of such data (GDPR). Digital marketing is not just a space within the digital market that allows companies to interact with their customers. Digital marketing allows for an equal and personalized relationship with the customer, but compared to conventional marketing, digital marketing enables this relationship at a lower cost and with a higher satisfaction rate on both sides [4]. It aims to improve communication, improve disclosure and improve the relationship between the customer, and the company, enhancing value co-creation, involving the company, the customer, and the supplier in a collaborative web [4]. Therefore, digital marketing stands out for its global reach, for the reputation and recognition given to the brand and its customers, and also for its cost-effectiveness.

One aspect of digital marketing, social media marketing, uses social networks to increase visibility on the Internet to promote and elevate products and services. The use of social networks increases the possibility of creating online communities, where it enhances and promotes the sharing of content, ideas, knowledge, and perspectives on products and services.

Social networks are described as the space constituted by a set of platforms that enable online interactions between people, companies, and brands, triggered by user actions and expressions: photographs, videos, texts, etc. [8].

Initially, the use of social networks was based solely on the ability to connect with friends, family, news and entertainment, and already in this capacity social networks provided unique opportunities for businesses to be closer to the consumer, making it possible to build a relationship between the business and the consumer [9]. Social media marketing works in a consumer-focused way. One of its key features is how campaigns, content, and communication are affected by analyzing consumer behaviors in connection with the brand/company in question. There is a transformation in the way these campaigns are communicated, created, distributed, and consumed by the target audience, the consumer. In a way the consumer is taking some of the conventional "marketing functions" out of the hands of the marketeers [10].

12.2.3 The Appearance of Social Media Influencers

Within social networks and social media marketing, new concepts, terms, techniques, and approaches have emerged. The social media influencer has become the latest phenomenon associated with social media marketing. A social media influencer impacts his/her followers that respect the influencer's opinion about something when they follow social media posts. In this way the influencer's opinion affects and influences the consumer's final judgment on whether or not to purchase the product/service that is mentioned or promoted [11].

There might not exist a commercial relationship between the influencer and the brand, but nowadays it is common to exist, initiated by either the influencer contacting the company and offering to advertise a given product or service or by the company contacting the influencer, sometimes through an intermediary, with an offer of sponsorship. Thus, the company sponsors the publications/content that will be generated by the influencer for a particular product or service, with the aim of increasing sales through the influence that the influencer has on the target audience that the brand intends to reach. This can cause increased sales, and increased communication between the company, the consumer, and the influencer. It also enables the emergence of engagement between these three entities, company, consumer, and influencer, through positive associations [12].

In the case of a commercial relationship, the contract between the influencer and the company specifies the frequency and type of content to be published but usually leaves most of the creative freedom in the hands of the influencer. This is necessary in order for the sponsored posts to blend in, in the normal flow of the influencer's posts, which makes them much more effective.

This process between the company and the influencer creates a link with electronic word of mouth marketing. That marketing method is based on a two-step flow of communication model. The model was developed by Lazarsfeld who stated that information does not flow directly from the source to the final recipient; instead the recipient gets it secondhand from an "opinion leader" that influences the recipient. This theory and model were first applied in the United States in the 1940 presidential election. In that context, Lazarsfeld studied how people are subject to different opinions from various "opinion leaders" (in this case candidates), and how that would cause them to interpret the same government news and projects in different ways. The conclusion drawn from this study indicated that voters created their views through conversations, where they shared similar perspectives to those they believed to be most fit for the presidency. Lazarsfeld concluded that people tend to create their opinions based on the opinions of others. These "others" are individuals whom we believe, meet, and relate to in some way [13]. After the initial investigation, Katz and Lazarsfeld expanded the model to include "opinion leaders" in general. This made the two-step flow communication model applicable outside the field of politics, and present in most of our perspectives and contexts [14].

In 2000, Gladwell presented the law of the few, classifying the people that influence and make information flow as "connectors", "mavens" and "salesmen" [15]. In 2003, Watts postulated that it is not just a few that governs influence but that instead many people play a critical role [16]. In 2016, Hilbert et al. adapted Lazarsfeld model to the study of social networks. By quickly expressing our personal opinion on everything, social networking platforms provide "opinion leaders" an easy way to spread their opinions on any topic, subject, event, project, product, or service [17].

Through the two-step flow communication model, influencers have great ease in influencing the opinion of their target audience (consumer). Influencer endorsement, consists of communicating and publishing content sponsored by a brand or company.

This sponsored content involves daily or weekly publications about a particular product or service that a company or brand intends to advertise using the influencer as a communication channel. Social media influencers are paid to promote a brand or product to their followers on one or more of the social platforms used by the influencer. In this way, followers/audience will associate the brand with the influencer, eventually transferring the influencer's opinions to the brand or product being promoted [18].

Influencer endorsement is the new term that has replaced celebrity endorsement. Despite this statement, not all influencers are considered celebrities. There are categories within this new medium, where influencers are placed at different levels depending on the number of followers they have on their digital platforms of choice. Veirman, Cauberghe and Hudders in 2017 distinguished and identified two groups of influencers. These two groups are differentiated by their popularity or number of followers. The concept of "followers" on social networks corresponds to the set of people who watch and interact with the content that the influencers post on their social networks [19]. Depending on the number of followers it is possible to distinguish between macro and micro-influencer, considering macro influencers those that have over 500,000 followers and micro-influencers those below 500,000 followers on their social platforms. In 2018, InflowLabs considered that it was necessary for a more detailed categorization of social media influencers, thus defining four levels: Nano (below 1,000 followers), Micro (below 10,000), Macro (below 100,000), and Mega influencer (above 100,000) [20]. The preferred influencer platforms are Instagram and YouTube.

12.3 Key Factors Governing the Influencers Impact

The influencers' impact is a function of many factors, including the influencer's personality, perceived authenticity, and how well the influencer matches the product being promoted. Although the impact of the sponsorship can be measured it might be difficult to find the best match for a given product because there are many subjective factors at play.

12.3.1 Authenticity and Sponsorship Disclosure

It is the blurring of the line between what is a genuine opinion and noncommercial endorsement of a product versus a paid endorsement that makes influencer marketing so powerful [21]. It is considered that around half of the users rely on recommendations from friends and half on recommendations from influencers. A study made by Nielsen Catalina Solutions concluded that influencer marketing can annually originate more than one order of magnitude (11 times larger) return on investment (ROI) than other more traditional forms of marketing [22]. Still this blurring of the line made the United States Federal Trade Commission publish in 2015, and then continuously update, a set of guidelines for the disclosure of this kind of endorsement [23]. The guidelines specify that a tweet, YouTube vlog, blog, or Instagram post for which the influencer has any connection or relationship with a company is considered a promotion and that such relationships must be disclosed, and if that is not done it can originate a large fine ($250,000 dollars). Other countries followed suite, and the rules for the use of most social platforms started suggesting or enforcing the use of #advertisement, #ad, #sponsored, #sp, #partnership tags. Still in 2014, a court in Germany considered that the addition of the hashtag "#sponsored" was not sufficient, as some people might not understand that it meant that the post was being sponsored [24]. This might seem strange but in the study described in [24], it was also found that in some instances the disclosure that it was an advertisement did not always make the persons seeing it understand that they were facing a promoted post. One might ask if adding that kind of tag is not enough to show that the post is a form of advertising, what would be enough? It seems necessary to look into the factors that increase people's awareness of influencer marketing and to identify what would be sufficient for people to understand that they are seeing paid content.

The collaboration with brands raises questions about the authenticity of the posts, namely, that even if the influencer seems passionate about a given topic in truth that passion might be somehow lacking in authenticity. Referring as much as possible to facts and not lying about the existence of partnerships, or about the real interest in the product being promoted, it is necessary to avoid the risk of rejection and alienating most of the followers.

A framework for authenticity management is proposed in [25] based upon a strategy of more or less passion and a strategy of more or less transparency, thus creating four possible paths: fake authenticity (little passion and transparency), fairytale authenticity (passionate but with little transparency), disembodied authenticity (transparent but with little passion), and absolute authenticity (passionate and transparent). Influencer marketing can be seen as a form of product placement because it involves integrating the brand message with the editorial media content created by the influencer. It is the fact that this media content might be suggested by the brand but in fact, is tuned and created by the influencer that gives the influencer message a ring of authenticity. But at the same time hiding advertising in entertainment is considered a deceptive communication practice, that is only avoided if the sponsorship is disclosed.

A study of the effects of product placement, sponsorship disclosure, type of influencer and the interplay with consumer responses concluded that non-sponsorship disclosure had an attenuated effect on "larger" celebrities and a more negative effect on micro-celebrities, as it affected message credibility [24]. A participant in another study [26] stated that "*If they are sponsored and they say something slightly negative, ..., I will trust them more than even the ones that weren't sponsored, because you are sponsored yet you choose to be honest*". This is good news for influencers because it means that if the promotion is artfully made and rings authentic then the disclosure can cause no harm while at the same time increases brand awareness.

There are some mixed results concerning whether micro-celebrities would generate more positive consumer responses than normal celebrities. In the study described in [24] it was shown that purchase intention was higher for normal celebrities and consequently lower for micro-celebrities. A possible explanation is that this was caused by a limitation of the study in that the participants might not be real followers of the micro-celebrity and on the other hand they probably knew and trusted the normal celebrity present in the study. We can conclude that the interplay between credibility, authority, match between influencer lifestyle and product being promoted, and its effects on purchase intention can be quite fragile.

Although consumers can try to explain why they trust the influencers that they follow, much of that influence occurs in the subconscious mind. It is not easy for us to identify the true causes behind our behaviors, much of our behavior is commanded by unconscious processes that we are not aware of. It is a well-known fact in marketing that not all our purchasing decisions come from a rationale reflection on the specific merit of the products. The advance of neuroscience allows researchers to better measure implicit cognition, including different attitudes, affects, and specific memories. By using functional magnetic resonance imaging (fMRI) we can get insights into the role of emotion and how different cognitive processes affect product choice and brand selection [27]. Influencing purchase decisions is somewhat tricky, there is a fine line between creating digital content that is agreeable and content that feels artificial and forced. On the other hand, studies made with teenagers showed that it is much easier to make them switch or alter their brand choice than it is to make them remain loyal to a given brand [28].

12.3.2 Achieving a Close Fit with the Brand

In influencer marketing, it is felt that authenticity is the biggest determinant of success. Still likeability, authority in the subject area being presented and promoted, and they fit with the brand are also an issue [29].

The match between an influencer and a given brand entails risks for both parties. If an influencer is associated with a given brand then a small indiscretion or out of character attitude in a weak moment in the influencer real life might be caught not only by the paparazzi industry but also by the camera of any passer-by with a smartphone. This is the same as when political leaders deviate from the policies that they promised their constituents and later on are punished by their voters for having abandoned their promises. The influencers self-branding through social media is based upon attention and narrative. Creating a compelling narrative attracts audiences, because they are inspired, instructed or relate with the influencer profile and lifestyle, but every influencer has to fight in order to stand out in the attention economy [30].

The effectiveness of electronic word of mouth marketing in the context of a given group of people can be considered a function of the personal characteristics of the social media influencer (personality, empathy, knowledge, skills) combined with social influence dimensions (who one is, what one knows, whom one knows). In the study reported in [31] these two dimensions of individual capital and social capital were the basis for the creation of nine categories of social influencers: Experiencer, Helpful friend, Reciprocal mercenary, Silent expert, Occasional influencer, Superspreader, Narrative expert, Macchiavellist, and Top influential. By conducting 480 interviews within the fashion industry it was possible to create a reduced topology with only three categories: The fashion superspreader, The narrative fashion expert, and The helpful friend. This reduced set highlights the differences that can occur between influencers' style and reach. A previous study this time in terms of automotive buying behavior identified four categories: The occasional influencers, The narrative experts, The social leaders, and The social followers [32]. This means that there are quite different motivational drivers that originate different types of social leaders and followers and that there are several factors to take into consideration when trying to combine an influencer with a given product, social situation, or culture. For the influencer to be able to create purchase intentions it is necessary for the influencer to be skilled enough to strike a chord with the followers in spite of them knowing that the product is being actively promoted by the influencer. But it is not only the influencers that need to achieve a good match between the brand values or the product being promoted and the influencer perceived expertise and lifestyle. In the movie industry, it is well-known that it is necessary to achieve a good match between the movie's star and the movie's genre or story [33]. Depending on the type of brand some influencers might not be appropriate. For instance, cooperating with influencers that have a large number of followers might not be the best marketing choice for promoting divergent and alternative lifestyle products, as this might lower the brand's perceived uniqueness and consequently the attitude toward the brand [19].

12.3.3 Measuring Fitness and Impact

To measure the influencer's impact, it is necessary to use difficult to measure criteria, such as the personality, authenticity, quality, or spontaneity of the digital content, together with factors that are easy to measure such as the number of page views, search engine rank, number of likes, followers, or engagement rate.

The engagement rate formula can be slightly different depending on the social media platform. Generically, we want to calculate the engagement rate by publication, or by grouping posts by theme or time range, or for all publication activity. The engagement rate is the number of audience interactions divided by the number of followers multiplied by 100 (so that the engagement rate is a percentage). Audience interactions consist of the sum of likes and comments on Twitter and Instagram. On Facebook, the audience interactions are the sum of likes, comments, and shares. LinkedIn uses several formulas to evaluate the engagement rate, one of them consists of summing up the number of comments, likes, and clicks, divided by the number of impressions (how many people viewed your updates), all multiplied by 1000. On YouTube, the total number of likes, dislikes, and comments to all the videos in the channel are added up and divided by the number of channel subscribers on the selected time range. As new features are continuously added, in particular in the case of Instagram, presumably the audience interactions will start counting other forms of interaction such as, for, e.g., views and shares. Analytics constructed by the social media platforms already include parameters that allow a qualitative evaluation of the engagement, for example, how many views and for how many seconds a video is viewed, how many clicks a photo receives, side by side with whether or not it was liked. The recent social network platforms crackdown into the use of social media bots to create fake accounts caused the removal of millions of fake accounts with a corresponding impact on the number of followers of some accounts, making them closer to reality.

From a more subjective point of view the exact placement of the product, the way how it is presented and how it catches the attention of the followers so that they might become consumers of that product is in itself an art form, trying to promote the product but without becoming too forceful [34–36].

12.4 Study of Portuguese Influencers

This section presents an overview of some of the Portuguese influencers studied in [37]. A limitation of that study was that although it evaluated the influencer's characteristics and engagement rate it did not try to account for subjective issues. To find an influencer that represents a good fit for a given brand can sometimes be easy and obvious, but in other cases it might be much more difficult as it entails evaluating subjective qualities. In that context in [38] it is suggested the use of the Q-sort technique. Q-sort is a methodology for the systematic study of subjective issues, such as the person's viewpoints, personality, beliefs, or attitudes [39, 40].

Catarina Filipe is a 24-year-old micro-influencer in fashion, beauty, and lifestyle with 224 k Youtube /165 k Instagram followers. Her Youtube account focuses on cosmetic product review videos, weekly vlogs, fashion, and personal matters. Her Instagram account has photos of her day outfits, travels, people close to her, or cosmetics. Her engagement rate is 3.05% in Youtube and 10.06% in Instagram (6.55% average).

Inês Rochinha is a 25-year-old micro-influencer in fashion, beauty, and lifestyle with 267 k Youtube/233 k Instagram followers. Her Youtube account focuses on cosmetic product review videos, weekly vlogs, fashion, and decoration. Her Instagram account has photos of her day outfits, decoration, or cosmetics. Her engagement rate is 1.60% in Youtube and 11.44% in Instagram (6.52% average).

Sofia Barbosa is a 19-year-old micro-influencer in fashion, beauty, and lifestyle with 270 k Youtube/298 k Instagram followers. Her Youtube account focuses on cosmetic product review videos, weekly vlogs, travels, and personal matters. Her Instagram account has photos of her day outfits, travels, cosmetics, or gastronomy. Her engagement rate is 2.12% in Youtube and 15.14% in Instagram (8.63% average).

Nuno Agonia is a 39-year-old influencer in technology with 1188 k Youtube (mega influencer)/291 k Instagram (micro-influencer) followers. His Youtube account focuses on laptop, smartphone, and smart TV reviews, and in his travels. His Instagram account has photos of new smartphone and laptop models, with family or during his travels. His engagement rate is 2.30% in Youtube and 5.81% in Instagram (4.06% average).

Mariana Cabral (Bumba na Fofinha) is a 31-year-old micro-influencer in entertainment with 170 k Youtube/269 k Instagram followers. Her Youtube account focuses on funny videos, travels, and personal matters with humor blended in. Her Instagram account has photos of her travels and pictures with some of her jokes in the caption. Her engagement rate is 9.32% in Youtube and 9.36% in Instagram (9.34% average).

Ricardo Fazeres (RicFazeres) is a 40-year-old in gaming with 705 k Youtube (macro-influencer)/199 k Instagram (micro influencer) followers. His Youtube account has videos about gaming. His Instagram account has photos with family and friends. His engagement rate is 0.44% in Youtube and 7.59% in Instagram (4.01% average).

What can be inferred from the examples above is that, as expected, a high number of followers does not always correlate into a high engagement rate. It is also relevant that RicFazeres achieves a much higher engagement rate in Instagram than in Youtube, and that in spite of his Youtube account presenting contents in gaming (where he is a macro-influencer) while his Instagram account has mainly photos with family and friends, something that in theory is much less relevant to his followers. This highlights another feature of influencers. Besides their ability to market third parties' products and services some influencers leverage their social media currency to market themselves and profit from their follower's willingness to have a more intimate relationship with them. Mobile Apps or subscription websites, such as `onlyfans.com`, which currently has 70 thousand creators uploading content for 7.5 million registered users with an average of 25 thousand new users registering daily [41], provide special content that their followers crave and are willing to pay for, ranging from the ability to see their influencer in a more intimate setting, to having access to a personalized phone number to text the influencer, or having access to influencer produced R-rated or even NC-17-rated content. Still this intimacy and paying captive audience is not without risks, as it can easily backfire and might be difficult to control [42, 43]. And, as was to be expected, virtual influencers with no human counterpart are already on the market [44].

12.5 Conclusion

Traditional celebrity endorsement came from the film, music, sports, or TV shows. An alternative form of celebrity started with the rise of reality TV stars, that can be seen as the precursors of social media stars. It is the perceived real life of those nontraditional stars and the deep connectedness that their followers feel toward them that results in higher purchase intentions of the products they endorse, that is, the consumers personally identify with the influencers and try to imitate them. The main characteristic of micro-celebrities is that they are seen by their followers as being closer to ordinary people, and living more or less ordinary lives. It can be said that social media influencers are simultaneously ordinary people and quite famous at least within the limits of their followers' group. A comparison of the trustworthiness (the degree of honesty, reliability and dependability) of Instagram Celebrities versus traditional celebrities found that consumers consider the Instagram celebrity more trustworthy and better able to connect with their audiences, achieving a higher degree of intimacy and creating a form of benign envy (desire to acquire the possessions or status of someone else but without hostile feelings toward that person), and affinity between the follower and the followee [45].

Initially, some of the social media platforms (e.g., Instagram), while trying to get onboard as many users as possible, did not allow brands to integrate promotional content into the platform, this created a need for advertisers to find a way into the platforms and a motivation for supporting the platform users that could fuse their product or brand into the posts [46]. This was one of the initial motivations for the appearance of influencers' social media marketing. And now they are here to stay.

References

1. Minuzzi, G., Larentis, F.: Marketing: definições, aplicações, tendências e desafios do profissional. RAIMED—Revista de Administração IMED **4**(1), 80–97 (2014). (in portuguese)

2. Kotler, P.: Marketing Management, 15th edn., Pearson (2016)
3. Sabbag, K.K.: Marketing Tradicional e Marketing Digital: evolução ou mudança?, Universidade Federal do Paraná (in portuguese) (2014)
4. Oliveira, J., Crucial, E.: Marketing Digital—Guia do Empresário, CTCP—Centro Tecnológico do Calçado de Portugal. https://www.ctcp.pt/docs/docmediagaleria/522x5j.pdf
5. Recuero, R.: Redes Sociais na Internet, Editora Sulina (in portuguese) (2009)
6. Santos, M.I.S.: Impacto das redes sociais na conversa entre as marcas e o consumidor, Universidade Católica Portuguesa (in portuguese) (2017)
7. Edosomwan, S., Prakasan, S., Kouame, D., Watson, J., Seymour, T.: The history of social media and its impact on business. J. Appl. Manage. Entrepreneurship (2011)
8. Pachiţanu, A.: Social media instruments' use and importance for the marketing communications mix—an exploratory analysis on companies' in Romania. J. Emerg. Trends Market. Manage.
9. Kelly, L., Kerr, G., Drennan, J.: Avoidance of advertising in social networking sites—the teenage perspective. J. Interact. Advertising **10**(2), 16–27 (2010)
10. Tsai, W.-H.S., Men, L.R.: Motivations and antecedents of consumer engagement with brand pages on social networking sites. J. Interact. Advertising **13**(2), 76–87 (2013)
11. Saito, K., Teramoto, T., Inoue, A.: How Opinion Leaders Are Made by the Social Media, Special Topic: The Future of Marketing—Expanding the Possibilities of Marketing, AD Studies, vol. 52, pp. 14–19 (2015)
12. Sudha, M., Sheena, K.: Impact of influencers in consumer decision process: the fashion industry. SCMS J. Indian Manage. July–September **2017**, 14–30 (2017)
13. Lazarsfeld, P.F., Berelson, B., Gaudet, H.: The People's Choice—How the Voter Makes Up His Mind in a Presidential Campaign. Duell, Sloan and Pearce, New York (1944)
14. Katz, E., Lazarsfeld, P.F.: Personal Influence: The Part Played by People in the Flow of Mass Communications. The Free Press, Glencoe (1955)
15. Gladwell, M.: The Tipping Point: How Little Things Can Make a Big Difference. Little, Brown and Company (2000)
16. Watts, D.J.: Six Degrees: The Science of a Connected Age. Norton, W. W (2003)
17. Hilbert, M., Vásquez, J., Halpern, D., Valenzuela, S., Arriagada, E.: One step, two step, network step? Complementary perspectives on communication flows in twittered citizen protests. Soc. Sci. Comput. Rev. **35**(4) (2016). https://doi.org/10.1177/0894439316639561
18. Lim, X.J., Radzol, A.R. bt M., Cheah, J.-H., Wong, M.W.: The impact of social media influencers on purchase intention and the mediation effect of customer attitude. Asian J Business Res. **7**(2), 19–36 (2017). https://doi.org/10.14707/ajbr.170035
19. Veirman, M.D., Cauberghe, V., Hudders, L.: Marketing through instagram influencers: impact of number of followers and product divergence on brand attitude. Int. J. Advertising **36**(5), 798–828 (2017). https://doi.org/10.1080/02650487.2017.1348035
20. InflowLabs: What are Nano, Micro, Macro and Mega Influencers and Why You Should Care. https://inflowlabs.com/2018/01/09/nano-micro-macro-mega-influencers-care/
21. Audrezet, A., de Kerviler, G., Moulard, J.G.: Authenticity under threat: when social media influencers need to go beyond passion. In: 2017 Global Fashion Management Conference at Vienna, pp. 170–172 (2017). https://doi.org/10.15444/gfmc2017.03.01.05
22. NCS (2016). Sales Effect Study: Influencer Marketing, Nielsen Catalina Solutions. http://www.tapinfluence.com/roi/
23. Federal Trade Commission: The FTC's Endorsement Guides: What People Are Asking (2017). https://www.ftc.gov/tips-advice/business-center/guidance/ftcs-endorsement-guides-what-people-are-asking

24. Ewers, N.L.: #sponsored—Influencer Marketing on Instagram: An Analysis of the Effects of Sponsorship Disclosure, Product Placement, Type of Influencer and their Interplay on Consumer Responses, Master Thesis, University of Twente (2017)
25. Audrezet, A., de Kerviler, G., Moulard, J.G.: Authenticity under threat: When social media influencers need to go beyond self-presentation. J. Business Res. Elsevier (2018). https://doi.org/10.1016/j.jbusres.2018.07.008
26. Dreifaldt, N., Drennan, T.: Does Your Brand Need Online Influencer Marketing? It Depends; A mixed method approach to country of origin, brand image, and online influencer marketing on consumers' purchase intention online, Master Thesis in Business Administration, School of Business. Society & Engineering, Mälardalen University (2019)
27. Young, A.M., Hinesly, M.D.: Identifying Millennials' key influencers from early childhood: insights into current consumer preferences. J. Consumer Market. **29**(2), 146–155 (2012). https://doi.org/10.1108/07363761211206393
28. Martin, C.A., Bush, A.J.: Do role models influence teenagers' purchase intentions and behavior? J. Consumer Market. **17**(5), 441–454 (2000)
29. Moore, A., Yang, K., Kim, H.M.: Influencer Marketing: Influentials' Authenticity, Likeability and Authority in Social Media, International Textile and Apparel Association, ITAA Annual Conference Proceedings, Poster 126 (2018)
30. Khamis, S., Ang, L., Welling, R.: Self-branding, 'micro-celebrity' and the rise of Social Media Influencers. Celebrity Stud. **8**(2), 191–208 (2017)
31. Wiedmann, K.P., Hennigs, N., Langner, S.: Spreading the word of fashion: identifying social influencers in fashion marketing. J. Global Fashion Market. Bridging Fashion Market. **1**(3), 142–153 (2010). https://doi.org/10.1080/20932685.2010.10593066
32. Langner, S., Hennigs, N., Wiedmann, K.: Social persuasion: targeting social identities through social influencers. J. Consumer Market. **30**(1), 31–49 (2013). https://doi.org/10.1108/07363761311290821
33. Liu, A., Liu, Y., Mazumdar, T.: Star power in the eye of the beholder: a study of the influence of stars in the movie industry. Mark. Lett. **25**(4), 385–396. https://doi.org/10.1007/s11002-013-9258-x
34. Glucksman, M.: The rise of social media influencer marketing on lifestyle branding: a case study of Lucie Fink. Elon J. Undergraduate Res. Commun. **8**(2), Fall 2017, 77–87
35. Dahlqvist, J., Preiksaite, S.: How Competing Brands are Being Communicated Through Influencer Marketing: A Qualitative Study Aimed to Explore How Consumers Evaluate an influencer communicating competing brands, Bachelor Thesis in International Management, Jönköping University
36. Gunnarsson, L., Postnikova, A., Folkestad, A.: Maybe Influencers Are Not Worth The Hype: An Explanatory Study on influencers' characteristics with perceived quality and brand loyalty, Bachelor Thesis, Linnaeus University
37. Oliveira, M.: Social Media Marketing e os Influencers. Master Thesis in Information Technology, Communication and Multimedia, University Institute of Maia—ISMAI (in portuguese) (2019)
38. Freberg, K., Graham, K., McGaughey, K., Freberg, L.A.: Who are the social media influencers? A study of public perceptions of personality. Public Relat. Rev. **37**(1), 90–92 (2011). https://doi.org/10.1016/j.pubrev.2010.11.001
39. Brown, S.R.: Political Subjectivity: Applications of Q Methodology in Political Science, Yale University Press (1980)
40. Brown, S.R.: A primer on Q methodology. Operant Subject. **16** (3/4), 91–138 (1993)
41. Parham, J.: When Influencers Switch Platforms—and Bare It All. WIRED 19.08.2019 (2019). https://www.wired.com/story/culture-fan-tastic-planet-influencer-porn/

42. Heck, S.: I Broke the Official Jeremy Renner App By Posting The Word "Porno" On It (2019). https://theconcourse.deadspin.com/i-broke-the-official-jeremy-renner-app-by-posting-the-w-1837804181
43. Barrett, B.: Jeremy Renner's App Developer: This is a Freak Situation. WIRED 06.09.2019 (2019). https://www.wired.com/story/jeremy-renner-app/
44. Ellis, E.G.: You Are Already Having Sex With Robots. WIRED 23.08.2019 (2019). https://www.wired.com/story/you-are-already-having-sex-with-robots/
45. Jin, S.V., Muqaddam, A., Ryu, E.: Instafamous and social media influencer marketing. Market. Intell. Plann. (2019). https://doi.org/10.1108/MIP-09-2018-0375
46. Woods, S.: #Sponsored: The Emergence of Influencer Marketing, University of Tennessee Honors Thesis Projects (2016). https://trace.tennessee.edu/utk_chanhonoproj/1976

Chapter 13
Potential and Advantages of Social Shopping in Portugal

Joana Ferreira, Vitor Santos(✉), and Pedro Malta

NOVA IMS, New University of Lisbon, Lisbon, Portugal
M2015076@isegi.unl.pt,
{vsantos,pmalta}@novaims.unl.pt

Abstract. Social networks are present in our day-to-day and have a strong influence in almost all our personal and professional aspects. Among several benefits and challenges brought by Social Networks, we cannot ignore their potential in what e-commerce is concerned. Social shopping allows potential buyers to obtain information and opinions issued by other buyers with similar profiles, on products and services they are planning to acquire. This way, instead of allowing a brand or product to create in the community users the need or will to buy, it is the community itself, through social networks, to self-influence in the sense of defining what are the products and services they want to be marketed and in what way, thus obtaining better prices and reaffirming its identity. To accompany the acceptance of cybernauts of these new communication spaces, like for instance Facebook, and understanding whether these users are as well motivated in participating and taking advantage of these platforms as online shopping places, seems to be a relevant issue. In the present article, we investigate how Social Shopping affects the Portuguese consumers in their purchases.

13.1 Introduction

In the last decade, the World Wide Web has witnessed the emergence of Web 2.0 services, such as video sharing, blogging platforms, social networking, and social bookmarking [1].

Social networks have conquered their space in our society's day-to-day, they continue to steadily grow and have nowadays a strong influence on people's decision-making. We know their behavior is influenced by several aspects, where the surrounding environment and contact list are paramount [2, 3].

Due to the number of social networks users and potential viral effect, these have become an attractive way in divulging products and services. The interest in knowing how to maximize the use of social networks has been a target in multiple studies in several areas of knowledge, from psychology, sociology, management, and economics, being the marketing area the one trying to use these platforms heavily, in order to promote products and services.

Among several benefits and challenges brought by social networks, we cannot ignore their potential in what e-commerce is concerned. To accompany the acceptance

Á. Rocha et al. (eds.), *Marketing and Smart Technologies*, Smart Innovation, Systems and Technologies 167, https://doi.org/10.1007/978-981-15-1564-4_13

of cybernauts of these new communication spaces, like for instance Facebook, and understanding whether these users are as well motivated in participating and taking advantage of these platforms as online shopping places, seems to be a relevant issue.

In this article, we investigate how social shopping affects the Portuguese consumers in their purchases. The main objective of this research is to contribute to understanding the potential benefit of social networks for consumers, in particular by enabling them to purchase better and more economical goods and services.

The present article, apart from this introductory section, is structured in four other sections: Social Shopping, Consumer Behavior in Online Shopping, Study, and Conclusions and future work.

13.2 Consumer Behavior in Online Shopping

Traditional ways of understanding consumer behavior, segmenting markets, defining policies, and marketing strategies have undergone significant changes throughout the twenty-first century.

According to Almeida et al. [4], the influence of these changes is due to the opening of markets, changes in capitalism and, mainly, the way new technologies have changed the way people work, relate and buy, creating synergies between consumers and organizations. Indeed, the Internet has changed the way consumers interact with each other and with businesses, using a direct relationship, eliminating intermediaries in purchasing, and communication processes. The change of focus has generated new perspectives, which are outlined by the use of the Internet as a vehicle of communication and a means of interactivity.

According to Kim et al. [5], Internet commerce is witnessing a steady growth rate, which makes it a shopping mode of choice for many consumers. The e-commerce environment allows consumers to search for information and buy products or services through direct interaction with the online store [6]. A consumer relies above all on the appearance of cyberspace, such as photographs, images, information quality, and product video clips rather than actual experience [7, 8]. Thus, shopping in the online store is like shopping in a mall through a catalogue, in which case the consumer cannot have a direct relationship with the products (touch or smell, for example) [9]. Therefore, the promise of e-commerce and online commerce depends to a large extent on user interfaces and how they interact with other digital media [10, 11]. In addition, the characteristics of reporting, browsing, and ordering in interactive shopping are considered more important in building trust than e-commerce [12, 13].

In recent decades, marketing experts have developed the notion that it is not only the demand for commercial products that can be influenced by marketing approaches but also the demand for social care services, lifestyles, and even the demand by political leadership [14]. This underpins the theory of marketing, known as social marketing, whose main purpose is to encourage people to take actions that, according to Scrivens [14], will be beneficial to themselves, rather than simply boosting business profits. Social marketing can be defined as "the design, implementation and control of programs that influence the receptivity of social ideas and considerations of product planning, pricing, communication, distribution and marketing research" [15].

Dholakia et al. [16] define virtual communities as social spaces in a digital environment that allow the formation of groups and stimulate continuous means of communication. These can also be defined as diverse consumer groups that meet and interact online for the purpose of achieving personal goals that are common to members. On the other hand, some authors characterize virtual community as an aggregation of individuals or business partners that interact around a common interest, where interaction is supported and/or mediated by technology and guided by some protocols and standards [4]. Virtual communities are supported by trust, commitment, and structure of consumer relationships with the organization and with each other [4]. Consumers who are part of virtual communities are active members and tend to be loyal and committed to the products and brand, giving the company a unique opportunity to interact with this select group [4].

13.3 Social Shopping

Boyd and Ellison define social network sites as being web-based services, that allow individuals to build a public or semi-public profile within a limited system, to articulate a list of other users with whom they share a connection, to watch and go through the personal connections lists, as well as those made by people within the system [17].

Social networks have proven to be very useful and powerful beyond their initial competences. An example of it is their use in politics, teaching, marketing, and in many other activities and areas of collaboration and cooperation. In this context, a new area arises known commonly by social shopping [18].

Shopping has always been a social experience and social networking allows consumers to interact with individuals—many of whom are likely strangers [19].

Social shopping is a way of electronic commerce based on social networks, where «friends» influence others to buy [3]. Social shopping uses technology to mime social interactions found in malls and physical stores. On another hand, social shopping is also a way for a company to assess the market and detect consumers' tendencies on a product and/or a service. In the present day, there is a social network created exclusively for this purpose, called Frugar (http://my.frugar.com.br). There are also plugins for social networks, as Facebook, for this purpose.

Frugar is a Brazilian social shopping network where one can find opinions of the members about products and services one can eventually acquire. It is a social network whose objective is to be a reference to people wanting to purchase something. One just needs to register to Frugar and create a list of products, contributing to the development of the community by commenting on them. It is also possible to add friends or meet new people with similar taste and follow them to see what they buy or what comments they are making. The objective of Frugar is to have 1 million users during the first year of existence.

Some virtual shops use social networks API's in order to interact with the customer, in the form of social shopping, like Levi's Friends Store. Levi's created a social shopping channel, destined essentially to an audience with ages between 18 and 34. The channel is meant to be complementary to Levi.com electronic platform. Levi.com electronic commerce platform visitors can see invites to use Facebook Connect and «Like» the brand, or in jargon «to become a fan».

The customers' preferences are expressed by the «brand ambassadors» through a simple «Like» telling your friends these are your favorite jeans. For instance: «John likes 501 Original Jeans». The «Like» will show in Facebook feeds that take the customer directly to the page with product details. Levi's Friends Store is filled with products where users and friends have manifested themselves via «Like» [20].

Naked Pizza held a promotion only divulged in Twitter. The result was a 15% increase in sales in a single day. Palm, needing to increase its smartphone device developed an application for Facebook. Users of that application registered and received news on the product. A message was sent to around 70 thousand registered users, and in 3 h, 3 thousand had read them. The return of this action in sales was made clear [21].

«Sony Electronics» chose to use social networks as a blog, which allowed to maintain contact between the user and the company. In the end, this company was reachable in other social networks as YouTube, Flickr, and Twitter. Through these networks they could convey information on their products to consumers [22].

The constructer «Tecnisa» uses Twitter to divulge their services. An example of success due to the use of this tool was the sale of an apartment by users who followed the company via Twitter [21].

In Portugal, «Tap» and «Coca-Cola» show themselves in social networks using these with the objective of showing their customers their products/services and promotions [23]. «PT» uses Web 2.0 tools to make available online support to their customers, allowing in a quick and effective way the enlightenment on questions [24]. «Sol Mélia» group resorted to Web 2.0 tools to get more information on their clients, and thus keep updated and prepared according to their demands. To do so, the group assigned several employees the task of maximizing the use of social networks use, in order to make business and getting to know the clients better [24].

13.3.1 A Social Shopping Model Example

There are several social shopping models. The model presented below consists of a social shopping approach, where the community itself determines what they are interested in buying (basket) and, in some situations, when [25]. This way, instead of letting a brand or product create the need or will to buy on the users, it will be the community, through a network effect to self-influence, in order to define what they want to buy, by assuring exemption related to brands, products, and services. The management of the community has as major tasks to propose products and services, and to handle the whole process, from the client's choice to delivery and due payment. Such a community allows two important things: on one hand, it is the very community that defines the basket of products and services they mean to acquire, translated in a specification for potential suppliers and ulterior contractual agreement; on another hand, after a while, it will be possible, resorting to data mining techniques, to create a concept that effectively represents the community.

The model is based on a member's community, interested in a certain set of products, come together to determine in a certain time period, a basket of products and services to be acquired.

The community starts by proposing a relatively vast basket. The members then point out their preferences and make comments on the products and services in the basket, which will influence the choice of other members.

When the moment or predetermined date is reached, they define a product basked for that period of time, based on the community members' preferences. These products choice process can be achieved based on the comment's analysis or simply through vote.

When the basket is set, so is its cost, and a period of ordering/payment is opened. At the end of this period, the orders are handed to the social shopping store suppliers and to the logistic operator, in charge of product pick & pack. This process is illustrated in Fig. 13.1.

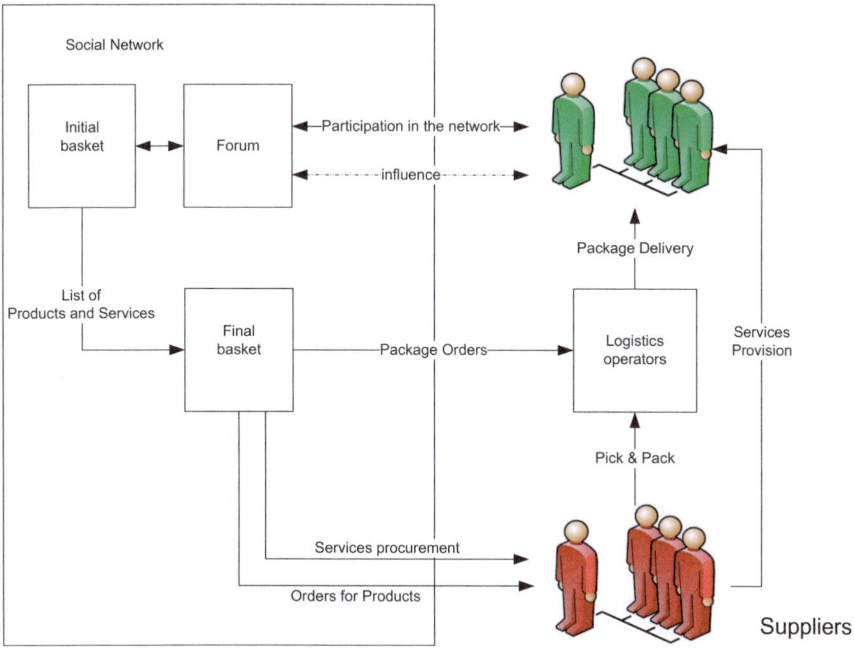

Fig. 13.1. Social shopping model with self-defined shopping basket

At the same time, the community management will have to ensure total exemption related to brands or products intrusion, that tries to influence directly or indirectly the array of the products in the basket.

13.4 Study

Based on the research objectives and a questionnaire was constructed and executed. The target population of this study was Portuguese consumers who use social networks and shop online. The questionnaire was available for answer from February 12, 2018 to March 4, 2018. The first two questions on the questionnaire are considered eliminatory because they ask if they are a social media user and shop online. The next three questions focused on sociodemographic factors such as gender, age, and complete educational level of the respondent. The questions in the questionnaire were divided into six categories, which are called: consumption triggers, impulse purchases, social media, social networks, social and business networks, and social shopping. The questionnaire obtained 198 answers.

13.4.1 Sample Characterization

The female gender predominated in the responses to this questionnaire with 59.7% as opposed to the male gender which obtained 40.3%. The age group with the largest number of respondents is between 26 years and 36 years old has a higher percentage than the others with 38.3%. Then we have the age group from 37 years to 45 years with 26%, from 18 years to 25 years with 19.5% and, finally, over the age of 45 presents the value of 16.2%. In terms of education, the "Bachelor" has the highest percentage with 45.5% in relation to the others; however, the option "Master" is second with 34.4%. The remaining ones have a small percentage, "Secondary Education" with 11.7%, "Doctorate" with 5.2% and, finally, 3.2% for the option "Basic Education".

13.4.2 Descriptive Analysis

Given the answers obtained, it was concluded that:

The three most common factors that consumers consider are "Online Payment Security" (99 replies), "Quality of Product Reviews" (73 replies), and "Product Characteristics Detail" (70 answers), respectively.

The three most important factors that consumers have in mind when completing their purchase are "Price" (132 replies), "Quality" (112 replies), and "Utility and Use" (111 replies). By contrast, "Humor" (4 replies), "Fun" (9 replies), and "Pleasure" (20 replies) are the factors that are least considered by the user when searching for a product.

13.4.3 Correlational Analysis

Impulse purchases in e-commerce

Table 13.1 mirrors the positive correlation of variable X with the variables K, Q, S, T, W, Y, and Z. The correlations that are most relevant to the study are those of the value 0.323, between the variables X and K, and 0.325, between variables X and Y. Thus, impulse purchases in e-commerce are shown to increase when confidence in online payment systems also increases. However, "friends" play a significant role in impulse purchases in e-commerce, because due to their positive correlation of 0.325,

Table 13.1. Correlation of variable X with the variables K, Q, S, T, W, Y, and Z

Spearman's variables correlation		K variable	Q variable	S variable	T variable	W variable	Y variable	Z variable
		Do you trust online payment systems?	Do product reviews help you decide if you make a purchase?	Are you part of any online community? (ex: OLX, eBay, LinkedIn, etc.)	Is the information presented on social media totally reliable?	If a "friend" follows a page they are likely to follow it too?	If a "friend" made a purchase at the company, is it likely to make it?	Are you likely to buy a product if your "friend" advises you?
X variable	Feel that you make impulse purchases in e-commerce?	0.323	0.167	0.171	0.164	0.243	0.325	0.207

impulse purchases increase if a “friend has made a purchase in the company” also increases. Also, with a positive correlation of 0.207, impulse purchases increase if the “friend” advises the product to increase as well.

On the other hand, impulse purchases are unrelated to the variables Q, S, and T, “Do product reviews help you decide if you make a purchase?”, “Are you part of an online community? (e.g., OLX, eBay, LinkedIn, etc.),” and “Is the information presented on social media fully reliable?” respectively.

Table 13.2 shows impulse purchases in electronic commerce by socioeconomic variables.

Table 13.2. Impulse buying in e-commerce by socioeconomic variables

Spearman's variables correlation		K variable	Q variable	T variable
		Gender	Age range	Level of education
X variable	Feel that you make impulse purchases in e-commerce?	−0.102	−0.129	0.162

13.4.4 Analysis by Association Rules

The analysis by association rules was performed in SPSS Modeler, targeting impulse purchases in electronic and traditional commerce. For this analysis the Apriori algorithm was used.

For the analysis of the impulse buying model in electronic commerce, the following variables were defined as consequent and antecedent:

- Consequent: “Impulse purchases in e-commerce”;
- Antecedents:

  1. “Do product ‘reviews’ help you decide if you make a purchase?”;
  2. “Is the information presented on *social media* totally reliable?”;
  3. “If a ‘friend’ follows a page they are likely to follow it too?”;
  4. “If a ‘friend’ made a purchase at the company, is it likely to make it?”;
  5. “Confidence of friends’ comments on social networks”;
  6. “Are most purchases ‘recommended’ by your friends?”.

From the execution of the Apriori algorithm, the following results were obtained:

Table 13.3 shows that there are strong support relationships of value 75.49%. Thus, it can be assumed that impulse purchases in e-commerce happen together when consumers identify “Product reviews help them decide whether to make a purchase,” “The information presented on social media is completely reliable,” “If a “friend” follows a page, is it likely to follow it too?” and “Are most purchases “recommended” by your friends?” with 100% confidence.

Table 13.3. E-commerce impulse buying rule

Consequent	Antecedents	Support %	Confidence %
Impulse purchases in e-commerce	"Do product 'reviews' help you decide if you make a purchase?"	75,49	100
	"Is the information presented on social media totally reliable?"		
	"If a 'friend' follows a page they are likely to follow it too?"		
	"If a 'friend' made a purchase at the company, is it likely to make it?"		
	"Are most purchases 'recommended' by your friends?"		

Table 13.4 shows 56.373% supportive relationships, i.e., impulse purchases in e-commerce always happen in conjunction with "Do product 'reviews' help you decide if you make a purchase?" "Information Are Social Media Presentations Completely Reliable?" "If a 'friend' following a page is likely to follow it too?" "If a 'friend' made a purchase at the company is likely to do so?, "Confidence of friend comments on social networks, and "Most purchases are 'recommended' by your friends?" with 100% confidence.

Table 13.4. Impulse purchases in e-commerce

Consequent	Antecedents	Support %	Confidence %
Impulse purchases in e-commerce	"Do product 'reviews' help you decide if you make a purchase?"	56,373	100
	"Is the information presented on social media totally reliable?"		
	"If a 'friend' follows a page they are likely to follow it too?"		
	"If a 'friend' made a purchase at the company, is it likely to make it?"		
	"Trusting friends' comments on social networks"		
	"Are most purchases 'recommended' by your friends?"		

Subsequently, for the analysis of the impulse buying model in the traditional commerce, the following variables were defined as consequent and antecedent:

- Consequent: "Impulse buying in traditional commerce";
- Antecedents:

  1. "Do product 'reviews' help you decide if you make a purchase?";
  2. "Is the information presented on *social media* totally reliable?";
  3. "Most purchases are 'recommended' by your friends?".

From the execution of the Apriori algorithm, the following results were obtained and presented in Table 13.5.

Table 13.5. Impulse buying association rule in traditional commerce

Consequent	Antecedents	Support %	Confidence %
Impulse buying in traditional commerce	"Do product 'reviews' help you decide if you make a purchase?"	75,49	100
	"Is the information presented on social media totally reliable?"		
	"Are most purchases 'recommended' by your friends?"		

From Table 13.5 it can be concluded that all respondents who make impulse purchases in traditional commerce have 75.49% support for "Product reviews help you decide if you make a purchase" "Product information presented on social media is totally reliable" and "Are you likely to buy a product if the 'friend' advised you?" with 100% confidence.

13.5 Conclusions and Future Work

Based on the analysis performed, it is possible to conclude that the Portuguese consumer is influenced by the recommendations of "friends" to make their purchases, both online and in person. The following conclusions were obtained for impulse purchases in e-commerce:

- These happen mostly in the same company that the "friend" ran it;
- These happen mostly when the "friend" advises the purchase of a product and on the page that was advised;
- Confidence in online payment systems increases when e-commerce impulse purchases also increase;
- Participation in online communities does not influence impulse purchases in e-commerce, as well as reviews of a product and the information presented on social media.

About Social Shopping in Portugal we can conclude the following:

- "Friends" recommendations are used to support impulse buying in both traditional and e-commerce;
- Being part of an online community is not about impulse buying in e-commerce;
- Confidence in online payments has strong support with impulse buying in e-commerce.

In this article, we intent to contribute to the debate on social networks and electronic commerce, by executing a study about the potential and advantages of Social Shopping in Portugal. As future work, we aim to extend the study to other countries and social contexts.

References

1. Corrocher, N.: The adoption of Web 2.0 services: an empirical investigation. In: Technological Forecasting & Social Change, vol. 78, pp. 547–558 (2011)
2. Jackson, M.O., Yariv, L.: Diffusion, Strategic Interaction, and Social Structure (2010)
3. Junsang, L., Sharon, E.: Factors affecting couples' decisions to jointly shop. J. Bus. Res. **64**, 774–781 (2011) (Julho 2011)
4. de Almeida, S.O., Mazzon, J.A., Dholakia, U.M., Muller Neto, H.F.: Os efeitos da participação em comunidades virtuais de marca no comportamento do consumidor: proposição e teste de um modelo teórico. Revista de Administração ContemporâNea. Curitiba. **15**(3), 366–391 (2011)
5. Kim, D.J., Ferrin, D.L., Rao, H.R.: A trust-based consumer decision-making model in electronic commerce: the role of trust, perceived risk, and their antecedents. Decis. Support Syst. **44**(2), 544–564 (2008). https://doi.org/10.1016/j.dss.2007.07.001
6. Park, C., Kim, Y.: Identifying key factors affecting consumer purchase behavior in an online shopping context. Int. J. Retail Distrib. Manag. **31**(1), 16–29 (2003). https://doi.org/10.1108/09590550310457818
7. Kolesar, M.B., Galbraith, R.W.: Internet research a services-marketing perspective on e-retailing: implications for e-retailers and directions for further research. Internet Res. (2003). https://doi.org/10.1108/10662240010349444
8. Lohse, G.L., Spiller, P.: Electronic shopping. Commun. ACM **41**(7), 81–87 (1998). https://doi.org/10.1145/278476.278491
9. Spiller, P., Lohse, G.L.: A classification of Internet retail stores. Int. J. Electron. Commer **2** (2), 29–56 (1997)
10. Griffith, D.A., Krampf, R.F., Palmer, J.W.: The role of interface in electronic commerce: consumer involvement with print versus on-line catalogs. Int. J. Electron. Commer **5**(4), 135–153 (2001). https://doi.org/10.1080/10864415.2001.11044219
11. Hoque, A.Y., Lohse, G.L.: An information search cost perspective for designing interfaces for electronic commerce. J. Mark. Res. **36**(3), 387 (1999). https://doi.org/10.2307/3152084
12. Alba, J., Lynch, J., Weitz, B., Janiszewski, C., Lutz, R., Sawyer, A., Wood, S.: Interactive home shopping: consumer, retailer, and manufacturer incentives to participate in electronic marketplaces. J. Mark. **61**(3), 38 (1997). https://doi.org/10.2307/1251788
13. Reynolds, J.: eCommerce: a critical review. Int. J. Retail Distrib. Manag. **28**(10), 417–444 (2000). https://doi.org/10.1108/09590550010349253

14. Scrivens, E.: The role of social marketing in influencing consumer behaviour. Soc. Policy Adm. **18**(2), 179–188 (1984)
15. Kotler, P., Zaltman, G.: Social marketing: an approach to planned social change. J. Mark. **35** (3), 3 (1971). https://doi.org/10.2307/1249783
16. Dholakia, U.M., Bagozzi, R.P., Pearo, L.K.: A social influence model of consumer participation in network- and small-group-based virtual communities. Int. J. Res. Mark. **21** (3), 241–263 (2004). https://doi.org/10.1016/j.ijresmar.2003.12.004
17. Boyd, D.M., Ellison, N.B.: Social network sites: definition, history, and scholarship. J. Comput. Mediat. Commun. **13**(1), article 11 (2007). http://jcmc.indiana.edu/vol13/issue1/boyd.ellison.html
18. Social shopping: Disponível em: http://en.wikipedia.org/wiki/Social_shopping (2011)
19. Sanjukta, P., Koesler, K.: Cultural influence on consumers' usage of social networks and its' impact on online purchase intentions. J. Retail. Consum. Serv. **18**(2011), 348–354 (2011)
20. Levi's 'Likes' Facebook a Whole Lot: Launches 'Friends Store' on Jeans: http://www.clickz.com/clickz/news/1710543/levis-likes-facebook-whole-lot-launches-friends-store-jeans-site (2011)
21. MIND, Marketing Internet Digital: http://www.marketinginternetdigital.com.br/blog/redes-sociais-e-cases-de-sucesso/ (2009)
22. Web 2.0 como ferramenta de negocio- CIO – Estratégias de negócio e TI para líderes corporativos Disponível em: http://cio.uol.com.br/gestao/2009/08/18/casos-de-sucesso-a-web-2-0-como-ferramenta-denegocio/ (2009)
23. Publico: Os trabalhadores da minha empresa têm Facebook. E agora? Disponível em: http://www.publico.pt/Economia/os-trabalhadores-da-minha-empresa-tem-facebook-eagora_1430670 (2010)
24. Turismo: sol, praia e uma estratégia dá muito jeito: http://agulha.camorim.eu/tag/redes-sociais (2010)
25. Santos, V., Gonçalves, R.: Social shopping model proposal. In: The 7th Mediterranean Conference on Information Systems, Adopting Emergent Knowledge and Technologies to Develop Innovative Information Systems (CloudWisdom), (MCIS), [ISI]—Guimarães, Portugal, 8–10 Sept 2012

Chapter 14
The Role of Digital Influencers on Buying Intention

Rita Silva[1], Ana Sampaio[2], and Pedro Rodrigues[1(✉)]

[1] ISMAI, Maia University Institute, Maia, Portugal
{catiasteixeiraramos,pmrodrigues}@gmail.com
[2] CEDTUR ISMAI, Maia University Institute, Maia, Portugal
asampaio@ismai.pt

Abstract. This paper aims to understand the role of digital influencers in consumer and brand loyalty. In line with this, a study was performed with 261 respondents over 16 years old and users of the social networks evaluating trust in influencers, user-generated content, perceived risk and buying intention with specific scales. All collected data was treated with SPSS statistical software and also Smart PLS software performing structural equation model analyses. The results presented the UGC influence on the purchase intention and also the trust in the influencers.

14.1 Introduction

The goal of this work is to understand if digital influencers have a significative impact on buying intention of others. This appreciation of the impact of digital influencers on the buying intention should be evaluated based on the trust that the consumers have in this influencer as well as Perceived Risk and User-Generated Content (UGC). Thus, this paper aims to answer the following questions:

- Do consumers rely on digital influencers?/Do consumers feel any risk when they buy something suggested by a digital influencer?/To what extent does the UGC influence consumers' buying intentions?
- Do consumers create greater affinity with brands when they adopt UGC?
- How should brands act in relation to digital influencers?

In order to have answers, the study will be direct to all users of social networks.

14.2 Marketing 1.0, 2.0, 3.0–4.0

Marketing is a discipline that has evolved a lot since its inception. Although marketing has been present for a long time now, compared to other management areas is still considered relatively new.

Á. Rocha et al. (eds.), *Marketing and Smart Technologies*, Smart Innovation, Systems and Technologies 167, https://doi.org/10.1007/978-981-15-1564-4_14

In the age of marketing 1.0, marketing was about selling the factory products to everyone who wants to buy them [1]. The objective was to make products available to consumers who were passive and had no opinion.

In the early 1990s, in the Information Age, Marketing 2.0 emerged, consumers were more informed, aware, educated and with the possibility and capability of comparing similar products. At this stage, attention is paid to the customer and to the need to satisfy their desires. Marketing tools became more interactive, and the content was more dynamic combining social and technical aspects [2].

The following era was named Marketing 3.0 and it was value-oriented. Consumers are no longer simple customers and marketers treat them as full human beings: with mind, heart and spirit [1].

Marketing 4.0 marks the era of transition from traditional marketing to digital marketing [3]. Digital marketing is characterized by any form of marketing that is done through new technologies, using channels of distribution of databases to reach consumers [4]. In a high-tech world, people are looking for proximity, so in this age, products become more personalized and less mass-skilled. In this new era of marketing 4.0, there have been a number of changes that radically modified the world and the way brands operate, so horizontal, inclusive, and social forces have triumphed over vertical, unique and individual forces [5]. Today, the market is becoming more inclusive, with social networks contributing to the elimination of geographical and time barriers, and what matters most is collaboration. At the social level, inclusiveness is seen, for example, in fair trade, acceptance of women's capacity and diversity of employment. Another concept that is no longer vertical is consumer trust. This concept is now horizontal because today's consumers believe less in marketing campaigns and more in their friends, family or Facebook. Therefore, consumers should be seen as peers and friends of the brand [5] and the brand must be sincere and authentic.

14.3 User-Generated Content

UGC refers to content created by users who want to share their ideas with other users through different online media. Online platforms such as YouTube, Wikipedia, Facebook or Trip Advisor are some examples of these platforms where users provide material to add content to them. UGC is the material that a platform obtains from its own users [6] and can be the creation of videos for YouTube, or a review of a restaurant on TripAdvisor.

There are different motivations that drive users to create content [7], state that these motivations can be divided into two categories: intrinsic, when people do it because it is interesting, enjoyable or satisfying; and extrinsic, when people expect some kind of reward for doing something.

14.4 Perceived Risk

Perceived risk involves two elements: uncertainty and consequences. Consumer behaviours involve risk because the buying action "will produce consequences that we can't anticipate and others that may be unpleasant" [8]. According to [9] risk is defined as an attribute of an alternative decision reflecting the variance of its possible results. When it comes to perceived risk, nowadays is closely associated with online shopping. Reference [10] show seven types of risks that consumers experience when they buy online: Financial risk; Performance risk; Risk of time; Risk of delivery; Risk of privacy; Psychological risk and Social risk.

14.5 Buying Intention

Buying intention is the consumer's preference to buy a product or service. There are several factors that affect consumer intention in selecting the product and the final decision depends on consumer intention and external factors [11]. These factors are the consumer's knowledge of the product or brand, the perception of other consumers, the attractive design of the packaging and the approval of celebrities.

14.6 Trust in the Influencer

With the emergence of online marketing, Internet users/consumers and social networks personal owners are in contact with digital influencers. A digital influencer is an individual who attracts an online audience and influences the behaviours, opinions and values of others [12]. However, for digital influencers to be credible they must also be reliable. Trust is very important, and it is also more difficult in the online environment than face to face [13]. Thus, trust in the online environment is also at risk, and therefore vulnerability. Trust is affected by different subjective circumstances and depends on the context of the situation.

14.7 Research Methodology

Following the literature review, it was possible to elaborate the hypotheses of this investigation presented in Fig. 14.1.

The approach of this study is quantitative, and it is inspired by the positivist paradigm. The quantitative approach represents a set of processes that is sequential and probative. Each stage precedes the next and you cannot skip steps [14]. Quantitative research occurs in the "external reality of the individual" [14].

This research also has a descriptive methodology. Descriptive studies seek to specify the properties, characteristics and profiles of people, groups, units, processes, objects or any other phenomenon that is subject to analysis [14].

In this study, the data collection method used was the Questionnaire, elaborated in *Google Forms*. In this case, there are questions with two possible answers to choose

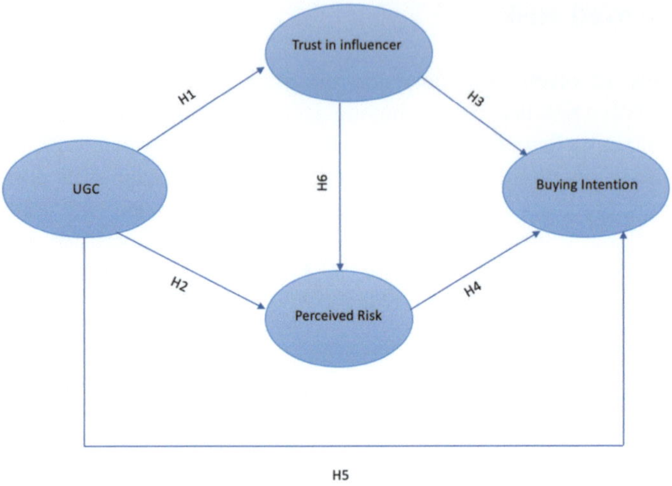

Fig. 14.1. Conceptual model

from and questions with more than two options. Then, there are other questions on specific scales used to measure the variables on study: Trust in influencer, Perceived Risk, UGC and Buying Intention. The scale used was the Likert scale where each construct is measured with five points ranging from totally disagree to fully agree.

14.8 Data Analysis

14.8.1 Reliability Analysis and Factor Analysis

In order to evaluate the validity of EFA, KMO values (Kaiser-Meyer-Olkin) we used the reference values of [15]. Regarding the variables, Trust in the influencer, Perceived Risk and Buying Intention, the results have medium values. Concerning UGC, the result was positive. In all variables Barlett's tests have a *p*-value of less than 0.005, demonstrating that the variables are correlated—see Table 14.1.

Table 14.1. KMO values

Construct	KMO	Bartlett test (*p*-value)	Test explained variance (%)
Trust in influencer	0.717	0.000	44.52
Perceived risk	0.647	0.000	66.09
Buying intention	0.659	0.000	61.64
UGC	0.851	0.000	58.78

14.8.2 Convergent and Discriminant Validity

Discriminant validity is the measure of the level at which a structure in a measurement model differs from other structures. It is an indicator of a low relationship between the items that form a construct and other items that form another construct [16]. Convergent validity indicates that the relationships between items that constitute a construct are high. The Cronbach value is calculated for each dimension separately. Values above 0.7 indicate that the internal reliability of the scale used is sufficient. In this study, Cronbach values are in agreement with the reference values. As for the values of composite reliability, when equal or beyond the limit of 0.7 show that there is composite reliability [16], which is the case in this investigation. Considering that AVE value should be greater than 0.50 or 0.50 [16], in this study all the constructs have values with a stroke greater than 0.50, except in the Trust in the influencer construct that is slightly lower (0.44). Finally, for RHO_A, values must be equal or greater than to 0.70, and in this investigation, all constructs have values that are in agreement with the reference values.

14.8.3 Structural Equation Model (SEM)

The validated and statistically significant hypotheses are H1, H3 and H5. UGC variable has a large positive influence on Trust in the influencer (H1: $\beta = 0.513$ $p < 0.05$) and also has a strong positive influence on the Buying intention (H5: $\beta = 0.614$ $p < 0.05$). With less significance, Trust in influencer that influences Buying intention is also highlighted (H3: $\beta = 0.174$ $p < 0.059$). The hypotheses H2 ($\beta = 0.95$ $p > 0.05$), H4 ($\beta = 0.012$ $p > 0.05$) and H6 ($\beta = 0.081$ $p > 0.05$) will no longer be considered in this study because they prove not to be statistically significant. As for the size of the effects (F2), the standard values are—0.03—low effect; 0.15—mean effect, 0.35—high effect. Thus, there are low effects on H2, H3, H4 and H6. Hypotheses H1 and H5 have high effects—see Fig. 14.2.

14.8.4 Hypothesis Validation

In Table 14.2 is presented the hypothesis validation.

14.9 Discussion

The sample of this study consists of 261 individuals and the majority of them are female. The sample is relatively young because the majority of respondents are in the 16–25 age group (54%) followed by 22% of the respondents in the range of 26–35 years. Data showed that most of the individuals are graduated, professionals, working on behalf of others and have an average monthly remuneration between € 501 and € 1000. All the respondents use frequently social networks; however, the three most used are Facebook, Instagram and Youtube. This data goes in accordance with the data of the QualiBest Institute which shows that the preferred social networks are: Facebook with 92%, Youtube, 90%, Instagram 73% and Twitter with 31%.

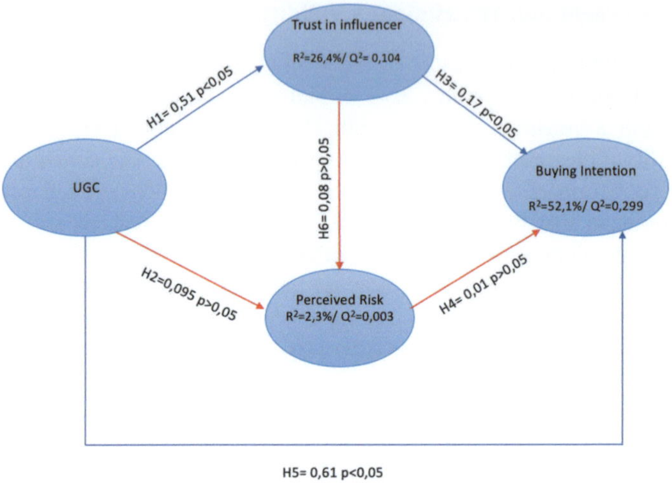

Fig. 14.2. Model in study

Table 14.2. Hypothesis validation

Hypothesis	Value	Validation
UGC has a positive impact on Trust in influencer	0.513	Validated
UGC has a positive impact on perceived risk	0.095	Not validated
Confidence in influencer has an influence on buying intention	0.174	Validated
Perceived risk has influence on buying intention	0.012	Not validated
UGC has an impact on buying intention	0.614	Validated
Trust in influencer has influence on perceived risk	0.081	Not validated

This study evaluated and analysed four variables, all of them assigned to a conceptual model. Regarding the variable confidence in the influencer, the results indicate that UGC has a 51% impact on Trust in the influencer and this shows that consumers rely more on their peers and on the content generated by an anonymous users than on the digital influencers and in the public figures of traditional ads. This is in line with what [5] said, that consumers nowadays believe less in marketing campaigns and more in their friends, family or Facebook.

Regarding the variable Trust, the sample showed to be afraid to answer because there are many answers with "I don't agree or disagree". Overall, respondents were neutral and did not fully show whether or not they relied on digital influencers.

Concerning Perceived Risk, respondents generally do not feel a lot of risk when buying something recommended by digital influencers. Overall, individuals feel the risk of buying something just because it was suggested by a digital influencer, mainly because they may have been paid so rather than because it may not live up to expectations. Therefore, for the respondents the greatest risk is the psychological risk that has to do with fears that are in the minds of the consumers when buying a product [17].

With regard to Buying Intention, the sample showed that individuals do not feel that digital influencers impact their purchases. Individuals refer that there are products and services that they do not know and only get to know when they were spoken by digital influencers. Respondents show that they do not buy products or services just because they are spoken by digital influencers they trust, which shows that, while being a digital influencer for whom respondents are fond of, knowledge and trust, they do not buy something they do not need. In fact, according to a study by Stackla, only 8% of consumers say that content created by influencers influences their purchasing decisions. The data show that UGC influences 61% the Buying Intention, that is, UGC influences more the Buying Intention than any digital influencer, where the influence is only 17%.

Regarding the UGC variable, although in the survey results the sample said they did not agree with many of the statements and thus did not feel influenced to buy just because it is user-created content, in the scale evaluation the contrary was found. This may have happened for two reasons: first, respondents do not like to feel or do not admit that their attitude has been manipulated or influenced; second they are not aware of the influence of UGC in their Buying Intention and since they do not have this awareness, they say they do not agree. However, the evaluation of the scale shows that the UGC is the variable that has the most influence on the other variables and influences in 51% Trust in the influencer and in 61% Buying Intention. A report by the Turn To Hearing the Voice of Consumer-UGC and the Commerce Experience revealed that 90% of US consumers report that user-generated content is the most influential part of their purchasing decisions and this is in line with the data obtained.

In conclusion, this study found that the sample relies far more on content created by users than on digital influencers and public figures in traditional ads. Yet they were quite neutral about trust. Respondents feel more of a risk perceived by influencers being paid to advertise a product or service than by the quality of that product or service.

14.10 Conclusions

This research is meant to understand the impact of digital influencers on the consumer and buying intention.

As far as digital marketing is concerned, consumers are looking for a differentiated and appropriate product for themselves. So, the consumer is currently rigorous with regard to their purchases. Although digital influencers are a good strategy for bringing brands closer to the consumer, consumers still do not feel totally confident and determined to buy into this new form of advertising, as far as the sample of this study is concerned. Regarding Perceived Risk, the biggest risk felt by the sample is the risk that digital influencers are advertising a product just because they are paid for it and not because they really like that product. As for UGC, in fact, it has a great influence on the Intention to buy (61%), although the respondents do not feel that they are influenced, the scale evaluation shows us that UGC has more influence on the buying intention than any digital influencer. UGC shows an influence of 51%, in the trust in the influencer, and trust in the influencer only impacts 17% in the buying intention.

While there is a large sample of this study following digital influencers, there is still a long way to go for digital influencers and the brands themselves to create mechanisms to implement trust and get consumers to believe and actually buy products or services advertised by them.

References

1. Kotler, P., Armstrong, G.: Principles of Marketing, 13th edn. Pearson Education, New Jersey (2010)
2. Erragcha, N., Romdhane, R.: New faces of marketing in the era of the web: from marketing 1.0 to marketing 3.0. J. Res. Mark. **2**, 137–142 (2014)
3. Carvalho, S.: Marketing: evolução e tendências. Revista Ciêntifica Multidisciplinar Núcleo do Conhecimento **01**, 5–20 (2018)
4. Todor, R.: Blending traditional and digital marketing. Bull. Transilv. Univ. Bras. **9** (2016)
5. Kotler, P., Kartajaya, H., Setiawan, I.: Marketing 4.0: Moving from Traditional to Digital. Wiley, New York (2017)
6. Luca, M.: User-generated content and social media. Obtido de Havard Business School. http://people.hbs.edu/mluca/UGC.pdf (2015)
7. Mckenzie, J.: User generated content: overview, current state and context. Peer Rev. J. Internet **17** (2012)
8. Chen, Y., Yan, X., Fan, W.: Examining the effects of decomposed perceived risk on consumer online shopping behavior: a field study in China. Inzinerine Ekonomika—Eng. Econ. **26**, 315–326 (2015)
9. Gefen, D., Rao, V.S., Tractinsky, N.: The conceptualization of trust, risk and their relationship in electronic commerce. The need for clarifications. In: Proceedings of the 36th Hawaii International Conference on System Sciences (HICSS 03), Hawaii, USA (2002)
10. Ariff, M., Sylvester, M., Zakuan, N., Ismail, K., Ali, K.: Consumer perceived risk, attitude and online shopping behaviour; empirical evidence from Malaysia. Conf. Ser. Mater. Sci. Eng. (2012)
11. Younus, S., Rasheed, F., Zia, A.: Identifying the factors affecting customer purchase intention. Glob. J. Manag. Bus. Res. Adm. Manag. **15**(2) (2015)
12. Lampeitl, A., Åberg, P.: The Role of Influencers in Generating Customer-Based Brand Equity and Brand-Promoting User-Generated Content. Department of Business Administration, Lund University, School of Economics and Management (2017)
13. Bauman, A., Bachmann, R.: Online consumer trust: trends in research. J. Technol. Manage. Innov. **12**(2) (2017) (Santiago)
14. Sampieri, R., Collado, C., Lucio, P.: Metodologia de la Investigación. 6° edición (2014)
15. Maroco, J., Silva, D., Rodrigues, A., Guerreiro, M., Santana, I., Mendonça, A.: Data mining methods in the prediction of Dementia: a real-data comparison of the accuracy, sensitivity and specificity of linear discriminant analysis, logistic regression, neural networks, support vector machines, classification trees and random forests. BMC Res. Notes **4**(1), 299 (2011)
16. Civelek, M.: Essentials of Structural Equation Modeling. Zea E-books, Lincoln, Nebraska (2018)
17. Dowling, G.: Perceived risk. In: Earl, P.E., Kemp, S. (eds.) The Elgar Companion to Consumer Research and Economic Psychology, pp. 419–424. Edward Elgar, Cheltenham (1999)
18. Coutinho, C.: Metodologia da Investigação em Ciências Sociais e Humanas: Teoria e Prática. Almedina (2011)

19. Ferreira, F.: O Papel dos Influenciadores Digitais no processo de Intenção de Compra dos Seguidores. Dissertação de mestrado. IPL, Escola Superior de Comunicação Social, Lisboa, Portugal (2018)
20. Margalina, V.: Manual para la creación de modelos de ecuaciones estructurales y análisis de resultados con SmartPLS (2016)
21. Pappas, N.: Marketing strategies, perceived risks, and consumer trust in online buying behaviour. J. Retail. Consum. Serv. **29**, 92–103 (2016)
22. Santos, S.: Métodos qualitativos e quantitativos na pesquisa biomédica. J. Pediatr. **75**, 401–406 (1999)
23. Swanson, R., Holton III, E.: Research in Organizations Foundations and Methods of Inquiry, I. 1st edn. Berrett-Koehler Publishers (2005)

Chapter 15
Driving Factors of Consumer Irrationality in Omnichannel Consumer Behaviour

Jurgita Radzevičė(✉) and Jūratė Banytė

KTU University, K. Donelaičio 73, LT-44029 Kaunas, Lithuania
jurgita.radzevice@ktu.edu

Abstract. The rapid enhancement in information and communication technology has brought a paradigm shift in the retailing market and consumer behaviour. Consumer behaviour in the omnichannel retailing context becomes increasingly more complex, as omnichannel consumers strategically choose where to gather information and buy products: online or offline. Insights into the drivers of channel choice need to be found. Consumers do not think of channels in isolation but rather combine them and make decisions based on their mood and lifestyle demands (Blázquez in Int J Electron Commer 18(4):97–116, 2014 [4]). Contributing factors of irrational consumers behaviour identification that determine their decision is treated as a new approach promoting the development of a standard of cognition of consumer behaviour. Paper aims to substantiate driving factors of consumer's irrationality in the context of omnichannel consumer behaviour. Guided by the Stimulus-Organism-Response (S-O-R) model, consumer irrationality and omnichannel behaviour literature, the study developed a theoretical framework that explores three types of driving factors, including price orientation, product category and convenience (Stimulus), consumer irrationality (Organism) and omnichannel behaviour, which is described through information search and purchase decision (Reaction).

15.1 Introduction

Traditionally, knowledge of consumer behaviour is associated with the environment in which he takes decisions. The inertial development of technologies is a key factor in promoting the evolution of commerce channels and consumers migration between them, without time and space limitations [20]. Under these conditions, consumers are described as "always-on", as communication with the retailer during consumer journey takes place even at several points of sale both physical and virtual [11]. Digital interactions influence 36 cents of each dollar spent in the retail stores, with an estimated value of over one trillion USD. The web already influenced 60% of retail sales in 2018 [40]. Therefore, creating an omnichannel strategy for an omnichannel consumer is critical for businesses if company wants to leverage the business potential of this interconnected customer behaviour [23].

Omnichannel's domination is one of the most recent coups in retail, with implications for both practice and theory [31]. While most studies focus on specific behaviour in omnichannel context [18], in the competitive omnichannel environment

Á. Rocha et al. (eds.), *Marketing and Smart Technologies*, Smart Innovation, Systems and Technologies 167, https://doi.org/10.1007/978-981-15-1564-4_15

retailers need to understand what drives the behaviour of omnichannel consumers [14]. Our first task is to prove, that an integrative approach considering various driving factors is necessary for a better understanding of how consumers behave and how they choose channels in the inertial uptake of technologies.

Consumers make decisions based on rational criteria or a wizard of emotions, which may lead to different experiences from consumer perceptions [22]. Xinhui and Han [41] research results indicated that the attitude and intention of consumer, the marketing promotion, and the guidance of consumer emotion have an obvious significance on consumer irrational purchase behaviour. Van der Horst and Matthijsen [37] have confirmed that only 5% consumer behaviour is conscious and planned, and the remaining 95% involves irrational actions. Consumer emotions are rarely discussed or examined in the extant literature of omnichannel, representing a major gap in the body of knowledge about behaviour in omnichannel context [34]. Hence becomes relevant assumption of research viability on consumers irrationality in omnichannel consumer behaviour.

Therefore, the aim of this study is to substantiate the driving factors of consumer irrationality in the context of omnichannel consumer behaviour. Contribution of this study authors to consumer behaviour research is enhanced by theoretical framework, which is developed based on the adaptation of the Stimulus-Organism-Response (S-O-R) model, consumer irrationality, and omnichannel behaviour literature. Authors settled conceptual model with irrationality as an Organism, its causes, as decisive factors in omnichannel context—as Stimulus and omnichannel consumer behaviour as a Result.

15.2 Theoretical Background

15.2.1 Concept of Omnichannel Consumer Behaviour

From scientific point of view, omnichannel behaviour is a new subject of consumer behaviour, in which investigation is testified by several studies. Omnichannel consumers are evolved multichannel users who want to use all channels (physical store, catalogue, call centre, e-shop, and mobile devices) at the same time, not in parallel. The idea of channel evolution is further explored by Mosquera et al. [27], who claim that omnichannel is the last step in the evolution of channels, where the aggregation of physical and digital environments is creating whole consumer experience. The use of multi-, cross-, and omnichannel terms in scientific literature is not clear. According to Beck and Rygl [3], different authors use the term “multichannel retailing” to define channels that are integrated, nonintegrated, interacting, or individual. In their view, the main differences between these concepts are the extent of user interaction and channel integration controlled by the retailer. Carried out scientific literature analyses suggest that omnichannel’s term in a common source is described in a similar way: it emphasizes consumer-friendly time, multiple devices, different platforms, continuity, integrity and consistent user experience. Despite extensive research having been conducted on deciphering the determinant of consumers’ channel preference, past studies tended to treat online and offline channels as competing alternatives.

Counter, omnichannel behaviour represent an unprecedented opportunity in channels synergies in synchronizing their interactions with consumers across multiple touchpoints [6]. With the proliferation of mobile technologies and social media, consumer journey has become more complex: the simultaneous use of different channels has accelerated the emergence of new types of consumer behaviour, such as showrooming and webrooming. These are two types of consumer behaviour, where physical and virtual channels change roles during the decision-making process [16]. Referring to traditional definitions, showrooming consists of searching for product information offline and purchasing online [39]; webrooming consists of searching for product information online and purchasing offline [12]. So essential stages for the particular omnichannel behaviour defining remains information search and purchase, which will be hereinafter used in this study.

15.2.2 Omnichannel Consumer Behaviour Driving Factors

The examination of driving factors is central to understand consumer behaviour [35]. A large amount of studies on determining the drivers of channel choice across the stages of the decision-making process have received attention in multichannel research for two decades [18, 21, 33]. According to Marmol and Fernandez [24], the most highlighted are stock availability, personalization, atmosphere, price and convenience. However, none of these existing studies has ever investigated drivers entirely, systematically combining them and analysing which of these variables drive consumer behaviour in omnichannel context most. To examine what influences consumers' decision for different omnichannel behaviour, the following Table 15.1 gives the summary of the newest scientific literature of driving factors in omnichannel shopping context. The study builds on and complements analysed literature by systemising, deriving and proposing essential drivers that may explain the different omnichannel behaviours.

Studies, investigating driving factors, are built on different theoretical foundations like information processing and uncertainty reduction theory [33], Person-Object-Situation paradigm [18], Critical lens of value co-destruction [8]. Driving factors of channel choice mostly are examined in one context, and only few studies revealing them in both omnichannel inherent behaviours [24, 32]. The most common drivers in analysed literature are price orientation, convenience, need for touch, availability, product category, perceived risk, security and trust. This study complements and advances the theoretical foundation of Daunt and Harris [8], by proposing established structure for showrooming to adjust for whole omnichannel shopping context and hereinafter to assign driving factors to product characteristics (product category, price orientation and availability) consumer characteristics (convenience, need for touch and perceived risk) and channel characteristics (security and trust).

15.2.3 Consumers Irrationality Phenomenon in Omnichannel Consumer Behaviour Context

Despite the growing role of big data analytics within retail, over three quarters (76%) of data experts admit that consumers are fundamentally irrational when making purchases

Table 15.1. Driving factors of omnichannel consumer behaviour

Author(s)	Shopping context	Driving factors	Research type	Method
Santos and Gonçalves [33]	Webrooming	Information attainment Price comparison orientation Consumer empowerment Need for touch Perceived risk Confidence	Empirical	263 respondents, personal and online surveys
Kim et al. [21]	Webrooming	Price consciousness orientation Convenience orientation Shopping enjoyment orientation Impulse buying orientation	Empirical	700 respondents, online survey
Heitz-Spahn et al. [18]	Omnichannel	Personal characteristics: objective and domain-specific Product categories: utilitarian and hedonic	Empirical	750 respondents, online survey
Daunt and Harris [8]	Showrooming	Product characteristic: technological speed of change, acquisition value, price, availability Consumer characteristic: product involvement, in-store shopping savviness, Internet savviness Channel characteristic: trust in in-store sales employees, trust in online stores, value of in-store shopping, value of online shopping		
Marmol and Fernandez [24]	Webrooming, Showrooming	"Touch and feel" product Convenience orientation Price consciousness orientation, avoid shipping costs Availability of the product Personal interaction with sales employees	Conceptual	30 reports from consulting companies
Nesar and Bin Sabir [28]	Omnichannel	"Touch and feel" product Return policy Advantage of in-store discount Convenience Online store availability	Empirical	378 respondents, questionnaire
Hilken et al. [19]	Omnichannel	Personal characteristics: need for touch, mental imagery abilities, product use experience	Conceptual	

(*continued*)

Table 15.1. (*continued*)

Author(s)	Shopping context	Driving factors	Research type	Method
Reid et al. [32]	Webrooming, Showrooming	Price criteria Convenience Time savings Tactile selection criteria Financial implications Inconvenience of unsuitable products	Empirical	109 respondents, online survey
Dohana et al. [9]	Showrooming	Consumer characteristics: involvement, prior knowledge, perceived risk, price consciousness Internet usage, access device usage, certain demographic variables	Empirical	500 respondents, online survey

[7]. This highlights the importance of irrational consumer perception as a consumer behaviour variable cognition in retailing and determines the extension of research for improving the standard of consumer behaviour awareness by extending the limits. Existing research into irrationality in consumer decisions reveals fragmented findings often lacking conceptual grounding and impossible to apply in practice. An irrational consumer behaviour study is usually the result of an analysis of the assumptions that induce irrationality of consumers. Initial analysis of the various dimensions of consumer irrationality are analysed by various authors reveals that the interpretation of irrationality in the course of time became increasingly fierce. According to Elliott [10] irrational consumer is a consumer whose experience relies on emotions. He is inconsistent and governed by unconscious desires. Irrational consumer is open, passive, and highly vulnerable by external factors influencing purchasing decisions. According to Martin and Morich [25], popular descriptions of the brain's unconscious processes include "emotional", "irrational" and "unconscious" terms, that minimize the importance, richness and complexity of mental processes that are not available to unconscious inspection.

Frasquet et al. [13] investigated how companies can design digital channels to evoke desired emotions. Findings from this study provide a new empirical support for the proposition that consumers responses based on emotions can be utilized to guide company digital strategy for building digital channel relationships with customers. Consumers' perceptions and behaviours are often moderated by situations that consumers have at that moment [15]. In the context of the investigated phenomenon, it is important to note that when discussing the omnichannel environment it is stated that omnichannel consumers are more impulsive and so demand a faster purchasing process [36]. As the relationship between irrationality and impulsiveness is already determined in several studies [17, 29], it becomes relevant assumption of research viability on consumers irrationality in omnichannel consumer behaviour. In order to cognize this phenomenon in more advance, it is important to analyse the reasons for consumer irrationality in the omnichannel context. As omnichannel behaviour driving factors are already revealed, it is essential to find out which of them are relevant to the impact of consumer irrationality.

15.3 Conceptual Framework

The theoretical framework of the present study is developed based on the adaptation of the Stimulus-Organism-Response (S-O-R) Model [26], consumer irrationality and omnichannel behaviour literature. The authors developed a theoretical framework that explores individual perceptions in irrationality traits and influencing factors to the consumer as they relate to irrationality-based omnichannel information search and purchase decisions, as illustrated in Fig. 15.1. Conceptual model settled with Irrationality as an Organism, its reasons, as decisive factors in omnichannel context—as Stimulus and omnichannel consumer behaviour as a Result.

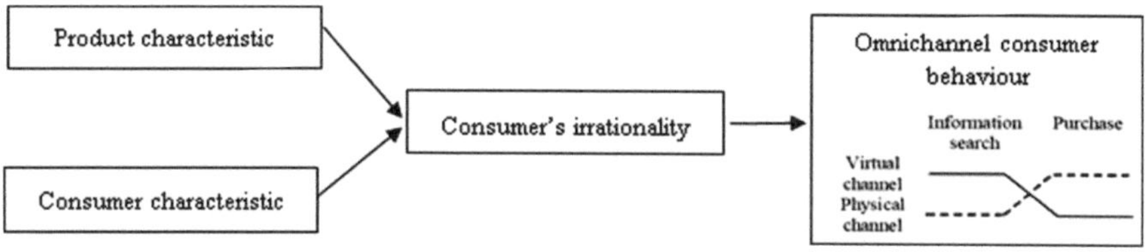

Fig. 15.1. Conceptual framework

Within the S-O-R framework, the influence of identified factors (stimulus) on behaviour is manifested through the consumers irrationality. Authors in 15.2.2. part distinguished three characteristics and assigned them as the most common driving factors of omnichannel consumer behaviour. After analysis, it was found that irrationality is associated with driving factors attributed to product and consumer characteristics. According to the current scientific literature of consumer irrationality analyses, it was found that irrationality is associated with driving factors attributed to product and consumer characteristics. The most identifiable bundles with irrationality are found in price orientation, product category and convenience driving factors. More detailed analysis is given below.

Authors distinguished price orientation as one of product characteristics driving factors. The studies that have been carried out so far suggest that the contrast between consumer rationality and irrationality is most obvious in price perception [2]. According to Asamoah and Chovancova [1], consumer perceptions are inseparable from the perception of price. Price perception is directly influenced by perceived expectations, the systematic information and incentives of previous knowledge. According to Fernández et al. [11], the showrooming consumer, at least in terms of hedonic products, is more likely to purchase products of a higher value and price. Omnichannel consumers have specific characteristics that make them special: on average, they spend more money than conventional shoppers [38]. Thus, price perception may positively impact consumer irrationality in the omnichannel context:

P1a. There will be a positive relationship between price orientation and consumer irrationality in the omnichannel context.

The next product characteristic factor—product category can also influence consumer channel choice behaviour. Kim et al. [21] conducted an exploratory analysis using the search and experience product classification. They found out that customers with different omnichannel behaviour driving factors make different choices of search channel. For instance, offline consumers who are high on price consciousness orientation are more likely to choose an online search channel when purchasing search products that when shopping experience products. Heitz-Spahn et al. [18] research results show that consumers' channel choice behaviours tendencies may be based on utilitarian and hedonic product categories. Yadav and Varadarajan [42] state that omnichannel consumers are the most valuable segment for hedonic product categories. Park et al. [30] study reveal that web browsing has a positive effect on impulse buying for hedonic product category on shopping websites, as mentioned above, relationship between irrationality and impulsiveness is already determined in several studies [17, 29]. Therefore, it is predicted that product category will positively impact consumer irrationality in the omnichannel context.

P1b. There will be a positive relationship between product category and consumer irrationality in the omnichannel context.

Kim et al. [21] defined convenience as the degree to which consumers can easily and quickly perform shopping activities. They assert, when searching for information, mobile devices have higher convenience than stationary ones. In omnichannel driving factors analyse convenience orientation is one of the most common factors and is mentioned by several authors [21, 24, 28, 32]. According to Dohana et al. [9] study, access device usage is assigned to consumer characteristics, they state that consumers using mobile devices tend to showroom more frequently than those who use other types of devices. Mobile omnichannel individuals are more impulsive and so demand a faster purchasing process [36]. Therefore, to explore the relationship between mobile device usage and consumer irrationality in the omnichannel context, the following proposition is formed.

P2. There will be a positive relationship between convenience orientation by using mobile devices and consumer irrationality in the omnichannel context.

The second element of consumer's irrationality leads to the third element of omnichannel behaviour. Organism refers to internal processes that intervene between stimulus characteristics to the person and behaviours. These processes consist of perceptual, physiological, emotional and cognitive activities.

Lastly, responses are the outcome or final action toward organism, such as consumers' behavioural reactions. Omnichannel consumers do not think of channels in isolation but rather combine them and make decisions based on their mood and lifestyle demands [4]. In spite of extensive research having been conducted on long-term omnichannel customer-retailer relationships, only few of them contributes to omnichannel research by examining digitalization related to individual purchase trips and consumer reactions to omnichannel-based stimuli [5]. Therefore, to explore the consumer impact on their omnichannel behaviour, the following proposition is formed.

P3. Consumer irrationality has a positive impact on their omnichannel behaviour.

15.4 Discussion

Consumer irrationality and omnichannel consumer behaviour have analysed quite often in consumer behaviour research, but combination of these two phenomena are still lacking scientific attention. Existing research into irrationality in consumer decisions reveals fragmented findings often lacking conceptual grounding and impossible to apply in practice. An irrational consumer behaviour study is usually the result of an analysis of the assumptions that induce irrationality of consumers. Although studies examining driving factors of channel choice in omnichannel context have been conducted in the past [18, 21, 33] but specific research which of these variables drive consumer behaviour in omnichannel context most is lacking. Authors of this study distinguished product, consumer and channel characteristics and assigned to them the most common driving factors of omnichannel consumer behaviour. According to the current scientific literature of consumer irrationality analyses found that irrationality is associated with driving factors attributed to product and consumer characteristics. The most identifiable bundles with irrationality are found in product characteristics—price orientation [11, 38] and product category [30, 42], in consumer characteristics—convenience [36] driving factors. As revealed, the proposed S-O-R model builds up a few testable propositions which call for empirical verification. Future scope of the study—mobile devices penetration and in-store usage are expected strongly affect omnichannel behaviour of the consumers [21, 24] but this has not been addressed. An investigation to examine driving factors of customers irrationality to collect information in on channel and make purchase in another, or wise versa, can be conducted to enhance the understanding of—consumers' omnichannel behaviour. Nevertheless, the abovementioned limitations provide scope for carrying out future research for enhancing understanding of the consumer irrationality in omnichannel behaviour.

References

1. Asamoah, E.S., Chovancova, M.: The influence of price endings on consumer behaviour: an application of the psychology of perception. Acta Universitatis Agriculturae et Silviculturae Mendelianae Brunensis **59**(7), 29–38 (2014)
2. Banyte, J., Rutelione, A., Gadeikiene, A., Belkeviciute, J.: Expression of irrationality in consumer behavior: aspect of price perception. Eng. Econ. **27**(3), 334–344 (2016)
3. Beck, N., Rygl, D.: Categorization of multiple channel retailing in Multi-, Cross-, and Omni-Channel retailing for retailers and retailing. J. Retail. Consum. Serv. **27**, 170–178 (2015)
4. Blázquez, M.B.: Fashion shopping in multichannel retail: the role of technology in enhancing the customer experience. Int. J. Electron. Commer. **18**(4), 97–116 (2014)
5. Blom, A., Lange, F., Hess, R.L.: Omnichannel-based promotions' effects on purchase behavior and brand image. J. Retail. Consum. Serv. (2017) (Article in press)
6. Chen, Y., Cheung, C.M.K., Tan, C.W.: Omnichannel business research: opportunities and challenges. Decis. Support Syst. **109**, 1–4 (2018)
7. Clicktale Homepage. https://www.clicktale.com/company/in-the-media/consumers-are-fundamentally-irrational-when-they-shop/. Accessed 12 Jan 2019
8. Daunt, K.L., Harris, L.C.: Consumer showrooming: value co-destruction. J. Retail. Consum. Serv. **38**, 166–176 (2017)

9. Dohana, W.D., Shin, H., Katsumata, S.: Influence of individual characteristics on whether and how much consumers engage in showrooming behaviour. Electron Commer. Res. **18**, 665–692 (2018)
10. Elliott, R.: Existential consumption and irrational desire. Eur. J. Mark. **31**(3/4), 285–296 (1997)
11. Fernández, N.V., Sanzo-Perez, M.J., Vázquez-Casielles, R.: Webroomers versus showroomers: are they the same? J. Bus. Res. **92**, 300–320 (2018)
12. Flavián, C., Gurrea, R., Orús, C.: Choice confidence in the webrooming purchase process: the impact of online positive reviews and the motivation to touch. J. Consum. Behav. **15**(5), 459–476 (2016)
13. Frasquet, M., Descals, A.M., Ruiz-Molina, M.E.: Understanding loyalty in multichannel retailing: the role of brand trust and brand attachment. Int. J. Retail Distrib. Manag. **45**(6), 608–625 (2017)
14. Frasquet, M., Molla, A., Ruiz, E.: Identifying patterns in channel usage across the search, purchase and post-sales stages of shopping. Electron. Commer. Res. Appl. **14**, 654–665 (2015). https://doi.org/10.1016/j.elerap.2015.10.002
15. Grewal, D., Marmorstein, H., Sharma, A.: Communicating price information through semantic cues: the moderating effects of situation and discount size. J. Consum. Res. **23**(2), 148–155 (1996)
16. Guardia, F.R., Luna-Nevarez, C.: "Showrooming" in consumer electronics retailing: an empirical study. J. Internet Commer. **16**(2), 174–201 (2017)
17. Harnish, R.J., Bridges, K.R., Karelitz, J.L.: Compulsive buying: prevalence, irrational beliefs and purchasing. Int. J. Ment. Health Addict. **15**(5), 993–1007 (2016)
18. Heitz-Spahn, S., Yildiz. H., Belaud, L.: Investigating the factors driving channel choice and retailer choice in an omni-channel environment. https://www.cairn.info/revue-projectique-2018-3-page-43.htm. Accessed 25 June 2019
19. Hilken, T., Heller, J., Chylinski, M., Keeling, D.I., Mahr, D., Ruyter, K.: Making omnichannel an augmented reality: the current and future state of the art. J. Res. Interact. **12** (4), 509–523 (2018)
20. Juaneda-Ayensa, E., Mosquera, A., Sierra-Murrillo, Y.: Omnichannel customer behaviour: key drivers of technology acceptance and use and their effects on purchase intention. Front. Psychol. **28**(7), 11–17 (2016)
21. Kim, E., Libaque-Saenz, C.F., Park, M.C.: Understanding shopping routes of offline purchasers: selection of search channels (online vs. offline) and search platforms (mobile vs. PC) based on product types. Serv. Bus. **13**, 305–338 (2018)
22. Klaus, P., Maklan, S.: Towards a better measure of customer experience. Int. J. Mark. Res. **55**(2), 227–246 (2013)
23. Lazaris, C., Vrechopoulous, A., Fraidaki, K., Doukidis, G.: Exploring the "Omnichannel" shopper behaviour. In: Conference Paper. International Service Research Conference (2018)
24. Marmol, M., Fernandez, V.: Trigger factors in brick and click shopping. Intang. Cap. **15**(1), 57–71 (2019)
25. Martin, N., Morich, K.: Unconscious mental processes in consumer choice: toward a new model of consumer behavior. J. Brand Manag. 1–23 (2011)
26. Mehrabian, A., Russell, J.A.: An Approach to Environmental Psychology. The Massachusetts Institute of Technology, USA (1974)
27. Mosquera, A., Olarte-Pascual, C., Juaneda-Ayensa, E.: Understanding the customer experience in the age of omni-channel shopping. Icono 14 **15**(2), 166–185 (2017)
28. Nesar, S., Bin Sabir, L.: Evaluation of customer preferences on showrooming and webrooming: an empirical study. Bi-annu. Refereed J. **8**(1), 50–67 (2016)

29. Olsen, S.O., Tudoran, A.A., Honkanen, P., Verplanken, B.: Differences and similarities between impulse buying and variety seeking: a personality-based perspective. Psychol. Mark. **33**(1), 36–47 (2016)
30. Park, E.J., Kim, E.Y., Funches, V.M., Foxx, W.: Apparel product attributes, web browsing, and e-impulse buying on shopping websites. J. Bus. Res. **65**(11), 1583–1589 (2012)
31. Piotrowicz, W., Cuthbertson, R.: Introduction to the special issue information technology in retail: toward omnichannel retailing. Int. J. Electron. Commer. **18**(4), 5–16 (2014)
32. Reid, L.F., Ross, H.F., Vignali, G.: An exploration of the relationship between product selection criteria and engagement with 'show-rooming' and 'web-rooming' in the consumer's decision- making process. Int. J. Bus. Glob. **17**(3), 364–383 (2016)
33. Santos, S., Gonçalves, H.M.: Multichannel consumer behaviors in the mobile environment: using fsQCA and discriminant analysis to understand webrooming motivations. J. Bus. Res. **101**, 757–766 (2019)
34. Sit, K.S., Hoang, A., Inversini, A.: Showrooming and retail opportunities: a qualitative investigation via a consumer-experience lens. J. Retail. Consum. Serv. **40**, 163–174 (2018)
35. Solomon, M.R., Bamossy, G., Askegaard, S., Hogg, M.K.: Consumer Behaviour: A European Perspective, 3rd edn. Prentice Hall, Upper Saddle River, New Jersey (2006)
36. Torrico, P.R., Cabezudo, S.J., San-Martín, S.: Tell me what they are like and I will tell you where they buy. An analysis of omnichannel consumer behavior. Comput. Hum. Behav. **68**, 465–471 (2016)
37. Van der Horst, F., Matthijsen, E.: The irrationality of payment behaviour. DNB Occas. Stud. **11**(4) (2018). https://www.dnb.nl/en/. Accessed 14 Jan 2018
38. Venkatesan, R., Kumar, V., Ravishanker, N.: Multichannel shopping: causes and consequences. J. Mark. **71**(2), 114–132 (2007)
39. Verhoef, P.C., Kannan, P.K., Inman, J.J.: From multi-channel retailing to omni-channel retailing: introduction to the special issue on multi-channel retailing. J. Retail. **91**(2), 174–181 (2015)
40. Whisby Homepage. http://www.whisbi.com/info/omnichannel/solution/retail/index.html/. Accessed 21 Apr 2019
41. Xinhui, C., Han, D.: Meta-analysis of consumer irrational purchase behavior based on Howard-Sheth mode. J. Bus. Retail Manag. Res. **10**(3), 69–80 (2016)
42. Yadav, M.S., Varadarajan, R.: Understanding product migration to the electronic marketplace: a conceptual framework. J. Retail. **81**(2), 125–140 (2005)

Chapter 16
Event-Based Marketing: A Trendy and Emotional Way to Engage with the Public

Amalia Hulubei (Georgescu)(✉) and Silvia Avasilcai

"Gheorghe Asachi" Technical University, 29 D. Mangeron Bd, T1, Iaşi, Romania
amalia.georgescu@tuiasi.ro

Abstract. Events impact people directly and make use of multiple tools in order to offer unique experiences. Events can generate the right emotions that make people interact more with the brand or buy from a certain organization. For creating an event and having sustained success, organizations must adapt constantly and introduce new media and access new instruments in the implementation of their communication strategy. This article outlines the conceptual investigation of the impact and attitude formation in the context of event-based marketing and public relations strategies conducted by an organization. The impact of events on the attitude toward the brand or the organization is very important today, due to the experiences created for the public and the ability to interact with them through various channels and tools. This article introduces the necessary terminology from the industry of events and project management and outlines the new trends in event planning and communication. It shows how events work, how can they be planned, and the best channels to be used to promote them.

16.1 Introduction

Everything happening in our lives is an event. We interact through the years with a different stimulus, different persons, objects, places, foods, and brands. We experience different types of emotions, gather memories, and speak about them with our family and friends. We want to believe that everything happening to us is important and that, though we do not always realize it, there were specific factors, actions, persons, and attitudes that influenced us to make decisions, to act in a specific way, to purchase various things, and to determine our results in life.

From this point of view, the same thing happens when a company is set up. The cycle of human life is in a way predefined and cannot be prolonged too much, but a company's cycle of life can be constantly reactivated and renewed, due to a various number of factors like the vision, strategy, process reinvention, human resources, marketing and public relations campaigns, technology, and innovations management. These factors could help a company or a brand exist for hundreds of years and impact many generations over its existence.

Á. Rocha et al. (eds.), *Marketing and Smart Technologies*, Smart Innovation, Systems and Technologies 167, https://doi.org/10.1007/978-981-15-1564-4_16

For a company to impact generations, to have a brand relevant to the market, to be able to respond to the demand, it has to think strategically and permanently generate innovative products or services, invest in new technology, and promote them through marketing and public relations (PR) management plans. These plans can be renewed periodically or annually and have to be relevant to the market the company serves.

Nowadays, marketing and PR campaigns have to work better together and engage with their stakeholders on different levels, being present on as many channels as possible, off-line and online, communicating strategically and translating an organization's mission and vision into reality. Therefore, marketing and PR are linked together and consider any communicational activity of great importance.

The businesses are competing on the world stage. To survive and become noticeable, they must strive to reach world-class standards. There are no more national or geographic boundaries to protect a company from its competitors. Nevertheless, customers today are traveling more and more, are well-informed, and are quick to make value judgements on the performance of a company or of a brand [1].

So, campaigns and events must be used to achieve and implement efficiently and effectively the goals or the mission of the company. Why events? Because consumers do not buy today only a product that they truly need, but they buy value or different benefits associated with the products and the services. They buy emotions, experiences, lifestyles, entertainment, a form of identity, and personal happiness. That is why events engage with the consumer at the highest level of emotions, better than any advertising campaign or press release [2].

The various aspects of event-based marketing strategies have been analyzed in many ways, from a theoretical and practical point of view. This article explains the effectiveness of these communication tools and shows the effects of events on creating a positive attitude toward the involved brands, engaging their consumers/stakeholders in creating emotions and experiences that can lead to financial and nonfinancial success in the company [1].

Events, as a strategically planned activity, have the power to create or modify a certain attitude toward a brand, but it is not clarified why and how because every experience generates an emotion that is personal and different from individual to individual.

Various models and studies have been conducted to estimate the influence of using events in marketing strategies, on attitude formation, and emotional response. These studies have brought into attention important aspects to take into consideration and demonstrate the importance of events in influencing the attitude toward the brand (its products and services) and toward the communication activities conducted by an organization. These studies have also shown that events have the highest potential to emotionally influence human decisions in experiencing, connecting with, and buying a specific product or service [1, 3–5].

Emotions are mediators and moderators for a consumer's positive or negative response when interacting with a stimulus. Emotions can manifest differently depending on the type of information. For example, the advertisements seen daily by a person can have informational or transformational impact, thus generating a specific answer back from the viewer toward the promoted brand or product [6, 7].

In order to better control these emotions and generate the right answer of the public toward the brand, marketers, and public relations, specialists tend to use more and more online and off-line events, as a tool to impact more, to communicate better, and to win consumers' decision in experiencing and buying their products and services.

16.2 Events: Definition, Types, and Purpose

According to the Business Dictionary, the term event is defined as an "occurrence happening at a determinable time and place, with or without the participation of human agents. It may be a part of a chain of occurrences as an effect of a preceding occurrence and as the cause of a succeeding occurrence" [8].

Events have been organized since the old times. The first recordings of organizing big public events were in Antic Greece in 776 BC, with the Olympic Games, and then, in Rome with the famous games organized at the Circus Maximus. Since then, events have grown in complexity, form, purpose, and access.

Nowadays, generating events is one of the main activities in a marketing and public relations strategy, irrespective of the type of organization: nonprofit entities, medical, academic, military, political, administrative or financial institutions, entrepreneurial, and other types of companies, freelancing or private activities.

Philip Kotler defines events, from an experiential marketing point of view, as occurrences designed to communicate particular messages to target audiences [9]. Many other definitions have been given to events, depending on its primary objective: to market a product/service, idea, place or person, to communicate a specific message to a targeted audience and influence its decisions. So, according to Woods, it becomes clear that all events can be seen as marketing events. An event is a live "occurrence" with an audience. If an audience exists, then a message or experience is being shared, transmitted, generated and, therefore, all events can potentially communicate something [10].

There are different types of events and they can be classified as shown in Table 16.1, having in mind: the type or characteristic of the event, the place, the audience, the environment, or the instruments that want to be used for promoting it.

All events have in common one basic element: the gathering of people for a specific purpose [11]. No matter what type of industry it is, the events are one of the most efficient tools for promoting a brand, a product, or a service of an entity. Events can be organized anytime during the life cycle of the product. They can be adjusted to any technological or market changes, to any requirements and can incorporate different types of multimedia and physical resources.

The primary advantage when organizing an event is that the organization can create a personal relationship with the public, has the control over the transmitted message, and has the highest degree of involvement and emotions of the public, the star of the event being in itself, the promoted brand or the company.

When planning and organizing a specific event, there are some important characteristics that any specialist should take into consideration: involvement, interaction, immersion, intensity, individuality, innovation, and integrity. These seven aspects are called the 7 "I"s of an event [13].

Table 16.1. Various types of events [11, 12]—personal adaptation

Criteria	Event types	Examples
Type/Character	Economical and business events	Conferences, Teambuilding events, Trainings, Meetings, Product launch events, Celebrations, Fairs, Networking events, Press conferences, Forums, TV shows, Stores and branches opening events, Public speaking events, Board meetings, Workshops, Products samplings
	Educational/Scientific	Conferences, Congresses, Colloquy, Olympics, Contests, Trainings, Forums, Conventions, Classes
	Political/ Diplomatic	Meetings, Official visits, Election Campaigns, Rallies
	Military	Parades, Strategic Trainings, Special Meetings
	Religious	Church events, Camps, Concerts, Conferences, Trainings
	Social/Environmental	Fundraising events, Awareness and CSR campaigns, Environmental events, Press conferences
	Cultural and creative industries	Arts and creative events, Entertainment shows, Book launching, Book or creative fairs, Concerts, Fashion shows, Lifestyle events, Exhibitions
	Sports	Specific games, Olympics, Contests for professionals of amateurs, Demo events, Camps, Tournaments
	Gastronomical/Agricultural	Wine and food tasting, Menu launchings, Food and agricultural fairs, Cooking shows, Cultures and vineyards visits
	Special-private events	Birthdays, Parties, Anniversaries, Celebrations, Retreats, Weddings, Funerals
Place	Indoor	In restaurants, hotels, conference centers, hub rooms, office buildings, private places, museums, libraries, theaters, arenas or sport facilities, and tents or other creative places
	Outdoor	In parks, roads, gardens, forests, streets, plains or mountains, zoos, sports courts, and beaches
	National	
	International	
Audience	Internal	Employees
	External	Stakeholders (consumers, suppliers, buyers, general public, competitors, mass media), general public, online audience
Environment/Instrument/Way	Off-line	As shown above, billboards, posters, TV/radio streaming
	Online	Social media events, webinars, YouTube streaming, livestreaming, and Internet pages

Any event is direct and emotional, that is why the implication part is crucial. If an event does not connect to the market and establishes no real interaction with its stakeholders and ambassadors, that it is not an event, but advertising on another public relation action. An event affects and appeals to all senses, so its immersion should be taken into consideration when planning an event, even in its most insignificant details (a welcome smile, an unexpected surprise, a funny moment, lights and sounds effects, and thank you card could be simple ideas that make your participants feel special). Intensity means being memorable and impacting the participants at the highest level, finding ways to remain in their minds, and talking about the event long after it ended.

Every event has to be unique, original, offer individuality, a sense of importance by attending it, and new opportunities to interact with the products, services, or with the VIPs/brand ambassadors. Having an innovative event means to launch news, new products, present specific innovations or include new technology, sounds and lights, special effects, decorations, styling or even chose a different, and unusual place for the venue (hubs, old buildings, parks, gardens, museums, forests, and stadiums).

Last but not the least, every event, based on its main purpose, has to offer real benefits for its participants, give true information, offer the opportunity to interact with the products and the company, ask questions, and receive feedback. This means that the company, through events, becomes more open to the market, is ready to interact with various groups, and shows its true values and principles. This gives a sense of integrity and veracity.

These 7Is of an event (involvement, interaction, immersion, intensity, individuality, innovation, and integrity), should become mandatory when creating and planning an event.

16.3 Event-Based Marketing Strategy: Why Events?

Events are an important and indispensable tool in any marketing and PR strategy and are considered to be part of the mix of communication strategies because they send a specific message to the targeted public [2].

Nowadays, consumers are tired of aggressive commercials and classical marketing communication. They want appreciation for the money paid for certain products and services. They want the right to choose, to experience, to feel and want to be free or at least to believe they are free in their decisions. Because of social media development, customers like and want to recommend different products, services and places, according to their experiences and feelings. Having this in mind, an organization must be more and more connected to its stakeholders, be involved in society, and bring its contribution.

Organizations must collaborate, communicate, innovate, access communication channels that are emotional, have maximum impact on people, and integrate new technologies into the strategic communication process.

The use of events in generating strategic marketing and PR programs helps organizations generate unique, memorable experiences, thus helping them develop the experiential marketing programs by engaging their potential and current customers at an emotional level.

For many consumers, brands are an image of a certain good, accompanied by an emotion that could be positive or negative, according to the experience the user had with that specific brand. Organizations can generate more positive emotions and impact the users' experience if they create specific events and occasions that allow them to interact more with the products, the services, and the employees.

Events, as experiential marketing, are extremely useful to marketers because they allow companies to reach and have direct contact with their potential and current customers [14]. An event can make use of all senses, thus offering a much durable impact, more than any commercial or advertising [2].

Having a real time with the customer, where the organization can talk to them face to face, is very important. This can be done by connecting them directly to their brand throughout the experiences which they can generate during the events. Furthermore, events are a useful technique for developing stronger inbound connections [15]. But also, the online is very important, being an important tool user nowadays, that is why it is of extreme importance to integrate the online platforms in promoting and generating the events. This can also motivate your followers and attendees to engage more deeply with your event during and after it takes place, making it easily accessible through the online platforms.

Before an off-line event takes place, it is better to connect it to online platforms. If there is not a dedicated website to promote the event, the event's platforms can be a user or, for example, a created Facebook event. Here, marketing specialists will be posted daily updates about what is about to happen, the schedule, the VIPs or the speakers invited, the partners, the ticketing, facilities offered, and where the event will take place. There can be made Twitter updates, with little reminders, teasers or shared links leading up to the event without being too intrusive. You can also use LinkedIn and Instagram for promoting your event in the online zone. These platforms help you "build momentum and anticipation for the event as the time comes closer" [15].

During the event, the most used way to interact with your virtual public that could not attend it is by live tweeting, live Facebook updates or livestreaming. Livestreaming during events allows companies to engage with their audience and make them feel like they are a part of the live event though there are a limited number of tickets available or the public could not attend the event for any other reason. In addition, it offers an added level of engagement for those who are actually attending the events, posting about it, sharing the livestreaming link, and inviting their friend to watch it. It is recommended to create a unique hashtag for the event and publicize it to both attendees and people following online so that everyone can easily follow and take part in the action, by using the same information when they are posting about the event.

After the event takes place, the marketers need to continue connecting with the audience by posting thank you messages to them and the partners, to post photos and allow the public to tag themselves. Also, posts can be made with short summarizing movies and announce future events. If the format of the event allows it, attendees can be encouraged to upload their own photos, giving them a chance to share their personal experience, creating a connection between attendees, the event, and the brand. At the same time, if the event has a dedicated website or a blog, there can be written articles about the event or capture links from other websites that published information about the events which can be made public by posting them and saying thank you to the mass media partners or influencers that wrote about it. There can also be a summary newsletter or an infographic of the main aspects made regarding the event. Infographics use visuals to make information easy to understand, present complex information quickly and clearly, and tell the story easier instead of using text and graphics [16].

16.4 The Architecture of Events: From Idea to Execution

Event marketing is the most customized form of marketing, usable in almost any industry, filling the strategy of direct marketing and PR programs. It is very strong in launching products and services, in attracting attention and generating awareness. It offers direct consumer contact and it can easily embrace any other form of communication that is based on the principles of project management.

Organizing an event asks for very good project management skills and requires knowledge, experience, and a dedicated team that will work as small departments for managing all the aspects of an event planning and implementation part.

The team responsible for organizing an event will need an interdisciplinary approach and knowledge of public relations, protocol rules, project and resources management, administration of goods and logistics, legislation concerning contracts, mass media relations and communication channels management, marketing, graphic design and advertising, accounting and finances management, commercial, sales management, and human resources coordination. If an organization does not have the right team to handle the event, it is advisable to access the services of a specialized event-organizing company or an event planner.

Being a marketing and PR tool, an event has to answer five questions: what and why?; who and with whom?; when?; where?; and with what and how?. If the organization answers all these questions and makes a plan, then it has all the chances to succeed in its event.

Management of event or event management is the process throughout the duration an event is planned, prepared, and produced. Successful event management asks for specific steps to be taken into consideration: the definition of the purpose and the objectives of the event, the type of the event, the establishment of the key performance indicators, the targeted public, budget and the written plan, a date and a location of the event, sponsors or partners, audio–photo–video resources, catering or restaurant services, list of speakers or special guests, the complete program setup, sales and marketing, the promotion of the event, and the follow-up after the event has been organized.

A successful event needs to be very well planned, to have as much marketing and PR as possible in order to announce to everyone (according to the targeted public). To attract many participants, an organization has to promote its event on all possible and appropriate channels (online and off-line) and make use of the social media channels (Facebook, Instagram, Twitter, and many others) to announce its updates.

A well-promoted event brings people and this is an important key performance indicator: the number of participants. If people are not coming, it means you did not promote it enough or it was unattractive to the public.

Choosing the right channels to promote you event means to start promoting it at the right time (at least 30–120 days before it takes place, especially if it is a public event), where your public is by sending invites, sending e-mails, newsletters, giving fliers and promotionals, putting posters, billboards, creating ads, launching the event's web page or the Facebook event, posting updates and teasers, making mass media announcements, blogging about it or asking for other websites and influencers to write about your event and start spreading the word (or creating the buzz) about your event [15].

There are nine important steps in planning an event, as shown in Fig. 16.1.

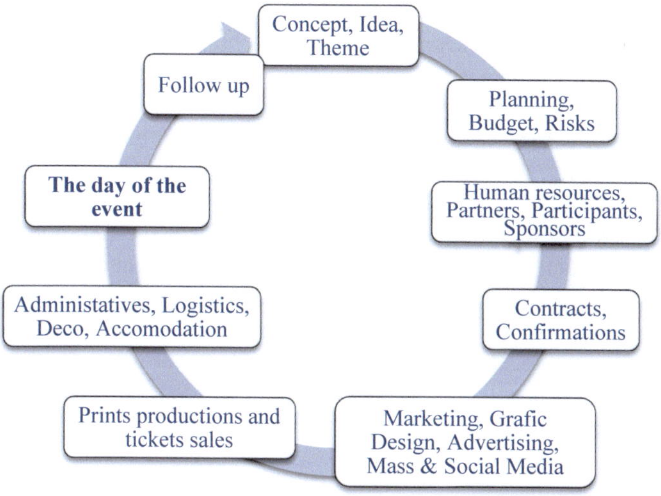

Fig. 16.1. The steps in planning and organizing an event

The first step takes into consideration the research part, the idea brainstorming, the concept, and the theme of the event. Here, it is very important to establish the type of event and even the place where the event will take place. The second step is to start planning, estimating the costs (even the possible revenues), and put the key performance indicators for the success of the event. At this point, a Gantt diagram is recommended to be created in order to establish the team, the resources needed for each activity, and the deadlines. This will help the event management team keep track of the activities and of what has been done or not.

The third step asks for the human resources team to be established and to give specific roles to each person (there could be employees, a specialized company, or a team of volunteers). At this stage, the team or the owner of the event also establishes the list of guests, speakers, participants, and the sponsors or partners that could be part of the event. The next step, the fourth one, will start contracting the venue, will look for suppliers for different aspects regarding the event, and will look for partners or sponsors to support the event.

The production of marketing materials, promotional, collaterals, the graphic design of all materials off-line and online, according to the chosen channels, is the next step. After the image and the specific design is established, the first materials are printed and spread to the targeted public (flyers, posters, invites, covers, and billboards). At this point, the marketing and sales team are in action and promote the event or sell tickets if the event requires this.

At the same time, the seventh step in event management, works simultaneously with the previous ones, because the event management team has to take care of administrative details, logistics on the venue, infrastructure and easy access to the venue, scenography, rehearsals, decorations, manage potential risks, accommodation for special guests, and marketing materials.

The day of the event takes place as planned (but it is better always to have a plan B, in case something unplanned happens), with every person in the team knowing what to do based on the plan and the general schedule. During the event, it is recommended to have photo posts and livestreaming if the event allows this.

After the event, the follow-up part is very important. Here, the organizer posts the official photos and videos of the events, sends thank you messages to the partners and the participants, posts links of reviews, and announces the future events if planned.

Some important key performance indicators of the event can be: specific objective, the concise plan and work delegated to the team, there was lots of PR and advertising around the event, things were well organized, invited persons attended the event, the venue us appropriate, there was at least one emotional moment, there was excellent feedback, the speakers had what they needed, the food was good, unexpected occurrences were handled competently and there is a plan for repeatability and expandability [17, 18].

16.5 Conclusions

Events, from the strategic communication point of view, are the ones that put the consumer or the public in the first place. Events are great means to manage a better relationship with the stakeholders, control the message and the feedback.

Events offer notoriety and visibility for the organization and can attract a great number of prospects, nowadays being a trend in marketing due to its impact and results in the young generation. If they are well planned, though in the beginning they are cost generators, events can bring lots of revenues during or after they take place, having a direct impact on the turnover of the organization.

Because the classical advertising is losing ground and the new business models are more oriented toward PR and nonconventional advertising (word of mouth, viral posts and videos, blogging, events, and pseudo-events), organizations are encouraged to access more the benefits of events in their marketing and PR strategy.

This review and small analysis over the benefits and the management of events can help organizations to update their strategy, to understand their marketing potential, and to be bolder in their actions toward their public, in order to connect better, longer, and with great impact.

References

1. Tum, J., Norton, P., Wright, N.J.: Management of Event Operations. Elsevier/Butterworth-Heinemann Publishing House (2006)
2. Schäfer-Mehdi, S.: Event-Marketing, p. 45. Cornelsen Verlag Scriptor Gmbh & Co. KG, Berlin (2005)
3. Weihe, F., Mau, G., Silberer, G.: How do marketing-events work? Marketing-events and brand attitudes. In: Diehl, S., Terlutter, R. (eds.) International Advertising and Communication. DUV, pp. 199–216 (2006)
4. Gwinner, K., Eaton, J.: Building brand image through event sponsorship: the role image transfer. J. Advert. **28**(4), 47–57 (1999)
5. Hoek, J., Philip, G., Michelle, J., Orsman, D.: Sponsorship and advertising: a comparison of their effects. J. Mark. Commun. **3**(March), 21–32 (1997)
6. Bagozzi, R.P., Gopinath, M., Nyer, P.U.: The role of emotions in marketing. J. Acad. Mark. Sci. **27**(2), 184–206 (1999)
7. Didier, C.: Les effets automatiques du parrainage télévisuel sur la marque: étude de la mésattribution de la familiarité, du transfert sémantique et de l'influence des émotions déclenchées par le programme. Recherche et Applications en Marketing **15**(1), 39–60 (2000)
8. Business Dictionary. http://www.businessdictionary.com/definition/event.html. Accessed 04 July 2019
9. Kotler, P.: Marketing Management, 11th edn., p. 576. Prentice Hall, Upper Saddle River, NJ (2003)
10. Wood, E.H.: Evaluating event marketing: experience or outcome? J. Promot. Manag. **15**(1–2), 247–268 (2009)
11. Kilkenny, S.: The Complete Guide of Successful Event Planning: With Companion CD-ROM. Revised 2nd edn., pp. 33–39. Atlantic Publishing Group, Inc., Florida (2011)
12. Oprea, D., Mesnita, G., Alexa, M.: Managementul evenimentelor: personale, organizationale, internationale, pp. 20–22, Polirom, Iasi (2016)
13. Wood, E.H., Masterman, G.: Event marketing: experience and exploitation. Paper Presented at the Extraordinary Experiences Conference: Managing the Consumer Experience in Hospitality, Leisure, Sport, Tourism, Retail and Events, Bournemouth University, 3–4 Sept 2007
14. Sorenson, L.: 7 Online Techniques to Improve Offline Events. https://blog.hubspot.com/blog/tabid/6307/bid/27621/7-online-techniques-to-improve-offline-events.aspx. Accessed 04 July 2019
15. How to host a successful Facebook event. https://www.compukol.com/how-to-host-a-successful-facebook-event/. Accessed 05 July 2019
16. Sterenberg, M.: Write Like a PR Pro: Why Writing is Still Crucial in Digital and Visual World. The Ohio State University, pp. 110–111, open text book. https://ohiostate.pressbooks.pub/writelikeapro/ (2017). Accessed 05 July 2019
17. Soard, L.: 7 Characteristics of successful events. https://attendee.events/7-characteristics-successful-event/. Accessed 06 July 2019
18. Reber, J.P.: The 7 Elements of A Successful Special Event. http://jamesreber.com/ProducingEvents/ElementsSpecialEvent.html. Accessed 06 July 2019

Chapter 17
The Application of Social and Viral Marketing and the Management of Social Networks to Promote the European Defence Conscience

Manuel A. Fernández-Villacañas Marín(✉)

Higher Technical School of Aeronautical and Space Engineering, Technical University of Madrid. Colonel Spanish Air Force (R), Madrid, Spain
manuel.fernandez@mm-ppm.com

Abstract. During the last two decades, the EU's political-military external influence has been increasing. However, the crisis in the Ukraine, the emergence of jihadist terrorism, the inadequate management of the migration problem, the departure of the UK, the proliferation of Euroscepticism, and the reappearance of ultra-nationalism are some of the problems that show its weakness in general and in terms of defence and security. In order to renew the political impulse to build Europe, as a necessary condition, it is essential to strengthen the European identity in general and the previous construction of a European Defence Conscience. The methodologies that are considered essential to achieve these objectives are the strategic management of the social image and the communication, especially in an increasingly technological and hyper-connected environment. The application of the social and viral marketing and the social network management should be its main instruments of action. In order to renew the political impulse to build Europe with adequate capabilities for its defence, it is essential to strengthen the European identity in general and the construction of a European Defence Conscience through the socio-political construction of the European Identity of Defence and Security. These targets require to design and implement an ambitious strategic plan of social and viral marketing, which would be led by the European Commission and developed by all EU Member States.

17.1 Introduction

Today's risks and threats, and consequently the problems of defence and security, are much broader than they were in the past due to their global impact, the characteristics and possibilities of the new digital technologies being applied, as well as the intelligent weaponry currently employed. This international framework moderately implies the usual threat of territorial invasion, as well as the significant presence of new, diverse set of threats, such as international terrorism, cyberattacks, corruption, drug trafficking, illegal arms trafficking, global organized crime or trafficking of human beings, and so on, all of which are supported by anonymous groups willing to achieve their objectives

Á. Rocha et al. (eds.), *Marketing and Smart Technologies*, Smart Innovation, Systems and Technologies 167, https://doi.org/10.1007/978-981-15-1564-4_17

at the cost of insecurity and international social destabilization. In this new era, in which structural change and globalization are permanent aspects, it is necessary to contemplate the risks and threats we will face during the third decade of the twenty-first century, and the most effective instruments we will require to accomplish it [1].

Consequently, in the first place, these new circumstances recommend the incorporation of technological and dynamic strategic management models capable of providing sufficient response capacity in the increasingly reduced time margins available. All these factors introduce philosophical and conceptual changes that are particularly intense in the field of defence and security, as well as in communicative relations within the social environment. In the second place, the Western societies are suffering from a considerable weakness as a consequence of the high-tech lifestyle, where the population is concentrated in large nuclei that are highly vulnerable to any kind of attack against them. In addition, there is a definitive lack of instruments on behalf of the governments to take action in the new international scenarios, once the traditional instruments have proved insufficient and ineffective against the new threats. Most of the new risks are a direct or indirect responsibility of new non-state entities (although in many cases they are supported by certain States), which use elements of unconventional combat, but with a great capacity for destruction and destabilization. It is clear that diplomacy becomes useless in such circumstances and the powerful Western armies, created and designed to fight against other armies, are inadequate to combat threats such as international terrorism, organized crime or illegal immigration networks. Thirdly, globalization implies modifications and changes in the benchmark elements of a European identity, some of which are difficult to reconcile with a number of more traditional European States and, consequently, appear to generate potential confusion for the citizens. The traditional defence culture was precisely the main guarantor for the maintenance of their national identities. It appears that they must be modified, enriched as well as adapted [2–4].

Objectives, Methodology and Contents

In view of the above, it is necessary to plan and strengthen the very elements of communication as well as the awareness of the new corporate realities of defence and security structures. In order to do so, it is essential to define a strategic model of social communication capable of overcoming the isolation and lack of social integration that the military, police and judicial systems regularly experience, and to respond flexibly and quickly to the turbulent changes that the very conception of a global environment may establish. In this way, the strategic management of the social image and communication takes on special prominence, especially in an increasingly technological and hyper-connected environment, and the application of social and viral marketing models of social network management should be its main instruments.

In relation to the methodology, the research has been developed under a systemic conception through a systematized deductive process (top-down) that is complemented by a qualitative research carried out on institutional communication, and the use of social networks in the Spanish Armed Forces and the author's experience as such are included in his cited bibliography (bottom-up). The rest of the bibliographic review has served to synthesize the essential concepts and approaches on the investigated problem, reviewing all this with a level of detail that is considered sufficient, looking for its original aspects to guarantee objectivity and avoid interpretive biases.

In order to achieve the objectives, the work has been structured in three sections, after which conclusions are presented. Let us first address a synthesis of the current international defence and security scenario. Secondly, the essential aspects of the construction of the European defence are going to be analysed. Finally, the keys to the improvement of strategic communication and institutional image through the application of social and viral marketing within the social networks will be studied.

17.2 The New World Order and the Current International Scenario of Defence and Security

The structural change of the twenty-first century and the emergence of a series of new, intertwined and integrated global risks and threats make it necessary to review the traditional approaches in setting up the current international defence and security scenario. The polysemic concept of the New World Order, which has been in continuous evolution for many years, can currently be identified by means of a set of characteristics, which, without exhaustive pretensions, are listed as follows [5]:

- Globalization in all its forms emphasized a transition from a bipolar order to a multipolar World Order.
- A recent emergence of "populism" on a global level is a consequence of the fact that traditional democratic systems of representation are undergoing a rapid evolution as a consequence of the socio-economic crisis, which affects traditional political parties which are unable to attract or represent the most popular classes.
- Complexity, uncertainty, hostility, dynamism, as well as continuous metamorphosis, competition and diversity.
- The crisis of values, as well as intense and continuous social change.
- The new concept of collective security, whose emphasis lies on the individual as the subject of security.
- The emergence of new dimensions of sovereignty, beyond identity, such as energy, food and environmental sovereignty.
- The continuous technological revolution, induced by ICTs, which definitely turn technology into a source of "military-police advantage".
- The assignment of new missions in collaboration with the military forces and the police forces is a consequence of the fact that defence and security have a tendency towards full and rapid functional integration.

After the Cold War, a multidirectional and multifaceted panorama of risks and uncertainties arose, giving rise to a complicated international situation. The world faces risks arising mainly from social tension generated by economic imbalances, population growth, democratic deficits, environmental aggressions and confrontation between different cultures. After the disintegration of the Soviet Union, the states that have recovered their freedom of action abroad and are integrated or closer to the EU have found themselves in a process of incorporation into the international community that has been fraught with difficulties. In this transition towards freedom, democracy and the market economy, both of these nations and of others located in other regions, there is a source of instability that has even led to the disappearance of the environment of trust, security and cooperation that was created at the end of the last century.

On the other hand, the North American unipolar power has been installed in a worldwide order that is neither accepted nor recognized. However, the other global participants and powers, in particular the Russian-Chinese alliance and the EU, are preparing to compete or stand out in the future global hegemony, giving way to a multiple dispute or to diverse and successive scenarios of hegemonic rivalry, at different scales and levels and with different manifestations that are not merely military.

17.3 The Defence and Security European Identity Construction

Apart from some previous initiatives, the political process of the construction of Europe began after the Second World War, as a unity strategy for the reconstruction of a devastated Europe, generating a process of economic integration through the Treaty of Rome of the European Economic Community. Years later, the process of political, economic and monetary integration was achieved with the entry into force of the Treaty of Maastricht in 1993. In the field of defence and security, after an arduous and unfinished process, which was constantly hampered by the national interests of certain European countries, above all the United States, it was the 1998 Franco-British summit in Saint-Malo that led to the promotion of a common European defence. In the same way, the conflict in Kosovo in the spring of 1999 generated the unanimous perception of European leaders regarding European shortcomings in the politico-military field. Faced with the failure of the 2004 draft of the European Constitution and in order to unblock the situation, the Treaty of Lisbon was signed at the end of 2007, which implied important modifications to the Constitutive Treaty of the EU (Maastricht) and the European Economic Community (Rome). Since then, Europe's political-military influence has been rapidly increasing, among other things, by its decisive role in the Iran crisis, by the military presence of European countries in Afghanistan or by its outstanding deployment of forces in the former Yugoslavia, in which its intervention was decisive for the maintenance of peace and the secession process.

However, regarding the Ukrainian crisis, the irruption into the European territory by Jihadist terrorist activism, the inadequate management of the migration problem, the fragmentation of the EU as a consequence of the exit of the United Kingdom and with it the induced NATO crisis as the basis of European collective defence, the proliferation of Euroscepticism and the reappearance of ultra-nationalism are some of the problems that show the EU's immense weakness in terms of defence and security. For all these reasons, among others, the EU has decided to renew its political drive to build a Europe of Defence, especially in terms of its industrial and technological base, which will make the industrial defence and security capabilities of the Member States more effective, efficient and integrated. In order to facilitate all of this, and insofar as it is directly related to aspects of the national identity of each of the nations, it is necessary that, in conjunction with the other initiatives, a genuine European Defence and Security Identity be built and encouraged.

The sociological concept of European identity is, to say the least, very complete. Although it is true that most of the Union's inhabitants feel European to a greater or lesser degree, it is no less true that most of its citizens consider their national or regional affiliation more strongly. It is obvious that the integration of Europe requires a European identity that needs to be built and promoted, and for that there is no linguistic or cultural homogeneity. To begin with, it is clear that European identity cannot be based on an impossible cultural standardization, nor should it ever be built against an imaginary "other one". On the contrary, in a liberal democracy, citizens must be loyal and identify not with an impossible common cultural identity, but with values and principles that fully guarantee their rights and freedoms. It will be the concept of multiple identity that constitutes the basis on which the European identity can be constituted, this characteristic being one of the most defining features of the European culture as a result of the interactions between multiple tendencies, in opposing menus, through dynamic relations between faith and reason, certainty and doubt, empiricism and rationalism, science and philosophy, religiosity and secularity and so on, as examples of some of the tensions on which the European cultural heritage is based. Rather than starting from an effort at cultural standardization, as explained above, it is necessary to be clear that the elements of European identity derive from a shared historical experience and its richness must be found in the interaction of different conceptions. This is the concept of pluriculturalism, understood as a necessity for mutual cultural enrichment, whereby cultures need each other in order to transform, develop and survive. It is not only a question of developing a tolerant attitude in this respect but of accepting the cultural reality of the other as an integral part of one's own cultural claim.

To sum up, as far as European identity is concerned, it seems essential to find a framework to support the process of European integration, that is, a soul that contributes to the citizens' finding attractiveness and warmth within the European project, which permanently runs the risk of lacking the necessary citizen support with a consequent legitimacy deficit. Without the capacity to construct the collective imagination that culture provides, there can be no social or political community. At the same time, the difficulty that arises from the disparity between the elements that form part of the European project, and not only between States but also between minorities, peoples and regions, is obvious. This concern is exacerbated by the prevailing uncertainty in an increasingly multicultural world, where certainties have disappeared, where the references have ceased to be untouchable dogmas, and where nowadays there are no territories but a geographical network and migratory flows.

From a political point of view, since the Treaty of Maastricht in 1992, the European Defence and Security Policy has sought to bring together the efforts of some European nations belonging to NATO, to constitute what was once called the European Defence and Security Identity, thereby attempting to express the will to act in this area with greater autonomy vis-à-vis its hitherto principal transatlantic ally. It was a new approach aimed at shifting more responsibility towards Europeans by facilitating a rebalancing of defence and security. In this way, within the transatlantic relationship, those allies that could not act together militarily would have difficulty in cooperating politically and economically. In addition, maintaining a fragmentation based on technology, in which Europe would provide the soldiers while the United States would continue to contribute with its technological potential, should be avoided. All of this posed as an undefined common objective, which would easily reflect a deep basic disagreement. According to this socio-political conception of the European Defence and Security Identity, the obstacles that have arisen regarding its development are clear, and the opposition of all those nations try to maintain their national sovereignty intact, especially in matters of foreign policy, as well as that of the United States, which tries to maintain and even increase their influence and supremacy over Europe.

If one starts from the concept of the *European identity* (institutional, multiple and diverse), and adds on the one hand, *security* (as a multidimensional term that encompasses the environment, terrorism, economy aspects, politics, organized crime, etc.), and, on the other hand, *defence* (supported by a strong military component firmly rooted in the sovereignty of each Member State), we find ourselves with a very complex concept that explains why the steps taken towards achieving the European Identity of Defence and Security have been very small and hindered by everyone. (To study in greater depth, see [2] and [6].).

17.4 Generating a Solution in the Field of Marketing and the Social Networks Management

17.4.1 Applying Social and Viral Marketing

To begin with, the social marketing was defined initially by Kotler and Roberto as the technology of managing social change that includes the designing, implementing and controlling programmes aimed at increasing the acceptance of an idea or social practice in one or more targeted adoptive groups [7]. Social marketing appeared as a non-profit extension of marketing that had a very specific general object, the voluntary social change, and which was implemented by means of complex, personal and anticipatory exchange relations that focus on specific ideas and social causes [8]. Costa and Vila considered that marketing at the beginning had two areas oriented towards society, which were clearly defined as follows: one area was made up of the distribution of wealth and intermediation, while the other sought to explain rationally the functioning of markets and systems [9]. Góngora, Schneider, Bins and Mauceri consider that social marketing originated as an adaptation of traditional marketing (marketing of products and services), but focused on the management and resolution of social conflicts [10]. Subsequently, social marketing was considered as a means to reach more people and generate a positive change in those socially rejected behaviours and practices, such as tobacco consumption or environmental pollution [11].

In recent years, social marketing is defined as a set of marketing strategies focused on creating value for customers, so that their welfare and community in general is guaranteed, turning citizens into consumers of social goods [12, 13], seeking that society adopts beneficial behaviours and moves away from those that are harmful, with a narrow relationship with legal, technological, economic and informational factors [14]. Currently, the purpose of social marketing is to improve the quality of life of people through initiatives and actions that allow achieving a sustainable social transformation [15, 16].

In general, as part of social marketing, we define cultural diffusion as the phenomenon through which a set of contents defining a culture is transmitted to a target population, seeking to produce certain favourable cognitive pregnancies in the individuals; that is to say an introspection of their contents. The result is the socialization of subjects, a process through which they are absorbed into the culture of their society and incorporated into it as constituent members of their identity. Fundamentally, socialization consists in learning by means of which the individual adapts himself to his social group and makes his norms, rules, beliefs, values and so on, which orient his conduct.

On the other hand, "word of mouth" marketing can be defined as an interpersonal form of communication between consumers about their personal experiences with a brand or product [17]. Bluntly, but not concretely, it is considered to be thousands of times more powerful than conventional marketing [18] and a very important force to model consumer attitudes and behaviours. The internet has enabled the development of new marketing strategies based on the transmission of similar marketing messages through internet-based channels. During these transmissions, information passes between individuals without involving the original source of the message, which spreads like a virus, infecting guests. We can define viral marketing as the marketing strategy that exponentially increases the recognition of a brand or product, through the media, whether it is traditional or recently created [19]. In the internet, "word of mouth" communication becomes a much more effective [20], penetrating and fast tool compared to traditional "word of mouth" communication [21], and yet the content of the message remains largely unchanged. It is also much more focused in that intermediate recipients transmit the messages in their social sphere, where their influence is more critical [22].

As a principle, the process of cultural diffusion and promotion has a totally natural and autogenic origin; it is, above all, the human groups themselves, originated in a natural way, who are in charge of the diffusion of the integral elements of their cultures to all their members on a regular basis. However, technically advanced societies have added to the traditional agents of socialization and cultural transmission (the family, the school, affinity and friendship groups, the church, etc.) other apparently anonymous and undirected mechanisms, such as radio, the press, television, video, the Internet and, above all, the social networks [23]. Thus, a great deal of information loaded with cultural values and traits is transmitted through these impersonal media. An analysis of the content of the messages transmitted will show us that the degrees of objectivity and respect for the autonomy of the socialized agent vary greatly. There are individuals and groups with restricted interests and opinions, who wish to inculcate them in others in order to achieve their own purposes and from the large networks to which they belong. They do not even hesitate to use false information [24]. Evidently, the phenomenon of socialization, consciously manipulated by those who wish to benefit from it, is not new. However, what is the vast media landscape that today's world is becoming accessible to those who are in a position to control it.

The planning of the social and viral marketing process is summarized in Fig. 17.1.

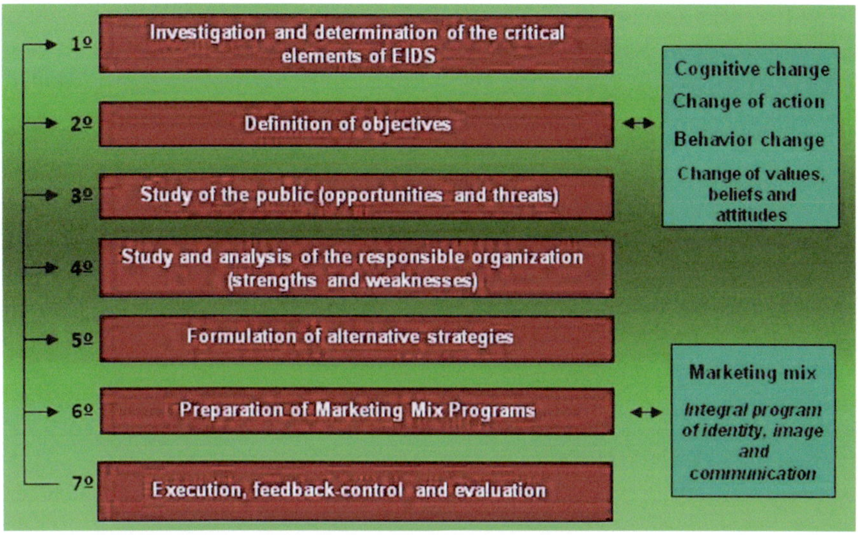

Fig. 17.1. The planning process of social and viral marketing to promote the European identity of defence and security [2]

17.4.2 Improving the Strategic Communication and the Strategic Image Management

The European Defence and Security institutions ought to carry out a global programme of European identity, both with regard to these institutions and the national identities of the Member States, communication and social image [25]. In this mission, the interpersonal communication "word of mouth" that has been analysed is critical, especially through social and viral marketing initiatives [26]. When a message is sent to a target population and affects a group of individuals, the communication process does not end there, but instead produces an added system of induced communication. The effective management of the strategic communication system should make it possible to radiate and animate the messages considered necessary, as well as detect and eliminate those that are harmful to the intended purposes.

However, the application of a strategic approach that we are developing also coincides with the management of its institutional image [27]. From this perspective, the organization is understood as a global corporate system in which it is not possible to separate any of its basic functions, and which is managed under various formal policies and consists of three specific areas: real identity, image and corporate communication. Strategic image management is an administrative function rather than a communication function. It consists of translating the strategic scheme into an image-based scheme in which the social perception of the organization is involved. Organizations must carry out a global programme of identity, image and communication, a set of actions aimed at building real identity, transforming it into messages of communication and behaviour, and disseminating it among strategically selected audiences [28].

The European Identity of Defence and Security is not going to remain static over time but rather it is going to be essentially dynamic. This changing factor will not always be voluntary. In some cases, there will be spontaneous changes of identity, in other words, generated by the very impact on the social environment. In other cases, the change will be planned and directed. In this sense, the planning of the evolution of the European identity of defence and security must take into account the evolution of the social environment in order to choose a specific objective for the future to be achieved, by incorporating aspects to be corrected as a consequence of the unplanned change into the design of the scheduled change. There will be a certain dynamic distancing between the identity reality and its social image, which will depend on the changes generated in reality and on the changes operated in the image. When changes in identity reality are oriented towards the accomplishment of a given desired reality, tending towards a certain ideal or utopian reality, the existing dynamic distancing must tend to be the minimum possible, that is, it must tend to have a full adaptation between identity reality and image [29] (Fig. 17.2).

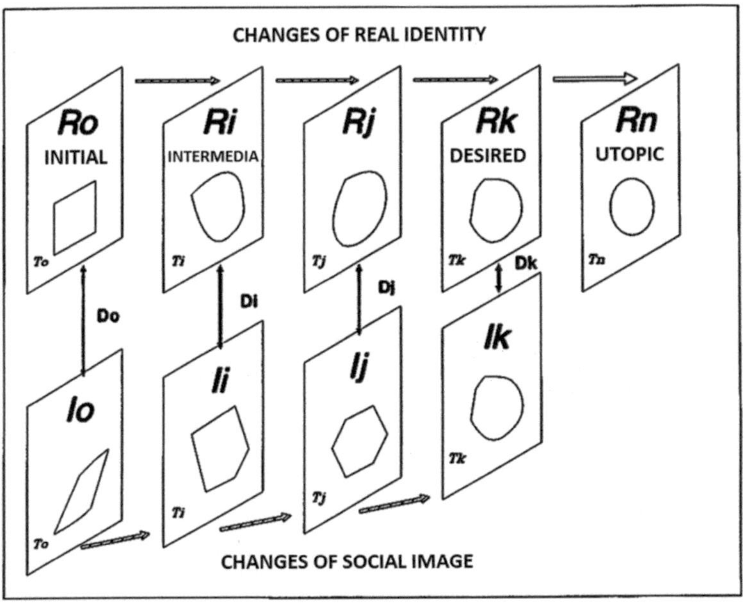

Fig. 17.2. The strategic management of European identity of D&S, social image and communications [29]

The qualitative research carried out on institutional communication and the use of social networks in the Spanish Armed Forces has allowed to contrast three dilemmas about it in Public Administrations. The first dilemma, the bureaucracy and public institutions are perceived as inefficient bureaucracies, in which there is usually no culture of informing, and in which transparency is usually understood as a response to requests for information from the "users". The second dilemma, the new digital media: there is intention and even need to use the new digital media as endorsement of a modern image, but it is not known how to become successful in communication, sending messages without a recipient clearly and exclusively to replicate official information. Third dilemma, clarity; the institutional communication account is used as if it were a personal account and very often the social media value the messages of public institutions in social networks as official statements.

The main conclusions of the study include: That social networks are the communication channel with the greatest capacity for interaction and immediacy; that the messages must use a natural and close language, different from the traditional one, oriented to a well-defined recipient, but they must be managed as official communication; the achievement of the maximum social interaction must be sought; it must always be kept in mind that everything that is written has consequences; and finally, public institutions in general and military and police forces in particular, must bear in mind that, beyond their great communicative capacity, the usefulness of social networks will depend above all on a critical aspect: the message [25].

17.5 Conclusions

In order to renew the political impulse to build Europe with adequate capabilities for its defence, it is essential to have adequate legitimacy and social support, to strengthen the European identity in general and the construction of a European Defence Conscience through the socio-political construction of the European Identity of Defence and Security.

In order to do so, it is essential to define a strategic model of social communication capable of overcoming the isolation and lack of social integration that the military, police and judicial systems regularly experience, and to respond flexibly and quickly to the turbulent changes of a global environment.

In this way, the strategic management of the social image and communication take on a special role, especially in an increasingly technological and hyper-connected environment, and the application of the social and viral marketing models of social network management is its main instrument of action.

These targets require to design and implement an ambitious strategic plan of social and viral marketing, which should be led by the European Commission and developed by all EU Member States.

References

1. Departamento de Seguridad Nacional: Estrategia de Seguridad Nacional de España 2017. Un Proyecto compartido de todos y para todos. Presidencia del Gobierno de España, Boletín Oficial del Estado, Madrid (Spain) Dec 2017

2. Fernández-Villacañas, M.A.: La construcción sociológica de la Identidad Europea de Seguridad y Defensa: La gestión estratégica de la imagen institucional y del marketing social. Ministerio de Defensa de España, SEGENTE, Madrid (Spain) Sep (2015)
3. Fernández Díaz, A., et al.: La Nueva Economía de la Defensa en un Nuevo Orden Mundial. Ministerio de Defensa de España, Madrid (Spain) Dec (2017)
4. García Arnáiz, J.: Inauguración de las Jornadas Aeroespaciales de Economía de la Defensa del Ejército del Aire. Ejército del Aire de España y UNED, Madrid (Spain) March (2017)
5. Fernández-Villacañas, M.A.: La Nueva Economía de la Defensa en un Nuevo Orden Mundial. Revista de Aeronáutica y Astronáutica, Madrid (Spain) June (2017)
6. Fernández-Villacañas, M.A.: La construcción sociológica de la Identidad Europea de Seguridad y Defensa, pp. 25–30. III Jornadas de Seguridad y Defensa, Universidad Politécnica de Cartagena (Spain) (2017)
7. Kotler, Ph, Roberto, E.: Social Marketing: strategies for changing public behavior. The Free Press, New York (USA) (1989)
8. Kotler, Ph., Andreasen, A.: Strategic Marketing for Nonprofit Organizations, 4th edn., Englewoods Cliffs, Prentice-Hall, New Jersey (USA) (1991)
9. Costa, G., Vila, M.: El societal marketing: Marketing de valor social y económico. Boletín de Estudios Económicos, **70**(215), 201–216, Universidad de Deusto, Bilbao (Spain) (2015)
10. Schneider, G., Bins, F.: Marketing social: abordagem histórica e desafios contemporâneos. Revista Brasileira de Marketing, **13**(3), 125–137. http://www.redalyc.org/articulo.oa?id=471747341008 (2014)
11. McAuley, A.: Reflections on a decade in social marketing. J. Soc. Mark. **4**(1), 77–86 (2014). https://doi.org/10.1108/JSOCM-09-2013-0062
12. Kotler, P., Armstrong, G.: Fundamentos de Marketing, Decimoprimera edn. Pearson Educación, Naucalpan de Juárez (Mexico) (2013)
13. Pykett, J., Jones, R., Welsh, M., Whitehead, M.: The art of choosing and the politics of social marketing. Policy Stud. **35**(2), 97–114 (2014). https://doi.org/10.1080/01442872.2013.875141
14. Santesmases, M.: Marketing: Conceptos y estrategias, Sexta edn. Ediciones Pirámide, Madrid (Spain) (2012)
15. Kennedy, A.M., Parsons, A.: Social engineering and social marketing: why is one "good" and the other "bad"? J. Soc. Mark. **4**(3), 198–209 (2014). https://doi.org/10.1108/jsocm-01-2014-0006
16. Saunders, S., Barrington, D., Sridharan, S.: Redefining social marketing: beyond behavioural change. J. Soc. Mark. **5**(2), 160–168 (2015). https://doi.org/10.1108/JSOCM-03-2014-0021
17. Richins, M.L.: Negative word-of-mouth by dissatisfied consumers: a pilot study. J. Mark. **47** Winter, SAGE (USA) (1983)
18. Silverman, G.: The secrets of word of mouth marketing: how to trigger exponential sales through runaway word of mouth. American Marketing Association, New York (USA) (2001)
19. Kirby, J., Marsden, P. (eds.): Connected Marketing: The Viral, Buzz and Word of Mouth Revolution. Butterworth-Heinemann (Elsevier), Oxford (UK) (2006)
20. Miller, R., Lammas, N.: Social media and its implications for viral marketing. Asia Pac. Relat. J. **11**, University of Newcastle, Callaghan (Australia) (2010)
21. Heim, S.: Viral marketing: establishing customer relationships by 'word-of-mouse'. J. Electron. Mark. **10**(3) (2000)
22. Woerndl, M., Papagiannidis, S., Bourlakis, M., Li, F.: Internet-induced marketing techniques: critical factors in viral marketing campaigns. Int. J. Bus. Sci. Appl. Manag. **3** (1), University of Kent (UK) (2008)

23. Túñez-López, M., Sixto, J., Guevara-Castillo, M.: Redes sociales y marketing viral: repercusión e incidencia en la construcción de la agenda mediática. Palabra Clave **14**(1), Universidad de La Sabana (Colombia) June 2011
24. Blázquez, M.: El problema de las noticias falsas: detección y contramedidas. XV Seminario Hispano-Mexicano de Investigación en Biblioteconomía y Documentación, 'México DF (México), 16–18 May 2018
25. Fernández-Villacañas, M.A.: Reflexiones sobre nuevas estrategias de comunicación social para el fomento de la cultura de la seguridad y defensa. Jornada de Economía y Tecnologías para la S&D, Universidad Politécnica de Cartagena (Spain) 26 April 2017
26. Aguado, G., García, A.: Del Word-of-mouth al Marketing viral: aspectos claves de la comunicación a través de redes sociales. Revista Comunicación y hombre, nº 5, Universidad Francisco de Vitoria (España) (2009)
27. Capriotti, P.: Planificación estratégica de la imagen corporativa. Barcelona: Ed. Ariel Comunicación, Barcelona (Spain) (1999)
28. Sanz de la Tajada, L.A.: Integración de la Identidad y la Imagen de la empresa. Desarrollo conceptual y aplicación práctica. Colección Empresarial e Innovación, ESIC, Madrid (1994)
29. Fernández-Villacañas, M.A.: La gestión estratégica de la imagen institucional de las Fuerzas Armadas. 5º Congreso de Investigación y Educación en Estudios de D&S REDES 2002 "Defence and Security Challenges in the 21st Century: Continuity or Change", CHDS, US National Defence University, Brasilia (Brazil) 7–10 August 2000

Chapter 18
A Survey on Centennials' Expectations of Mobile Operators

Mirjana Stojanović(✉), Marko Projović, Lazar Živojinović, Dušan Barać, and Zorica Bogdanović(✉)

Faculty of Organizational Sciences, University of Belgrade, Jove Ilića 154, 11000 Belgrade, Serbia
stojanovic.p.mirjana@gmail.com, markoprojovic@gmail.com, {lazar,dusan,zorica}@elab.rs

Abstract. To attract and retain consumers born in the era of social networks and grown up with the smartphones requires marketing and customer relationship management to be reshaped and adapted to their needs, habits, and aspirations. Digital natives expect digital customer experience through all the stages of the customer journey. Besides the challenges common for all industries, for mobile operators this additionally means to satisfy the customer segment which has the most demanding requirements in terms of mobile network and services quality but perceive it as a commodity. This paper addresses the topic of relationship between Generation Z and mobile operators, leveraging on the survey made among post-Millennials in Serbia. It presents their preferences and expectations and outlines the most important implications this might have on mobile operator's business.

18.1 Introduction

Numerous sources are using the term 'Centennials' to refer to the generations born in the period from the mid-1990s until the early 2010s. The alternative names used for this demographic group are 'post-Millennials', 'Generation Z' or 'Gen Z', 'iGeneration', and so on. Centennials have grown up with the internet. They are familiar with the technology and accustomed to multiple information sources, with messages bombarding them from all sides. The companies that want to reach them as the potential customers must take into consideration their specific characteristics, lifestyle, and attitudes when deciding on the appropriate marketing and communication strategy, as well as product and services for this particular segment [1]. Generic answer that new generation of digital customers requires digital marketing and digital channels of interactions is not sufficient. To engage with this 'always on' generation requires a serious shift to be made, both in marketing and customer relationship management in many industries [2]. Telecommunications are not the exception.

Being born into the social media era, post-Millennials are intensively interacting with friends/family via mobile devices such as smartphones [3]. Although mobile phones are their device of choice, always close by, and within reach, they are usually not

Á. Rocha et al. (eds.), *Marketing and Smart Technologies*, Smart Innovation, Systems and Technologies 167, https://doi.org/10.1007/978-981-15-1564-4_18

heavy users of the traditional services provided by mobile operators, as voice or SMS. On the other hand, availability and the speed of internet access are very important to them [4]. Furthermore, they prefer their smartphone manufacturer vs. their mobile operator, and the biggest gaps in satisfaction appear among data-heavy users [4].

The subject of this paper is to analyze the relationships of the mobile operators with their users from Generation Z and to better understand the expectations and communication preferences of the young subscribers. It could be a good starting point for the operators to customize their marketing strategy and their offering intended to this customer segment. This paper is organized as follows. Section 18.2 provides an overview of the existing findings related to the Centennials as the specific customers' segment that is targeted. Section 18.3 gives the perspective of mobile operators, together with an overview of the ongoing initiatives worldwide. Section 18.4 explains the main research questions which are further evaluated in the survey conducted among the post-Millennials in Serbia, presents the findings and analyses the results obtained. Section 18.5 concludes the paper with conclusions, practical implications on mobile operators, limitations, and suggestions for future work.

18.2 Who Are the Centennials?

In 2001, Marc Prensky [5] introduced the term 'Digital Natives' for the kids that have grown up with the digital technology and have had computer games, email, the internet, cell phones, and instant messaging as the integral parts of their lives. Although often used for Millennials, this term is even more appropriate for the generation that followed them, that is, Centennials or Generation Z.

Different researches have confirmed that Generation Z spends more time online than any other demographic segment. Besides being connected, they also find important to create and share the content over social media [4]. Sharing the content, ideas, or opinions via social networks is perceived as significant for visibility inside the community and for the reputation [4]. Peer acceptance is very important to them, they need to belong [1]. However, unlike the Millennials' mentality of broadcasting anything and everything to these platforms, Gen Z is more selective. They are shifting to a mentality of sharing only specific stories to specific people on specific social channels [6]. Additionally, being more familiar with the technology, they are more aware of the security and privacy issues [3].

To reach them as consumers demands appropriate marketing strategy. Using digital marketing might not be enough, or in some cases can even trigger negative response (e.g. banners can be perceived as annoying and intrusive) [7]. This is particularly true for any type of advertising that can be perceived as interrupting of the current activities [6]. Centennials want to get required information quickly and easily, and to achieve this, most of them use YouTube on their smartphones [3]. As purchasers, they expect transaction to be fast, simple, and reliable. They will lose interest if a page takes more than a few seconds to load, or a web form is to complex and not optimized for smartphone display [2].

This generation values authenticity and 'realness' [1]. Thus, Centennials as consumers are keen to interact with others on social networking sites to exchange information and opinion about products. Peer judgment and evaluation is valued and perceived as more credible than information provided by organizational sources or traditional media, that is, recommendation from other consumers is highly influential [7, 8]. Companies that want to increase and improve engagement of this consumer segment must consider them as partners, not as a target audience [9].

The other findings that are relevant to be mentioned here are that post-Millennials care about society and the environment [7] and positively react to cause-based marketing [10]. They are very proud of being recognized by friends and family as the experts in new technologies [4]. Additionally, they significantly influence the purchasing decisions and spending of other family members [11].

18.3 Centennials from the Mobile Operators' Perspective

As for the people, the term 'digital natives' can be used for the companies born in the same period, like Google, Facebook, Amazon, and so on. For them, an online marketplace, communication with the users through the different digital channels, and also the internal business processes that are supporting digital strategy are natural and being present from the beginning of their operations. On the other side, companies established before the digital era that want to adapt to the digital economy must pursue the process of digital transformation.

In case of mobile operators, according to the industry-agreed Digital Maturity Model by TM Forum [12], transformation or improvement (depending on the current maturity status and wanted position) is needed across five key dimensions that are: customer, strategy, technology, operations, and finally, culture, people, and organization [12]. Transformation related to the customer aims to provide an experience where customers view the organization as their digital partner using their preferred channels of interaction [12]. A segment of young, tech-savvy customers or 'digital natives' is particularly important for this transformation journey. Getting closer to this customer segment and engaging with them in a more innovative way is one of the primary objectives for the operators, especially in mature markets with increased competition and those with high percentage of Millennials/Centennials [13].

Following the communications, purchasing, and lifestyle preferences of the 'digital natives' consumer segment, operators already embraced the new ways of interacting with them and understood the importance of the effective self-service, omnichannel strategy, and the necessity of using social media. Digital Maturity Model by TM Forum [12] distinguishes four broad categories related to customer (customer engagement, customer experience, customer trust and perception, and customer insights and behavior) and the social media can and should play important roles in all of them. Besides the use of social media as the marketing channel, the operator can use it as a channel for customer care. Effective customer service response through social media can improve the trust and perception and finally, social media can be a very valuable source of information from which to develop sophisticated customer insights [13]. Communication that is perceived as the effective by this demographic segment often is described as the 'tap and text' phenomenon [2]. In the context of communication with the mobile provider, this would mean that the user can simply drop a message with the question/query to the mobile provider without interrupting whatever activity he/she is currently engaged in. This requires new approach to the user interfaces and many operators worldwide already introduced chat-bots to fulfill this type of requests [2]. Although most of the operators worldwide accepted social media as the preferred way of communication with the new generations of customers, only a few of them really made a radical change in the interaction and the relationship with this segment. The limitations in the legacy systems, products, and business processes, in most of the cases, impose the limited engagement of the youngest consumer segment, despite the operators' presence in all popular social networks.

Example of the operator who made the real step forward is GiffGaff, a sub-brand of O2 (Telefonica) launched in the UK in 2009 [13, 14]. GiffGaff addresses the users of its services as the community members rather than traditional customers. Their marketing slogan is 'network run by its customers' as the users actively contribute to sales, marketing, product development, and customer services. Community members are promoting the brand and offerings and recruiting new users, as a replacement of traditional marketing and sales department. Customer service requests (questions and complaints) are mainly answered satisfactorily by the other users, instead of traditional customer care. Some of the users are even making video tutorials for common queries. In addition, GiffGaff is leveraging the creative potential of its users which are proposing ideas for innovation that are commented upon and rated by other members, before being implemented. Users that are contributing are receiving in return, twice per year, incentives in the form of cash, airtime, or contribution to charity [13, 14].

A similar example is Jawwy launched in Saudi Arabia by Saudi Telecom Company (STC) [13, 15]. In both cases, operators are targeting youngest consumer segment through a dedicated sub-brand and leveraging much more on the collaborative potential of social media. Such an approach allows the experiments in order to create a new revenue stream and possibly a platform on which to build a new ecosystem and at the same time protects the value of the main brand and the contracted service level of its subscriber base. On the other side, it might require significant investments to maintain two parallel infrastructures [13]. Important question in this context is if such a solution would be applicable for both Millennials and Centennials, or the generational differences would require different approach.

18.4 Survey and Results

The survey made in this study examined the perception and preferences of the Centennials in Serbia about their relationship with mobile providers, as well as their expectations and readiness for new services. The main research questions evaluated in the survey are the following:

- What are the Centennials' decision criteria when selecting mobile operators/offers?
- How satisfied they are with the selected mobile operator and its current offering?
- How ready they are to engage on social media in relation to the operator's brand?

- How ready they are to consent with the further processing or disclosure of the personal data that mobile operators have about them?
- What do they expect from the new technologies, both in terms of improvements that operator can apply and in the domain of new services?

18.4.1 Data Sample

This survey was performed in the period from October 2018 till February 2019 among the population born between 1993 and 2003. This study examines the opinion of 221 participants, mainly coming from the Faculty of Organizational Sciences and the Faculty of Mining and Geology, University of Belgrade, and from two professional schools in Belgrade providing secondary education in the field of economy and transport. The main demographic data about the participants are summarized in Table 18.1.

Table 18.1. Main demographic data

Variable	Values	Frequency	Percentage
Age group	<16	3	1.4
	16–20	159	71.9
	21–25	55	24.9
	>25	4	1.8
Occupation	University student	118	53.4
	Secondary school student	102	46.2
	Employee	1	0.5
Gender	Female	107	48.4
	Male	114	51.6

18.4.2 Survey Methodology and Instruments

The questionnaire created for this research was anonymous and consisted of two parts. The first one was focused on demographic information, while the second part contained 29 questions, derived from the five main research questions listed above. Out of them, 28 were answered using the five-point Likert scale and grouped as follows:

- Decision criteria when selecting mobile operators/offers (DC1-DC5)
- Current satisfaction with the mobile operator (CS1-CS6)
- Interest to engage on social media in relation to the operator's brand (SE1-SE3)
- Motivation factors and/or remuneration for the engagement above (MF1-MF3)
- Readiness to consent with the further processing or disclosure of the personal data that mobile operators have about subscribers (PD1-PD3)
- Improvements in the business processes and in the customer relationships that the operator could make leveraging on the recent technology developments (BI1-BI3)
- New services that the operator could offer in the near future (NS1-NS5).

The last question was an open-ended question (OE1) for suggestions and ideas for new offers and services.

18.4.3 Results and Analysis

In addition to the basic data presented above in Sect. 18.4.1, describing the sample evaluated in this survey, demographic part of the questionnaire collected the other information relevant for the profile of respondents as the mobile operators' customers. In the context of the expectations that Centennials have from their mobile providers and from the mobile technology in general, it is particularly important to consider for how long they have been using a mobile phone. As per the results, 85.1% of the respondents have got their own mobile device when they were between 7 and 12 years old.

The majority of Centennials still economically depends on their parents or guardians. This fact is reflected in the answers to the question related to the cost of mobile subscription. Another important conclusion from Table 18.2 is that 53.8% of respondents spend more than 3 GB of mobile data per month, while 3.3 GB was the average mobile data usage in Serbia in Q3 2018 [16].

Table 18.2. Centennials as mobile subscribers

Variable	Values	Frequency	Percentage
Who is bearing the cost of your mobile subscription?	Paid by parents/guardians	183	82.8
	Paid from my own pocket	36	16.3
	Paid by the company I am working for	1	0.5
Your monthly usage of mobile data is:	<3 GB	66	29.9
	3–10 GB	92	41.6
	>10 GB	27	12.2
	I do not know	36	16.3

Results of 28 questions answered using the five-point Likert scale are analyzed on the total respondents' base, and are also within the six biggest demographic groups (female, male, university students, secondary school students, the population in the age between 16 and 20, and population in the age between 21 and 25).

The tables in the sub-chapters show the average value of the results for the total respondents' base, standard deviation, and the confidence intervals for the confidence level of 95%. The results obtained in the corresponding demographic groups were further compared using the t-test (female vs. male, age group '16–20' vs. '21–25', and university students vs. secondary school students). As per this comparison, there is no significant difference between the genders. Particular questions for which t-test showed a significant difference ($p < 0.05$) in comparison of the responses obtained in the age group '16–20' versus '21–25' or university students versus secondary school students are emphasized in the sub-chapters below.

Decision criteria when selecting mobile operators/offers. When selecting a mobile operator or particular package from its portfolio, Centennials find the economic criteria (DC2-DC3) as the most important (price-3.99, benefits and bonuses-3.79). Criteria

DC1–DC3 are particularly relevant for the older participants (results of the *t*-test based on the age groups were: $*p < 0.022$, $**p < 0.001$, and $***p < 0.001$). On the other side, traditional advertising (DC5) does not have a significant impact on this population (mean-1.92) (Table 18.3).

Table 18.3. Survey results: importance of the different decision criteria

	Mean	SD	CI
DC1*: The opinion of your friends or the comments of other users	2.90	1.35	0.18
DC2**: Price	3.99	1.23	0.16
DC3***: Benefits and bonuses you can get as their user	3.79	1.26	0.17
DC4: Information posted by the operator on their website or profiles and channels on the different social networks	2.65	1.27	0.17
DC5: Traditional advertising (TV, billboards, etc.)	1.92	1.11	0.15

Current satisfaction with the mobile operator. Overall satisfaction of the respondents is good (average rates are >3.4). The significant difference in the answers exists only between university and secondary school students for CS2, CS3, and CS6, where the results of the corresponding *t*-test were: $*p < 0.049$, $**p < 0.004$, and $***p < 0.049$ (Table 18.4).

Table 18.4. Survey results: satisfaction on the different aspects

	Mean	SD	CI
CS1: Quality and speed of mobile internet	4.31	0.92	0.12
CS2*: Quality of other services	4.19	0.93	0.12
CS3**: Prices	3.71	1.10	0.15
CS4: Benefits and bonuses I can get as their subscriber	3.44	1.19	0.16
CS5: Possibility of getting online, quickly and easily, all the information about the way of using the service, spending, etc., or of changing the tariff/package without going to the shop or calling a contact center	3.97	1.28	0.17
CS6***: Possibility of personalizing the offer (e.g. special rate plan for the numbers, sites, services that are important to me)	3.66	1.37	0.18

Interest in engaging on social media in relation to the operator's brand. While questions ES1–ES3 addressed general interest in different types of engagement, questions MF1–MF3 evaluated what would motivate them to accept this task. Contrary to what was expected, the results showed a low interest (<2.5) of Centennials in this kind of engagement. Interest is increasing (to 3.6) when this engagement is in charitable purposes (MF2). Another observation is that the reputation on social networks (MF3) is more important motivational factor for the age group between 16 and 20 than for the older population ($*p = 0.035$) (Table 18.5).

Table 18.5. Survey results: interest in engaging on social media and potential motivation

	Mean	SD	CI
ES1: To promote the brand and services of the mobile operator	2.36	1.49	0.20
ES2: To support other users (by answering their questions, preparing video-tutorials, etc.)	2.25	1.28	0.17
ES3: To share your ideas and proposals for new services, new tariff plans, or new ways of communicating with the end users	2.42	1.37	0.18
MF1: To get additional credit for mobile services	2.97	1.49	0.20
MF2: To secure that certain amount of money is paid for the charity	3.60	1.27	0.17
MF3*: To get a better reputation among your contacts on social networks	2.54	1.46	0.19

Readiness to consent with the further processing of the personal data. The mobile operator has a lot of information about subscribers' communications habits and the way they are using the mobile network. The data collected in such a way can be used in different scenarios, out of which three have been evaluated in the survey. According to the results, participants would rather allow the operator than a third party to further process their personal data. The willingness to consent to disclose the data to the third party is significantly lower in the older (21–25) age group (*$p = 0.032$) (Table 18.6).

Table 18.6. Survey results: readiness to consent with the further processing of the personal data

	Mean	SD	CI
PD1: To get an offer for new services from the operator, personalized according to your habits and needs	3.36	1.38	0.18
PD2: To get an offer for the rate plan which would be optimal for you instead of selecting one from the standard tariffs	3.64	1.31	0.17
PD3*: To give the consent to the operator to share your data with the third party, which will be used to offer to you its own product and services, and you will get in return certain bonuses and incentives	2.77	1.36	0.18

Expectations from the new technologies. Recent technology developments (5G, AI, IoT, blockchain, etc.) create different opportunities for the mobile operators. In this section, participants rated the importance of the potential enhancements in the business processes and in the customer relationships that the operator could make leveraging on that (BI1–BI3) and their interest in the new services that the operator could offer in the near future (NS1–NS6). Improvements related to the quality of service and personalized tariff are more important to this segment than the improvements in the way they are interacting with the operator (mean for BI1 = 2.75). Quality of service (BI3) is significantly more important for the age group 21–25 than for the younger respondents (*$p = 0.001$). Interest in new services is good (average rates in all cases >3.1), particularly for services for which the awareness in the local market has been already created (e.g. smart home). Considerable difference between the answers of the respondents from two age groups is registered only in case of the question NS4 (**p = 0.020) (Table 18.7).

Table 18.7. Survey results: expectations from the new technologies

	Mean	SD	CI
BI1: To have a personalized digital assistant (similar to Siri on iPhone) available, which will accept and answer on your voice commands and inquiries about the services/packages/tariffs	2.75	1.44	0.19
BI2: To have the operator applying automatically the tariff that is the most optimal for you and your usage of the network and services in the given time period	3.14	1.32	0.18
BI3*: To have the operator using artificial intelligence to prevent the issues in the network and to secure the guaranteed quality of service	3.36	1.37	0.18
NS1: To use your mobile phone for locking/unlocking the objects you are renting, and the rental fee to be included in the monthly bill you are paying to the mobile operator	3.12	1.45	0.19
NS2: To use your mobile phone to control future smart devices and the different sensors in your home	3.63	1.33	0.18
NS3: To enrich/complement the image/audio from the real world or the information available via mobile phone, by using augmented reality/virtual reality on your mobile device	3.26	1.31	0.17
NS4**: To enable on your mobile phone the access to the different multimedia content (e.g. movies, music, sports events, lectures, etc.) with the possibility of interacting with the other users following the same content	3.60	1.30	0.17
NS5: To monitor in real time the important indicators of your health (e.g. blood pressure, sugar level, etc.) and that in case of irregular values, you get the advice on how to react or get in contact with the doctor	3.57	1.29	0.17

In the last, open-ended question (OE1), respondents mainly suggested lower prices, unlimited consumption of data traffic, and faster data rates.

18.5 Discussion and Conclusions

Although some demographic studies have been reported in the industry, the topic of Centennials as the users of mobile technology has not been explored in detail in scholarly research. At the same time, customers, especially the 'digital natives' present one of the main pillars of the digital transformation of the mobile operator. Thus, the main contribution of this paper is in providing survey's results that could help in better understanding of this cohort and their expectations. This survey has confirmed several statements from the existing literature about Generation Z, but in some cases different results were obtained.

It is justified that this hyper-connected population uses the mobile data more than the average mobile users and that the faster mobile internet is one of the features they are most interested in [4]. As per the recent worldwide results [17], Gen Z shows even higher levels of adoption of the digital services than Millennials and continues to lead the shift in usage patterns. Their interest in new services, including mHealth, mEducation, mEntertainment, has been confirmed in this survey. As expected, results showed high importance of other users' opinion and almost complete irrelevance of the traditional advertising [7, 8]. Respondents confirmed that Generation Z values fast and easy access to the correct information [3]. Although the personalization is important for this segment, they are concerned about privacy and security of their personal data and want to know who is going to have the access to them [3, 6]. At the same time, survey confirmed on this particular data sample the general conclusion that the operators are among the most trusted organizations by consumers for handling personal data [18]. The hypothesis related to the readiness of this segment to take an active contribution in promoting operators on social media and co-creation of new offerings, based on [4, 9], has not been confirmed, except when it relates to humanitarian causes. This approves the thesis about social responsibility of the post-Millennials [7].

Following the results from this study, the mobile operators should do the following:

- Prepare the offers affordable to this segment, preferably with the clear bonus and reward policies, but avoid the trap that economic benefits could compensate poor coverage or quality of service.
- Stay focused on network quality and reliability, especially taking into consideration data-intensive applications (streaming, video, AR/VR, mobile gaming).
- Enable simple, fast, and reliable access via digital channels to all the information relevant for the end-users in all the stages of the customer journey (example 'tap and text' [2]). This has obvious implications not only on the operator's shops and contact center but also huge impact on all business support systems and processes that should secure consistency regardless of the channel used for customer interaction.
- Develop sophisticated customer insight from the digital interactions above. Take into consideration 'word of mouth' via social networks, that is, users' opinions, comments, complaints, and preferences related to the operator's services.
- Carefully consider the different aspects of personalization (recommendations, tariffs, services, etc.) and secure that the way of using personal data is properly documented and protected from misuse and that relevant consent has been obtained from the user.

The main limitations of this study are related to the data sample and the fact that respondents are still within the educational system. It would be interesting to compare the results obtained in this analysis with the opinion of the post-Millennials who have not enrolled to University. Furthermore, the readiness for the new services in mobile networks can be verified using the model described in the Unified theory of acceptance and use of technology in a consumer context, UTAUT2 model [19], or its modification, and the results can be compared between Millennials and post-Millennials as two most important customer groups in the coming period.

Acknowledgements. The authors are thankful to Ministry of Education, Science and Technological Development, grant no. 174031.

References

1. Williams, K., Page, R.: Marketing to the Generations. J. Behav. Stud. Bus. **3**, 37–52 (2011)
2. COMARCH: Putting the digital customer experience at the heart of telco omnichannel strategies. White paper (2018)
3. Crearie, L.: Millennial and centennial student interactions with technology. GSTF J. Comput. **6**(1), 1–10 (2018)
4. CommScope: The generation Z study of tech intimates. https://www.commscope.com/Insights/uploads/2017/09/Generation-Z-Report.pdf, last accessed 2019/07/19
5. Prensky, M.: Digital natives, digital immigrants Part 1. On Horiz. **9**, 1–6 (2001)
6. Fromm, J.: What marketers need to know about social media and Gen Z. https://www.forbes.com/sites/jefffromm/2016/12/19/what-marketers-need-to-know-about-social-media-and-gen-z/#6165278a7e65, last accessed 2019/07/19
7. Smith, K.: Longitudinal study of digital marketing strategies targeting Millennials. J. Consum. Mark. **29**, 86–92 (2102)
8. Cruz, I.: Centennials, the generation that has never known a world without the internet. https://www.bbva.com/en/centennials-generation-never-known-world-internet/, last accessed 2019/07/19
9. Fromm, J., Butler, C., Dickey, C.: How to engage Millennials: re-imagining the consumer as a partner, not a target audience, to increase engagement. J. Brand. Strat. **4**(1), 27–36 (2015)
10. Beckmann, M., Noll, F., Berndt, A.: Cause-related marketing (CM): the perspective of millennials (Abstract). In: Stieler, M. (eds.) Creating Marketing Magic and Innovative Future Marketing Trends. Developments in Marketing Science: Proceedings of the Academy of Marketing Science. Springer, Cham (2017)
11. Barkley: The power of Gen Z influence. htpp://www.millennialmarketing.com/wp-content/uploads/2018/01/Barkley_WP_GenZMarketSpend_Final.pdf., last accessed 2019/07/19
12. TM Forum: Digital maturity model project. https://www.tmforum.org/collaboration/dmm/, last accessed 2019/07/19
13. TM Forum: Targeting millennials: CSPs use social media and sub-brands to improve customer centricity. https://inform.tmforum.org/research-reports/targeting-millennials-csps-use-social-media-sub-brands-improve-customer-centricity/, last accessed 2019/07/19
14. The giffgaff community. https://community.giffgaff.com/, last accessed 2019/07/19
15. Jawwy from STC, https://www.jawwy.sa/en, last accessed 2019/07/19
16. RATEL: An overview of the telecom market 2018-Q3. https://www.ratel.rs/en/page/quarterly-reports, last accessed 2019/07/19
17. Meyer, M., Giles, M.: Mobile internet use mapping the age gap. GSMA Intelligence (2019). https://www.gsmaintelligence.com/research/2019/08/mobile-internet-use-mapping-the-age-gap/796/, last accessed 2019/08/27

18. van den Dam, R.: The trust factor in the digital economy: why privacy and security is fundamental for successful ecosystems. In: 14th ITS Asia-Pacific Regional Conference, Kyoto (2017): Mapping ICT into Transformation for the Next Information Society. International Telecommunications Society (ITS), Kyoto (2017)
19. Venkatesh, V., Thong, J.Y.L., Xu, X.: Consumer acceptance and use of information technology: extending the unified theory of acceptance and use of technology. MIS Q. **36**(1), 157–178 (2012)

Chapter 19
Mobile Applications at Music Festivals in Portugal

Daniel Fonseca[1], José Luís Reis[1,2(✉)], Sandrina Teixeira[1], and Marc K. Peter[3]

[1] ISCAP, Porto Superior Institute of Accounting and Administration, Research Unit CEOS.PP/ISCAP/P.PORTO, Matosinhos, Portugal
fonseca.danielmartins@gmail.com, jreis@ismai.pt, sandrina@iscap.ipp.pt
[2] ISMAI Maia University Institute, Research Unit CEDTUR/CETRAD, Maia, Portugal
[3] FHNW School of Business, Olten, Switzerland
marc.peter@fhnw.ch

Abstract. In Portugal, the number of music festivals as well as the number of participants has been increasing. It is essential to be able to provide the best possible experience to the participants of these same festivals. Based on a survey, this paper analyses the frequency of use of mobile applications of major music festivals in Portugal. In order to verify the factors leading to the adoption of applications, the Unified Theory of Acceptance and Use of Technology (UTAUT) model was used. The results show that less than half of the participants are using mobile applications, and that the main reason for not using is the lack of knowledge about the applications. Correlations of the analysed constructs of the UTAUT model were also tested, and correlations between the constructs were verified with the mediation of the age and gender moderators.

19.1 Introduction

Music festivals are one of the central elements of the tourist destination system with the aim of expanding the tourist potential of these destinations beyond the narrow focus on leisure tourism, providing authentic experiences [1]. The importance of integrating event visitors into the destination to increase satisfaction and improve experiences is a strategic imperative [2].

A study in Portugal [3] found that internet access on handheld devices accounted for 79% of internet users in 2017, and this indicator continues to show a growing trend. Of the portable equipment used to access the mobile internet, the mobile phone or smartphone is the most mentioned (76%), followed by the laptop (51%). Using smartphone apps that offer services to visitors of outdoor cultural events before, during and after the event can be a major marketing tool for event organisers [4].

This article aims to identify the drivers and utilisation level of mobile applications at music festivals in Portugal based on Unified Theory of Acceptance and Use of

Á. Rocha et al. (eds.), *Marketing and Smart Technologies*, Smart Innovation, Systems and Technologies 167, https://doi.org/10.1007/978-981-15-1564-4_19

Technology (UTAUT), a model utilised with the aim to understand if music festival participants know the mobile applications of the attended events and, if using them, what their expectations and needs are, what they expect to find in the official mobile application of the event and the reasons that influence their use.

As this work is exploratory, a quantitative approach was chosen based on the defined model: An online questionnaire was developed and distributed to smartphone holders who participated at music festivals in Portugal in 2018.

19.2 Music Festivals in Portugal

An event is a planned happening that will take place at a certain date, place and time previously announced. The promoter of an event always has, regardless of the event typology, a concrete goal to bring people together for a specific purpose [5].

Various types of events and festivals were part of human history throughout the ages. According to Getz [6], the events can be classified as: cultural, political, arts and entertainment, business, education and scientific celebrations, sports and recreational competitions, with festivals framed in cultural celebrations. Following Sarmento [7], festivals have the potential to constitute local political strategies for promoting leisure activities with the aim of attracting visitors and tourists. The author further concludes that the success of events depends on the innovative potential and creativity of programmes and organisations.

In Portugal, following the concept of the Woodstock festival of 1969, the Vilar de Mouros festival was born in 1971. Considering the study by Martinho and Neves [8], about 20 music festivals were accounted for in 1985, predominantly on the coast, and especially in the Algarve region. In 1999, in addition to the increase in the number of festivals, there was a greater diversity of the musical genre. Here, the rock genre has increased its appearance with festivals such as Noites Ritual Rock in Porto and the Paredes de Coura Music Festival [8].

The study by Guerra [9] states that in 2005, 115 music festivals were held in Portugal. According to data provided by APORFEST [10] in its annual report for 2017, 272 music festivals were held in Portugal, reflecting a 9.2% growth over the previous year. In 2017, 38 festivals had their first edition, and 71.4% of the festivals were paid admission. As for their size, 57.2% of the festivals were small (meaning less than 1000 visitors) and only 12.6% large. Lisbon, Porto, Braga and Funchal were the locations with the most festivals [10].

For all the festivals held during 2017, over 2.5 million visitors were recorded, a figure equivalent to one quarter of the resident population of Portugal [10].

19.3 Mobile Marketing

Mobile marketing consists of a set of practices that enable organisations to communicate and engage with the public in an interactive and relevant way through any mobile device or network [11].

19.3.1 Definition of Mobile Marketing

For Leppäniemi, Sinisalo and Karjaluoto [12], mobile marketing can be defined as the use of an interactive communication channel that provides consumers with geo-localised and personalised information that promotes goods, services and ideas that generate value for all stakeholders. However, sending non-relevant content may lead to negative consumer attitudes towards a particular product and/or service.

Huang and Symonds [13] confirm the main features of localisation, personalisation and interactivity of mobile marketing. Reaching or being sought by consumers involves creating activities that foster the building of relationships between the brand and the consumer, such as advertising, promotions and customer support, among other forms of interactive communication that are only possible because of a two-way or multi-directional communication using mobile devices as a medium [14].

19.3.2 Mobile Applications

Mobile applications or apps are software installed on top of the operating system of any type of mobile device (smartphones and tablets) with the ability to access online and offline features and content. Today, mobile applications have become increasingly important to consumers' daily activities. Smartphones and tablets give us the opportunity to connect to a global network of services, information and things at any time, regardless of one's location. An individual not only has the opportunity to search for information or use a service but is also able to generate and contribute information generated by himself/herself and disseminate it to the general public, thereby becoming a living member of a global community [4].

Apps have the potential to change the user experience of music festivals, especially as the festival market attracts participants who are most likely to be smartphone users interested in using the technology [2].

The mobile application marketplace allows developers to market and distribute apps, and users to search and install applications on a mobile device. For most use cases, both free and paid apps are available. Most of the mobile applications are programmed by third-party developers, such as companies wishing to advertise or extend their value proposition, or by simply offering an online product, or freelance programmers who sell their applications to generate revenue [15].

19.3.3 Mobile Marketing Tools

Shankar and Balasubramanian [14] state that mobile devices have three important features: geolocation, portability and wireless connectivity. Communication in mobile marketing exists through two communication types, namely push and pull. In push communication, information is sent to the individual through SMS, MMS, notifications, alerts or other advertising and content. Pulling requires users' initiative to take the first step by looking for the information they want. Some of the tools used are e-mail marketing, coupons or sending location-related information (geolocation), among others.

Finally, the Mobile Marketing Association [16] highlights mobile video, display or audio ads, a mobile website, mobile apps, response codes, mobile search marketing, SMS and MMS, location-based marketing and near field communications as important tools of mobile marketing. Table 19.1 summarises important mobile marketing tools.

Table 19.1. Mobile marketing tools adapted from [12–14, 16]

Tools	Source	Description
Mobile website	14, 15, 16	It is a version of a website that is specifically designed to be mobile compatible. Mobile websites offer an engaging and streamlined mobile experience that appeals to the mobile user
E-mail marketing	15	While e-mails are also utilised on other devices, they also support mobile marketing as most users have e-mail installed on their mobile devices
Mobile video, display or audio ads	14, 16	They are an effective way to attract new and potential customers with valuable content in different shapes and formats
Personalised services and content	13, 14	Personalisation is important as it tailors the communication to the user's needs and behaviour, thus leveraging the chance that the user will engage with the sender (i.e. website or app provider)
Coupons	15	Coupons support the engagement and interaction between users and the provider of goods and services as they provide an incentive for the user to take an action
Mobile apps	14, 16	These are software programmes that can be installed on a smartphone or tablet. Apps can be used by providers to educate, entertain, engage and/or sell products
Response codes	14, 16	There are several different types of response codes in the market, including QR codes, Microsoft tags, ScanLife and SPARQ codes. Similar to coupons, they push the engagement between the provider and the users
Mobile search marketing	16	Mobile search includes a set of techniques in order to be found through mobile search, including SEO (search engine optimisation) and SEA (search engine marketing)
SMS and MMS	14, 15, 16	Short Message Service (SMS) and Multimedia Messaging Service (MMS) are systems that allow a sender to distribute text or multimedia (graphics, video, audio) content to users
Location-Based Marketing (LBM)	13, 14, 15, 16	Similar to personalisation, LBM enables the distribution of context-based information to users. There are two subcategories of LBM: Location-based services such as Foursquare, SCVNGR and WHERE, and location-based advertising, which uses smartphone image ads to geographically target customers in a specific location
Near-Field Communications (NFC)	15, 16	Like BlueTooth, NFC uses a small smartphone chip to wirelessly connect to another chip placed for instance in a kiosk, point of sale or debit/credit card terminal. It supports LBM and personalisation

The brief review of the literature enabled the identification of the most important mobile marketing tools. They include the mobile website, ads, personalisation, mobile apps, SMS/MMS and location-based marketing. When organisations communicate to mobile users, these tools should be considered for effective mobile marketing strategies.

19.4 Conceptual Model and Research Hypotheses

The underlying conceptual model is the Unified Theory of Acceptance and Use of Technology (UTAUT) with its four central determinants/constructs of intention and use, and up to four key relationship moderators. The four constructs that play a significant role as direct determinants of user acceptance and use behaviour are: (1) performance expectation, (2) effort expectation, (3) social influence, and (4) enabling (or facilitating) conditions. These constructs are moderated by gender, age, voluntary use and experience. Labels used for constructs describe the essence of the construct and should be independent of any specific theoretical perspective [17].

Performance expectancy is defined as the degree to which an individual believes that using the system will help him or her to gain in the performance of their work; effort expectation is defined as the degree of ease associated with using the system; social influence is defined as the degree to which an individual realises that other important or influential people believe he or she should use the new system; and facilitating conditions are defined as the degree to which an individual believes that there is an organisational and technical infrastructure to support the use of the system.

Figure 19.1 represents the conceptual research model, which reflects the constructs mentioned and adapted from the UTAUT model [17], as well as the different defined hypotheses. The adapted model includes three of the constructs (performance expectation, effort expectation and social influence) that will influence the behavioural intent and will be moderated gender and age.

Based on the literature review and conceptual model, nine hypotheses were defined. Hypotheses H1, H2 and H3 are validated through the analysis of the answers obtained to specific questions of the:

- H1: Most music festival participants do not use music festival applications.
- H2: The moment when most participants install the application is before the music festival.
- H3: The moment when participants use the application the most is during the music festival.

Two hypotheses concern the influence of moderators on the correlation between the performance expectation and behavioural intention constructs:

- H4: Gender influences performance expectation in relation to behavioural intention.
- H7: Age influences performance expectation in relation to behavioural intention.

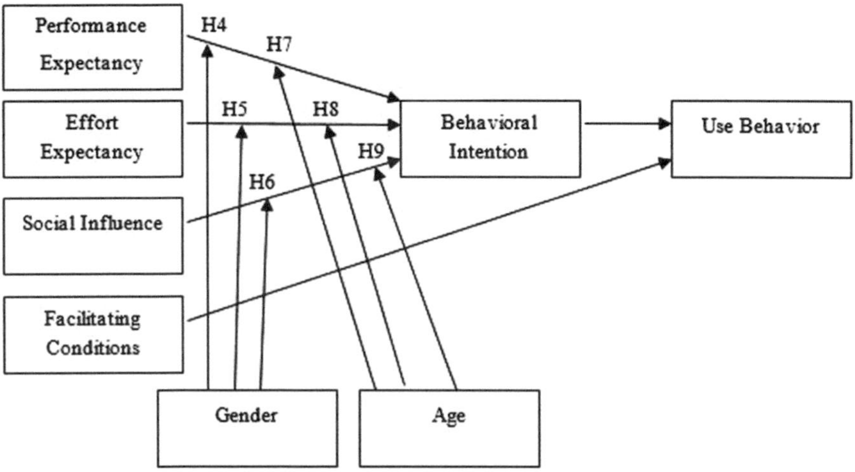

Fig. 19.1. Conceptual model of research adapted from [17]

Two hypotheses refer to the influence of moderators on the correlation between the constructs effort expectation and behavioural intent:

- H5: Gender influences the expectation of effort in relation to behavioural intention.
- H8: Age influences the expectation of effort in relation to behavioural intention.

Two hypotheses concern the influence of moderators on the correlation between the constructs social influence and behavioural intention:

- H6: Gender influences social influence in relation to behavioural intention.
- H9: Age influences social influence in relation to behavioural intention.

19.5 Methodology

This work is based on an exploratory, quantitative research approach, using an online questionnaire, addressed to mobile devices (i.e. smartphone) users who participated in music festivals in Portugal in 2018. The questionnaire was distributed on the social media platform Facebook in selected interest groups from 15 August 2018, via e-mail and SMS invitations. It is therefore a convenience sample.

The questions in the online survey are closed questions and mostly measured using a Likert scale, except for options where it was possible to indicate another value different from those available to choose from. The questions were created based on the model provided by Venkatesh et al. [17].

The questionnaire was intended for participants of music festivals in 2018 and owners of a smartphone. Given these conditions, the first question corresponded to whether a smartphone was owned, the second to whether the user participated in a music festival. The third question referred to an actual participation in 2018. These screening questions allowed the segmentation of those who did not own a smartphone and/or have not participated in music festivals in 2018, while still allowing them to complete the survey. The remaining questions focussed on the nine hypotheses.

19.6 Analysis and Discussion of Results

This section presents the results obtained through the online survey, the verification of the research hypotheses previously presented, as well as the discussion of the results obtained.

19.6.1 Size and Socio-Demographic Characterisation of the Sample

The total sample includes 208 respondents who own a smartphone and have participated in music festivals. Regarding the socio-demographic setup of the sample, 46.2% of respondents are male and 53.8% female. As for age, the most represented age group is 25–29 years (23.6%), followed by 35–39 years (22.6%) and 30–34 years (22.1%).

Regarding their academic qualifications, 47.1% of respondents have completed a degree, 26.4% have a master's degree, and 22.6% of respondents have completed the 12th grade. With regard to their employment status, 115 respondents are employees, representing 55.3% of the sample; 12% are self-employed, 10.6% are students and 10.1% are both part-time employees and students.

19.6.2 Participation in Music Festivals and the Use of Mobile Applications

Of the total sample, 56.3% participated in one or multiple music festivals in 2018; and 43.8% did not. At this point in the survey, those who did not participate in a music festival in 2018 were referred to the question about the perception of future behavioural intention in relation to mobile applications of music festivals.

The most visited music festivals in which the respondents participated in 2018 were NOS Primavera Sound (21.28%), NOS Alive (17.73%), Rock in Rio (16.31%) and MEO Marés Vivas (15.60%). The option "Other" represents several other festivals which did not have any significance in the overall data set, even though they represent 26.24% of participants.

Regarding the use of mobile applications from music festivals (Fig. 19.2), the focus topic of this research, 58.12% of respondents say that they do not use any applications, a result that meets the results obtained by APORFEST [10] in their study of the profile of festival visitors. Hence, only 41.88% of the respondents use a mobile application provided by a music festival, which is why hypothesis H1 (most music festival participants do not use music festival applications) is validated, albeit 42% of survey participants are using a festival's mobile app.

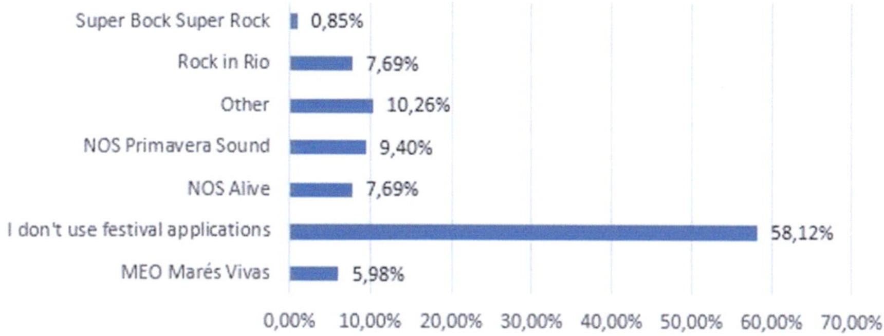

Fig. 19.2. Percentage of use of the official mobile applications in the music festivals

Regarding the way in which festival participants obtained information about the existence of the mobile application of a music festival, the event's official website as a source of information was the answer given by the majority of respondents and represents 34.7% of the sample, followed by social networks with 28.6%. Also noteworthy is the "through friends" response, which represents 16.6%; the festival's promotional materials (12.2%) and mobile app stores (8.2%).

In order to validate hypothesis H2, participants were asked at which point in time they installed the application, with 79.6% of respondents installing the application before the festival, and only 20.4% installing it during the festival. The installation after the festival was not chosen by any of the respondents. Hypothesis H2 can therefore be validated, since most participants install the application before the music festival.

Regarding the point in time when participants use the application most, the survey revealed that 85.7% of respondents do so during the festival, 14.3% say that they used the app before the festival most. Thus, the validity of hypothesis H3 is confirmed, as it is during the music festival when participants use the application the most.

19.6.3 Degree of Satisfaction with the Application Used

Regarding the degree of users' satisfaction matching the expectations of use of mobile applications, the survey identified that 40.8% of respondents are satisfied, 36.7% very satisfied, 20.4% of respondents say they are not very satisfied and only 2% are not satisfied with the mobile app. Overall, this is positive feedback for the app providers, as the majority of the sample (77.5%) is either satisfied or very satisfied.

19.6.4 Correlation Analysis

The Pearson correlation was used between the constructs of expectation of effort, social influence and expectation of performance with the behavioural intent, moderated by gender. The analysis shows that only for females, a correlation between all the constructs exist, as for males, the only correlation is between the behavioural intention and the social influence (Table 19.2). Data shows that the gender moderator exerts some influence on the correlation of the constructs, thus supporting hypotheses H4 (Gender influences performance expectation in relation to behavioural intention), H5 (Gender influences the expectation of effort in relation to behavioural intention) and H6 (Gender influences social influence in relation to behavioural intention).

Table 19.2. Pearson correlation between the constructs expectation of effort, social influence and expectation of performance with behavioural intent, moderated by gender

Gender			Social influence	Expectation of effort	Performance expectation
Masculine	Behavioural intention	Pearson correlation	348[a]	0.127	0.025
		Sig. (2 extremities)	0.005	0.319	0.847
		N	64	64	64
Feminine	Behavioural intention	Pearson correlation	0.373[a]	0.422[a]	0.433[a]
		Sig. (2 extremities)	0.001	0.000	0.000
		N	76	76	76

[a]The correlation is significant at the 0.01 level (2 extremities)

In order to verify the correlation of the constructs performance expectation, expectation of effort and social influence with the construct of behavioural intention, moderated by age group, the Pearson correlation was used as well.

As there were no answers obtained from the "under 18" and "over 50" age groups, and the "40–44 years" and "45–49 years" age groups obtained only two and three answers, respectively, no conclusion can be drawn from the data set. Of the remaining echelons, the only one that had a significant sample to perform Pearson's correlation was the age group "35–39 years" with 50 responses, and where significant correlation values were found (Table 19.3).

The highest correlation value was found between the constructs social influence and behavioural intent in the age group "25–29 years" with Pearson's correlation value of 0.654. Despite the low correlation values, it can be said that age (age group) is somehow a moderator of the presented constructs, thus validating hypotheses H7 (Age influences performance expectation in relation to behavioural intention), H8 (Age influences the expectation of effort in relation to behavioural intention) and H9 (Age influences social influence in relation to behavioural intention).

19.6.5 Reasons for Low Music Festival Mobile Apps Adoption

Finally, the data set provided insights to the reasons for a low adoption of mobile apps from music festival providers: 44.12% of respondents say that they do not use mobile apps from music festivals because they do not know that they exist, 16.18% of respondents indicate that they do not find the application useful, 11.76% respond that there are no advantages compared to the use of other festival communication channels, and 10.29% of respondents say that they cannot see any benefits in music festival

Table 19.3. Pearson correlation between the constructs expectation of effort, social influence and expectation of performance with "behavioural intention", moderated by age group

Age group			Expectation of performance	Expectation of effort	Social influence
18–24 years old	Behavioural intention	Pearson correlation	0,162	0,333	0,471
		Sig. (2 extremities)	0,565	0,225	0,076
		N	15	15	15
25–29 years old	Behavioural intention	Pearson correlation	0,352[a]	0,306	0,654[b]
		Sig. (2 extremities)	0,041	0,078	0,000
		N	34	34	34
30–34 years old	Behavioural intention	Pearson correlation	−0,094	0,091	−0,066
		Sig. (2 extremities)	0,585	0,597	0,704
		N	36	36	36
35–39 years old	Behavioural intention	Pearson correlation	0,436[b]	0,455[b]	0,400[b]
		Sig. (2 extremities)	0,002	0,001	0,004
		N	50	50	50

[a]The correlation is significant at the 0.05 level (2 extremities)
[b]The correlation is significant at the 0.01 level (2 extremities)

mobile apps. This might highlight the need for improved market communication about the existence of such mobile apps, and to improve the offering (features and/or content) of the said apps from music festival providers.

19.7 Conclusions

In mobile marketing, mobile websites, ads, personalisation, SMS/MMS, location-based marketing and mobile apps play an important role in the market communication mix of organisations, including music festival providers.

This study focused on mobile apps and their use and perception in the music festival industry. Despite the relatively low mobile application utilisation rate obtained in the results of the present study (only 41.88% of respondents who participated in music festivals in 2018 used the official application), the result is similar to that obtained in a study from 2017 [10].

As it is before the event that the participants install the festival applications, it is recommended that the availability of a mobile app shall be communicated before the event, motivating the use of the application. In addition, as most festival participants are using the apps during the event, it is recommended that during this period, the event provider might want to continuously update the contents (and potentially features) of the app.

The most mentioned factors for the non-use of mobile applications of music festivals are the lack of knowledge of the existence of such an app and the lack of a perceived value for an adoption (i.e. installation and use). Therefore, any festival marketing communications plan might benefit from improvements in relation to the dissemination and benefits of its use, as it is not sufficient to merely upload the mobile application to an app store. Here, the most important channels are the festival website, social media and word of mouth. Overall, music festival mobile apps deliver customer satisfaction when they provide value to the user.

References

1. Getz, D., Page, S.J.: Progress and prospects for event tourism research. Tour. Manag. **52**, 593–631 (2016)
2. Luxford, A., Dickinson, J.E.: The role of mobile applications in the consumer experience at music festivals. Event Manag. **19**(1), 33–46 (2015). https://doi.org/10.3727/152599515x14229071392909
3. INE, I. d. E.: Inquérito à Utilização de Tecnologias da Informação e da Comunicação pelas Famílias 2017 (2017)
4. Koukopoulos, Z., Koukopoulos, D.: Smart dissemination and exploitation mobile services for carnival events. Procedia Comput. Sci. **110**, 24–31 (2017)
5. Isidoro, A.M., Simões, M., Saldanha, S., Caetano, J.: Manual de Organização e Gestão de Eventos. Edições Silabo, Lisboa (2013)
6. Getz, D.: Event tourism: definition, evolution, and research. Tour. Manag. **29**(3), 403–428 (2008)
7. Sarmento, J.: Festivais de Música de Verão: artes performativas, turismo e território. Geo-Working Papers (13), 5–21 (2007)
8. Martinho, T., Neves, J.S.: Festivais de música em Portugal. Folha OBS **1**, 1–5 (1999)
9. Guerra, P.: Lembranças do último verão: Festivais de música, ritualizações e identidades na contemporaneidade portuguesa. https://portugalaoespelho.files.wordpress.com/2016/05/ficha_-lembrancas-ultimoverao.pdf (2016)
10. APORFEST: Annual Report 2017. Obtido de: http://www.aporfest.pt/ (2018)
11. American Marketing Association (AMA): Definition of marketing. https://www.ama.org/AboutAMA/Pages/Definition-of-Marketing.aspx (2013)
12. Leppäniemi, M., Sinisalo, J., Karjaluoto, H.: A review of mobile marketing research. Int. J. Mob. Mark. **1**(1), 30–40 (2006)
13. Huang, R.Y., Symonds, J.: Mobile marketing evolution: systematic literature review on multi-channel communication and multi-characteristics campaign. Paper presented at the Enterprise Distributed Object Computing Conference Workshops, 2009. EDOCW 2009. 13th (2009)
14. Shankar, V., Balasubramanian, S.: Mobile marketing: a synthesis and prognosis. J. Interact. Mark. **23**(2), 118–129 (2009)
15. Rakestraw, T.L., Eunni, R.V., Kasuganti, R.R.: The mobile apps industry: a case study. J. Bus. Cases Appl. **9**, 1 (2013)

16. Mobile Marketing Association (MMA): The Mobile Marketing Roadmap: How mobile is Transforming Marketing for Targeting Next Generation Consumers. https://www.mmaglobal.com/files/documents/mobilemarketingroadmap.pdf (2016)
17. Venkatesh, V., Morris, M.G., Davis, G.B., Davis, F.D.: User acceptance of information technology: toward a unified view. MIS Q. 425–478 (2003)
18. Mobile Marketing Association (MMA).: MMA updates definition of mobile marketing. http://mmaglobal.com/news/mma-updates-definition-mobile-marketing (2009)

Chapter 20
The Use of Consumer Neuroscience Knowledge in Improving Real Promotional Media: The Case of Worten

José Paulo Marques dos Santos[1,2(✉)], Hugo Ferreira[3,4], Joaquim Reis[4], Diana Prata[4], Sofia Pereira Simões[5], and Inês Drummond Borges[5]

[1] University Institute of Maia, Maia, Portugal
jpsantos@ismai.pt
[2] Faculty of Medicine, University of Porto, Porto, Portugal
[3] Faculty of Sciences, University of Lisbon, Lisbon, Portugal
[4] IBEB—Institute of Biophysics and Biomedical Engineering, Lisbon, Portugal
[5] Worten, Direcção de Marketing, Carnaxide, Portugal

Abstract. All those that work in markets research know, for long, that there is a distance between the intentions that consumers manifest overtly (e.g. in surveys or focus groups) and what they practise. One cause is that these methods assume the consumer as a full rational decision-maker, although most consumption acts arise in the non-conscious plan, where intuition and emotions play the most important roles. Traditional inquiry methods ground on the former while consumption decision processes live in the latter. The unfit is notorious. Consumer neuroscience aims to go beyond this hurdle by directly questioning the brain processes. We use in this study EEG, eye-tracking, EDA/SCR, and heart rate monitoring to measure differences in the redesign of the logo and communication materials of the largest appliance retailer in Portugal, Worten. The results allowed identifying and measuring ameliorations in consumers' emotional engagement in the TV commercials and media layout. Commercials tend now more towards approach than withdrawal and the flow of the eyes while searching for information in printed materials is more placid and smoother. These improvements contribute to enhance marketing communication with consumers.

20.1 Introduction

20.1.1 Why Consumer Neuroscience? (And Not Traditional Research Techniques)

Questioning overtly the consumers, either orally, digitally, or by writing, as in surveys and focus groups, figures among the most used methods in market research. This procedure roots on cognitive psychology, more specifically assuming that consumption acts are performed grounded on consciousness and deliberation. The consumer decides and the process is named "decision-making". The word "making" reflects the conscious

Á. Rocha et al. (eds.), *Marketing and Smart Technologies*, Smart Innovation, Systems and Technologies 167, https://doi.org/10.1007/978-981-15-1564-4_20

control of the situation and the optimal decision is calculated, as it followed an algorithm, that is reproducing the computer metaphor [1].

However, although their massive use, these methods encompass some important difficulties [2], for example, the difficulty that humans have in self-reporting their emotional states [3]. Consumers' emotional states are critical in marketing communication, either because one's actual emotional state influences the perception of the marketing stimuli, or because it is supposed that such individual emotional states move according to the marketer intention, which is known for long [4].

The contrast with the computer metaphor is evident. Nonetheless, one must recognise that it is far easier to deal with the explicit answers in surveys and focus groups than to access the processes that run in the brain which give rise to overt behaviours. Most of the time, these processes develop implicitly and, although the existence of research in these realms [4–6], the technique has not been deploying solutions to grasp such hidden decisional processes, yet.

Consumer neuroscience/neuromarketing and its range of neuroscientific techniques may help in understanding the intrinsic state and the dynamics of consumers' emotional states and other brain-based processes [3, 7–10].

20.1.2 The Challenge

Worten is the largest appliance retailer in Portugal and planned and performed a redesign of the logo among other improvements in the communicational media (e.g. printed leaflets) and shop interior architecture and customer experience. The present study focuses on the logo and the communicational media. Fig. 20.1 depicts both logos, being the previous one on the left and the redesigned on the right.

Fig. 20.1. Previous Worten logo on the left and the redesigned on the right

The challenge here is to measure eventual differences in brand perception with the aid of the neuroscientific techniques, focusing on the putative influence of emotions. Therefore, here we report the application of EEG,[1] eye-tracking, EDA/SCR,[2] and heart rate monitoring to real marketing communication media, like TV spots, printed leaflets,

[1] Electroencephalography.

[2] Electrodermal activity/Skin conductance responses.

and posters for mupi.[3] The aim of the study was to evaluate the post hoc consumers' reactions to the Worten logo redesign, mostly those which ground on emotions.

20.1.3 Detecting Emotions

Approach-Withdrawal Theory and EEG. EEG has been used to study the brain bases of human behaviour. The theoretical ground is the Approach-Withdrawal theory. In the 1990s and while studying happiness and disgust, it was found, with the aid of EEG, that there was asymmetry in brain waves depending on the cerebral hemisphere: happiness was characterised by left-side activation of the brain (focusing on the anterior temporal region), while disgust was characterised by the activation of the right hemisphere (focusing on the frontal and anterior temporal regions) [11]. The activation of the left hemisphere is related to happiness and with approach behaviour, whereas the activation of the right hemisphere is related to disgust and with withdrawal behaviour [12]. These findings have been extended to advertising research [13, 14].

The underlying rationale is: if one pretends to have an emotional influence on consumers by the means of the marketing communication, then the marketing communication (e.g. a commercial) should provoke some reaction in the brain and, if that reaction is emotion-based, then the hemispheric asymmetry is a proxy of such effect. Thus, one would have a metric that does not depend on consumer self-reporting that could measure the emotional effect provoked by the putative emotional content of the advertisement. Having that measurement, marketing communication may be successively tuned to match the desired emotional effect on the consumer. The objective would be to increase the left hemisphere activation in order to induce an approach behaviour in the consumer, acknowledging that such approach (happiness) brings benefits to the brand/product/service/point-of-sale [7].

Still in consumer studies, the use of EEG has been extended from frontal alpha asymmetry and Approach-Withdrawal theory, now including ERPs[4] and other bands [15]. For example, higher theta band power suggests preference, which is linked with the N200 [16]. The most intriguing point is that these EEG recordings were used to predict future choices and not merely present behavioural reactions.

Eye-Tracking. Eye-tracker consists of a computer screen and a device that detects whether the eyes are focusing on the screen surface by the means of infrared light. Therefore, while using an eye-tracker one may obviate asking where the consumer looked at, acknowledging that it would be a difficult task because some saccades may last less than a centisecond and the observer may not realise their existence, and also may miss the correct full path of the gazing. With the eye-tracker, every point gazed, the duration on the spot, and the full sequence are recorded.

Considering the market research field, eye-tracking has been used to improve advertisements [17, 18], labelling [19], and packaging [20].

[3] "Mobiliário urbano para informação" Street furniture for information.

[4] Event-related potential.

Electrodermal Activity/Skin Conductance Responses and Heart Rate Monitoring. While the electrodes in EEG measure the electric fields originated in the central nervous system (brain) activity, the reflexes of this central activity may also be detected in the peripheral nervous system, even the emotional activity. Putting it simply, EDA/SCR consists of two electrodes and an electric current flowing between them. Because sweating changes the skin conductance, such alterations interfere with the flowing of the electric current and the fluctuations may be recorded by the electrodes. Sweating is a consequence of the emotional behavioural reactions.

The same happens with the heart rate: arousal stimuli lead to increased heart rate and apathetic situations lead to decreased heart rate. Therefore, heart rate may be used to measure the arousal content of stimuli, like a commercial.

It was shown that humans anticipate EDA/SCR signals even before they get aware of risky choices [21], that is, EDA/SCR signals reflect the brain processes that support decisions well before one is conscious about them, if s/he will ever be conscious about them. EDA/SCR was already used to analyse the implications of TV commercials visualisation and heart rate monitoring as well [7].

20.2 Methods

20.2.1 General Structure

The targeted stimuli for this experiment consisted of two versions of (previous and redesigned logo) TV commercials, the first page of print leaflets, and screen versions of mupi posters. However, the single presentation of materials of Worten marketing communication could introduce a severe bias. Therefore, the materials from Worten were mixed with marketing communication materials of other brands.

Twenty-two TV commercials and the two commercials of Worten were pseudo-randomly sequenced (the two commercials of Worten were forced to be separated, one in each half) in order to simulate a common TV break. The Supplementary Materials include the exhibited videos of the two TV commercials (files Worten_SpotTv_previous and Worten_SpotTv_redesigned). Before visualising the commercials sequence, a two-minute video showing a water cascade with calm music was presented (file Baseline in Supplementary Materials). The purpose of the presentation of this video was twofold: on the one hand, to reduce anxiety derived from the cap montage (between 20 and 30 min) and the unusual technical apparatus; on the other hand, to have a measure of the intrinsic frontal alpha asymmetry of each individual when stimulated with a neutral stimulus. The EEG signal measures are referenced to this baseline as explained in the section Data Analysis.

Besides the printed leaflets and mupi posters of Worten, the subjects also saw similar materials of three other competitors, so subjects could not anticipate which brand was being studied. Figure 20.2 depicts the two first pages of printed leaflets and Fig. 20.3 depicts two posters for mupi, in either case figuring the previous on the left and the redesigned on the right. Images were shown for 8 s.

All the subjects accomplished with the three tasks: first, passive visualisation of the TV commercials while the subjects' electric brain signals were recorded with EEG, the eye gaze was recorded with the eye-tracker, and the peripheral nervous system signals (electrodermal activity and heart rate) were recorded; secondly, passive visualisation of the printed leaflets while the eye gaze was recorded with the eye-tracker, and thirdly, passive visualisation of mupi posters while the eye gaze was recorded with the eye-tracker. All the marketing materials were the real ones, used in the company communication.

20.2.2 Montage

All the three sets of videos and still images were displayed in a 24″ 144 Hz Asus VG248QE screen and were sequenced with the software Gazepoint Analysis Professional Edition v.3.2.1. The eye-tracker was a Gazepoint GP3, installed below the screen, 65 cm away from the subject's eyes and with 60 Hz sampling frequency.

Fig. 20.2. Printed leaflets (first page): previous on the left and the redesigned on the right

Fig. 20.3. Mupi posters: previous on the left and the redesigned on the right

The EEG montage consisted of an actiCHamp (Brain Products GmbH) 64-channel amplifier (although only 32 channels were used in the experiments) and actiCAP 32 (Brain Products GmbH) caps encompassing 32 active electrodes arranged according to the 10–20 international system configuration [22]. The impedance was kept below 5 k Ω and the sampling rate was 1 kHz.

In order to synchronise EGG and eye-tracking recordings, a 10 pixels square, alternating between black and white, was added to the bottom left of all the videos and images. The coding provided by the alternating black and white square is read by a photocell which generates a trigger signal that is recorded together with the EEG signal.

The electrodermal activity was recorded with the respective sensor of BITalino (Plux Wireless Biosignals SA) and the heart rate was registered with the photoplethysmography sensor of the same manufacturer.

20.2.3 Subjects and Instructions

A total of 56 individuals were recruited to participate in this study, 30 females and 26 males, with 35.8 years as average age, ranging from 18 until 59 years recruited in the two biggest cities in Portugal. All of them were Portuguese natives, with absent neurologic or psychiatric disorders, with no dependence on alcohol or drugs, nor administrating psychotropic medication, absence of depressive states, without metal parts in the skull, nor lesions in the scalp. It was ensured that vision and audition were sufficient for normal comprehension of the stimuli.

Subjects were also screened for being habitual buyers of technology, consumer electronics, or household appliances.

The instruction passed to the subjects was just to passively look at the screen avoiding excessive head movement and, very importantly, eye blinks.

20.2.4 Data Analysis

The EEG signal analysis largely followed the procedure already published [7, 23], including band-pass filtering and the use of ICA to remove artefacts. The power spectral density (PSD) of the alpha band (8.0–12.0 Hz) in electrodes $F3$ and $F4$ was used to calculate the asymmetry between the two brain hemispheres composing the Approach-Withdrawal Index (AWI)

$$\mathrm{AWI} = \left(\frac{\mathrm{PSD}_{F3} - \mathrm{PSD}_{F4}}{\frac{\mathrm{PSD}_{F3} + \mathrm{PSD}_{F4}}{2}} - \frac{\mathrm{PSD}_{basF3} - \mathrm{PSD}_{basF4}}{\frac{\mathrm{PSD}_{basF3} + \mathrm{PSD}_{basF4}}{2}} \right) \times 100$$

where PSD_{F3}, PSD_{F4}, PSD_{basF3}, and PSD_{basF4} are the power spectral density of the EEG signal recorded in, respectively, the electrode $F3$ during the visualisation of the commercial, the same for electrodes $F4$ and $F3$ during the visualisation of the baseline, and the same for electrode $F4$, all calculated following the Welche procedure [24].

For the TV commercials, the eye-tracker data consisted of sequence of the subjects' gaze spotting on the screen, while ROIs[5] were defined for the printed materials, which allowed to measure the number of ROI's views, time to the first spotting, ROI's stay time, number of revisitors, and revisit's average time.

20.3 Results

20.3.1 TV Commercials

The results of the TV commercials data analysis are summarised in the following graphs. Figures 20.4 and 20.5 depict, respectively, the AWI time course for the previous Worten TV commercial and the redesigned. The blue line is the AWI time course, the green line is the all-time-points median, the light blue area is the inter-quartile range (IQR), and the black crosses signal the statistically significant differences. For the black crosses, graphs are complemented with the respective video frames having the eye gazes spotting overlaid. The Supplementary Materials include two videos, each one encompassing the sequence of eye gaze spotting overlaying the TV commercial (files Worten_SpotTv_previous_fixationmap and Worten_SpotTv_redesigned_fixationmap).

The AWI of the previous TV commercial is characterised by an initial (0–4 s) tendency to mark on the approach side, which is immediately followed by a decrease towards the withdrawal side (4–7 s), and then again towards the approach side, and staying there stable until around 12 s. So far, both peaks and troughs have significant

[5] Region of interest.

points. From the 12 s until the end of the commercial, the AWI stays in the withdrawal side peaking significantly negative twice. Both human elements (in the beginning and in the end) attract the eye gaze but with opposite valences. Initially, when the camera focuses on the female face the AWI rises to a maximum and decreases after the plan is back until a trough. First, her face and then her body attract the eye gaze. The female element in the end is a famous actress and her appearance is accompanied by a significant decrease towards the withdrawal side. In the first video frames, the eye focuses on her body and then change to her face but, both cases lead to significant troughs.

In the redesigned TV commercial, each of the two simultaneous scenes (living room and kitchen) in the opening attracts the eye gaze and it is possible to follow the flipping gazes between them. The next scene encompasses several elements (TV set, text with the model, price, logo with discount, and logo with warranted minimum price) and the eye gazes distribute among them. It is possible to verify that, in parallel, the AWI of the redesigned TV commercial is around the median during this period (until 8 s). Then, the movement of the price leads to a trough in AWI. The next scenes with the logos are accompanied by three peaks in the approach side until 14 s (the peak generated by the gazing over the football player in the TV set and the dynamic logos twice) and then until the end of the commercial the AWI returns to values around the median.

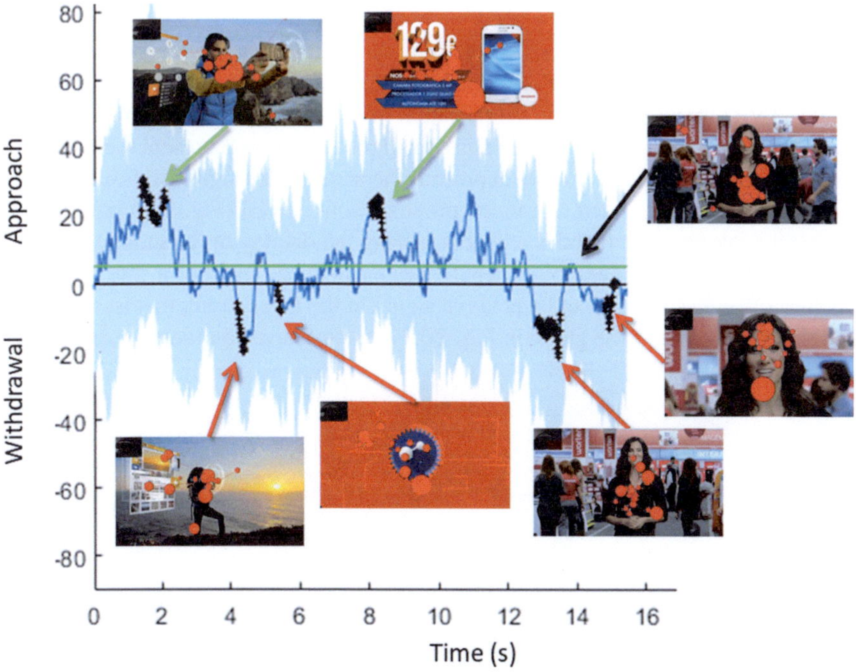

Fig. 20.4. Approach-withdrawal index time course for the previous Worten TV commercial

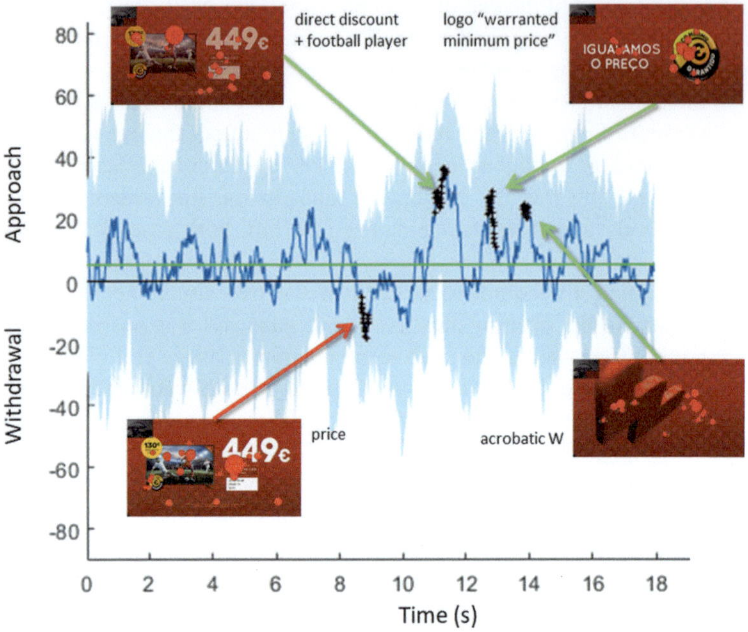

Fig. 20.5. Approach-withdrawal index time course for the redesigned Worten TV commercial

Figures 20.6 and 20.7 depict, respectively, the variation in the heart rate (left graph) and the variation in the EDA/SCR signal (right graph) during the visualisation of the previous TV commercial and the visualisation of the redesigned TV commercial. In all graphs, the blue line is the signal (median), the light blue areas represent the interquartile range (IRQ), and the black crosses represent the statistically significant differences.

The variation in the heart rate in the previous commercial (Fig. 20.6 left) exhibits a significant peak between 6 and 8 s, that is, during the display of the mobile phone, and variation in the EDA/SCR exhibits a negative signal between 8 and 10 s, that is, during the dynamics of price showing, and a positive signal in the last second, that is, the actress's face feeling the screen. The graph in Fig. 20.7 does not have significant points.

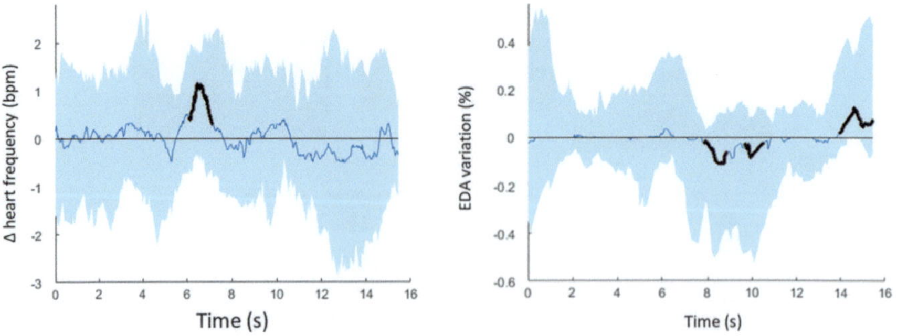

Fig. 20.6. Variation in the heart rate (left) and of EDA/SCR signal (right) during the visualisation of the previous Worten TV commercial

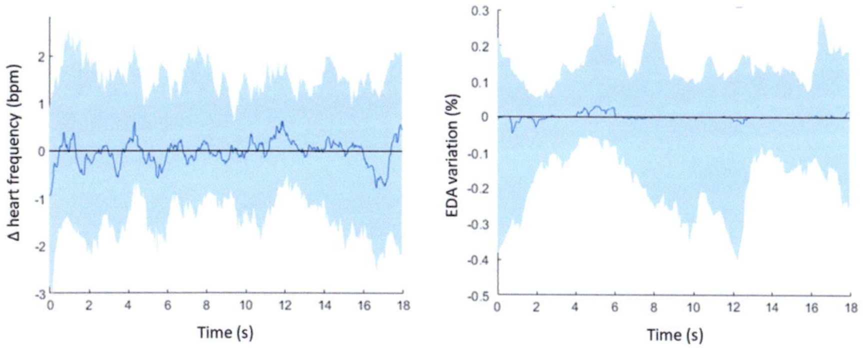

Fig. 20.7. Variation in the heart rate (left) and of EDA/SCR signal (right) during the visualisation of the redesigned Worten TV commercial

20.3.2 Printed Leaflets

Figure 20.8 depicts the eye gaze path among the elements that compose the first page of the printed leaflets, both for the previous version (left) and the redesigned one (right). Tables 20.1 and 20.2 complement with numeric group results: viewers (the total number of subjects that focused their gazes in the ROI until the time limit), first viewed (the mean time that the ROI was first gazed), time viewed (mean duration of the gaze in the ROI), revisitors (number of subjects that returned to the ROI), and revisits (mean revisits to the ROI). The Supplementary Materials include two videos containing the eye-tracker fixation maps for the previous and the redesigned printed leaflets (files Worten_Leaflet_previous_fixationmap and Worten_Leaflet_redesigned_fixationmap).

It is possible to verify in Fig. 20.8 that the gaze flowing in the previous printed leaflet (left) is more erratic than in the redesigned (right). The exceptions are steps 8 and 9 (discount logos), but they were seen by a minority of subjects. If these steps are removed from the analysis the gaze flowing in the redesigned leaflet is even cleaner. The human female element in the previous leaflet takes the first gaze, while the brand receives the primacy in the redesigned leaflets. The products (washing machine and TV set) in the redesigned leaflet were seen for more time than in the previous version (1.36 and 1.43 s vs. 0.51 and 0.39 s, respectively). The same happened with their prices (0.46 and 0.42 s vs. 0.40 and 0.32 s) and the total number of viewers (52 and 50 vs. 44 and 38).

Fig. 20.8. Eye gaze flowing among the elements in the first page of printed leaflets in the previous (left) and redesigned (right). Numbers represent the sequence in time

20.3.3 Mupi Posters

Because some elements change from the previous mupi posters to the redesigned version, it is not possible to do a full comparison. In any case, some elements maintain, and it is possible to assess the effects caused by the introduction of new elements. Figure 20.9 depicts the considered ROIs in both mupi posters and Table 20.3 summarises the results.

Table 20.1. Eye-tracker results for the visualisation of the first page of previous Worten printed leaflets. Total viewers are 56

	Girl	Brand	Text	Laptop	Motto	Price laptop	Washing machine	Promo	TV	Price TV	Price wash. m.	Disclaimer
	1	2	3	4	5	6	7	8	9	10	11	12
Viewers	53	47	50	47	45	46	44	10	38	19	23	7
1st viewed (s)	0.78	0.93	1.57	1.99	2.23	3.03	3.83	4.07	4.57	4.82	4.86	5.79
Time viewed (s)	0.46	0.66	1.12	0.71	0.48	0.52	0.51	0.15	0.39	0.32	0.40	0.31
Revisitors	38	32	44	41	33	31	22	3	21	11	8	3
Revisits	2.03	2.25	3.39	2.76	2.73	2.52	1.91	1.33	1.67	1.36	2.13	1.00

Table 20.2. Eye-tracker results for the visualisation of the first page of redesigned Worten printed leaflets. Total viewers are 56

	Brand	Text	Logo W	Logo min. price	Washing machine	TV	Price wash. m.	Discount TV	Discount wash. m.	Price
	1	2	3	4	5	6	7	8	9	10
Viewers	48	47	39	45	52	50	30	20	19	30
1st viewed (s)	0.83	1.35	1.83	1.86	2.05	2.63	3.85	4.00	4.50	4.57
Time viewed (s)	0.43	0.78	0.22	0.54	1.36	1.43	0.46	0.25	0.18	0.42
Revisitors	38	32	18	22	42	40	11	7	5	22
Revisits	2.21	2.03	1.78	1.91	2.05	2.35	2.73	2.86	2.00	1.77

Fig. 20.9. ROIs in the mupi posters in the previous (left) and in the redesigned (right) versions

In both cases, price is the first viewed element and viewed buy more subjects. It is also possible to verify that the posters' reading is done from top to bottom. Consequently, one remarkable fact is that the brand (in the bottom in both cases) is the last seen element, with fewer subjects viewing (approximately just half of the cohort), for less time and fewer revisitors and mean revisits. The new logos (discount and minimum price) are noticed by almost all subjects and with important viewing duration.

Table 20.3. Eye-tracker results for the visualisation of the mupi posters. Total viewers are 56

	Previous				Redesigned				
	Price	Motto	Charact.	Brand	Price	Discount logo	Charact.	Min price logo	Brand
	1	2	3	4	1	2	3	4	5
Viewers	53	41	49	23	49	49	44	47	26
1st viewed (s)	0.43	2.69	3.40	3.62	1.35	1.55	2.87	3.00	4.14
Time viewed (s)	1.16	0.45	0.64	0.16	0.58	0.87	1.05	0.64	0.14
Revisitors	50	26	37	9	39	40	39	34	9
Revisits	3.72	2.31	2.73	1.67	2.62	2.75	3.31	1.56	1.56

20.4 Discussion

Considering the redesigned TV commercial, the paired analysis of the EEG and eye-tracking results allow to conclude about the efficacy of the dynamic elements (either the acrobatic W and the spinning logo announcing the guaranteed minimum price), and the still photo figuring in the TV set with the football player. This is evident that these elements contribute to approach, which happens in the middle of the commercial. On the other hand, when the subjects' gaze targets the price (which was previous), the signal moves to withdrawal, suggesting the involvement of rational processing, which compromises and harms the emotional engagement.

The final of the previous TV commercial is significantly in the withdrawal side. This compromises the advertisement efficacy as it may leave an unwanted flavour. On the contrary, the last 8 s in the redesigned commercial are mostly on the approach side, in fact peaking there for three times. Although the slope in the very end is decreasing, there is an improvement in the commercial. One may even speculate that the redesigned commercial could end by the 14th second (i.e. just after the last peak), maximising the approach effect and leaving an interesting flavour.

The visual ergonomics of the printed communication materials also benefited from the eye-tracking study. The new arrangement of the printed leaflets allowed achieving 68% revisits in the logo area, and 57% revisits in the motto area, which means that the focus is now being on marketing elements and not on secondary components, as in the previous version. Here, one may also speculate about the inclusion (or not) of certain elements in the layout. For example, in the redesigned printed leaflet, one may now consider removing steps 8 and 9 (yellow spots announcing extra discounts) in order to improve the eye gaze flowing (i.e. minimising eye flipping between the two sides of the leaflet). In either case, this is now testable and assessed with objective metrics.

While the element price is dominant in mupi posters, the brand receives about half of the attention (and it is always the last element to be seen). The use of eye-tracking detected this fact, which may be problematic because the consumer may be prompted by the advertisement but may not memorise which company is beyond that offer. The positioning of the brand logo is to be studied in future versions of mupi posters. Returning to the element price, price is strategic in marketing communication and this is clearly achieved and ratified.

There is one limitation in mupi posters that must be considered: while real mupi posters dimensions are around 2A0 (printed surface width is 1160 mm, height 1710 mm) and are installed close to the ground, the reproductions visualised in this study (on the screen) are much smaller and are at the eye level. The consequences of this rescaling are not fully known but real brand logo visualisations can be even worse because the poster is closer to the ground.

20.5 Conclusions

It is evident from the results of this study that now marketers have techniques to address the consumers' emotional engagement and other non-consciously accessible processes, exceeding the limitations of traditional inquiry procedures like surveys and focus groups. Now, it is possible to go beyond. For example, the decision to stop (or not) the redesigned commercial in the 14th second may now be taken using the metrics yielded by the neuroscientific techniques. Other fine tunings are possible too (like deciding to include the yellow spots with extra discounts in the redesigned printed leaflet). It is even possible to design the advertisements with specific profiles, improving their efficacy.

Communication media visual ergonomics may also benefit from neuroscientific and biometric techniques as it results from the analysis of the eye-tracker results in printed leaflets and mupi posters. There were potential problems raised by the technique which were not detected previously, but it was also possible to confirm that the strategic promotional guidelines are effective.

In fact, improving communication with the consumers interest both companies and consumers, that is, markets and consumer neuroscience already started giving tangible contributions.

References

1. Østergaard, P., Jantzen, C.: Shifting perspectives in consumer research: from buyer behaviour to consumption studies. In: Beckmann, S.C., Elliott, R.H. (eds.) Interpretive Consumer Research—Paradigms. Methodologies & Applications. Copenhagen Business School Press, Copenhagen (2002)
2. Walla, P., Mavratzakis, A., Bosshard, S.: Neuroimaging for the affective brain sciences, and its role in advancing consumer neuroscience. In: Fountas, K. (ed.) Novel Frontiers of Advanced Neuroimaging. InTech (2013)

3. Chamberlain, L., Broderick, A.: The application of physiological observation methods to emotion research. Qual. Mark. Res.: Int. J. **10**, 199–216 (2007). https://doi.org/10.1108/13522750710740853
4. Havlena, W.J., Holbrook, M.B.: The varieties of consumption experience: comparing two typologies of emotion in consumer behavior. J. Consum. Res. **13**, 394–404 (1986). https://doi.org/10.1086/209078
5. Mehrabian, A.: Framework for a comprehensive description and measurement of emotional states. Genet. Soc. Gen. Psychol. Monogr. **121**, 339–361 (1995)
6. Mehrabian, A., de Wetter, R.: Experimental test of an emotion-based approach to fitting brand names to products. J. Appl. Psychol. **72**, 125–130 (1987). https://doi.org/10.1037/0021-9010.72.1.125
7. Vecchiato, G., Cherubino, P., Maglione, A.G., Ezquierro, M.T.H., Marinozzi, F., Bini, F., Trettel, A., Babiloni, F.: How to measure cerebral correlates of emotions in marketing relevant tasks. Cogn. Comput. **6**, 856–871 (2014). https://doi.org/10.1007/s12559-014-9304-x
8. Ariely, D., Berns, G.: Neuromarketing: the hope and hype of neuroimaging in business. Nat. Rev. Neurosci. **11**, 284–292 (2010). https://doi.org/10.1038/nrn2795
9. Smidts, A., Hsu, M., Sanfey, A.G., Boksem, M.A.S., Ebstein, R.B., Huettel, S.A., Kable, J. W., Karmarkar, U.R., Kitayama, S., Knutson, B., Liberzon, I., Lohrenz, T., Stallen, M., Yoon, C.: Advancing consumer neuroscience. Mark. Lett. **25**, 257–267 (2014). https://doi.org/10.1007/s11002-014-9306-1
10. Plassmann, H., Venkatraman, V., Huettel, S., Yoon, C.: Consumer neuroscience: applications, challenges, and possible solutions. J. Mark. Res. **52**, 427–435 (2015). https://doi.org/10.1509/jmr.14.0048
11. Davidson, R.J., Ekman, P., Saron, C.D., Senulis, J.A., Friesen, W.V.: Approach-withdrawal and cerebral asymmetry: emotional expression and brain physiology: I. J. Pers. Soc. Psychol. **58**, 330–341 (1990). https://doi.org/10.1037/0022-3514.58.2.330
12. Davidson, R.J.: Anterior cerebral asymmetry and the nature of emotion. Brain Cogn. **20**, 125–151 (1992). https://doi.org/10.1016/0278-2626(92)90065-T
13. Silberstein, R.B., Nield, G.E.: Measuring emotion in advertising research: prefrontal brain activity. IEEE Pulse **3**, 24–27 (2012). https://doi.org/10.1109/MPUL.2012.2189172
14. Ohme, R., Reykowska, D., Wiener, D., Choromanska, A.: Application of frontal EEG asymmetry to advertising research. J. Econ. Psychol. **31**, 785–793 (2010). https://doi.org/10.1016/j.joep.2010.03.008
15. Lin, M.-H., Cross, S.N.N., Jones, W.J., Childers, T.L.: Applying EEG in consumer neuroscience. Eur. J. Mark. **52**, 66–91 (2018). https://doi.org/10.1108/EJM-12-2016-0805
16. Telpaz, A., Webb, R., Levy, D.J.: Using EEG to predict consumers' future choices. J. Mark. Res. **52**, 511–529 (2015). https://doi.org/10.1509/jmr.13.0564
17. Pieters, R., Wedel, M., Batra, R.: The stopping power of advertising: measures and effects of visual complexity. J. Mark. **74**, 48–60 (2010). https://doi.org/10.1509/jmkg.74.5.48
18. Hutton, S.B., Nolte, S.: The effect of gaze cues on attention to print advertisements. Appl. Cogn. Psychol. **25**, 887–892 (2011). https://doi.org/10.1002/acp.1763
19. Ares, G., Giménez, A., Bruzzone, F., Vidal, L., Antúnez, L., Maiche, A.: Consumer visual processing of food labels: results from an eye-tracking study. J. Sens. Stud. **28**, 138–153 (2013). https://doi.org/10.1111/joss.12031
20. Hurley, R.A., Ouzts, A., Fischer, J., Gomes, T.: Effects of private and public label packaging on consumer purchase patterns. Packag. Technol. Sci. **26**, 399–412 (2013). https://doi.org/10.1002/pts.2012
21. Bechara, A., Damásio, H., Tranel, D., Damásio, A.R.: Deciding advantageously before knowing the advantageous strategy. Science **275**, 1293–1295 (1997). https://doi.org/10.1126/science.275.5304.1293

22. Jurcak, V., Tsuzuki, D., Dan, I.: 10/20, 10/10, and 10/5 systems revisited: Their validity as relative head-surface-based positioning systems. NeuroImage **34**, 1600–1611 (2007). https://doi.org/10.1016/j.neuroimage.2006.09.024
23. Vecchiato, G., Cherubino, P., Trettel, A., Babiloni, F.: Neuroelectrical Brain Imaging Tools for the Study of the Efficacy of TV Advertising Stimuli and their Application to Neuromarketing. Springer-Verlag, Heidelberg (2013)
24. Welch, P.D.: The use of fast Fourier transform for the estimation of power spectra: a method based on time averaging over short, modified periodograms. IEEE Trans. Audio Electroacoust. **15**, 70–73 (1967). https://doi.org/10.1109/TAU.1967.1161901

Chapter 21
The Influence of Thinking Styles on Perceived Price Fairness: An Experimental Study

Sayanti Shaw(✉)

Department of Management and Marketing, Faculty of Management and Economics, Tomas Bata University in Zlin, Mostni 5139, 76001 Zlin, Czech Republic
shaw@utb.cz

Abstract. Customers expect pricing practices to be fair, they are specifically sensitive to price increase considered unfair or unacceptable, which eventually influences both behavioural intentions and emotions. Substantial investments are deployed in building positive associations with consumers, which can be subverted by the incident of price increase leading to unfairness perceptions. It is significantly important to mitigate the perceived unfairness perceptions, given the stakes involved. This article investigates the role that styles of thinking (holistic versus analytic) plays in perceived price fairness. The findings of this research show that styles of thinking play a significant role in influencing perceived price fairness. When exposed to negative and/or undesirable events such as price increase, styles of thinking influence cognitive attribution, which, in turn, influences price fairness perceptions. This research adds to the growing literature in price fairness by exploring the influence of thinking styles on perceived price fairness that has not been examined before. These findings suggest that strategies to influence styles of thinking could be useful in managing perceived unfairness perceptions. Marketers and managers could utilise marketing-management strategies based on findings of this research to maintain positive consumer's perception of price, thereby will improve firms' performance, in terms of, value, profitability and sales.

21.1 Introduction

A product/service's price has a significant influence on consumers' perceptions and behaviours. Thus, pricing as a managerial decision is important and at the same time challenging [1]. Not only economic motives but also subjective preferences and perceptions play a significant role in consumers' price perceptions and purchase decisions. Just as understanding of what makes a price perceive as unfair is important, likewise understanding of what makes a price appear fair is also equally important. For determining fairness perceptions, customers have to go through a series of cognitive steps. It is important for marketers as well as managers to understand the cognitive process that consumers go through while determining a price to be fair. This is because of the fact that making a price/price increase appear to be fair or lowering unfairness perceptions

Á. Rocha et al. (eds.), *Marketing and Smart Technologies*, Smart Innovation, Systems and Technologies 167, https://doi.org/10.1007/978-981-15-1564-4_21

has been widely proven in literature for leading to a higher level of consumer satisfaction and loyalty [2, 3].

Fairness has been defined as 'a judgment of whether an outcome and/or the process to reach an outcome are reasonable, acceptable, or just' [4]. Previous research has shown that price fairness perceptions influence the behaviour of consumers. It is very crucial for organisations to know when consumers perceive price and price change to be unfair or fair. Consumers evaluate the way products'/services' prices are set and make perceptions about price fairness of the product/service. Perceived price fairness has a positive influence on the consumer's intent to purchase [5], consumer satisfaction [6], consumer loyalty [3] and consumer's attitude towards the seller [1]. On the other hand, perceived price unfairness can lead consumers to feel the urge to protect themselves from unfair treatment, via return of goods/services, complaints and negative word of mouth [7]. Perceived price unfairness can also lead to punishments by consumers in the form of switching to competing firm or reprisal in court [8].

21.2 Literature Review

Price Fairness Perception Price fairness perception is identified as a key element of consumers' responses to prices [9]. As fairness depends on the outcome, thereby the price fairness perception is subject to what or who is responsible for that outcome. The increase of price incited by internal costs or internal factors is perceived as more unfair (less fair) than those driven by external costs [10, 11] or external factors (e.g. inflation). When people meet certain types of events especially negative or undesirable, they deduce the events' cause(s) and depending on cognitive attribution (how the causes are attributed), different types of emotional responses are generated, which thereby lead to their reactions to the events. In the evaluation of price fairness, the consumers' perceived motive concerning the seller's choice of price increase play a crucial role [12]. Firms who are bound to raise prices may enhance consumers' price fairness perceptions by revealing the increase of price and giving a proper explanation corresponding to the extent of increase [13]. Customer-oriented business strategies are increasingly gaining importance [14]. In today's world of a dynamic market characterised by continuous price changes, consumers can end up paying different prices for the same product, in spite of the fact that the seller is the same [15]. The discriminatory price-setting strategies (such as uniform vs. differential pricing, posted vs. auction pricing) have their influences on price fairness perceptions [16].

The perceived price fairness has its impact on customer attitudinal and behavioural outcomes. The process of fairness may lead to perceptions of price fairness or unfairness, which thereby results in positive or negative consequences. Perception of price fairness encourages buyers to buy the product/services. On the other hand, unfairness perceptions may ensue to negative consequences such as buyers may leave the relationship, spread negative word of mouth, or engage in other behaviours that can be harmful to the seller.

Attributional Approach to Price Fairness In order to comprehend an individual's fairness perceptions, the understanding of his/her attributions of cause and responsibility is required. Attribution theory has been used to study the effect of locus of causality, controllability and temporal stability on perceived price fairness [11, 17–19]. Based on any or all the attributional dimensions, consumers can infer the cause(s) of an event/action [20]. In the case of internal (external) locus of causality, price increases are perceived as less (more) fair. Similarly, controllable (uncontrollable) price increases are perceived as less (more) fair [11]. Thereby provide evidence of contextual–situational dependence of price fairness perceptions. Hence, it can be said that customers perceive a price increase caused by external contextual and/or situational factors to be more (or less) fair than a price increase driven by the internal object and/or disposition-based factors [10, 11, 18].

When faced with negative and/or undesirable events like price increase, consumers are likely to engage in the cognitive attribution process. Cognitive attribution is defined as a cognitive process that infers the cause(s) of an event or others' behaviour, which in turn leads to behavioural intentions or consequences [12, 19]. Cognitive attribution effects distributive price fairness and procedural price fairness. Subject to the understanding of three cognitive attribution dimensions (i.e. locus of causality, controllability and temporal stability) outcome evaluation engenders positive/negative emotions [12], which thereby is connected to behavioural intentions [21]. Previous literature has recognised locus of causality and controllability as the key causal dimensions influencing cognitive attributions and consequent behaviours [10, 18].

Holistic and Analytic Style of Thinking Holistic style of thinking (or mode of cognition) is defined as 'involving an orientation to the context or field as a whole, including attention to relationships between a focal object and the field, and a preference for explaining and predicting events on the basis of such relationships' [22]. On the other hand, analytic style of thinking (or mode of cognition) 'involves a detachment of the object from its context, a tendency to focus on attributes of the object to assign it to categories, and a preference for using rules about the categories to explain and predict the object's behaviour' [23, 24]. Previous literature has shown many instances of the fact that the style of thinking has significant impacts on consumers' perceptions [25–27].

Analytic thinkers tend not to pay attention to contextual and/or situational determinants of event/behaviour and assign causality of the event/behaviour to the object/individual. On the contrary, holistic thinkers are more likely to concentrate on the field along with considering external contextual and/or situational influences as significant determinants of event/behaviour [28, 29].

Holistic Versus Analytic Style of Thinking and Attribution Tendency Thinking styles (holistic vs. analytic) are likely to dictate the individuals' level of considering contextual and/or situational factors while drawing attributions [30]. Psychologists have observed that people with diverse thinking styles (holistic vs. analytic) differ in placing importance on external and/or internal factors in cognition. The holistic and analytic style of thinking can influence individuals' cognitive process of making causal attributions, i.e. cognitive attribution to a behaviour/event. Taking into consideration contextual and/or situational factors, consumers are more prone to attribute blame/responsibility to sources outside the firm (external attribution) than sources inside the firm (internal attribution) [29, 31]. Holistic thinkers are more inclined in attributing causes to broader context and emphasis more on interrelationships between person/event and context and/or situation, i.e. tendencies of external attribution [32]. Analytic thinkers favour attributing causes to internal object/disposition based factors by ignoring the role of contexts and situations, i.e. tendencies of internal attribution [29].

In the process of cognitive attribution, holistic thinkers were found to use context- and/or situation-based information more than did analytic thinkers [29, 32, 33]. These attributions apart from allowing consumers to envisage and sway their environment, can also determine customer satisfaction, perceptions, behavioural outcomes (such as complaining, demanding refunds), emotions (such as anger) and brand evaluations [34].

21.3 Research Objective

Fairness or unfairness perceptions depend on consumers' subjective perceptions relying on cognitive reasoning [10]. In pricing literature, cognitive attribution has already been well-recognised as a significant factor influencing perceived price fairness. However, a surprising gap in the literature is the identification and assessment of the underlying factors that influence consumers' cognitive attribution process in the context of perceived price fairness. This research tries to extend the existing price fairness literature by incorporating cognitive factor i.e. styles of thinking, which is expected to have a significant influence on perceived price fairness. It is also assumed that the relationship between the styles of thinking and perceived price fairness can be explained through the specific path of cognitive attribution. Hence, this study points to a novel and important avenue for research that takes into account the interrelationships between perceived price fairness and styles of thinking.

21.4 Methodology

21.4.1 Hypothesis

Attribution theory asserts that individuals are prone to seek for causal explanations of an event, especially in case of surprising, undesirable, and/or negative event [35]. Negative and/or undesirable events, for instance, price increase, are more likely to incite causal reasoning. Styles of thinking (holistic vs. analytic) can affect the cognitive process of making causal attributions, i.e. cognitive attribution to a behaviour/event. Holistic thinkers are more prone to consider external factors in addition to internal factors, whereas analytic thinkers are inclined to focus exclusively on internal factors [29, 33, 36]. During the cognitive attribution process, consideration of external (internal) factors diverts (attribute) blame from (on) company and thus holistic (analytic) thinkers are less (more) likely to revise their brand evaluations in a negative manner [18, 31, 37]. Based on the above explanations, we would like to propose that styles of thinking influence price fairness perceptions. When exposed to negative and/or undesirable incidents such as price increase, styles of thinking (holistic vs.

analytic) influence cognitive attribution, i.e. consideration of external or/and internal factors. Moreover, this cognitive attribution would thereby influence price fairness perceptions. This leads to the following hypothesis:

H_1: Activating holistic versus analytic thinking enhances perceived price fairness. To observe the effect of styles of thinking manipulation on perceived price fairness through the specific path of cognitive attribution, it is hypothesised that.

H_2: Cognition attribution mediates the relationship between styles of thinking manipulation and perceived price fairness.

21.4.2 Conceptual Framework

In accordance with the identified research problem, the following conceptual framework of perceived price fairness (see Fig. 21.1.) is developed. In order to achieve specific objectives of the research, the conceptual framework would be analysed via an experimental study. Figure 21.1 illustrates the conceptual framework for the current study.

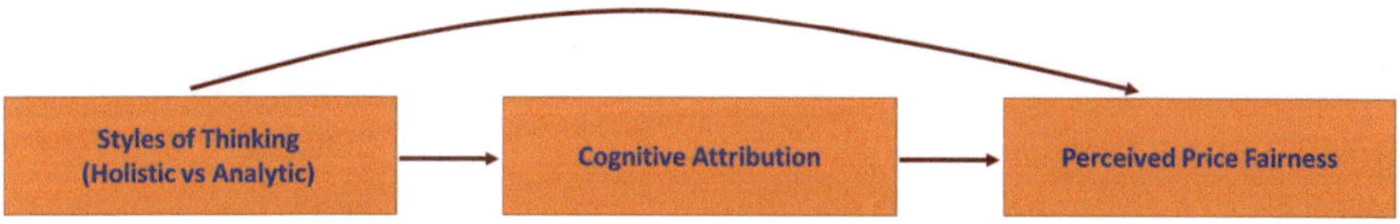

Fig. 21.1 A conceptual framework of perceived price fairness. *Source* The author's research

21.4.3 Research Methods

H_1, H_2 are examined via an experimental study. Participants were randomly assigned to one of the three experimental conditions, i.e. holistic thinking condition, analytic thinking condition and control condition. Following previous studies [26, 29, 38], thinking styles were manipulated using a well established technique of showing participants a black-and-white picture in which line drawings of 11 smaller objects are embedded. One hundred and thirty-eight students of a public university in India participated in this study. Participants were provided a hypothetical scenario in the context of the restaurant's price increase. After reading the scenario, participants completed cognitive attribution and perceived price fairness measurement scales along with certain demographic questions. Figure 21.2 depicts the current study's research methods.

All the measurement scales that were used in this study were adopted from the literature and modified according to the concerned study. Perceived price fairness was assessed via six-item, five-point Likert scale [1, 10]. An example of items included in perceived price fairness measurement—the price increase is acceptable. Cognitive attribution was measured using five-item, bipolar rating (semantic differential) scale from 1 to 7 [10]. An example of items included in cognitive attribution measurement—the cause(s) of price increase is something inside/outside the restaurant.

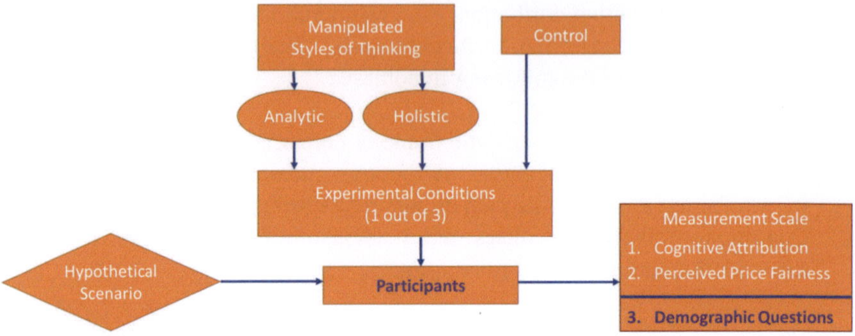

Fig. 21.2 Research methods of the current study. *Source* The author's research

21.4.4 Data Analysis

The data collected from the study were analysed via statistical analysis/test techniques. The following table summarises the techniques.

Table 21.1 List of statistical analysis/test techniques, description and software used in the research. (*Source* The author's research)

Analysis/Tests	Description	Software
Cronbach's Alpha	Measure of scale reliability	SPSS
One-Way Anova	Determine whether there are any statistically significant differences between the means of two or more independent groups	
Multiple regression	Use to predict a dependent variable based on multiple independent variables	
Pearson's correlation	Determine the degree of relationship between two variables	
Sobel test	Determine the statistical significance of mediation effect	The calculation for the Sobel test QUANTPSY. ORG)

21.5 Results and Discussions

According to the reliability analysis, the Cronbach's alpha values (0.849) and (0.946) suggest that items used to measure cognitive attribution and perceived price fairness had acceptable and reliable internal consistency respectively. A one-way ANOVA was performed to test the effect of styles of thinking manipulation on cognitive attribution. As expected, the effect of thinking styles manipulation was significant, $F(2, 88.72)$ =

24.55, $p < 0.0005$. The effect of the thinking styles manipulation was decomposed by creating two dummy contrasts (holistic thinking vs. control; analytical thinking vs. control) and the effects of the two contrasts on cognitive attribution were assessed by using multiple regression. As shown in Table 21.1, cognitive attribution was higher in the holistic thinking condition than the control condition, $b = 0.919$, $t(135) = 3.72$, $p < 0.0005$. In addition, cognitive attribution was lower in the analytic thinking condition than the control condition, $b = -0.915$, $t(135) = -3.77$, $p < 0.0005$. In short, the thinking styles manipulation had the expected effect on cognitive attribution.

Table 21.2 Means with standard deviations. (*Source* The author's research)

Measure	Holistic thinking condition	Control condition	Analytic thinking condition
Cognitive attribution	4.86 (1.25)a	3.94 (1.04)b	3.03 (1.24)c
Perceived price fairness	3.81 (1.46)a	3.16 (0.76)b	2.51 (0.96)c

Notes Means with standard deviations in parentheses. Means that do not have the same subscript were significantly different from each other

The effect of styles of thinking manipulation on perceived price fairness was significant, $F(2, 83.70) = 13.66$, $p < 0.0005$. Again, the effect of the thinking styles manipulation was decomposed by assessing the effects of the two dummy contrasts (holistic thinking vs. control; analytical thinking vs. control) on perceived price fairness by using multiple regression. As shown in Table 21.2, compared to the control condition, higher perceived price fairness was observed in the holistic thinking condition, $b = 0.651$, $t(135) = 2.85$, $p = 0.005$, and lower perceived price fairness was observed in the analytic thinking condition, $b = -0.645$, $t(135) = -2.88$, $p = 0.005$. To show that the effect of styles of thinking manipulation on perceived price fairness through the specific path of cognitive attribution, the hypothesis that the cognition attribution would mediate the relationship between styles of thinking manipulation and perceived price fairness was tested. As already demonstrated, the thinking styles manipulation had a significant effect on both cognitive attribution and perceived price fairness. As expected, the cognitive attribution was positively associated with perceived price fairness ($r = 0.707$, $p < 0.05$). Furthermore, when both the thinking styles manipulation and cognitive attribution were included as predictors of perceived price fairness in a multiple regression analysis, the cognitive attribution was a significant predictor of perceived price fairness, $b = 0.574$, $t(134) = 9.17$, $p < 0.0005$, and effect of the thinking styles manipulation became non-significant, $b = 0.124$, $t(134) = 0.656$; $p = 0.513$; $b = -0.120$, $t(134) = -0.648$, $p = 0.518$. The results of the Sobel test ($z = 3.54$, $p < 0.0005$; $z = 3.58$, $p < 0.0005$) were consistent with the hypothesis that cognitive attribution mediated the effect of the holistic versus analytic thinking on perceived price fairness.

21.6 Conclusions

The results of the experimental study provide support for our research hypotheses. Our findings confirm that styles of thinking have an important influence on perceived price fairness. When exposed to negative and/or undesirable incident such as price increase, holistic thinkers are likely to consider contextual and/or situational factors, i.e. external factors while deducing causes of the event, leading to higher cognitive attribution and perceived price fairness. In contrast, analytic thinkers are likely to ignore contextual and/or situational factors while deducing causes of the event, leading to lower cognitive attribution and perceived price fairness. The current research would be able to provide marketers and managers with guidance on the way to communicate a price increase, so that consumers are likely to concede the price increase and perceive it as fair, or at least less unfair.

Acknowledgements. The author would like to thank the Faculty of Management and Economics at Tomas Bata University in Zlin. This work was supported by the Internal Grant Agency of Tomas Bata University in Zlin under project number IGA/FaME/2018/015.

References

1. Chung, J.Y., Petrick, J.F.: Measuring price fairness: development of a multidimensional scale. J. Travel Tour. Mark. **32**, 907–922 (2015)
2. Han, H., Hyun, S.S.: Customer retention in the medical tourism industry: impact of quality, satisfaction, trust, and price reasonableness. Tour. Manag. **46**, 20–29 (2015)
3. Kasiri, L.A., Cheng, K.T.G., Sambasivan, M., Sidin, S.M.: Integration of standardization and customization: impact on service quality, customer satisfaction, and loyalty. J. Retail. Consum. Serv. **35**, 91–97 (2017)
4. Bolton, L.E., Warlop, L., Alba, J.W.: Consumer perceptions of price (un) fairness. J. Consum. Res. **29**, 474–491 (2003)
5. Bettray, J., Suessmair, A., Dorn, T.: Perceived price fairness in pay-what-you-want: a multi-country study. Am. J. Ind. Bus. Manag. **7**, 711 (2017)
6. Gorondutse, A.H., Hilman, H.: Mediation effect of customer satisfaction on the relationships between service quality and customer loyalty in the Nigerian foods and beverages industry: Sobel test approach. Int. J. Manag. Sci. Eng. Manag. **9**, 1–8 (2014)
7. Xia, L., Monroe, K.B., Cox, J.L.: The price is unfair! A conceptual framework of price fairness perceptions. J. Mark. **68**, 1–15 (2004)
8. Jin, L., He, Y., Zhang, Y.: How power states influence consumers' perceptions of price unfairness. J. Consum. Res. **40**, 818–833 (2013)
9. Reavey, B., Suri, R.: Determining fairness: a cognitive process of price fairness situations. In: Proceedings of the 2010 Academy of Marketing Science (AMS) Annual Conference, p. 177 (2015)
10. Chung, J.Y., Petrick, J.F.: Price fairness of airline ancillary fees: an attributional approach. J. Travel Res. **52**, 168–181 (2013)
11. Vaidyanathan, R., Aggarwal, P.: Who is the fairest of them all? An attributional approach to price fairness perceptions. J. Bus. Res. **56**, 453–463 (2003)

12. Somervuori, O.: Profiling behavioral pricing research in marketing. J. Prod. Brand Manag. **23**, 462–474 (2014)
13. Rothenberger, S., others: Fairness through transparency: the influence of price transparency on consumer perceptions of price fairness. Work. Pap. CEB. 15 (2015)
14. Shaw, S., Chovancová, M., Bejtkovský, J.: Consumer behaviour and warranty claim: a study on czech consumers. Econ. Sociol. **10**, 90–101 (2017)
15. Li, W., Hardesty, D.M., Craig, A.W.: The impact of dynamic bundling on price fairness perceptions. J. Retail. Consum. Serv. **40**, 204–212 (2018)
16. Haws, K.L., Bearden, W.O.: Dynamic pricing and consumer fairness perceptions. J. Consum. Res. **33**, 304–311 (2006)
17. Bhowmick, S.: Essays on three price judgments: price fairness, price magnitude, and price expectation (2010)
18. Pallas, F., Bolton, L.E., Lobschat, L.: Shifting the blame: how surcharge pricing influences blame attributions for a service price increase. J. Serv. Res. 1094670518755314 (2017)
19. Young Chung, J., James, F.: A conceptual framework of perceived price fairness: an attributional approach (2016)
20. Chung, J., Petrick, J.: A conceptual framework of perceived price fairness: an attributional approach. In: TTRA Annual Conference Proceedings (2009)
21. Dominique-Ferreira, S., Vasconcelos, H., Proença, J.F.: Determinants of customer price sensitivity: an empirical analysis. J. Serv. Mark. **30**, 327–340 (2016)
22. Nisbett, R.E., Peng, K., Choi, I., Norenzayan, A.: Culture and systems of thought: holistic versus analytic cognition. Psychol. Rev. **108**, 291 (2001)
23. Cheng, H.-Y., Zhang, S.-Q.: Examining the relationship between holistic/analytic style and classroom learning behaviors of high school students. Eur. J. Psychol. Educ. **32**, 271–288 (2017)
24. Monga, A.B., John, D.R.: What makes brands elastic? The influence of brand concept and styles of thinking on brand extension evaluation. J. Mark. **74**, 80–92 (2010)
25. Monga, A.B., John, D.R.: Understanding cultural differences in brand extension evaluation: the influence of analytic versus holistic thinking (2009)
26. Lalwani, A.K., Shavitt, S.: You get what you pay for? Self-construal influences price-quality judgments. J. Consum. Res. **40**, 255–267 (2013)
27. Hossain, M.T., Bagchi, R.: How cognitive style influences the mental accounting system: role of analytic versus holistic thinking. J. Consum. Res. (2018)
28. Dikici, A.: Relationships between thinking styles and behaviors fostering creativity: an exploratory study for the mediating role of certain demographic traits. Educ. Sci. Theory Pract. **14**, 179–201 (2014)
29. Monga, A.B., John, D.R.: When does negative brand publicity hurt? The moderating influence of analytic versus holistic thinking. J. Consum. Psychol. **18**, 320–332 (2008)
30. Choi, B.-J.: The influence of cultural thinking style on consumer cognitive complexity underlying wine purchase decisions. J. Appl. Bus. Res. **32**, 1257 (2016)
31. Monga, A.B., Hsu, L.: How consumers' styles of thinking can control brand dilution. GfK Mark. Intell. Rev. **10**, 40–45 (2018)
32. de Oliveira, S., Nisbett, R.E.: Culture changes how we think about thinking: from "human inference" to "geography of thought". Perspect. Psychol. Sci. **12**, 782–790 (2017)
33. Monga, A.S.B., Williams, J.D.: Cross-cultural styles of thinking and their influence on consumer behavior. Curr. Opin. Psychol. **10**, 65–69 (2016)
34. Song, L., Swaminathan, S., Anderson, R.E.: Differences in customers' online service satisfaction across cultures: the role of thinking style. J. Mark. Channels. **22**, 52–61 (2015)
35. Maxwell, S., Mayer, V.F., Stamer, H.H., Diller, H., Avila, M.G.: Reactions to a price increase: what makes it seem fair. Glob. J. Manag. Bus. Res. (2013)

36. Hollebeek, L.D.: Individual-level cultural consumer engagement styles: conceptualization, propositions and implications. Int. Mark. Rev. **35**, 42–71 (2018)
37. Monga, A.B., John, D.R.: The influence of brand concept and styles of thinking on brand extension evaluation. ACR Eur, Adv (2011)
38. Chinchanachokchai, S., Noel, H.: The devil is in the details: when holistic thinkers react negatively to incongruent information. Contemp. Manag. Res. **11**, 25 (2015)

Chapter 22
Disruptive Technologies or Big-Bang Disruption: A Research Gap in Marketing Studies

Mohammad Bagheri(✉)

IPAM, Portuguese Marketing Institute, Porto, Portugal
moe.bagheri@universidadeeuropeia.pt

Abstract. The term disruption has widely been used across industries to conjure the innovation concept and support such discussions in different contexts, but what disruptive technology means and what characteristics of disruptive technologies or innovations are. Based on Christensen's (Christensen, The Innovator's Dilemma When New Technologies Cause Great Firmsto Fail. Harvard Business School Press, Boston, Massachusetts, 1997, [1]) theory of disruptive innovation and big-bang disruption of Downes and Nunes (Harv Bus Rev, 44–56, 2013, [2]), this paper deeps dive into the concept of disruptive technology and big-bang disruption to discover the main features of such disruptions. It looks at the two concepts to uncover the characteristics of such disruptions and signify their differences. It reveals a research gap to be further studied and proposes empirical research to validate whether the traditional technology adoption of Rogers is applicable to disruptive innovations, or the newly proposed model of Downes and Nunes (Harv Bus Rev, 44–56, 2013, [2]), is more valid. The missing link is the lack of having the right marketing strategy for products and services which were born based on the theory of Big-Bang Disruption. Hence, the paper looks at the academic literature and presents some of the most relevant scholars' outlook on disruption concept. It proposes the methodological strategies and the approach for the future study, which is positivism. It discusses the research design that is "case study" and its method of measurement and concludes by the expected contribution.

22.1 Introduction

22.1.1 Presentation and Delimitation of the Topic

Bower and Christensen [3] raised a question about the reason behind the failure of the established firms in making a proper technological investment for the future customer's demand, while they strongly–and successfully–invest in the technologies to retain their present customers. The authors answered the question by highlighting the fact that the essential reason is that leading companies surrender to one of the most general, and valued, management dogmas, which is staying close to the customers that have astonishing power in guiding a company's investments [3].

Á. Rocha et al. (eds.), *Marketing and Smart Technologies*, Smart Innovation, Systems and Technologies 167, https://doi.org/10.1007/978-981-15-1564-4_22

There are several cases such as IBM Minicomputers, Xerox's Tabletop Copiers, and Bucyrus-Erie Hydraulic Excavators that companies listened to their customers and offered them the performance of the product they needed, but finally, the companies were hit by the same technologies their customers led them to ignore [3].

The strong well-known companies are constantly developing technologies for their current customers to meet their needs for the forthcoming performance required, however, they failed to detect and offer technologies for the small customer segment, which their needs are not the same as the mainstream customers [3]. Lots of firms had to learn it the hard way to ignore the threats of new technologies that do not primarily meet the needs of mainstream customers [3].

Disruption is done especially when incumbents concentrate on refining and enhancing the current products and services for their most profitable and demanding customers, which lead to exceeding the needs of some customer segment and overlooking other segments [4]. Consequently, the new entrants focus on offering the disregarded segments and obtain a grip through a better functional offer at a reduced price [4]. In this manner, while the incumbents are after gaining more profit from the current customers, they are inclined not to actively respond to the movements from the new entrants [4]. Therefore, the entrants move toward the top of the market by responding to the mainstream customer needs leading to the time when the mainstream customers start using and taking up the new offerings, which is the moment that disruption occurs [4].

Bower and Christensen [3] emphasized that the technologies that destroy the incumbents are not necessarily new or difficult from technology viewpoint, instead (1) they normally offer different performance attributes, which are not favored by the current customers, and (2) the preferred technology characteristics of the present customers advance so quickly that the companies with new technologies can sweep the established markets. While the old-established firms–incumbents–aim to enhance the products and services for their mainstream customers who are mostly profitable segment, they surpass the need of their most demanding segment, which makes them to neglect the needs of other customers [4]. That is how the companies that offer cheap substitutes to the products of incumbents, capture new low-end customers, and then they gradually move upmarket to pick off higher end customers [2].

As a crucial aspect in competitive strategy, the uprising identification of disruptive innovation has made the researchers scrutinize the appearance of disruptive innovation [5]. Wan et al. [5] argued that the condition of which disruptive innovations in a firm would most likely arise from are the company's resource allocation processes, its organizational structure and culture. One of the factors that can stimulate disruptive innovations in emerging markets is that the changes in product design and business models can considerably reduce the costs and improve the value-for-money attribute [5]. These changes are frequently being perceived as the prerequisite for exposing the mass-market customer segments with inadequate disposable income in the emerging markets [5].

Wan et al. [5, p. 95] argued about the existence of two preconditions for a market disruption: (1) "performance overshoot on the mainstream attributes of the existing product"; (2) "asymmetric incentives between an existing healthy business model and the potentially disruptive business model," which were documented through case study in various industries.

Although disturbing the status quo, not every disruptive technology has the potential to be disruptive and alter the business or social landscape and rearrange the value pools. Tellis [6] asserted that in contradiction to the popular belief that only small new firms introduce new technologies, the incumbents along with new entrants launch new technologies. The old-established firms (i.e. Procter and Gamble) are constantly innovating to be able to accommodate the emerging market and they are afraid that their competitors reach there prior to them–good examples are Microsoft and Intel [6]. One can notice that old-established firms are sometimes even eager to cannibalize their current assets to grasp the future potential, which is contrary to the practice of the dominant incumbents that ignore the technological change and new entrants by sticking to the current customers and market success [6]. Tellis [6] stated that based on his research findings, the author did not observe the straight linear pattern with a constant slope in the examples of disruptive technology of Christensen. So, this study will investigate the outlook on the disruptive technologies and big-bang disruption. It is recommended to use a case study approach to validate whether the theory of disruption is applicable to the newly introduced technologies refereed to as disruptors.

22.1.2 Justification of the Study

Christensen et al. [4] claimed that there are too many wrong interpretations of disruption by the people who did not read a serious book or article within this area. That is the reason why the term "disruption" has been utilized so loosely and repeatedly to conjure the innovation concept to support such discussions [4]. For instance, several researchers and writers referred to "disruptive innovation" as any situation that an industry is restructured, which the previous successful incumbents stumble, but that is a very broad usage of the term and concept [4]. Christensen et al. [4, p. 4] described disruption as "a process whereby a smaller company with fewer resources is able to successfully challenge established incumbent businesses."

The term "Disruptive Technology" was first coined in 1997 in a book titled "The Innovator's Dilemma" by Clayton M. Christensen, which is any type of technology that emerges to displace the old-established ones [7].

The issue of amalgamating a disruptive innovation with any breakthrough innovation, that changes the market landscape and competitive patterns, is that various kinds of innovations need different strategies [4]. The authors indicated that being a follower as a disruptive innovator or being defensive against a disruptive challenger could not be practiced in every firm in a shifting market [4]. If we are careless in labeling or do not incorporate the views from the research and experience into the main theory, then this makes managers selecting the incorrect tools for their settings and lead them in having less chance of success, which overall dilutes the practicality of the theory over time [4].

In a disrupted market, the businesses with substantial impacts are technology-oriented and information-intensive businesses. It is critical that new businesses prepare themselves for the upcoming changes in the market landscape. According to the McKinsey report, there are twelve technologies that will disrupt and transform life, business and the global economy with the economic potential impact between \$14 trillion and \$33 trillion a year in 2025 [8]. The report specified that the said value was based on in-depth studies and the application of the twelve technologies in creating consumer surplus from better products and lower prices, cleaner environment and better health [8].

Since the US Taxi market was hit by the introduction of Uber and its business model has been extended to various businesses and industries, there have been many discussions about Uber being a disruptive innovation. Consequently, Christensen et al. [4] raised the question whether Uber is a disruptive innovation or not. The authors argued that according to the theory of disruptive innovation, Uber is not a disruptive innovation because its financial and strategic achievements do not qualify the firm as genuinely disruptive even though it has almost always been described like that due to two main reasons [4]. *Reasons number one* is that disruptive innovations derive from low-end or new-market footholds, which was not the case in Uber [4]. Disrupters usually start by getting the appeal from low-end and under/unserved consumers then move on to the mainstream, however, in Uber's case it went all the way on the reverse direction, by first constructing a market position in the mainstream market and afterward engaging the generally unnoticed segment [4]. *Reason number two* is that disruptive innovations do not become popular with mainstream customers until quality catches up to their standards [4]. That is how disruptive theory differentiated disruptive innovation from sustaining innovations, that the latter is more of enhancing the current products for the existing customers of incumbents–that can be incremental advances or major breakthroughs–whereas the disruptive innovations are primarily presumed menial by the customers of incumbents, when they are not keen in switching to the new product offerings only because of its lower price, rather they prefer to postpone it up until the quality rise up to par then they switch to the new technologies with less price, which is the secret of disruption decreasing the price in the market [4].

Christensen et al. [4] described the diagram as contrasting the product performance with the customer demand axis, which are simply red lines demonstrating improvements of products and services throughout the time versus the blue lines that illustrated the inclination of customers to pay for performance, which incumbents target the upper red line offering products with high quality and they ignore the customers of low end, leaving the chance of lower end segment for new players that offer low-priced products and services targeted for lower red line, and gradually enhance their product performance and move toward the top of the market and challenging the incumbents and their market control.

Nunes and Downes [9] argued that the normal pattern of technology adoption of Rogers does not apply to the big-bang disruption because they have only two stages of Trial Users and Vast Majority. Accordingly, this paper looks at the concept of disruptive technology and innovation to uncover the characteristics of such disruptions and signify its differences with big-bang disruption. It further investigates these two concepts to realize their pattern and to test them in an empirical context.

22.1.3 Problem Formulation

Referring to the brief look at the literature, it can be observed that disruptive technology is a topic which has captured the attention of academia and practitioners alike. One cannot ignore that the storm of change will embrace most of the industries abruptly, in other words, it might appear as a bang. Accordingly, it could be inspected that the upcoming years will witness the appearance of new disruptive technologies targeted at offering new products and services at different verticals of the industry. Hence, there is a strong need to market these offerings in the best possible way. Thus, this study is set to investigate the methods of which market disruptors educate, inform and persuade the users to take up the new product and services. The missing link is the lack of having the right marketing strategy for products and services which were born based on the theory of disruptive technology.

The normal pattern of technology adoption of Rogers does not apply to the big-bang disruption, given that they have only two stages of Trial Users and Vast Majority [9]. The big-bang disruption model is condensed, it has some trial users at the initial stage and then an acceptance of a great majority of customers [9]. The big-bang disruption can have a new product cycle that is simplified into three basic stages: development, deployment, and replacement [2].

Consequently, the research gap is to validate whether the traditional technology adoption of Rogers is applicable to disruptive innovations, or the newly proposed model of Downes and Nunes [2] is more valid. The research opportunity recognized is to empirically test the concept of big-bang disruption. This will assist marketers in selecting the right marketing strategy in the new product life cycle for products and services built upon the theory of big-bang disruption. The missing link is the lack of having the right marketing strategy for products and services which were born based on the theory of Big-Bang Disruption.

22.2 Literature Review

22.2.1 Current Research Related to the Topic Under Research

In the face of continually evolving technologies, companies need to have strategies in place to balance between sustaining technology and disruptive technology. Since technology can progress faster than demand, some suppliers offer customers more than what they require or are willing to pay. Based on the essential elements of the theory of Christensen, the sustaining strategy is the innovations that help incumbent companies to win higher margins by selling better products to their best customers via simple, incremental engineering improvements [7]. On the contrary, disruptive innovations appeal to customers who are not interested in the incumbents, it engages simple adoption of known technologies [7]. Danneels [10] described disruptive technology as a technology that changes the bases of competition by altering the performance metrics by which companies compete. New companies that demonstrate disruptive starts by successfully targeting the overlooked market segment, which will gain them a ground in offering better functionality usually at a lower price [4].

These new entrants then gradually move toward the top of the market to deliver the required performance of the incumbents' mainstream customers, while protecting their uniqueness which made their early success [4]. The time that the mainstream customers adopt the new offerings in mass scale is the time when the disruption has already ensued [4]. Typically, disruptive technologies are cheaper, simpler and more convenient and reliable than the established technologies [11]. Although disruptive technologies originally underperform incumbents in serving the mainstream market, they ultimately replace the established technologies [10].

One of the reasons that disruptive technologies are not well accepted by the majority of the customers is that at the initial stage they do not meet the minimum requirements along the performance metric, which is most appreciated by the mainstream segment [10]. The products built upon the notion of disruptive technology primarily satisfy the needs of the niche segments that value the higher dimension of performance on which the disruptive technology excel [10]. A new market disruption is basically an innovation that allows people with less money, training and skill to begin consuming and using a product [12]. In 2003, in the new book written by Christensen and Raynor, the term "disruptive technology" is substituted with "disruptive innovation" supposedly to broaden the theory's applicability [10]. Christensen has argued that since disruptive technology at the beginning stage merely serves a minimal market segment with low-profit margin, it is overlooked by incumbents that are serving more profitable and attractive market segments [10].

As technology develops and matures in the marginal market, it finally facilitates its performance and fulfills the needs of the upmarket segments and disrupts the whole market [10]. Tellis [6] critically analyzed the Christensen's thesis and summarized his notion of disruptive technology into five premises: (1) initially disruptive technology underperforms the dominant technology in the principal dimensions appreciated by the typical customers; (2) but it has other features that are valued by marginal customers, which happened to be cheaper, simpler, and more convenient than the dominant technology; (3) it is not well received by the general and most profitable customers, so they are commercialized in emerging and minor market segments and in terms of financial investment they are not welcomed with the incumbents; (4) the new technology improves its performance to reach a point that it satisfies the required standard performance of the mainstream market; (5) and that is the time when the new disruptive technology replaces the dominant technology and the new firm displaces the incumbent in the mainstream market.

For the incumbents and the established firms, it is a choice to advance their current products and stick to the current market and customers using sustaining strategies or gaining new segments of the market by implementation of new technologies and adopting new business models using disrupting strategies [13]. The examples of disruptive innovations are the expensive mainframe computers being replaced with personal computers, IBM stepping into the PC market, Netflix converting from sending out rental DVDs by post to a new business model of streaming on-demand videos to its customers [13]. Simply put, disruptive innovations do not compete; they create a new sphere for themselves.

Disruptors target the bottom of the market with unproved unpolished products that are usually priced low, moving under the radar not to be detected, whereas, the incumbents are content with their profit margins of the current customer and they are slow to recognize the threats, which are gradually capturing their market and reshaping the entire market landscape and industries [13]. Some of the successful examples of these disruptions are classified ads "Craigslist", long-distance calls "Skype", record stores "iTunes", research libraries "Google", local stores "eBay", and newspapers

"Twitter" [13]. Several of the close future disruptions with a highly influential impact on the current market players are on the way, Google is working to reinvent cars as autonomous vehicles; Amazon is developing the new concept of delivery and reinventing shopping experience using drones, the 3D printing is disrupting manufacturing, distance learning, and virtual universities are delivering education over the Internet and health care is using the technology of health information system [13].

Christensen was accused of cherry-picking the example of disruptive technology to back his notion and theory, and due to this matter, all his disruptive technology case studies were successful [14]. The critical fact about some of the criticisms is that, even with the repeated usage of the "disruptive innovation" term, the theory's concepts are not always applied in a consistent manner [15].

It is not unusual for the expression to be employed with a radical meaning that is distant from what was intended by Christensen.

These market disruptors did not come from rivals in the same industry or even with similar business models, or targeted unprofitable or underserved market segment, nor did enter from the bottom of a mature market and then moved up the market [2]. They appeared in a way that consumers in every segment simultaneously and massively are defected by their kind of innovation which changes the rules [2]. As Downes and Nunes [2, p. 16] indicated "the change will come not over time but suddenly, not with a whimper but with a bang." Consequently, one can observe the replacement of mature products with new technologies that are introduced by the disruptors that come from nowhere and instantaneously be everywhere [2]. These game-changers are called "big-bang disrupters" that are unplanned and unintentional and do not follow conventional strategic routes or normal patterns of market adoption [2].

Nonetheless, the disruptors do not necessarily come from the same industry or with the same business model, neither they would capture the underserved market segment nor the bottom of the market [2]. As referred by Downes and Nunes [2], big-bang disruptors once launched it is hard to fight such disruptions, and it affects the whole consumer segment. They simply do not create dilemmas for innovators, they trigger disasters [2]. Good examples of big-bang disruptors are Angry Birds: downloaded over a million times in the first 24 h and more than 200 million times within 7 months; CampusBookRentals: offering affordable rates to rent books for students; Khan Academy in education; Pandora and Spotify in radio and recorded music; Skype and FaceTime in voice and video calling; Square in mobile credit-card processing; and finally Sony PlayStation that was sold only at $299 with hundred games compared to the arcade pinball machines that cost up to $7,500 [2].

When the big-bang disrupters launch in the market they bring a lot of surprises–they are up, and then fade away–which is due to their three main characteristics that Downes and Nunes [2, p. 5] described as: (1) *unencumbered development*: means that these innovations are "often born of rapid-fire, low-cost experiments on fast-maturing, ubiquitous technology platforms" in hackathons events in Silicon Valley that is not intended to disrupt the businesses, but some businesses are just collateral damage (i.e. Twitter was born in a hackathon in 2006); (2) *unconstrained growth:* meaning big-bang disrupters do not follow the traditional classic bell curve of Everett Rogers [16], so these types of innovation have an adoption curve closer to a straight line that heads up and then falls quickly (i.e., upon the launch of iPad everyone wanted to have one, even a millionaire); (3) *undisciplined strategy*: means that big-bang disrupters do not follow the strategic Discipline of Market Leaders introduced by Treacy and Wiersema [17] which stated to pursue only one dimension of low cost (operational excellence), constant innovation (product leadership), or customized offerings (customer intimacy), but big-bang disrupters rather have better performance at a lower price with greater customization which competes with the conventional products on all the abovementioned three value principals from day one (i.e., Smartphones killing the portable navigation tools such as Garmin and TomTom that lost 70 and 85% of their market capitalization) [2].

It is important to highlight that the big-bang disruptions have a different pattern of adoption compared to the traditional technology adoption that was theorized by Rogers [2]. Although disruptive products and services have apparent benefits, not everyone will adopt it instantly. Diffusion of Innovations (DOI) theory of Everett Rogers described how, over time, an idea or product gains momentum and diffuses or spreads through a specific population or social system [16]. Everett Rogers identified some personality traits of consumers that indicate how customers accept and adopt a new innovation over time [16].

The normal pattern of technology adoption of Rogers does not apply to the big-bang disruption, because it only has two stages of Trial Users and Vast Majority [9]. The big-bang disruption model is compressed, it has some trial users at the preliminary stage and then an acceptance of a great majority of customers [9].

The big-bang disruption's three basic stages of development, deployment, and replacement [2] needs to be empirically studied to confirm whether the traditional technology adoption of Rogers is applicable to disruptive innovations, or the newly proposed model of Downes and Nunes [2] is more valid.

22.2.2 Research Proposition

Referring to the above look at the literature and the identified gap, this study is set to empirically investigate the theory of big-bang disruption.

The research objectives are namely:

1. To empirically analyze and understand the factors that impact the adoption of products and services born based on disruptive technology and big-bang disruption.
2. To unveil the above behavior to marketers for selecting the right marketing strategy to increase the adoption rate.

Taking these two objectives into account, there are two research questions which need to be investigated:

1. What factors impact the adoption of products and services born based on disruptive technology and big-bang disruption?
2. How market disruptors utilize disruptive technologies in their favor to accelerate the adoption of their products and services?

22.3 Methodological Strategy

22.3.1 Methodological Approach

Making the proper choice between qualitative, which is investigating theories by the means of numbers and large samples, and quantitative methodology, that is using words and meaning in smaller samples to construct theories, is of concern to many researchers [18]. While some researchers adopt only one methodology, the others recommend having both even if the results of the two methods may be opposing for the same aspect [18]. Having said that, the main concern for investigators is not selecting the right methodology but recognizing the paradigms of the research [18]. As Perry, Riege and Brown [19] indicated, a paradigm is a general conceptual framework that a researcher may work within, and it also can be considered as the "basic belief system or worldview that guides the investigator" [20, p. 105].

A paradigm commentates on a researcher's understanding of the nature of existence that is apart from any logical debates, as each paradigm is rational within its particular constructed logic, hence there is no objective ground for choosing a paradigm [18]. Accordingly, all the things that a researcher can do is to work within a paradigm that is consistent with his or her own presumptions, which cannot be tested on any empirical or logical ground [18]. For this reason, the future researcher could choose the positivism paradigm to apply in this proposed research and investigate the findings.

22.3.2 Research Design

This paper grounded its research on Christensen's theory of disruption [1] and the newly proposed theory of big-bang disruption of Downes and Nunes [2]. The methodologies used in Christensen's theory were mainly qualitative and case studies. Danneels [10, p. 250] also indicated that "the historical case studies in The Innovator's Dilemma are considered best as wonderfully rich empirical data used for theory-building purposes." Consequently, the proposed research paper should also use qualitative and case studies methodology to find the answers to its questions.

One of numerous ways of conducting social science research is the case study, which other approaches can include experiments, surveys and histories to name a few [21]. "Each strategy has peculiar advantages and disadvantages, depending upon three conditions: (a) the type of research question, (b) the control an investigator has over actual behavioral events, and (c) the focus on contemporary as opposed to historical phenomena [21, p. 1]. Said that, there are three main features on when utilizing case study is the ideal approach: (1) when there is a need to answer the research questions of how and why; (2) when the researcher does not have any possibility or control over on the events in the research; (3) when the research concentrates on a phenomenon in real-life context [22], which can be named explanatory type of case study aggregated by two other types of case studies exploratory and descriptive.

Case study is an empirical investigation concentrating on a current issue in its real-life situations of which the limitations between the issue itself and its context are not very distinct, that are applicable for investigating complex social phenomenon through the design of a single case study or multiple case studies, using qualitative, quantitative, or both methods to answer the research questions [21]. As recommended by Yin [21] and implemented and tested by Christensen [1] for disruptive technology, the proposed research paper should use case study to find the answers to the research questions.

By answering the questions of why and how, discovering the theoretical propositions such as narrowing down the scope of the study and finding the links between the issues, realizing the unit of analysis to be in parallel with the research questions and previous studies, building a logical reasoning between the data and the research proposition, setting proper criteria to analyze and interpret the research findings bring about a loose theory and a blueprint of the case under investigation [21].

In terms of selecting the type of case study, whether explanatory, exploratory or descriptive to be used depends on having a rich theoretical framework, that explanatory design is best suited when having the richest theories; and in terms of case selection, one should use developed preliminary theory as an outline to compare the features and findings of the case that indicates the underlying theoretical propositions and/or conceptual framework; and finally in terms of generalizing the results of the case study, it depends on the proper development of the study's design and its initial theory [21].

Since the theories behind disruptive technology/innovation and big-bang disruption are not rich enough–in terms of being young theories and being tested in various studies and industries–in order to use the explanatory case study, the proposed research could use the exploratory type of case study to reach its intended goal.

Based on Johnson and Christensen [22], the research methodology needs to be exploratory, having a mixed approach of qualitative and quantitative. The quantitative approach is to enhance the qualitative method. In order to collect primary data, in-depth interviews need to be conducted along with distributing questionnaires to the targeted population.

The study can have a cross-country analysis to compare the companies that offer technology-oriented products and services that meet the criteria of disruptive technology and big-bang disruption. The selected countries could be the United States and Portugal. The reasons for such choices are namely: (1) the US being the market leader in introducing and launching disruptive technologies; (2) Portugal being one of the fast-moving forward countries in having tech startups, which is evident by being the organizer and center of attention in Web Summit 2018. The interviews could be conducted with the tech companies in the US and Portugal to validate the theories under investigation. The interview questions should be extracted and adapted from the relevant literature. Since the theory of Big-bang disruption is still very young (2013), and to the best of the author's knowledge no empirical study has been found that tested the theory, this concept needs more in-depth investigations in an empirical context.

Following the interviews, the questionnaire that is developed based on the literature should be distributed among American and Portuguese consumers to test the big-bang disruption and check the take-up rate of two selected disruptive technologies.

Since the theory claims that big-bang disruption has two stage of Trial Users and Vast Majority, the questionnaire should be focused on testing whether such algorithm and patterns of behavior is detected from the collected data from American and Portuguese consumers.

Based on these two sources of data (case studies of companies and questionnaires among the end users), the data needs to be validated against the academic literature, and it should be interpreted and triangulated.

22.3.3 Measurement

Considering the number of cases and unit of analysis, one should consider: (1) case characterizes the topic under investigation in an empirical manner; (2) the unit of analysis, which is the actual source of information, can be different aspects such as individual, organizational documents, capital flow of a company or even a policy; (3) a single case study is the right choice when the case represents a critical test to existing theory resulting from an unusual or unique happenings; (4) multiple case study is appropriate when a replication logic is required to demonstrate support for theoretically similar findings, and contrasting the results for predictable reasons; (5) in terms of replicating case numbers, it can be contingent with the desired certainty and richness of the underlying theoretical propositions [21].

Accordingly, the proposed paper could use multiple case studies to uncover the theory of disruptive technology and big-bang disruption and their take-up rate, so the unit of analysis could be one or two technologies that are considered disruptive or big-bang disruption based on their criteria and features described in the literature.

Yin [21] argued that the units of analysis can have either a holistic designs, which comprises of a single unit of analysis with the objective of surveying the global nature of the phenomenon when there are no logical subunits, which this method has the trap of abstractness; or embedded designs, which contains multiple units of analysis with the possibility of having main units and subunits at different levels in search for detecting consistent patterns of evidence across the units within the case under investigation. Since disruptive technology and big-bang disruption have different subunits and it is not only one main unit, the proposed study should use embedded designs to study the concepts and uncover the results.

22.4 Expected Contribution

It is important to highlight that disruptive technologies start by engaging the low-end market segment and underserved consumers, then move toward the mainstream market, however, Uber is an exception case because it started just in the opposite direction by first forming a position in the mainstream market and then serve the ignored and disregarded market segment [4]. Christensen [4] claimed that based on the theory of disruption Uber is an outlier, which made it difficult for the authors to have a universal way to explain such nonconforming results. In actual fact, the old-fashioned disruption impersonated as the innovator's dilemma, but the big-bang disruption is the innovator's disaster, which will keep the managers of any industry in cold sweat in the years ahead [2]. Nevertheless, the most successful innovators in the close future are the ones who could ride on the right aggregation of other people's technologies [2].

The normal pattern of technology adoption of Rogers does not apply to the big-bang disruption, given that they have only two stages of Trial Users and Vast Majority

[9]. As it was mentioned earlier, the big-bang disruption model is compressed, therefore it has some trial users at the initial stage and then an acceptance of a great majority of customers [9]. The big-bang disruption can have a new product cycle that is simplified into three basic stages: development, deployment, and replacement [2]. Consequently, the future paper could empirically investigate and validate whether the traditional technology adoption of Rogers is applicable to disruptive innovations, or the newly proposed model of Downes and Nunes [2] is more applicable to such disruptions.

The contribution of this research is realizing a research gap in the marketing field to be further studied. It will contribute to the marketing studies and practices in selecting the right marketing strategies for introducing and promoting the new technologies in the process of the new product life cycle for products and services built upon the theory of disruptive technology or big-bang disruption. From the theoretical point of view, it will validate the implication of theories under testing by using the case study approach and adds value to the academic literature on the confirmation or rejection of such theories in real-life.

From the managerial point of view, the proposed research will assist unveiling the adoption pattern of products and services built upon disruptive technology or big-bang disruption, which can be utilized for recognizing the most suitable marketing strategies to target the end users of such products and services. Hence, through this research and its case studies, the managers will have a chance to observe the success or failures of the case companies under investigation and analyze the implemented strategies, which will empower them in selecting and implementing the right strategies in their own firms. Accordingly, this scientific investigation will help managers decide on selecting the most suitable marketing strategies for their disruptive products and services, and realize the fact that either the life cycle of their product or service has been shortened and follow the proposed theory of Downes and Nunes [2] or it follows the Rogers [6]'s traditional technology adoption.

References

1. Christensen, C.M.: The Innovator's Dilemma When New Technologies Cause Great Firms to Fail. Harvard Business School Press, Boston, Massachusetts (1997)
2. Downes, L., Nunes, P.F.: Big-bang disruption. Harv. Bus. Rev. 44–56 (2013)
3. Bower, J.L., Christensen, C.M.: Disruptive technologies: catching the wave. Harv. Bus. Rev. **73**, 43–53 (1995). https://doi.org/10.1016/0024-6301(95)91075-1
4. Christensen, C.M., Raynor, M.M., McDonald, R.: What is disruptive innovation? Harv. Bus. Rev. **93**, 44–53 (2015). https://doi.org/10.1353/abr.2012.0147
5. Wan, F., Williamson, P.J., Yin, E.: Antecedents and implications of disruptive innovation: evidence from China. Technovation **39–40**, 94–104 (2015). https://doi.org/10.1016/j.technovation.2014.05.012
6. Tellis, G.J.: Disruptive technology or visionary leadership? J. Prod. Innov. Manag. **23**, 34–38 (2006). https://doi.org/10.1111/j.1540-5885.2005.00179.x
7. Christensen, C.M., Johnson, M.W., Rigby, D.K.: Foundations for growth: how to identify and build disruptive new businesses. MIT Sloan Manag. Rev. **43**, 22–31 (2002)

8. Manyika, J., Chui, M., Bughin, J., Dobbs, R., Bisson, P., Marrs, A.: Disruptive technologies: advances that will transform life, business, and the global economy. McKinsey Glob. Inst. 1–164 (2013)
9. Nunes, P.F., Downes, L.: Big-Bang disruption–the innovator's disaster. J. High-Perform. Bus. Innov. Accent. Outlook. (2013)
10. Danneels, E.: Disruptive technology reconsidered: a critique and research agenda. J. Prod. Innov. Manag. **21**, 246–258 (2004)
11. Christensen, C.M.: The Innovator's Dilemma: When New Technologies Cause Great Firms to Fail. Harvard Business School Press, Boston, Massachusetts (2000)
12. Christensen, C.M., Raynor, M.M.: The Innovator's Solution: Creating and Sustaining Successful Growth. Harvard Business School Press (2003)
13. Walker, A.: What disruptive innovation means (2015). http://www.economist.com/blogs/economist-explains/2015/01/economist-explains-15
14. Cohan, P.S.: The Dilemma of the innovators Dilemma: clayton Christensen's management theories are suddenly all the rage, out are they ripe for disruption. Ind. Stand. **10**, 2000 (2000)
15. Tadao Kawamoto, C., Giovinazzo Spers, R.: A systematic review of the debate and the researchers of disruptive innovation. J. Technol. Manag. Innov. **14**(1), 73–82 (2019)
16. Rogers, E.M.: Diffusion of Innovations, vol. 12. New York (1995)
17. Treacy, M., Wiersema, F.: How market leaders keep their edge. Fortune **131**, 52–57 (1995)
18. Sobh, R., Perry, C.: Research design and data analysis in realism research. Eur. J. Mark. **40**, 1194–1209 (2006). https://doi.org/10.1108/03090560610702777
19. Perry, C., Riege, A., Brown, L.: Realisms role among scientific paradigms in marketing research. Irish Mark. Rev. **12**, 16–23 (1999)
20. Guba, E.G., Lincoln, Y.S.: Competing paradigms in qualitative research. Handb. Qual. Res. 105–117 (1994). http://www.uncg.edu/hdf/facultystaff/Tudge/Guba%20&%20Lincoln%201994.pdf
21. Yin, R.K.: Case Study Research—Design and Methods. Sage Publications, Thousand Oaks (1994)
22. Johnson, B., Christensen, L.: Educational Research: Quantitative, Qualitative, and Mixed Approaches. Sage Publications (2008)

Chapter 23
The Influence of Brand Experience in City Marketing

António Coelho[1], Ana Sampaio[2], and Pedro Rodrigues[1](✉)

[1] ISMAI, Maia University Institute, Maia, Portugal
{catiasteixeiraramos, pmrodrigues}@gmail.com
[2] CEDTUR ISMAI, Maia University Institute, Maia, Portugal
asampaio@ismai.pt

Abstract. This study aims to understand the extent to which brand experience influences city marketing by investigating its influence on visitor-based brand equity (based on consumer-based brand equity scale), as well as on the City Brand Love (CBL). The investigation focuses on the visitors of the city of Matosinhos. For this purpose, a study was carried out with Internet users who meet the requirements of having already visited the city at least once, using later validated study scales of brand experience, consumer-based brand equity, and brand love. A questionnaire survey was answered by a total of 224 respondents. The data collected were treated using SPSS statistical software in order to perform the described analysis of the variables and exploratory factor analysis, and the smart PLS software for the confirmatory factor analysis and the Structural Equation Model (SEM). The results confirm the influence of brand experience in CBL and CBBE. It was also demonstrated a significant influence of the CBBE in CBL. The study shows that brand experience influences in a relevant way city marketing and consequently the economic performance of it.

23.1 Introduction

Experience Economy is a relatively recent term, emerging as a paradigm shift in the marketing of goods and products, aimed at a new way of capturing and retaining customers by different organizations [1]. Likewise, city marketing is a management practice whose research is recent [2], arising as a consequence of the phenomenon of globalization [3].

Local authorities have been searching for ways of creating value for their visitors and therefore their retention and the willingness to repeat the visit and share their experience with others, as well as encourage them to invest in the city. Nowadays, the cities are in a great competitive environment, reason why the differentiation through its endogenous resources, material, and immaterial patrimony becomes the greater and important challenge.

The brand experience emerges as a differentiating and catalyzing element for the economic performance of cities, bringing benefits such as the leveraging of declining brands and differentiation of competition, generating brand awareness and leading to

Á. Rocha et al. (eds.), *Marketing and Smart Technologies*, Smart Innovation, Systems and Technologies 167, https://doi.org/10.1007/978-981-15-1564-4_23

product acquisition and brand loyalty [4]. The creation of unique and memorable experiences has been shown to lead to loyalty [5], satisfaction [6, 7] and is vital to stakeholder assessment of the global brand value, thus generating greater potential for revenue generation. In addition, these high levels of satisfaction and loyalty can lead to a strong feeling and passion for the city.

In short, visitors of a certain city are essentially looking for well-being, pleasant, and memorable experiences that they can repeat later, and municipalities have been paying attention to this due to the positive consequences of promoting these feelings.

23.2 Marketing and Experience: Conceptualization

Marketing as a management philosophy aimed at creating value for customers is a process whereby "individuals get what they need and/or want for the creation, supply and exchange of valuable products with others" [8].

Kotler and Keller [9] argues that nowadays companies and brands should link a product or service to unique and interesting experiences for consumers, what Schmitt [10] defines as "perceptions, feelings and thoughts that consumers have when they find products and brands in the marketplace and engage in consumer activities, as well as the memory of such experiences."

Earlier in 1998, Pine and Gilmore presented a paradigm shift in economics denominated experience economy, distinguishing four stages in the progression of economic value, where the current one is the creation of experiences.

23.3 Experiential Marketing and Brand Experience

Although Pine and Gilmore's [1] work grew awareness about experience, Holbrook and Hirschman [11] already advocated a hedonic consumption perspective.

Experiential marketing emerges with the goal of differentiation, generate brand awareness, promote innovation, and produce acquisition and loyalty [12]. In fact, customers "want products, communications and campaigns that appeal to their senses, touch their hearts and with which they can relate and incorporate into their lifestyles" [10, 13].

Smilansky [14] defines experiential marketing as a "process of identifying and satisfying customer needs and aspirations profitably, engaging them through two-way communications that bring brand personalities to life and add value to the target audience."

The core of experiential marketing is the consumer experience [15], closely linked to a particular brand, defined as "sensations, feelings, cognitions, and behavioral responses evoked by brand-related stimuli that are part of a brand's design and identity, packaging, communications, and environments" [16].

Brakus et al. [16] have created a scale to measure and explore the dimensionality of experiences composed of four experiential dimensions: sensory, affective, intellectual, and behavioral.

Brakus et al. [16] stated that most of the research on experiences to date has focused on utilitarian product attributes and category experiences, not on experiences provided by brands. When consumers shop, they are exposed to the utilitarian features of a certain product as well as to distinct brand-related stimuli that constitute the major source of subjective, internal consumer responses, which authors refer as brand experience.

Brand experience leverages word of mouth [17–20], affects loyalty [21–23], and influences brand equity [24], brand image [17, 23], and the consumer identification with the brand [25].

Brand experience has been increasingly explored and relevant in place marketing, more specifically in destination management [26].

23.4 Place Marketing and City Marketing

Place marketing encompasses city marketing and refers to the "application of marketing tools to places such as cities, towns, regions and communities" [27], with the purpose of attracting investors, tourists, and residents [27], by applying strategies of promotion like launching favorable images [28].

The objective is to "communicate comparative advantages to target segments in a clearly understandable way" [29] and make them choose one place over another [30]. According to Kavaratzis [31], the application of city marketing is "dependent on the construction, communication and management of the City Image."

For a strategic management of a city in a marketing perspective, there must be an identification of the motives why such is needed to the town. After that, the values that highlight the city in the consumers' vision are defined. Then, it is drawn the identity, mission, and objectives of the organizations that are responsible for urban management, and finally, the strategy is operationalized through the marketing mix selection [32].

Tourism is highlighted as one of the sectors most associated with city marketing, whose professionals have been developing products and services to meet the needs of visitors [33]. The authors point out that "the tourism potential of the destination attracts managers and employees to new job opportunities."

23.5 Place/City Branding

As mentioned above, city marketing's main purpose is to manage the city image, which is the starting point for the development of city branding. Place branding alludes to the "development of brands for geographic locations, usually with the aim of triggering positive associations and distinguishing the place of others" [34], a construct that encompasses city branding.

23.6 Place Marketing and Place Branding: Comparison

In the scope of academic research, it has been difficult to distinguish the place marketing and place branding constructs, so it becomes relevant to distinguish them.

Place marketing is more embracing, including place branding itself, uses marketing tools to communicate, provides and exchanges urban deals, alludes to general management of the place, and its main objective is to offer value to customers and the community of a place. On the other hand, place branding aims at the development of brands for geographical locations and refers to the promotion of the place [34].

Place marketing is defined as "the measures taken, by actors appointed to govern a place, to improve the competitive image of that place with the explicit aim of attracting capital from elsewhere" [35].

Place branding is focused on attractions and location image that might attract events, tourists, investors, visitors (e.g., students or professionals), trade, and the category of people termed "influencers" [3].

23.7 Brand Love

Brand love is defined "as the degree of passionate emotional attachment a satisfied consumer has for a particular trade name" [36]. It is an effective concept that results from a long-term commitment and relationship between the consumer and the brand, being something deeper than satisfaction. In addition, brand love includes "a willingness to declare love and involves integration of the brand into the consumer's identity" [36].

23.8 City Brand Love (CBL)

Ardyan and Farida [37] defines city brand love as the "emotional attachment level passionate tourist and the happiness that exists in a certain city." The same author has shown that the attractiveness of a city influences city brand love and that city brand love influences in a positive and significant way repetition of the visit and word of mouth.

Two years later, Ardyan and Susanti [38] proves that a memorable city brand experience positively influences brand love.

23.9 Customer-Based Brand Equity (CBBE)

In order to assist management in brand-building efforts, Keller [39] designed a Customer-Based Brand Equity (CBBE) management model consisting in a pyramid with four steps and six blocks to shape how customers feel and think about a certain brand. The first step is the identity by creating brand awareness; the second step is the meaning which implies brand performance and imagery; the third step is the response of the customers toward the brand through judgments and feelings; and the fourth step is the relationship between customers and the brand implying brand resonance that is a deep, psychological bond.

Yoo and Donthu [40] designed a customer-based brand equity scale to process information and build trust in the purchasing decision, and increase the efficiency and effectiveness of marketing programs, profits, and brand extensions. Its multidimensional scale is made up of the dimensions of brand loyalty, perceived quality, brand awareness/associations, and brand global value.

23.10 Visitor-Based Brand Equity

The multidimensional scale developed by Yoo and Donthu [40] was adapted to investigate in this study the visitor-based brand equity of Matosinhos city.

23.11 Conceptual Model

Based on the literature review, a conceptual model was outlined with the hypotheses to be tested.

Figure 23.1 depicts the conceptual model regarding the present investigation.

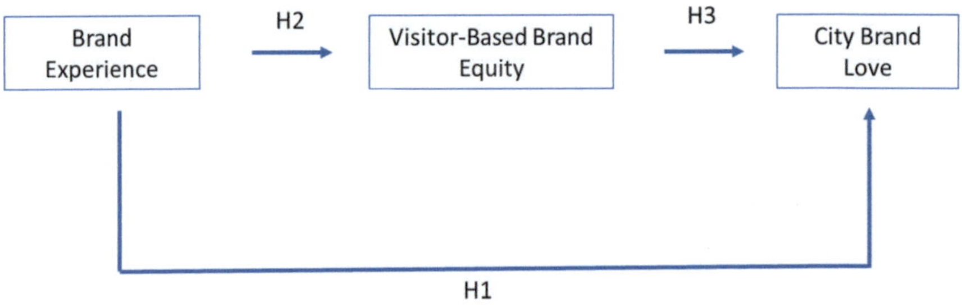

Fig. 23.1. Study's conceptual model

23.12 Research Methodology

In order to develop the present investigation, a bibliographical and documentary research was first done through scientific articles, books, magazines, newspapers, and statistical reports. Afterward, a set of data was collected through a questionnaire survey.

The research will follow a quantitative and descriptive methodology (deriving from a positivistic paradigm), using a questionnaire survey. Most of the questionnaire questions were closed. Its elaboration was done in Google forms online platform, and the collected data were interpreted using SPSS and Smart PLS.

It was applied the brand experience scale of Brakus et al. [16] with 12 items, measured by a seven-point Likert scale; the customer-based brand equity scale of Yoo and Donthu [40] with 14 items, used a seven-point Likert scale; and the brand love scale of Carroll and Ahuvia [36] of six items, measured by a five-point Likert scale.

23.13 Data Analysis

23.13.1 Sample Characterization

A total of 224 responses regarding the city of Matosinhos were obtained.

The majority were female (55.8%), from the age group between 16 and 25 years old (37.5%), single (56.7%), with secondary education as literary qualification (37, 1%), employed by others (57.6%) and has a monthly wage between 501 and 1000 euros (51.8%). It is also important to point out that 32.6% of the respondents live in Porto, 16.2% in Maia, and 10.3% and 9.4% in Gondomar and Vila Nova de Gaia, respectively.

23.13.2 Reliability and Internal Consistency

Using the SPSS statistical software, Cronbach's Alpha was verified. Good consistency was found in the study model, in its three constructs. Brand experience has a Cronbach's alpha of 0.889; the visitor-based brand equity of 0.938; and city brand love of 0.911.

23.13.3 Factor Analysis and SEM

With this consistency, exploratory factor analysis was applied. All variables showed a favorable KMO value and Bartlett's tests present a p-value of less than 0.005, demonstrating that the variables are significantly correlated. In addition, a single factor was extracted for each variable, and the proportion of variance explained by the factors extracted was as follows: brand experience (factor) with $\approx$69%; CBBE (factor) with $\approx$72%; and city brand love (factor) $\approx$71%.

Then, the reliability and viability of constructs were analyzed, and the AVE value is slightly below the ideal value in the brand experience construct (0.431) and is within the values recommended in the visitor-based brand equity (0.559), and love to the brand of the city (0.652) constructs. The values of the composite reliability are within the reference values (Brand experience = 0.892; CBBE = 0.944; City brand love = 0.918).

With the aid of the SmartPLS 3.0 program (2019 version), it was performed the Least Squares Analysis (PLS) to analyze the model of this study.

Figure 23.2 represents the path diagram of the structural equation model.

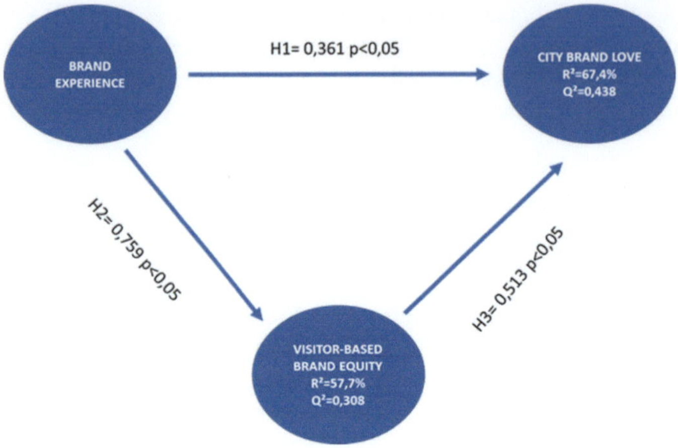

Fig. 23.2. Path diagram of the structural equation model

Based on the results shown (Table 23.1), all hypotheses of the present investigation were validated.

Table 23.1. Hypothesis validation

Hypothesis	Value	Result
Brand experience influences city brand love (CBL)	0.361	Validated
Brand experience influences visitor-based brand equity	0.759	Validated
Visitor-based brand equity influences city brand love (CBL)	0.513	Validated

23.14 Discussion

Brand experience is a relevant component of city marketing for visitors in Matosinhos, and the data suggest that visitors feel more satisfied in a sensorial point of view, as mentioned by Brakus et al. [16].

With respect to visitor-based brand equity, the sample demonstrates loyalty to Matosinhos, perceives an elevated quality of the city, remembers well its characteristics, is able to distinguish it from other cities, and shows neutrality regarding the preference for Matosinhos in relation to other cities.

Palusuk et al. [41] adds that brand love has the potential to increase consumer involvement, brand advocacy, commitment, and loyalty. This can be applied to city brand love. In this study, the visitors showed passion for the city and if we take into account what Ardyan and Susanti [38] said, and then we will have positive and significant influence on repetition of the visit and the word of mouth.

The model of this study presented high validity with values above the standard ones. The R^2 value showed that 67.4% of the variance of the city brand love variable is explained by the brand experience variable; 57.7% of the variance of the visitor-based brand equity variable is explained by the variable BE; and 67.4% of the variance of the variable city brand love is explained by the variable visitor-based brand equity.

In order to achieve the objectives of this study, three hypotheses were considered for the conceptual model. The results obtained in the Structural Equation Model (SEM) support all the hypotheses, leading to the conclusion that the brand experience influences city marketing through the visitor-based brand equity and city brand love.

Brand experience showed to have a statistically significant influence on city brand love with a positive coefficient of 0.361. These results are consistent with Ardyan and Susanti [38] who proved that a memorable city brand experience positively influences city brand love. Brand experience also influences visitor-based brand equity, since there is a link between the two variables, approved by a positive coefficient of 0.759. The literature studied is consistent with this information, being that the creation of experiences for different stakeholders positively affects their loyalty [21–23], perceived quality [16], and brand awareness [4]. Visitor-based brand equity produces a positive effect in the city brand love, and among these variables there is a positive coefficient of 0.513.

The different analyses and statistical tests carried out showed that the model of the present study is valid, reliable, and significant for the investigation.

Given the results presented, it is possible to say that those responsible for city management, namely, municipal entities and destination marketing organizations, should invest in the introduction of experiences in their strategies.

23.15 Conclusions

The objective of this research was to understand the influence of brand experience in city marketing through visitor-based brand equity and city brand love.

Brand experience is an essential tool to differentiate cities, generate brand awareness, build loyalty and fidelity of visitors, generate word of mouth, and help revitalize cities that are in a declining state.

This study demonstrated that brand experience positively influences city brand love.

Strategies promoting unforgettable experiences that touch the hearts of the visitors should be promoted, causing them to fall in love with the city and to have the predisposition to declare that love with others. These statements represent an intense communication through word of mouth that emphasize the satisfaction of different stakeholders and can lead to the attraction of new visitors without financial costs.

It has also been demonstrated that brand experience has a very significant weight in visitor-based brand equity, being that the creation and implementation of sensory (mainly), affective, behavioral, and intellectual experiences will add value to the city as a brand, and will lead to the loyalty of its visitors, making them return, sharing their experience, and making them potential residents; brand knowledge and its distinctiveness in relation to competition; and a superlative global brand value, leading visitors to prefer a certain city in relation to the rest.

The study confirms that visitor-based brand equity influences city brand love, so efforts and investments must be made in this area in order to outline strategies to increase brand value.

In sum, the present research demonstrates to what extent and how brand experience influences city marketing, proven that local authorities should invest in the creation of experiences for their visitors and in managing visitor-based brand equity, aiming city brand love.

23.16 Limitations and Future Research

In the first instance, a construct named economic performance was included, in order to evaluate the influence that brand experience has on it. However, it was not possible, given the scarcity of scales that measure the economic performance, and the same applies to the construct marketing of cities.

Regarding future lines of research, it is important to apply this model in the field of citizen-based brand equity and compare its results with visitor-based brand equity of this study.

References

1. Pine II, B., Gilmore, J.: Welcome to the experience economy. Harv. Bus. Rev. 97–105 (1998)
2. Lucarelli, A., Berg, P.: City branding: a state-of-the-art review of the research domain. J. Place Manag. Dev. **4**, 9–27 (2011)
3. Anholt, S.: Definitions of place branding—working towards a resolution. Place Brand. Public Dipl. **6**, 1–10 (2010)
4. Silva, M., Tincani, D.: Características e componentes do Marketing de Experiências: análise das ações realizadas pelo Itaú Unibanco no rock in rio 2011. Rev. Científica Eletrônica UNISEB, Ribeirão Preto **1**(2), 147–161 (2013)
5. Taghipourian, M., Bakhsh, M.: From single-stage loyalty to four-stage loyalty. Int. J. New Technol. Research. **1**, 48–51 (2015)
6. Alkilani, K., Ling, K.C., Abzakh, A.A.: The impact of experiential marketing and customer satisfaction on customer commitment in the world of social networks, pp. 262–270 (2012)
7. Wu, M., Tseng, L.: Customer satisfaction and loyalty in an online shop: an experiential marketing perspective. Int. J. Bus. Manag. **10**(1), 104–114 (2014)
8. Kotler, P.: Administração de Marketing: Análise, Planejamento, Implementação e Controle —5ª Edição. Atlas, São Paulo (1998)
9. Kotler, P., Keller, K.: Marketing Management, 14th edn. Prentice Hall, Upper Saddle River, New Jersey (2011)
10. Schmitt, B.: Experience Marketing: Concepts, Frameworks and Consumer Insights, pp. 55–112. Foundations and Trends® in Marketing (2010)
11. Holbrook, M., Hirschman, E.: The experiential aspects of consumption: consumer fantasies, feelings, and fun. J. Consum. Res. **9**(2), 132–140 (1982)
12. Azambuja, C., Bichueti, R.: Marketing de Experiência: Estratégias para impulsionar o Market Share e fortalecer a marca do energético Energy. Rev. Adm. Da UFSM **9**(1), 88–106 (2016)

13. Schmitt, B.: Experiential marketing. J. Mark. Manag. **15**(1–3), 53–67 (1999)
14. Smilansky, S.: Experiential Marketing, A Practical Guide to Interactive Brand Experiences. Kogan Page, London, UK (2009)
15. Moral, M., Alles, M.: Nuevas tendencias del marketing: el marketing experiential. Entelequia: Revista Interdisciplinar (14), 237–251 (2012)
16. Brakus, J., Schmitt, B., Zarantonello, L.: Brand experience: what is it? How is it measured? Does it affect loyalty? J. Mark. **73**(3), 52–68 (2009)
17. Sabiote, E., Ballester, M.: Marcas de experiencia: marcando la diferencia. Estud. Gerenc. Les **27**(121), 59–77 (2011)
18. Li, L., Lee, L.: Experiential consumption and customer satisfaction: moderating effects of perceived values. Int. J. Mark. Stud. **8**(5), 32–40 (2016)
19. Klein, J., Falk, T., Esch, F., Gloukhovtsev, A.: Linking pop-up brand stores to brand experience and word of mouth: the case of luxury retail. J. Bus. Res. **69**(12), 5761–5767 (2016)
20. Ong, C., Lee, H., Ramayah, T.: Impact of brand experience on loyalty. J. Hosp. Mark. Manag. **27**(7), 755–774 (2018)
21. Moreira, A., Silva, P., Moutinho, V.: The effects of brand experiences on quality, satisfaction and loyalty: an empirical study in the telecommunications multiple-play service market. Rev. Innov. J. 23–36 (2017)
22. Riza, A., Wijayanti, D.: The triangle of sensory marketing model: does it stimulate brand experience and loyalty? J. Bisnis Dan Manaj. **8**(1), 57–66 (2018)
23. Gavurova, B., Bacik, R., Fedorko, R., Nastisin, L.: The customer's brand experience in the light of selected performance indicators in the social media environment. J. Compet. **10**(2), 72–84 (2018)
24. Lin, Y.: Innovative brand experience's influence on brand equity and brand satisfaction. J. Bus. Res. **68**(11), 2254–2259 (2015)
25. Ngo, L., Northey, G., Duffy, S., Thao, H.: Perceptions of others, mindfulness, and brand experience in retail service setting. J. Retail. Consum. Serv. **33**, 43–52 (2016)
26. Barnes, S., Mattsson, J., Sørensen, F.: Destination brand experience and visitor behavior: testing a scale in the tourism context. Ann. Tour. Res. **48**, 121–139 (2014)
27. Eshuis, J., Klijn, E., Braun, E.: Place marketing and citizen participation: branding as strategy to address the emotional dimension of policy making? Int. Rev. Admin. Sci. **80**(1), 151–171 (2014)
28. Greenberg, M.: Branding New York: How a City in Crisis was Sold to the World. Routledge, New York (2008)
29. Iványi, T., Bíró-szigeti, S.: Smart city: an overview of the functions of city. Perspect. Innov. Econ. **18**(1), 44–57 (2018)
30. Boisen, M., Terlouw, K., Groote, P., Couwenberg, O.: Reframing place promotion, marketing territorial, and marca territorial—moving beyond conceptual confusion. Cities **80**, 4–11 (2018)
31. Kavaratzis, M.: From city marketing to city branding: towards a theoretical framework for developing city brands. Marca Territ. Public Dipl. **1**(1), 58–73 (2004)
32. Moreira, P. (11 de Julho de 2011) Gestão de Marcas Cidade: O Caso da marca Porto Turismo, p. 157 (2011)
33. Manhas, S., Manrai, L., Manrai, A.: Role of tourist destination development in building its brand image: a conceptual model. J. Econ. Financ. Adm. Sci. **21**, 25–29 (2016)
34. Vuignier, R.: Marketing Territorial and Marca Territorial: a systematic (and tentatively exhaustive) literature review. Hal Archives-ouverts (2016)
35. Niedomysl, T., Jonasson, M.: Towards a theory of place marketing. J. Place Manag. Dev. **5** (3), 223–230 (2012)

36. Carroll, B., Ahuvia, A.: Some antecedents and outcomes of Brand Love. Mark. Lett **17**, 79–89 (2006)
37. Ardyan, E., Farida, N.: Attribute attractiveness of city, city brand love and tourist behavior: the case of Jakarta. In: Innovation in Regional Public Service for Sustainability, pp. 101–103 (2016)
38. Ardyan, E., Susanti, A.: The effect of city brand love on tourist based city brand equity: city branding study in Indonesia. Int. J. Asian Bus. Inf. Manag. **9**(3), 44–60 (2018)
39. Keller, K.: Conceptualizing, measuring, and managing customer-based brand equity. J. Mark. 1–22 (1993)
40. Yoo, B., Donthu, N.: Developing and validating a multidimensional consumer-based brand equity scale. J. Bus. Res. **52**, 1–14 (2001)
41. Palusuk, N., Koles, B., Hasan, R.: "All you need is brand love": a critical review and comprehensive conceptual framework for brand love. J. Market. Manag. **35**(1–2), 97–129 (2019)

Chapter 24
The Human Senses as Consumption and Brand Identity Boosters

Cátia Ramos[1], Ana Sampaio[2], and Pedro Rodrigues[1(✉)]

[1] ISMAI, Maia University Institute, Maia, Portugal
{catiasteixeiraramos, pmrodrigues}@gmail.com
[2] CEDTUR ISMAI, Maia University Institute, Maia, Portugal
asampaio@ismai.pt

Abstract. Brands increasingly seek to stand out from the competition and reach beyond the minds of consumers. This research has as main objective to understand the importance of the senses and sensorial stimuli in the consumption and in the own identity of the brand. In order to consolidate this research, Lidl was the chosen brand. Throughout the study, the following variables will be studied: sensorial senses/senses, store environment, brand experience, and brand identity. It seeks to understand their influence relationships and in what way they can also influence consumption. This is a quantitative study where a questionnaire survey was conducted for 328 respondents. The collected data were processed using statistical software such as SPSS to perform the descriptive analysis of the variables as well as the exploratory factorial analysis and the Smart PLS for the confirmatory factorial analysis and the structural equations model (SEM). The results show that sensory stimuli have a positive and significant influence on the Store Environment as well as the Brand Experience and this is also positive and significant in Brand Identity. The influence that the Brand Identity exerts on the Brand Experience is also positive and, although smaller, the Store Environment also influences the Brand Experience positively.

24.1 Introduction

In the market, there are varied brands of different product categories. More and more brands are trying to stand out from the competition in order to become more attractive to the consumer. Why not do it through the senses? Marketing and branding are already being explored by a variety of brands, but there is still much room for maneuver, there is still much that can be explored. The exploitation of senses can make a great difference in the captivation of the consumer, in the memorization of the brand, product or service and in the emotional attachment to them. Nowadays, the use of the five senses is a very influential weapon in marketing to conquer the human being. The senses are linked to the feelings, and these are the weak point of the man. Emotions reward the purchase of products as well as an affective bond with the brand [1]. Thus, brands need to encourage consumers to like their products, services, and the brand itself by offering them a positive and differentiated experience in order to create an emotional bond with

Á. Rocha et al. (eds.), *Marketing and Smart Technologies*, Smart Innovation, Systems and Technologies 167, https://doi.org/10.1007/978-981-15-1564-4_24

the customer [2]. The consumer seeks not only a good or service to acquire, but memorable, personal, relevant, emotional, and sensory experiences. Therefore, consumers seek meaningful experiences [3].

The objective of the present investigation is to study and understand the influence of senses and sensorial stimuli on consumption through the store environment and the brand experience, as well as the influence of sensory stimuli on brand identity.

24.2 Marketing and Brand

Kotler [4] defines Marketing as the "*social and administrative process by which individuals and groups obtain what they need and/or desire by creating, offering, and exchanging valuable products with others.*" It is a philosophy of management oriented to the clients aiming at the creation of value for these and for the own organization. Later, this definition becomes more extensive, establishing marketing as a process of managing the relations of exchange between the individual or legal entity, differentiated from the competition and to the public [5].

Brito [6], regards the brand as an element of extreme importance since it acts as a link between companies and customers. The brand, according to Martins [7], is the union of tangible and intangible attributes, symbolized in logos that properly directed create influence and generate value. It is an integrated system that promises and delivers solutions desired by people. Lencastre [5] adds that the European Union has adopted a definition of a brand that focuses on all its requirements and figurative dimension. They are thus admissible as a brand "All signs which may be the object of a graphic representation …". Certain legislation of Member States, in particular, the Portuguese, already explicitly integrates "sounds" as one of the signs susceptible to this same representation.

24.3 Sensory Marketing

Sensory marketing aims to strengthen the relationship between a brand and its customers on a deeper level, involving the human senses [8].

Hultén [9] highlights it as a service process focused on strategies and sensorial stimuli with the objective of creating a multisensory brand experience, supporting the creation of the individual's identity through the mind and the senses to generate value and experiences for the consumer and mark it as an image. According to Khanna and Mishra [10], brand perception changes positively when the stimuli are consistent with other elements of the brand. One of the starting points for sensory marketing is that brands must offer customers multisensory brand experiences in a more emotional and personal way [11].

Senses:

According to Lindstrom [12], the senses are relevant to the behavior of buying and acquiring experiences, and Khanna and Mishra [10] add that sensory marketing aims to strengthen the relationship between a brand and its customers at a deeper level.

Vision:

According to Hultén [13], vision is the most dominant sensory system and most humans depend to a large extent on visible and tangible sensory signals that create attention for certain products in the environment. According to Raghubir [14], the consequences of visual cues include how vision interacts with other sensory stimuli received from the hepatic/tactile, olfactory, gustatory, and auditory to make judgments regarding touch, smell, taste, and sound.

Smell:

Smell is the most basic, most direct [12], and more primitive sense [15]. Moreover, it is the only sense with duality, it can detect smells both in the external world (just by resorting to the nose) as well as by feeling them in the mouth [16]. According to Lindstrom [12], fragrance is a deciding factor whenever a consumer buys or uses a product.

Touch:

The importance of touch has been recognized for centuries and according to Krishna [17], touch is the first sense to develop in the womb and the last sense that is lost with age. According to Lindstrom [12], touch is our connecting tool, when all else fails. Peck [18] and Klatzky [19] argue that by touch it is possible to explore properties that are not perceptible through another sense.

Audition:

According to Lindstrom [12], sounds create states of mind, emotions, and feelings and are fundamental to create the atmosphere of the story that the brand intends to tell. Sound may not be a central aspect of consumer decision-making, although it plays a valuable role for business and consumer [20]. According to Solomon [21], many aspects of sound affect people's feelings and behavior.

Taste:

According to Soloman [21], taste receptors contribute to the experience with many products, and people form strong preferences for certain flavors. When we think of taste, food comes immediately to mind. According to Krishna and Elder [16], the taste sensations that food provides us, play a vital role in a physiological, survival, social, and emotional perspective.

24.4 Consumer Behavior

Solomon [21] describes consumer behavior as the actions of a consumer in relation to an object. In turn, the same author argues that the consumer is the person who identifies a need or desire, makes a purchase, and/or discards a product. The same author states that consumer behavior is the study of the processes involved when individuals or groups select, purchase, use or dispose of products, services, ideas, or experiences to satisfy needs or desires.

24.4.1 Consumer Behavior and the Senses

There are several fundamental aspects between the connection of the senses and the behavior of the consumer. Learning is one of them. According to Solomon [21], this is a permanent change in behavior caused by experience.

24.4.1.1 Consumer Behavior and Vision

Hultén [13] states that consumers can be positively or negatively affected by visual stimuli without access to other information. According to Spence et al. [22], visual clues can trigger specific associations in consumers that can facilitate decision-making. However, according to Elder et al. [23], complementing the visual clues with other sensory cues can increase the influence of consumer images on perceptions, attitudes, and behaviors.

24.4.1.2 Consumer Behavior and Smell

According to Elder et al. [23], the presence of a perfume is generally recognized as having a positive impact on consumer attitudes, both in different environments and infused directly into the products. Herz [15] states that pleasant fragrances improved mood and relieved some of the symptoms associated with unpleasant physical conditions.

24.4.1.3 Consumer Behavior and Touch

Touch, whether between two products, two human beings, or a product and a human, seems to affect consumer behavior in significant ways [10]. Concerning the touch on consumer behavior, Hultén [11] states that there are individual differences in touching, and some people always prefer to touch the brand's products before buying it.

24.4.1.4 Consumer Behavior and Audition

Audition is a sense that requires no effort to operate. According to Elder et al. [23], people have no control over the auditory stimuli they perceive, although they have some control over the stimulus to which they are attentive. Elder et al. [23] affirm that the congruence of music has positive effects on product ratings. Meyers-Levy et al. [24] add that music can shape the behaviors that consumers exhibit.

24.4.1.5 Consumer Behavior and the Taste

The palate, according to Hultén [11], interacts with smell and touch, offering different taste perceptions. According to Yoon and Park [25], the palate is the one that exerts greater influence in loyalty.

24.5 Store Environment

Kotler [26] argues that the store environment has more influence than the product itself in the purchasing decision and this results in a competitive advantage over the competition. According to Kotler [26], space can evoke effects on consumers and in turn affect purchasing behavior. Solomon [21] agrees that the dimensions of the physical environment can significantly influence consumption.

24.5.1 Store Environment and the Senses

24.5.1.1 Store Environment and Vision

Vision is the most used sense in marketing and also the most stimulated by the environment. Choices of product colors, shapes, and materials are important to organize a sales outlet by implementing a promotional campaign for key success or failure factors that should be well understood by marketers to attract a consumer.

24.5.1.2 Store Environment and Smell

The use of scents in the store environment can be beneficial if it is congruent with the shopping environment [27]. Spangeberg et al. [28] have shown that environmental perfumes generally improve product evaluations.

24.5.1.3 Store Environment and Touch

According to Morales [29], consumers like to touch products but do not like other people to touch the product they want to buy. Consumers have a preference for environments where they are allowed to inspect and touch products [17, 30].

24.5.1.4 Store Environment and the Audition

In store environments, music is often used to change the environment of the shopping experience or even to change the pace of consumer behavior. Music can shape the behaviors consumers exhibit, a conclusion that comes from the idea that motion may be the most primordial sensory response to music [24]. Meyers-Levy et al. [24] state that slow-paced versus fast-paced background music changes the pace of purchases and the total volume of sales.

24.6 Brand Identity

In order to establish its own identity, it has to identify and differentiate its public, since it is through the brand that consumers identify the products or services and distinguish them from the competition [5]. Through Lindstrom [12], brands that create a connection with customers are stronger than those that can not. Identity is the tangible manifestation of your personality and it must project itself into everything the organization does, from how the organization behaves. According to Machado [31], the stronger the personality of the brand, the more human-centered and less product-centered, the easier it is for the consumer to associate words, phrases or phrases with the brand.

24.6.1 Brand Identity and the Senses

Acting in the customer's sense, a company can create a deeper and personal relationship with customers and thus with appropriate tactics, increase sales and achieve business goals [32]. A brand has to become a sensory experience that goes beyond what we see [12]. According to Khanna and Mishra [8], by appealing to the five senses, brands create strong memories in the minds of consumers.

24.6.1.1 Brand Identity and Vision

Hultén [11] argues that when applying a visual sensory strategy, the identity of a brand is visually expressed and becomes a distinctive resource. According to Nadanyiova et al. [32], colors and shapes are the first way to identify and differentiate. The colors are easy to store in the customer's memory.

24.6.1.2 Brand Identity and Smell

The identity of a brand can be expressed atmospherically, where the perfume is intended to create awareness in the consumer. According to Lindstrom [12], in brands, it is important that their smell is clear and distinct. Morrin and Ratneshwar [33] point out that the aroma can contribute to increase brand recall and brand awareness.

24.6.1.3 Brand Identity and Touch

The texture associated with the mark is often a neglected feature. According to Lindstrom [12], brands can take advantage of it, but many of them have not yet realized the advantages that can bring them.

24.6.1.4 Brand Identity and the Hearing

The identity of a brand can be expressed in the form of sound and become a distinctive feature [11]. Yorkston [20] states that music can be an excellent cross-promotion tool that can link brands to similar audiences and strengthen brand identity.

24.7 Brand Experience

Brakus et al. [34] conceptualize "*brand experiences as subjective responses, internal to the consumer (sensations, feelings and cognitions) and behavioral responses evoked by brand-related stimuli that are part of the design and identity of a brand, packaging, communications and environments*." For Shieh and Lai [35], Experiential Marketing is "*taking the essence of a product, expanding it into a set of tangible, physical, and interactive experiences that reinforce it.*"

24.7.1 Brand Experience and the Senses

According to Ding and Tseng [36], the use of different types of sensory experiences can improve the quality of customer service and consequently result in improvements in brand image and recognition in the long run. And, Hultén [11] argues that brands should offer customers branded multisensory experiences in a more emotional and personal way. Multisensory brand experiences support individual value creation and enable consumers to react to companies' initiatives [37].

24.8 Conceptual Model

See Fig. 24.1.

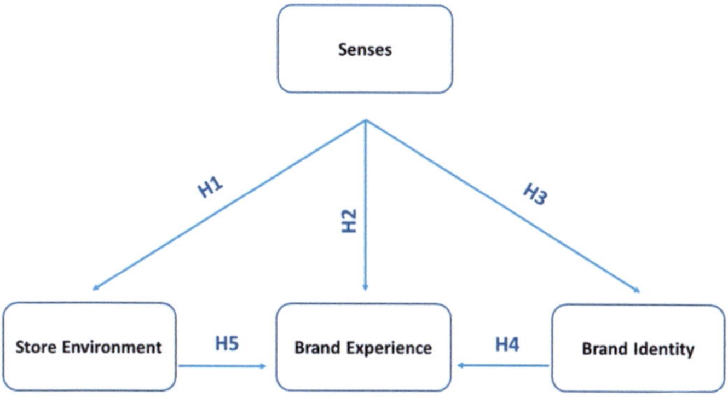

Fig. 24.1. Conceptual model of the study

24.9 Research Methodology

In order to develop the present investigation, a bibliographical and documentary research was done through scientific articles, books, newspapers, and statistical reports.

Subsequently, the investigation was reinforced by a set of data collected through a questionnaire survey.

The research follows a quantitative and descriptive methodology. The questionnaire used in the study will be composed of mostly closed questions. The elaboration of the same will be done in the online platform Google forms and the collected data will be interpreted in the statistical software SPSS and Smart PLS.

In order to carry out this questionnaire, we used the scales used in previous studies. The first scale presented is referring to the senses more concretely the vision and this one is composed of 6 items. The construction of this scale is based on the scales used by Nadiri and Gunay [38], Yoon [39], Peck and Childers [40]. The items were measured on a seven-point Likert scale ranging from “totally disagree” to “strongly agree”. There is also an additional question about the preference of space lighting based on a study by Areni and Kim [41]. Regarding olfaction, the scale presented is composed of 4 items, based on the scales presented by Nadiri and Gunay [38] and Peck and Childers [40]. The items presented were also measured on a seven-point Likert scale ranging from “strongly disagree” to “strongly agree”. The scale used for tact is based solely on the Peck and Childers scale [40] and is composed of 12 items that were measured on a Likert scale of seven points ranging from “totally disagree” to “strongly agree”.

Concerning the hearing, consumers’ preferences regarding music volume, music rhythm, and music familiarity were explored, based on the Smith and Curnow [42], Milliman [43] and Yalch and Spangerberg [44] scales, respectively.

The store environment is analyzed using two different scales. A scale has 3 items and the other has 4 items. The construct is analyzed according to two scales because they have different measurements. The scales presented were constructed based on the studies of Seughwan Lee et al. [45], Lin et al. [46]. The first items are measured on a five-point Likert scale, and the other items are measured on a seven-point Likert scale. Both range from "totally disagree" to "strongly agree".

It is also used the Brakus Brand Experience Scale [34] where 12 items are present that were measured on a seven-point Likert scale ranging from "totally disagree" to "strongly agree".

For the last scale presented in the questionnaire, this is referring to the brand identity and is based on the scale presented by Coleman et al. [47] and is measured using a seven-point Likert scale ranging from "totally disagree" to "strongly agree".

24.10 Data Analysis

In this section of the paper the analysis of data obtained in the work will be made, considering the responses obtained in the questionnaire.

24.10.1 Sample Characterization

A total of 328 validated responses were collected. The majority of respondents are female (72%), are in the 26–35 age group (35.7%) and they are single (64.3%). Regarding education, the majority have a Bachelor's Degree (46.6%), are employed by others (62.2%), and have a monthly remuneration of between 501 and 1000 euros (48.2%). It is also noted that 75% of respondents are located in the Porto district. Second, representatives of the city of Lisbon, representing 5.8% of the sample. In addition, 4% of the participants reside in the city of Viseu.

24.10.2 Reliability and Internal Consistency

Using the SPSS statistical software, Cronbach's alpha was verified. It is concluded that the Store Environment construct and the Senses construct have a very good consistency and the Brand Experience and Brand Identity constructs have a good internal consistency. The Store Environment has a Cronbach Alpha of 0.901; the Senses 0.918; the Brand Experience 0.755 and the Brand Identity 0.896.

24.10.3 Factor Analysis and SEM

With the consistency described above, the data were reduced through Exploratory Factor Analysis. It was found that the store environment variable presents a good KMO value as well as the brand experience variable and brand identity.

For the senses, they have an excellent KMO value. With regard to Bartlett's tests have a p-value <0.005, thus concluding that the variables are significantly correlated. Additionally, a single factor was extracted for each variable, the proportion of variance explained by the factors extracted were as follows: Store environment (factor) $\approx$ 64%;

Senses (factor) ≈ 38%; Brand Experience (factor) ≈ 45%, and Brand Identity (factor) ≈ 45%.

The reliability and viability of the constructs were analyzed, and not all the values of the Average Extracted Variance obtained were greater than or equal to that indicated. Some AVE values are slightly below the reference values as in the case of the Senses (0.314), Brand Experience (0.448), and Brand Identity (0.449) constructs. Since only the value obtained in the store environment construct is within the recommended values (0.639). For composite reliability, all values are within the reference values (Senses = 0.910; Store Environment = 0.925; Brand Experience = 0.848; Brand Identity = 0.922).

With the aid of the SmartPLS 3.0 program (2019 version), we performed the least-squares analysis (PLS) to analyze the model of this study. Figure 24.2 represents the path diagram of the Structural Equation Model.

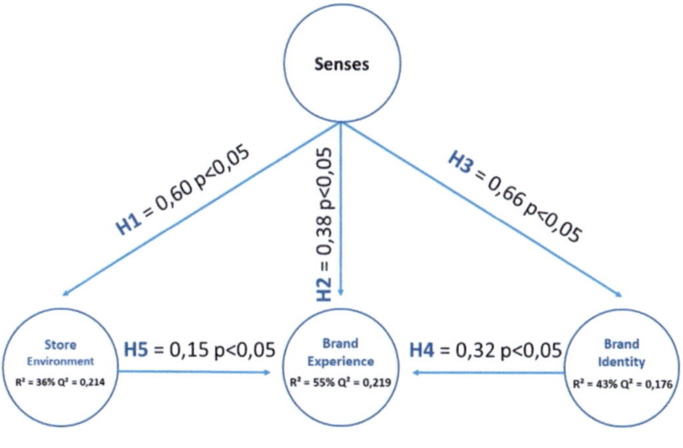

Fig. 24.2. Study's model

The hypotheses of the present study were all validated.

24.11 Discussion

The present study analyzed and evaluated four distinct variables and a conceptual model. The first construct is referring to the Senses and the sensory stimuli. Most respondents believe that the senses greatly influence decisions about buying a product or not. Within the senses, these were analyzed separately, leaving aside the palate, since the brand and its store environment do not resort to this sense in general.

Consumers generally agree that the brand relies on the use of appealing materials and services in order to capture the customer's attention. They also agree that the store's lighting is pleasant and adequate, but they do not consider it as an asset in the visual evaluation of the product. In general, the consumer has a neutral opinion regarding the intensity, pleasantness, and interest of the aroma of the brand stores. Consumers seem to prefer to feel pleasant smells. The consumer feels more confident to make a purchase after they smell the product. For tact, most consumers place more confidence in products that can be touched before the purchase, feel more comfortable to buy after examining the product physically and feel more confident if they make a purchase after touching the product. The brand does not feature any sound addition in its store environment. Despite this, most have a preference for sounds that create a relaxing and pleasurable atmosphere.

Regarding the second construct, the respondents have a positive impression and opinion of the brand stores. There are many consumers who claim to have made various purchases through the influence of the environment and also several times they buy unplanned items.

The third construct related to the brand experience presented a rather neutral position regarding all the components that make up the construct. In this way, consumers do not have a strong opinion regarding any of the aspects of the experience that the brand gives them.

The last construct explored, the brand identity, generates a great positive agreement in the consumer regarding the provision of services and support of the brand. In addition, the consumer feels that corporate identity is useful to make the organization recognizable. As far as the personality of the brand is concerned, the associations that the consumer makes are positive and favorable. But, contrary to this positive agreement presented, the vast majority of consumers do not associate a song to the brand.

The model referring to this study reveals high convergent validity with values above the standard ones. Based on the results obtained in R^2, it's observable that 36% of the store's Ambient variable is explained by the variable of the senses, just as the variable of the senses also explains in 55% the variable of the Brand Experience and in 43% the Brand Identity variable.

For the accomplishment and conclusion of this study, we considered five hypotheses represented by a conceptual model. The results obtained in the Model of Structural Equations (SEM) support all hypotheses. In this way, it can be concluded that Senses and sensory stimuli influence both the Store Environment, the Brand Experience, and the Brand Identity. It is still proven that the Brand Experience has a positive influence through the Brand Identity and the Store Environment itself.

Senses and sensory stimuli showed a statistically significant influence on the Store Environment with a coefficient of 0.604 showing that for each increased unit in the sensory variable leads to an increase of 0.604 in the Store Environment. Regarding the influence of the Senses in the Brand Experience, the results prove it through a positive coefficient of 0.379, that is, for each increased unit in the Senses variable leads to an increase of 0.379 in the Brand Experience. The senses also showed to have a statistically significant influence on the Brand Identity presenting a coefficient of 0.659, thus, for each increased unit in the variable of the directions leads to an increase of 0.659 in the Brand Identity. The Brand Experience is also influenced by the Brand Identity, presenting a coefficient of 0.317 and the Shop Environment through a coefficient of 0.148. For each unit increased in the Brand Identity variable leads to an increase of 0.32 in the Brand Experience and for each increased unit in the Store Environment variable leads to a 0.15 increase in Brand Experience.

The different analyses and statistical tests carried out showed that the model presented in this study is valid, reliable, and meaningful for the investigation. Given the results presented, it is concluded that there should be a higher investment in sensorial characteristics in the store environment. The brand already explores the senses, yet there is room for improvement. A key point that can be explored and proven to influence consumers and brands is the exploration of the auditory component. This can bring benefits to both parts. Meyers-Levy et al. [24] have shown that music can effectively shape the behaviors consumers exhibit, can add and increase brand value and consumer value, can make the visits more enjoyable and interesting and thus make the brand experience more interesting, pleasant, and pleasurable. Morrin et al. [33] state that there is considerable evidence to suggest that background music can positively impact mood and arousal levels of consumers as well as their overall affective response to the environment. This also explores the influence that the store environment has on brand experience.

In order to highlight the differentiated characteristics of brand identity and, above all, to highlight and more efficiently explore the sensorial characteristics of the brand, there must also be a greater investment. This is in line with what is advocated by Nadanyiova et al. [32] by acting in the customer's sense, a company can create a deeper and personal relationship with customers and thus with appropriate tactics, increase sales and achieve business goals. There is already a greater involvement of the brand with the consumer, but this needs to be further explored. The vast majority of consumers do not associate any music or jingle specifically to the brand, which proves that the brand has to continue investing in its sensorial characteristics so that they are a differentiator. Linsdstrom [12] states that many are the brands that underestimate the sound value. But music can contribute as much to brand communication as to its overall look. Brands that use songs that relate to their identity, through the author, are 96% more likely to be memorized. Lindstrom [12] adds that it is vital to keep alive the sensorial points of contact of the brand. And they must be improved because they give the brand its identity. The more distinctive characteristics of the brand are present and the more noticeable and recognizable to the consumer, it will also bring benefits to the brand experience. Sarquis et al. [48] argue that to achieve the desired brand identity, client experiences need to be planned, managed, and deliver relevant emotional and functional benefits.

The central point of all the investigation was the senses and the sensorial stimuli and it is indicated that if there is greater exploitation of the same, it can result for the brand. The brand can be evident and present a more pronounced identity, which will positively influence the experience that the consumer has with the brand.

24.12 Conclusions

This research aimed to understand to what extent the senses and sensorial stimuli influence the brand and its consumption through the holistic experience of the brand. We also sought to understand whether sensory stimuli influenced the store environment and, in turn, had an influence on consumption. Regarding the brand identity, it was sought to understand if the sensorial stimuli would also influence the same and in what way they could influence the brand experience.

Consumers are looking for brands to provide them with new experiences and the senses are relevant to pushing brands to another level. The senses and sensory stimuli help the brands to provide differentiated experiences that will prove to be advantageous for the brand itself, because by strengthening the connection with the customer this will become a profit for the company. It is concluded that the senses are an added value for the brands because they end up influencing the consumer and boosting consumption.

Consumers are looking for experiences, but it is also essential to provide the consumer with a space that is conducive to this. The way the brand presents and organizes the store is fundamental to provide a pleasant visit as well as the store's own exploration. Through sensory stimuli, branding will capture more consumer attention, which will get more involved in the store environment, will gain a better brand experience and will also boost consumption. The consumer is more attracted to environments where sensory stimuli are more exploited. In addition, consumers tend to rely more on products and brands that allow them to explore autonomously.

It would be pertinent for the brand to insert more strategies and sensory stimuli in store environments, since the brand under review does not explore the auditory component in the store environment. This can be something with which the brand can work in order to attract and please the customer even more. The generality of the consumer has a preference for atmospheres with pleasant sounds, and previous research has already proven that adding music to the environment influences the consumer and leverages consumption.

Brand Experience is an essential tool to differentiate brands, raise consumer awareness for the brand and also create customer loyalty. It is fundamental to promote innovation and this can effectively be in the exploration of the senses of the consumer. The senses influence the brand experience and this was also demonstrated in this study. Exploitation of sensory experiences can improve the quality of care and, as a consequence, result in improvements in brand image and recognition in the long term. The experiences that the brand offers to its consumer as well as multisensory must be emotional and personal.

With regard to Brand Identity, it is crucial that the consumer has a good brand image and there are several ways of doing so. Exploring the brand's sensory identity is important so that it stands out and differentiates itself from the competition. In this study, it was verified that the sensorial stimuli are an asset and positively influence the Brand Identity. In this way, the company only has to gain from greater exploration of the senses and sensorial stimuli. The investment of the brand must be superior in regard to evidencing its sensorial characteristics. There is already a greater involvement of the brand with the consumer, but this needs to be further explored. The vast majority of consumers do not associate any brand-specific music or jingle, which proves that the brand has to continue investing in its sensory characteristics so that they are a differentiator. Music can contribute to both brand communication and overall look.

It is thus a fundamental necessity to keep alive the sensorial points of contact of the brand. And they must be improved in a continuous way, since they give the brand its identity. As more differentiating features of the brand are present and these are more noticeable and recognizable to the consumer, it will also result in benefits in the brand experience.

References

1. Ramos, M.G., Braga, T.C.: Branding sensorial: a relação marca x consumidor criada pela ambientação das lojas, pp. 81–109 (2011)
2. Hinestroza, N., James, P.: The effects of sensory marketing on the implementation of fastfood marketing campaigns. J. Manag. Mark. Res. 1–11 (2014)
3. Andrés, A., Caetano, J., Rasquilha, L.: Gestão de Experience Marketing: Quimera (2006)
4. Kotler, P.: Administração de Marketing: Análise, Planejamento, Implementação e Controle, 5ª Edição. Atlas, São Paulo (1998)
5. Lencaste, P.: O livro da Marca. Publicações Dom Quixote, Lisboa (2005)
6. Brito, C.: Gestão relacional da marca. In Lencaste, P. (ed.) O livro da Marca, pp. 329–344. Publicações Dom Quixote, Lisboa (2005)
7. Martins, J.: Branding – Um manual para você criar, gerenciar e avaliar marcas. São Paulo, Global Brands (2006)
8. Khanna, P., Mishra, S.: Power of senses in branding and its impact on consumer. VSRD Int. J. Bus. Manag. Res. **2**(12), 20–623 (2012)
9. Hultén, B.: Sensory Marketing: Theoretical and Empirical Grounds. Routledge, New York (2015)
10. Krishna, A.: An integrative review of sensory marketing: engaging the senses to affect perception, judgment and behavior. J. Consum. Psychol. **22**(3), 332–351 (2012)
11. Hultén, B.: Branding by the five senses: a sensory branding framework. J. Brand Strateg. **6** (3), 281–292 (2017)
12. Lindstrom, M.: Brand Sense: Os segredos sensoriais que nos levam a comprar. Free Press, Nova Iorque (2010)
13. Hultén, B.: Sensory cues as in-store innovations: their impact on shopper—approaches and touch behaviour. J. Innov. Manag. 17–37 (2013)
14. Raghubir, P.: Visual perception—an overview. In: Krishna A. (ed.) Sensory Marketing—Research on the Sensuality of Products, pp. 201–217. Nova Iorque e Londres, Routledge (2010)
15. Herz, R.: The emotional, cognitive, and biological basics of olfaction—implications and considerations for scent marketing. In: Krishna A. (ed.) Sensory Marketing—Research on the Sensuality of Products, pp. 87–108. Nova Iorque e Londres, Routledge (2010)
16. Krishna, A., Elder, R.: The gist of gustation—an exploration of taste, food, and consumption. In: Krishna, A. (ed.) Sensory Marketing—Research on the Sensuality of Products, pp. 281–301. Nova Iorque e Londres, Routledge (2010)
17. Krishna, A., Morrin, M.: Does touch affect taste? The perceptual transfer of product container haptic cues. J. Consum. Res. **34**(6), 807–818 (2008)
18. Peck, J.: Does touch matter? Insights from haptic research in marketing. In: Krishna, A. (ed.) Sensory Marketing—Research on the Sensuality of Products, pp. 7–31. Nova Iorque e Londres, Routledge (2010)
19. Klatzky, R.: Touch—a gente tutorial with implications for marketing. In: Krishna, A. (ed.) Sensory Marketing—Research on the Sensuality of Products, pp. 33–47. Nova Iorque e Londres, Routledge (2010)
20. Yokston, E.: Auxiliary auditory ambitions—assessing ancillary and ambient sounds. In: Krishna, A. (ed.) Sensory Marketing—Research on the Sensuality of Products, pp. 157–167. Nova Iorque e Londres, Routledge (2010)
21. Solomon, M.: Comportamento do consumidor – Comprando, possuindo, sendo, 7ª edn. Bookman, Porto Alegre (2008)

22. Spence, C., Puccinelli, N., Grewal, D., Roggeveen, A.: Store atmospherics: a multisensory perspective. Psychol. Mark. **31**(7), 472–488 (2014)
23. Elder, R., Aydinoglu, N., Barger, V., Caldara, C., Chun, H., Lee, C., Mohr, G., Stamatogiannakis, A.: A sense of things to come—future research directions in sensory marketing. In: Krishna, A. (ed.) Sensory Marketing—Research on the Sensuality of Products, pp. 361–376. Nova Iorque e Londres, Routledge (2010)
24. Meyers-Levy, J., Bublitz, M., Peracchio, L.: The sounds of the marketplace—the role of audition in marketing. In: Krishna A. (ed.) Sensory Marketing—Research on the Sensuality of Products, pp. 137–156. Nova Iorque e Londres, Routledge (2010)
25. Yoon, S., Park, J.: Do sensory ad appeals influence brand attitude? J. Bus. Res. **65**, 1534–1542 (2012)
26. Kotler, P.: Atmospherics as a marketing tool. J. Retail. **49**(4), 48–64 (1974)
27. Chebat, J., Michon, R.: Impact of ambient odors on mall shoppers' emotions, cognition, and spending: a test of competitive causal theories. J. Bus. Res. **56**, 529–539 (2003)
28. Spangenberg, E., Sprott, D., Grohmann, B., Tracy D.: Gender-congruent ambient scent influences on approach and avoidance behaviours in a retail store. J. Bus. Res. **59**(12), 1281–1287 (2006)
29. Morales, A.: Understanding the role of incidental touch in consumer behavior. In: Krishna, A. (ed.) Sensory Marketing—Research on the Sensuality of Products, pp. 49–62. Nova Iorque e Londres, Routledge (2010)
30. McCabe, D.B., Nowlis, S.M.: The effect of examining actual products or product description on consumer preference. J. Consum. Psychol. **13**(4), 431–439 (2003)
31. Machado, J.: A identidade da marca. In: Lencaste, P. (ed.) O livro da Marca, pp. 212–262. Publicações Dom Quixote, Lisboa (2005)
32. Nadanyiova, M., Kliestikova, J., Kolencik, J.: Sensory marketing from perspective of support tool for building brand value. Econom. Cult. **15**(1), 96–104 (2018)
33. Morrin, M., Ratneshwar, S.: Does it make sense to use scents to enhance brand memory? J. Mark. Res. **40**(1), 10–25 (2003)
34. Brakus, J., Schmitt, B., Zarantonello, L.: Brand experience: what is it? How is it measured? Does it affect loyalty? J. Mark. **73**(3), 52–68 (2009)
35. Shieh, H., Lai, W.: The relationships among brand experience, brand resonance and brand loyalty in experiential marketing: evidence from smart phone in Taiwan. J. Econ. Manag. **28** (2), 57–73 (2017)
36. Ding, C., Tseng, T.: On the relationships among brand experience, hedonic emotions, and brand equity. Int. J. Distrib. Log. Manag. **49**(7/8), 994–1015 (2015)
37. Moreira, A., Fortes, N., Santiago, R.: Influence of sensory stimuli on brand experience, brand equity and purchase intention. J. Bus. Econ. Manag. **18**(1), 68–83 (2017)
38. Nadiri, H., Gunay, G.N.: Na empirical study to diagnose the outcomes of costumer's experiences in trendy coffee shops. J. Bus. Econ. Manag. **14**(1), 22–53 (2013)
39. Yoon, S.: Antecedents and consequences of in-store experiences based on an experiential typology. Eur. J. Mark. **47**, 663–714 (2013)
40. Peck, J., Childers, T.L.: To have and to hold: the influence of haptic information on product judgments. J. Mark. **67**(2), 35–48 (2003)
41. Areni, S., Kim, D.: The influence of in-store lighting on consumer's examination of merchandise in a wine store. Int. J. Res. Mark. **11**, 117–125 (1994)
42. Smith, P.C., Curnow, R.: "Arousal hypothesis" and the effects of music on purchasing behavior. J. Appl. Psychol. **50**(3), 255–256 (1966)
43. Milliman, R.: Using background music to affect the behavior of supermarket shoppers. J. Mark. **46**(3), 86–91 (1982)

44. Yalch, R.F., Spangenberg, E.R.: The effects of music in a retail setting on real and perceived shopping times. J. Bus. Res. **49**(2), 139–147 (2000)
45. Lee, S.: Which senses matter more? The impact of our senses on team identity and team loyalty. Sport Mark. Q. 203–213 (2013)
46. Lin, K., Chang, C., Lin, Z., Tseng, M., Lan, L.: Application of experimental marketing strategy to identify factors affecting guests' leisure behaviour in Taiwan HotSpring Hotel. WSEAS Trans. Bus. Econ. **6**, 229–240 (2009)
47. Coleman, D., Chernatony, L., Christodoulides, G.: B2B service brand identity: scale development and validation. Ind. Mark. Manag. (40), 1063–1071 (2011)
48. Sarquis, A., Glavam, A., Casagrande, J., Morais, A., Kich, M.: Marketing sensorial na comunicação de marca: um ensaio teórico. Braz. J. Manag. Innov. **2**(3), 1–21 (2015)

Chapter 25
Geomarketing Based on Beacons BLE

Isaac Aleixo(✉), João Paulo Fernandes,
and Cláudia Jacy Barenco Abbas

Faculty of Technology, Department of Electrical Engineering, Campus
Universitário Darcy Ribeiro, Brasília – D.F., Brazil
{isaac.aleixo, joaopaulofb83}@gmail.com,
barenco@unb.br

Abstract. The breakthrough in the field of wireless technologies, such as Bluetooth Low Energy (BLE), the availability of large volume of customer data and the growing use of beacon messages, all of these contribute to the development of a solution that makes shopping more enjoyable and efficient for the customer with the real-time monitoring facilitating the business management. The objective of this paper is to present a BLE solution using beacon technology applied to Geomarketing. The study addresses the construction of an application that makes use of BLE technology.

25.1 Introduction

The purpose of this work is to develop a Geomarketing solution: an application where the user will be able to advertise and get ads based on their location. To understand the goals of the application it is necessary to understand the following key concepts: BLE —Bluetooth Low Energy, Beacons, and Geomarketing that will be described in the following subsections.

25.1.1 Bluetooth Low Energy

BLE stands for Bluetooth Low Energy [1]; BLE is a form of wireless communication designed especially for short-range communications. The BLE is intended for situations where battery life is preferred over high data transfer rates. For example, if a user wants to broadcast marketing campaigns in the vicinity of a phone, the amount of data that is required to transfer to another is extremely small so BLE-compliant beacons do the job quickly without depleting the battery.

A BLE beacon transmits packets of data at regular intervals; these packets of data are detected by applications or services pre-installed on nearby smartphones. This BLE communication triggers actions such as sending a message, promoting an application, or sending an advertisement.

To save energy and provide higher data transfer speeds, the entire BLE communication structure consists of 40 frequency channels, separated by 2 MHz. 3 of these channels are the main advertising channels, while the remaining 37 channels are secondary channels, also known as data channels. Bluetooth communication starts with

Á. Rocha et al. (eds.), *Marketing and Smart Technologies*, Smart Innovation, Systems and Technologies 167, https://doi.org/10.1007/978-981-15-1564-4_25

the first three advertisement channels and transfers data on secondary channels (see Fig. 25.1).

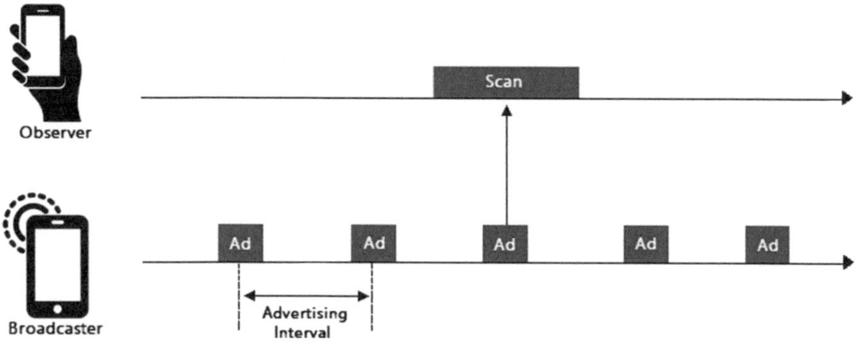

Fig. 25.1. BLE communication example [1]

25.1.2 Beacons BLE

Beacons are small Bluetooth transmitters [2], they transmit a unique identifier to electronic devices close enough to pick up their signal.

Beacons transmit small amounts of data via BLE packets, therefore having distance limitations, as a result, are often used for indoor localization technology, although beacons can also be used in outdoor environments as well.

The unique identifier and multiple bytes that are transmitted by the beacon can be used to determine the physical location of the device, track customers, or trigger a location-based action on the device, such as a notification when it arrives at a location.

Firstly, it is important to inform that beacons are used in BLE precisely because they allow for better energy savings and because they have periodic data transfer, not a continuous flow. These features make BLE ideal for broadcasting Beacons and advertisements. Therefore, it is ideal for the beacon-based Geomarketing project.

Beacons are small packets transmitted via BLE signals at a given interval. This range depends on the hardware capacity. On average, a BLE beacon can transmit BLE signals up to 80 meters. This beacon BLE signal is able to perform a specific action relevant to the location, such as showing an ad.

Other advertising methods are also possible with Google's beacons, URIBeacon, and Eddystone allows for a URI transmission mode. URI beacons transmit a URI that can be a link to a web page and the user will see that URI directly on their phone.

The beacons send an ID number through the BLE channels, approximately 10 times per second. A Bluetooth enabled nearby beacon handles this ID number. When an application or service pre-installed such as Google Nearby recognizes the ID number, it binds the beacon to an action, such as downloading an application or a marketing offer and displays it on your smartphone.

25.1.3 Geomarketing and Marketing with Beacons

Geomarketing is a marketing technique that relies on user location to define target audiences and better tactics to reach the end consumer at the right time and place [3]. Coupled with the increase in the use of mobile devices, tablets, cell phones and other devices, the information of the location and places visited by a person, if used efficiently by an algorithm or Artificial Intelligence (AI), can lend companies valuable information to assemble a strategy of effective marketing, delivering ads and offers that are more relevant to your end customer.

BLE Beacons are a convenient way for marketers to communicate with customers in a meaningful and personalized manner.

According to Beaconstac [4], social media marketing, such as Facebook and Twitter ads, also has below-average click-through rates for beacons. The average click-through rate for Facebook ads in 2018 is 0.119%, while the average click-through rate for Twitter ads is 1–3%. However, the Average Click-Through Rate for beacon campaigns is 2–4%. Well-executed signaling campaigns can generate a rate of 12–15%.

With this data, it is possible to notice that the technology of beacons used in ads is evolving rapidly.

25.1.4 Differences Between Bluetooth Classic and BLE

There are two main technologies within the specification Bluetooth: Bluetooth Classic and BLE also known as Bluetooth Smart. The main difference between the two is in the energy consumption in each case. However, there are other factors why Bluetooth Smart is being used for technological applications [5].

Bluetooth Classic consumes more power than BLE. Bluetooth Classic is ideal for applications that require continuous data flow, such as headphones. However, BLE is suitable for applications that work well with periodic data transfer, and therefore reduces a significant amount of battery usage. This makes BLE suitable for applications related to proximity marketing, which is the case with the Beacons in this project.

BLE can establish up to 20 connections simultaneously, it supports more simultaneous connections as it transfers small data packets and establishes fast connections. The Bluetooth Classic can only initiate 7 simultaneous connections.

25.2 Developing the Application

The application was developed using Java programming language and is limited for Android operational systems. To understand how the application was developed, it is necessary to comprehend the core tools used to develop the application: Android Studio, Google Developer Console, API Nearby Messages, which are shown in the following subsections.

25.2.1 API Nearby Messages

The API used for building the application is the Google Nearby Messages API [6] The API makes use of both Bluetooth, BLE, and Wi-Fi to make devices communicate, the server makes it easy to exchange messages between devices with the same pairing type encoding.

The exchange of messages between devices takes place as follows:

I. A publishing application makes a request to associate the message with a unique pairing code (token).
II. The publishing device uses a combination of Bluetooth, Bluetooth Low Energy, and Wi-Fi to make the token discoverable by nearby devices. The publishing device also uses these technologies to search for tokens from other devices.
III. A subscription application associates its signature with a token and uses a combination of the above technologies to send its token to the publishing device and to detect the publishing device token.
IV. When one side detects the token of the other, it informs the server.
V. The server makes it easy to exchange messages between two devices when both are associated with a common token, and the API keys used by the calling applications are associated with the same project in the Google Developers Console.

In order to integrate the application with Google Play services, you need to create a buildGoogleApiClient () method, both in the class responsible for publishing and in the class responsible for subscribing (see Fig. 25.2).

```
private void buildGoogleApiClient() {
    if (mGoogleApiClient != null) {
        return;
    }
    mGoogleApiClient = new GoogleApiClient.Builder( context: this)
            .addApi(Nearby.MESSAGES_API, new MessagesOptions.Builder()
                    .setPermissions(NearbyPermissions.BLE).build())
            .addConnectionCallbacks(this)
            .enableAutoManage( fragmentActivity: this,  onConnectionFailedListener: this)
            .build();
}
```

Fig. 25.2. buildGoogleApiClient() Method

25.2.2 Google Developers Console

To use the API Nearby Messages you must have a Google account, but for the end-user of the application is not necessary to have an account.

To use the API it is necessary to create a KEY API in the Google Developers Console [7], the KEY API is also used to restrict which applications can access the beacons with the messages published by the publishers. In this solution only the applications with the same KEY API can view the beacons.

The KEY API is configured in the AndroidManifest.xml file.

25.2.3 Publishing and Subscribing with Beacons

First, to make sure that only beacons will be published and subscribed; it is necessary to configure the strategy, the Google Messages API provides two strategies:

- Default: Default behavior is currently making publications and subscriptions, using all available sensors, to discover nearby devices regardless of distance.
- BLE_Only: The behavior of BLE Only is to make publications and subscriptions only using beacons; this is the strategy to be used for the application and to study the properties of beacons and BLE.

Once the strategy to be used has been chosen, you can now create the methods to publish and subscribe the beacons, the methods have been created in separate classes, a java class is responsible for publishing beacons and another java class responsible for beacons subscription.

Within the folder, "java" was created the classes: "TelaAnuncio" which stands for AdvertisingScreen and "TelaOferta" which stands for OfferScreen responsible for publishing and subscribing respectively.

Within the class the method private void publish () is created, within the publishing options there is the parameter.setStrategy, within the Strategy class of the Nearby Messages API, is the constructor for BLE ONLY.

And at the very beginning of the "TelaAnuncio" class, a variable is created to store this strategy (see Fig. 25.3).

```
private static final Strategy PUB_SUB_STRATEGY = (new Strategy.Builder()).zzjS(2)
        .setTtlSeconds(TTL_IN_SECONDS).build();
```

Fig. 25.3. Variable created to store strategy

With this done it is possible to create the publish () method, with the adopted strategy parameter (see Fig. 25.4).

To subscribe, you must first have a publisher in the range of the device, and if this condition is fulfilled, the device you want to subscribe to will perform the subscribe method, that is in the class "TelaOferta" which stands for OfferScreen.

Just as in the case of publication, it is necessary to define the strategy adopted for the subscription, and as previously only interested in subscribing in beacons, so the strategy used remains BLE ONLY.

In Fig. 25.5 there is an application diagram showing how each of the steps mentioned works.

From the home screen, you can make three main decisions: Create an ad; Publish ad; or View published ads.

```java
private void publish() {
    Log.i(TAG, msg: "Publishing");
    PublishOptions options = new PublishOptions.Builder()
            .setStrategy(PUB_SUB_STRATEGY)
            .setCallback(onExpired() → {
                    super.onExpired();
                    Log.i(TAG, msg: "No longer publishing");
                    runOnUiThread(() → {
                            mPublishSwitch.setChecked(false);
                    });
            }).build();

    Nearby.Messages.publish(mGoogleApiClient, mPubMessage, options)
            .setResultCallback((ResultCallback) (status) → {
                    if (status.isSuccess()) {
                        Log.i(TAG, msg: "Published successfully.");
                    } else {
                        logAndShowSnackbar( text: "Could not publish, status = " + status);
                        mPublishSwitch.setChecked(false);
                    }
            });
}
```

Fig. 25.4. Publish method

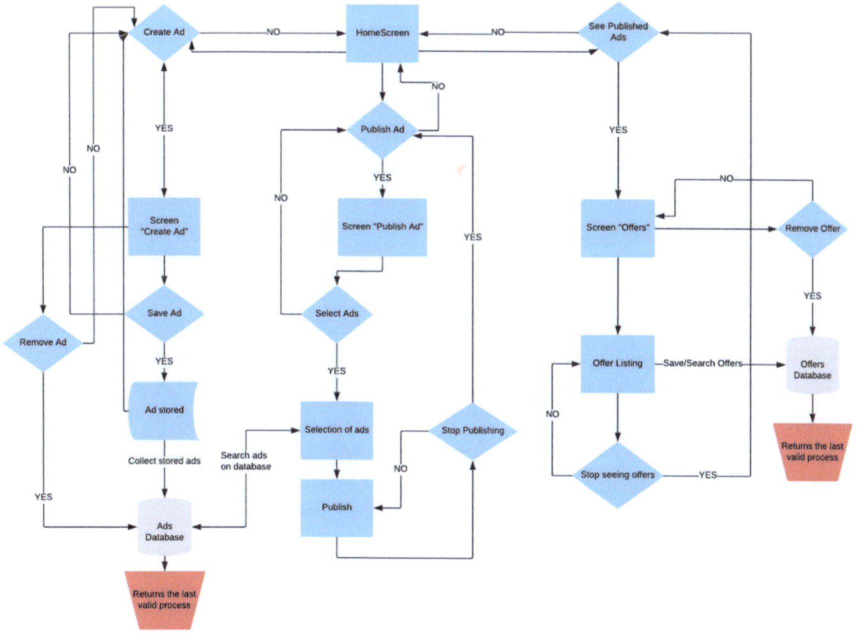

Fig. 25.5. Application flowchart

25.2.4 Create Advertisement

When the application user chooses to create an advertisement, they will be taken to the screen “TelaCriaAnuncio” which stands for “CreateAd”. In this class, the methods responsible for creating, storing and deleting an advertisement are defined.

25.2.5 Publish Advertisement

Before publishing, the application will send to the Google Developer Console the API KEY information, after this step it will be possible to publish to other applications that also have the same validated API KEY.

When the user of the application selects the option to publish, they will be taken to the screen “TelaAnuncio” which stands for “AdScreen.” In this class are defined as the methods responsible for selecting the ads to be published; the publication of the selected ads and to stop publishing.

In order to make the announcement, the user must select which ads created and stored in the ad database they want to post. If they want to start a new post with different ads, the current post must be stopped using the unpublish () method of the class, and a new post should be made with the desired announcements. From this screen, the user can stop publishing at any time using the unpublish () method.

25.2.6 View Published Ads

Before you begin viewing served ads, the application will send the Google Developer Console API KEY information, after which you can subscribe to other applications that also have the same validated API KEY.

When the user selects the option to view the ads that are being published, they will be taken to the screen “TelaOferta” which stands for “OfferScreen”, in this class are defined the methods responsible for listing and saving the ads that have been published by devices in which the application has been subscribed, as well as methods for deleting offers from the offer database and also stopping the subscription.

When subscribe is selected, the application will automatically start a scan of devices that are publishing within the range supported by the physical limitations of the device, up to 80 meters. Upon finding a device in range of the application will then store all ads sent by the advertiser’s beacon into an offer database.

The offers database is used to group offers by categories, so users can filter ads according to their needs as well as delete offers that do not attract their attention.

If the user wishes to receive ads, they may simply confirm the subscription through the unsubscribe () method.

25.2.7 App Appearance

Each home screen button is responsible for performing an action corresponding to Sects. 25.2.4, 25.2.5 and 25.2.6 described before (see Figs. 25.6 and 25.7).

Fig. 25.6. Home screen

The home screen buttons are:

- Criar Anúncio “CreateAd” (Sect. 25.2.4);
- Anunciar “Advertise” (Sect. 25.2.5);
- Ofertas “Offers” (Sect. 25.2.6);

25.2.8 Screen “CreateAd”

By clicking on the button “Create an Ad” the user can perform the actions explained in Sect. 25.2.4.

To create an ad the user must fill in the fields:

- Ad Name; (Nome do Anúncio)
- Product’s name; (Nome do Produto)

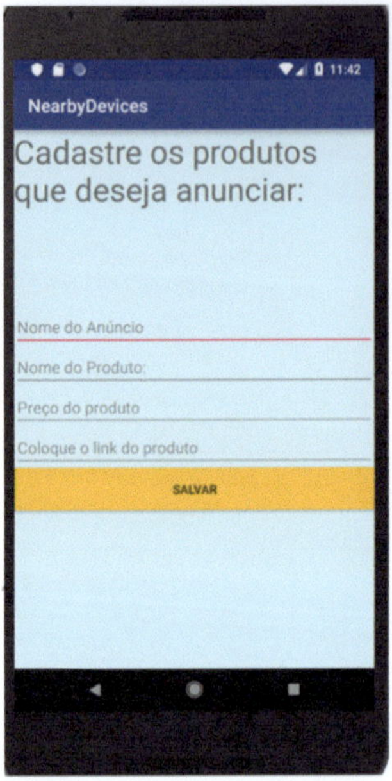

Fig. 25.7. Screen "Create Ad"

- Price of the product; (Preço do Produto)
- Link of the Product; (Link do Produto)

After filling in this data and clicking the save button, your ad will be available in the ad database and ready to be published.

25.2.9 Screen "PublishAd"

By clicking the button "Announce" the user can perform the actions explained in the Sect. 25.2.5.

On this screen the user must select the ads they want to post, and then activate the "toggle switch: Announce" to start posting. It is noteworthy that while toggle is active it will not be possible to modify the publications being made, in the case the user desires to make any alterations, they must then disable the toggle and modify the ads to be published.

25.2.10 Screen "Offers"

By clicking on the button to view the offers the user can perform the actions explained in subsection 25.2.6.

From this screen, you can view offers being published and store them in the offer database, as well as manage the offers saved in the database.

25.3 Conclusion

This paper presents a powerful method of employing beacon messages and BLE technology to further enhance advertisement strategies through the use of Geomarketing. The use of BLE and beacons is notable, as they can be used to innovate in multiple areas, such as in those of commerce, transit, home solutions, and other indoor applications, with the benefits of reduced battery expenditure and the possibility for multiple simultaneous connections.

The use of these technologies has yet many challenges in its horizon, however, this application and its functions serve as a first step in accomplishing future goals of the geomarketing sector, notably changing the way the consumer and seller dynamic functions.

This paper shows a powerful solution for Geomarketing using beacons messages and BLE technology.

BLE and beacons can be used to innovate on a lot of areas such as selling, car traffic, home solutions, and other indoor applications.

There are a lot of challenges on the future, and this work is only one step to accomplish the future goals in the Geomarketing area that is changing the way of the relationship between consumer and dealer.

References

1. UFRJ.: Bluetooth Low Energy. https://www.gta.ufrj.br/ensino/eel879/trabalhos_vf_2012_2/bluetooth/ble.htm
2. Beaconstac.: Bluetooth Beacon. https://www.beaconstac.com/what-is-a-bluetooth-beacon
3. Zaim, D.: Bluetooth Low Energy (BLE) Based Geomarketing System. https://ieeexplore.ieee.org/document/7772263
4. Bellafkih, M.: Developing a Geomarketing Solution. https://www.sciencedirect.com/
5. Meera, R., Archan, M., Rajesh K.B., Youngki, L.: Smartphones & BLE Services: Empirical Insights. http://smedia.ust.hk/james/projects/people_aware_smart_city_applications/paper/3.pdf
6. Developer Android.: Android. https://developer.android.com
7. Developers Google.: Nearby Google. https://developers.google.com/nearby/messages/android/

Chapter 26
Customer Acceptance of Shopping-Assistant Chatbots

Tiago Araújo[2] and Beatriz Casais[1,2,3(✉)]

[1] School of Economics and Management, Universtity of Minho, Braga, Portugal
bcasais@eeg.uminho.pt
[2] IPAM, Porto, Portugal
[3] CiTUR, Leiria, Portugal

Abstract. Due to the increased use of chat applications, it is expected that consumer buying behavior will change. Chatbots have been identified by companies as being an excellent opportunity to better exploit social networks and also get competitive differentiation through relationship marketing in the digital environment. This paper discusses through the Technology Acceptance Model (TAM), the adoption of chatbots as shopping assistants in e-commerce. A sample of 237 Portuguese respondents between 18 and 54 years old that answered a questionnaire evidenced that the dimensions that most influence their acceptance of chatbots were the compatibility and attitude towards mobile advertising. These results emphasize that there is a need to introduce better digital solutions to boost the results of companies. The investment in chatbots is thus an important factor of differentiation.

26.1 Introduction

Consumers are spending more time in the digital environment and companies need to implement a digital marketing strategy to meet consumers in the electronic marketplace. Leeflang et al. [16] point out great opportunities, such as the use of real-time feedback from consumers to compete effectively, the use of social networks to enhance relationships between company and costumers, and the analytics of the return on investments in digital marketing activities.

Digital marketing strategies have been following the inbound marketing approach. Dakouan et al. [5] define inbound marketing as a marketing strategy that refers to the creation of relevant and appropriate content to attract Internet users. It requires targeting the audience with content of extreme quality and relevance [22], exploring digital techniques, as search engine optimization or email marketing, the use of websites, social networks, and blogs, for example, in order to convert, close, and delight over time [5, 29]. Within the consumer interactions with the company, passing the different steps of the inbound funnel, the perspective of relationship marketing in user experience should be a central concern [18].

Considering that consumers tend to use all the shopping channels available, both in the offline and online environment, expecting an integrated service, the concept of

Á. Rocha et al. (eds.), *Marketing and Smart Technologies*, Smart Innovation, Systems and Technologies 167, https://doi.org/10.1007/978-981-15-1564-4_26

omnichannel rose and implied the development of techniques to integrate the dimension of relationship marketing in online selling services [1]. It is imperative to maintain a consistent experience across all available channels to avoid any negative impact in the consumer buying experience, overcoming any barriers and meeting all the consumer expectations [31].

The growing importance of mobile and social networks creates new challenges and opportunities for companies [24], for example, with artificial intelligence (AI) through chatbots [32]. However, the consumer acceptance of AI remains under research and this study proposed a discussion about the consumer acceptance of chatbots, in Portugal.

26.2 Theoretical Framework

In order to understand the digital consumer experience, the perception about the quality of service in shopping encounter and its acceptance of chatbots application, this section presents a review about the technology acceptance model (TAM), the online user experience and the state of the art about the application of chatbots in marketing.

26.2.1 Technology Acceptance Model

With the constant appearance of new technologies, it is necessary to measure the level of their acceptance by the users in order to obtain useful evidence for companies which are considering technological investments to gain business competitive advantages. The Technology Acceptance Model (TAM) was created by Davis [6] in order to better understand the factors that lead to the use of new technology. The model considers two dimensions. The first dimension states that users use certain technology according to the expectation that the use will facilitate the performance of a given task. This variable is called the perceived *utility*. The second dimension, perceived *ease of use*, concerns the user's perception of the usability and intuitivism of new technology [6]. These same dimensions were later sustained for the online environment by Chen and Teng [4].

The TAM model has been reinforced by several authors, as reported in Van der Heijden and Verhagen [27]. Legris et al. [17] reviewed the model and proved the existence of a positive relationship between attitude and behavioral intention and also the need to include change processes such as the *compatibility* with lifestyle, proposed by Moore and Benbasat [21] and included by Chen Gillensen and Sherrel [3] in their studies of consumer behavior in online stores.

Venkatesh [30] organized another model, called UTAUT integrating the dimensions of TAM with models. Kaasinen [15] modified the TAM model and applied the trust variable that proved to be very relevant in his study. This variable was later supported by Wu and Wang [33]. In order to build a model that allows the study of the acceptance of mobile services, Zarmpou et al. [34] maintained some TAM variables such as perceived ease of use, perceived utility and behavioral intention, and added new variables such as innovation, relationship, and functionality. Currently, it continues to be a reference model in the study of new technologies, although it presents slight modifications between studies.

26.2.2 Chatbots

Chatbots are a digital solution with a huge growth potential designed to enhance the e-commerce process by simulating an intelligent conversation with the customer via auditory or written methods [11]. There are two types of chatbots, those based on previously defined keywords and those with a more advanced programming that use algorithms of machine learning. The chatbot regulated by keywords is quite limited since it only responds to certain commands for which it was previously defined, while the chatbot that uses machine learning algorithms is able to perceive the user's own language, even if it contains errors. However, all simulate a conversation in order to provide the answers in the chat the consumers want to get [11] and ideally make unnoticeable to distinguish whether they are talking to a human or a chatbot [19]. According to the theory of social response, people tend to treat the computer as a social actor, and even attribute human characteristics, even though the machines do not have feelings, intentions, or human motivations [20], then relations created with the people behind the computers are different. Holzwarth et al. [14] points out that human–computer interactions can be enhanced by simply assigning an avatar to the bot.

Chatbots can be equated as shopping assistants in physical and online stores. Purchasing assistants are considered to be one of the greatest tools in traditional commerce, as they provide store support, provide additional information about products, and simplify the decision-making process [2], and this same role can be played by chatbots, in the digital world of the omnichannel integration of purchasing experiences. These purchase assistants may be designated by artificial software agents, external agents, autonomous agents, shopping bots, or shopbots [25].

With the improvement of machine learning techniques and decision-making skills, chatbots have become important elements in commerce [26]. They make possible the multitask, allowing to attend several customers at the same time, create instant leads, the starting of a conversation and the creation of customer–brand relationships, and allows for a better first impression, since it provides an instantaneous response by reducing the waiting time [7]. The distinction between chatbots and live chat is essentially based on the absence of the human touch and the possibility of providing customer service 24 h a day, however, the absence of the human touch divides experts' opinions. Eroglu et al. [10] argue that human presence should be absent from virtual stores since they are two totally different realities. Holzwarth et al. [14] points out that using chatbots makes the online store more credible and pleasant for consumers, potentiating the purchase.

Previous research has discussed the acceptance of live customer support, developing, for example, framework for e-service quality in this context [8]. Elmorshidy et al. [9] highlight that the customer intention to use chatbots in online purchase environments is influenced by the usefulness, ease of use, and attitude toward the technology.

26.2.3 Online User Experience

In order to better understand human–computer interactions, Petrie and Bevan [23] argue that the definition of usability should consider dimensions that fit the user experience: likeability, pleasure, comfort, and confidence. There are pragmatic aspects of interactive products that have a significant impact on the well-being of the user [13]. These authors define user experience as being the consequence of the interaction of the user with system design and the context in which the interaction takes place. User experience is a set of user perceptions and responses resulting from the use of a product, system, and service. In this sense, the service provided by live customer support chats is part of the user experience, affecting the dimension of the technology acceptance model connected to usability or ease of use.

26.3 Methodology

This research aims to discuss customer acceptance of chatbots in e-commerce experiences. In this section, we show the hypothesis developed, the process of data collection and the sample of the study, as well as the process for data analysis.

26.3.1 Purpose and Hypothesis

The research is based on the TAM model of Davis [6], maintaining some dimensions that have proved to be important in the adoption of new technologies, such as perceived ease of use, perceived utility, and behavioral and attitude-dependent variables. Van Eeuwen [28] developed an adapted conceptual model to research the acceptance of chatbots, adding variables of attitude towards mobile advertising, compatibility, and concern for privacy on the Internet.

In this sense, this paper integrates dimensions previously proposed in the literature of technology acceptance and in the specific case of chatbots and mobile commerce. Table 26.1 summarizes those dimensions.

Table 26.1. Summary of dimensions under analysis

Dimension	Authors
Perceived utility	Davis [6]
Perceived ease of use	Davis [6]
Compatibility	Chen Gillensen and Sherrel [3], Moore and Benbasat [21]
Concern about privacy on the internet	Moore and Benbasat [21]
Attitude towards mobile advertising	Van Eeuwen [28]
Attitude	Davis [6]
Behavioral intent	Zarmpou et al. [34]

Based on the conceptual model, where the variables are presented; perceived utility, perceived ease of use, compatibility, attitude toward mobile advertising, concern for Internet privacy, attitude and behavioral intention, six hypotheses are presented:

H1—Perceived utility positively influences attitude toward chatbots;
H2—Perceived ease of use positively influences attitude toward chatbots;
H3—Compatibility positively influences attitude toward chatbots;
H4—The concern for online privacy negatively influences the attitude toward chatbots;
H5—Attitude toward mobile advertising through chatbots positively influences attitude about chatbots;
H6—Attitude toward the adoption of chatbots positively influences behavioral intention.

Based on the research problem and its objectives, in this case, the acceptance adoption of chatbots in electronic commerce, a quantitative methodology is applied. With regard to interaction with chatbots, a rather simplified explanation of what this digital solution was presented in the questionnaire, with a video of 1-minute length in order to prevent misunderstandings.

26.3.2 Sample

This study obtained a sample of 237 respondents collected in Portugal in 2018. The majority of respondents are female, 140 (59.1%) and 97 are male (40.9%). The respondents' ages were given through intervals ranging from 18 to 54 years of age, however, no age limits were placed on the questionnaire. The age range that obtained the greatest number of responses, 139 more precisely, corresponded to the ages between 18 and 24 years (58.6%), 74 between 25 and 34 years (31.2%), 17 between the ages of 35 and 44 (7.2%) and only 7 people over 44 (3%). In terms of academic education, 50.2% were graduated, 30.4% have a master, 0.8% have a doctoral degree and the remain respondents have a high school or less.

26.3.3 Data Analysis

A descriptive analysis was made of the respondents in order to understand the purchasing behavior of the respondents the average of the responses attributed to each variable of the conceptual model applied.

The *t-test* and the ANOVA test were applied in order to understand if the sociodemographic characteristics and the buying behavior of the respondents influenced the dependent variables. In order to validate the conceptual model and the hypotheses placed, we conducted regression and correlation analysis.

26.4 Results

The results show that there are a large number of people have already bought online (90.7%) This number reinforces the importance of the digitization of companies and the introduction of new digital solutions to satisfy consumers' expectations and establish an omnichannel environment. In this section, we explain in detail the consumer knowledge and attitudes toward chatbots' interactions.

26.4.1 Knowledge and Behavior Intention About Chatbots Interaction

56.5% of respondents answered they did not know what chatbots were. 39.7% assumed they have never interacted with a chatbot and 21.9% answered they were not aware whether they had interacted with a chatbot or not.

In this study, two dependent variables are presented, one of which is the behavioral intention. It is possible to verify that the level of behavioral intention in relation to the use of chatbots falls mainly on levels 3 and 4 that correspond to a neutral and positive behavioral intention, respectively. Despite the positive trend, the average number of responses obtained is 3.38, which translates into a neutral behavioral intention, with a standard deviation of 0.987.

The t-test was analyzed in order to verify if this dependent variable changes based on the gender of the respondent, academic qualifications, the preferred device for purchases, the fact that they have already made purchases online, and also by be knowledge of what chatbots are. Regarding the gender of the respondents, it is possible to verify that there is no significant difference (sig. 0.370) between the values attributed to the behavioral intention by male (N = 97) and female (N = 140). In the academic level, two groups were created: the group with the lowest academic qualifications (N = 44) and the group with the highest academic qualifications. The respondents who considered the "other" option were classified with less academic qualifications group. Similarly to what happened in the gender analysis, the difference in the values attributed to behavioral intention is also not significant (sig. 0.343). Among users who prefer to use mobile equipment when making purchases (N = 76) and those who prefer to use the computer (N = 139), it is possible to perceive that there is also no significance (sig. 0.834). Finally, the difference in the values attributed between the group of respondents who have already made online purchases (N = 215) and those who never bought online (N = 22), there is a significative variance of 0.023. The knowledge or lack of knowledge about what is a chatbot also does not have a significant variance with respect to behavioral intention. Finally, in order to understand if there is an influence of the respondents' ages and residence in behavioral intention, an ANOVA test was conducted, and it was possible to verify that both characteristics have no significance (sig. 0.51 and 0.76, respectively).

26.4.2 Attitudes Toward Chatbot

It is possible to verify that the mean and the standard deviation in the attitude (N = 237 M = 3.48 SD = 0.97) is very similar to the results for the behavior intention (N = 237 M = 3.38 SD = 0.99). This result demonstrates that respondents have on average a neutral attitude with a positive trend regarding the use of chatbots in e-commerce. Attitude values below 1.5 are interpreted as being a very negative attitude, between 1.5 and 2.5 as negative, between 2.5 and 3.5 is considered neutral, between 3.5 and 4.5 positive, and results over 4.5 are considered as a very positive attitude.

The same tests previously mentioned in behavioral intention were also implemented to the variable of attitude towards the adoption of the *chatbot.* Any significant difference of results based on gender, academic qualifications, preferred device to make purchases, for the knowledge or not of what is a *chatbot*, nor for the fact of the respondent already had made online purchases were identified. The same was observed in the age and the district of residence that also did not present significance in the attitude (sig. 0.44 and 0.41, respectively).

There is a strong positive correlation (0.865) between the attitudinal variable and behavioral intention, with a significance value of less than 0.05.

26.4.3 Hypothesis Test

When performing the confirmatory factorial analysis with *varimax* rotation, the value of the Kaiser–Meyer–Olkin presents a value of 0.772. Thus, the KMO test obtained statistically valid values. The Bartlett sphericity test showed a significance lower than 0.01, so factorial analysis is appropriate. All items were analyzed through confirmatory factor analysis and verified in terms of reliability. The reliability analysis is also performed through Cronbach's alpha, and it is possible to verify that the reliability levels of all variables respect the minimum value of 0.70, thus proving that the questions present a good reliability.

Using regression analysis using the Stepwise method, it is possible to verify that only 3 of the 5 independent variables presented in the conceptual model significantly affect the dependent variable attitude toward chatbot. The variables indicated as significant are compatibility, attitude toward mobile advertising, and perceived utility, leaving out the concern for privacy on the Internet and perceived ease of use. The three variables, together, present a variance in attitude of 62% (sig. 0.003), and the variable that contributes the most is the compatibility with 52%, followed by the attitude toward mobile advertising and the utility perceived with 8.9% and 1.4%, respectively.

Analyzing the coefficients, it is possible to verify that the independent variables' compatibility, attitude toward mobile advertising and perceived utility have a significance value of less than 0.05. The regression analysis also allows us to verify that all the variables presented in the models indicated by stepwise affect the attitude-dependent variable in a positive way, being that the compatibility is the one that presents a higher value, considering as being the variable with a greater degree of prediction in the attitude, since for each unit increased in the compatibility, the dependent variable increases to 0.432, followed by the variables attitude toward mobile advertising and perceived utility. When analyzing the regression between the attitudinal variable and the behavioral intention it is possible to verify that for each increased level in the attitude, the dependent variable behavioral intention increases to 0.879 (Sig < 0.05).

A linear regression analysis was used in order to verify whether the hypotheses presented are supported or not. Table 26.2 shows the results of regression analysis.

H1—Through linear regression, it is possible to verify that the variable perceived utility positively influences the dependent variable attitude, explaining in 58% the total variance in attitude.

H2—Through linear regression, it is possible to verify that the compatibility variable positively influences the dependent variable attitude, explaining it in 72%.

H3—Through linear regression, it is possible to verify that the attitude variable toward mobile advertising positively influences the dependent variable attitude, explaining in 65% the total variance in attitude.

Table 26.2. Linear regression applied to attitude-dependent variable

	Non-standardized coefficients	Standardized coefficients	T	Sig.
	B	B		
Perceived utility	0.772	0.582	10,976	0.000
Compatibility	0.665	0.724	16,096	0.000
Mobile advertising	0.765	0.647	13,009	0.000
Perceived ease of use	0.604	0.509	9,074	0.000
Privacy concern	−0.012	−0.15	−0.225	0.822

H4—Through linear regression, it is possible to verify that the perceived ease of use positively influences the attitude-dependent variable, explaining in 51% the total variance in attitude.
H5—Through linear regression, it is possible to verify that the concern variable with privacy on the Internet negatively influences the dependent variable attitude.
H6—Through linear regression, it is possible to verify that the attitude variable positively influences the dependent variable behavioral intention, explaining in 87% the total variance in behavioral intention (Sig < 0.05).

In brief, all the hypotheses were accepted.

26.5 Conclusions

The high number of respondents unaware of chatbots may indicate a transition period and the need for greater dissemination of this digital solution. Both the attitude and behavioral intentions presented a neutral position of the respondents regarding the acceptance of chatbots, although with a slightly positive trend. These results may reflect the time of data collection, in an early stage of implementation of chatbots in Portuguese e-commerce.

It was possible to conclude that the sociodemographic variables such as gender, age, academic background, and district of residence do not have any type of influence on the attitude and intention behavior to adopt chatbots.

Attitude and behavioral intention are strongly correlated with each other. The fact that the respondent has already made purchases online is a significant variable that changes the behavioral intention of the respondent.

Contrary to the study of Van Euwen [28], the attitude toward mobile advertising through the use of *chatbots* did not negatively affect the dependent variable attitude. All variables, with the exception of the concern for online privacy, positively influence attitude, as proposed in the hypothesis. Compatibility is assumed to be the greatest predictor of attitude variation as previously noticed [28, 33], when people find *chatbots* compatible with their buying behavior and their lifestyle.

References

1. Barwitz, N., Maas, P.: Understanding the omnichannel customer journey: determinants of interaction choice. J. Interact. Mark. **43**(August), 116–133 (2018)
2. Bogdanovych, A., Simoff, S.: Implicit training of virtual shopping assistants in 3D electronic institutions. In: E-Commerce 2005 IADIS International Conference Proceedings, January, pp. 50–57 (2005)
4. Chen, M.Y., Teng, C.I.: A comprehensive model of the effects of online store image on purchase intention in an e-commerce environment. Electron. Commerce Res. **13**(1), 1–23 (2013)
3. da Chen, L., Gillenson, M.L., Sherrell, D.L.: Enticing online consumers: an extended technology acceptance perspective. Inf. Manag. **39**(8), 705–719 (2002)
5. Dakouan, C., Benabdelouahed, R., Anabir, H.: Inbound marketing vs. outbound marketing: independent or complementary strategies. Expert J. Mark. **7**(1), 1–6 (2019)
6. Davis, F.: Perceived usefulness, perceived ease of use, and user acceptance of information technology. MIS Q. **13**(3), 319–340 (1989)
7. Elmorshidy, A.: Benefits analysis of live customer support chat in E-commerce websites: dimensions of a new success model for live customer support chat. In: 2011 10th International Conference on Machine Learning and Applications and Workshops, vol. 2, pp. 325–329. IEEE (2011)
8. Elmorshidy, A.: Applying the technology acceptance and service quality models to live customer support chat for e-commerce websites. J. Appl. Bus. Res. **29**(2), 589–596 (2013)
9. Elmorshidy, A., Mostafa, M., El-Moughrabi, I., Al-Mezen, H.: Factors influencing live customer support chat services: an empirical investigation in Kuwait. J. Theor. Appl. Electron. Commer. Res. **10**(3), 63–76 (2015)
10. Eroglu, S.A., Machleit, K.A., Davis, L.M.: Atmospheric qualities of online retailing: a conceptual model and implications. J. Bus. Res. **54**, 177–184 (2001)
11. Gupta, S., Borkar, D., Mello, C.De, Patil, S.: An E-commerce website based Chatbot. Int. J. Comput. Sci. Inf. Technol. **6**(2), 1483–1485 (2015)
12. Gursoy, D., Chi, O.H., Lu, L., Nunkoo, R.: Consumers acceptance of artificially intelligent (AI) device use in service delivery. Int. J. Inf. Manag. **49**, 157–169 (2019)
13. Hassenzahl, M., Tractinsky, N.: User experience—a research agenda. Behav. Inf. Technol. **25**(March–April), 91–97 (2006)
14. Holzwarth, M., Janiszewski, C., Neumann, M.M.: The influence of avatars on online consumer shopping behavior. J. Mark. **70**(4), 19–36 (2006)
15. Kaasinen, E.: User acceptance of mobile services—value, ease of use, trust and ease of adoption, 566. VTT Publications (2005)
16. Leeflang, P.S.H., Verhoef, P.C., Dahlström, P., Freundt, T.: Challenges and solutions for marketing in a digital era. Eur. Manag. J. **32**(1), 1–12 (2014)
17. Legris, P., Ingham, J., Collerette, P.: Why do people use information technology? a critical review of the technology acceptance model. Inf. Manag. **40**(3), 191–204 (2003)
18. Lemon, K.N., Verhoef, P.C.: Understanding customer experience throughout the customer journey. J. Mark. **80**(November), 69–96 (2016)
19. Mctear, M.: Future and Emerging Trends in Language Technology. Machine Learning and Big Data, vol. 10341, pp. 38–49 (2017)
20. Moon, Y.: Intimate exchanges: using computers to elicit self-disclosure from consumers. J. Consumer Res. **26**(4), 323–339 (2000)
21. Moore, G.C., Benbasat, I.: Development of an instrument to measure the perceptions of adopting an information technology innovation. Inf. Syst. Res. **2**(3), 192–222 (1991)

22. Patrutiu-Baltes, L.: Inbound marketing—the most important digital marketing strategy. Bull. Transilvania Univ. Brasov **9**(2), 61–68 (2016)
23. Petrie, H., Bevan, N.: The evaluation of accessibility, usability, and user experience. In: The Universal Access Handbook, vol. 1, pp. 1–16 (2009)
24. Piotrowicz, W., Cuthbertson, R.: Introduction to the special issue information technology in retail: toward omnichannel retailing. Int. J. Electron. Commerce **18**(4), 5–16 (2014)
25. Redmond, W.H.: The potential impact of artificial shopping agents in E-commerce markets. J. Interact. Mark. **5**(1), 56–66 (2002)
26. Sotolongo, N., Copulsky, J.: Conversational marketing: creating compelling customer connections. Appl. Mark. Anal. **4**(1), 6–21 (2018)
27. Van der Heijden, H., Verhagen, T.: Online store image: conceptual foundations and empirical measurement. Inf. Manag. **41**(5), 609–617 (2004)
28. Van Eeuwen, M.V.: Mobile conversational commerce: messenger chatbots as the next interface between businesses and consumers. Master's thesis, University of Twente (2017)
29. Vargo, R.L.S.: Service-dominant logic—a guiding framework for inbound marketing. Mark. Rev. St. Gallen, 6–10 (2009)
30. Venkatesh, V.: Determinants of perceived ease of use: integrating control, intrinsic motivation, Acceptance Model. Inf. Syst. Res. **11**(4), 342–365 (2000)
31. Verhoef, P.C., Kannan, P.K., Inman, J.J.: From multi-channel retailing to omni-channel retailing. Introduction to the special issue on multi-channel retailing. J. Retai. **91**(2), 174–181 (2015)
32. Wirth, N.: Hello marketing, what can artificial intelligence help you with? Int. J. Mark. Res. **60**(5), 435–438 (2018)
33. Wu, J.H., Wang, S.C.: What drives mobile commerce? an empirical evaluation of the revised technology acceptance model. Inf. Manag. **42**(5), 719–729 (2005)
34. Zarmpou, T., Saprikis, V., Markos, A., Vlachopoulou, M.: Modeling users' acceptance of mobile services. Electron. Commerce Res. **12**(2), 225–248 (2012)

Chapter 27
Words, Frequency, and Understanding: Ranking Marketing Discipline Terms Using Machine Learning

Peter Vitartas(✉)

La Trobe University, Melbourne, Australia
p.virtartas@latrobe.edu.au

Abstract. This paper reports on research currently underway that aims to refine a list of terms that represent the marketing discipline in order for them to be used in the teaching of marketing at the tertiary level. The research also forms part of a project for an intelligent tutoring system. It draws terms from existing textbooks and prior research to produce a list of 593 terms. The terms are evaluated for presence using frequency and TFIDF across a corpus of 227 first-year marketing assignments. The results reveal a high proportion of term usage, although this was not across all terms. TFIDF provided additional insights into term usage among the selected terms evaluated across the corpus. It is believed such lists can be used to inform shortfalls in areas of assignment (e.g., concepts not being addressed) and in the development of intelligent tutor systems, which can provide feedback to students on topic relevance in assignments. Being able to measure discipline knowledge is an important step in being able to assure employers of a student's preparedness for work and is a key challenge for most Universities today.

27.1 Introduction and Background

The development of the web and new technology to assist learning has emerged as an important area in pedagogy, teaching, and learning [15]. The move to the web as a source of information and for teaching and learning has meant there has been an explosion in the volume of textual data that is available for analysis and the opportunity to explore all aspects of assessment. A broader area that has been founded is learning analytics which is described as the measurement, collection, analysis, and reporting of data about learners and their contexts [17].

Yet, the use of data mining and machine learning tools in education is still relatively nascent. One of the issues has been the accessibility of the tools and an understanding of the techniques used by data scientists. Recently, however, there have emerged a number of tools which, when integrated with friendlier computer programs, like Python and R mean access to these tools is more readily available [18]. Programs in natural language processing (NLP), for example, have facilitated the widespread use

Á. Rocha et al. (eds.), *Marketing and Smart Technologies*, Smart Innovation, Systems and Technologies 167, https://doi.org/10.1007/978-981-15-1564-4_27

of textual analysis of social media posts, websites, customer reviews, and other digital documentation [14, 16]. This has been supported with rapid developments in machine learning and data science, which have provided further tools and algorithms to assist in the processing of this explosion of data. Such tools are now available to the field of education [18].

In being able to evaluate the development of learning in a discipline, such as marketing, there needs to be a foundation upon which knowledge can be benchmarked. The teaching of discipline is often based on textbooks, readings, and documents, which summarize, introduce and explain concepts that are to be studied and learned by students. These texts provide one means of capturing information that forms the basis of discipline and the concepts that are incorporated into the knowledge base.

This paper reports on research currently underway that aims to refine a list of terms that represent the marketing discipline in order for them to be used in the teaching of marketing at the tertiary level. The research also forms part of a project for an intelligent tutoring system. The paper consists of six sections; the first introduces the concept and development of an ontology for the marketing discipline. The methodology used in identifying terms for the ontology is then discussed. In selecting terms, the paper proposes the use of a term frequency-inverse document frequency (TFIDF) score and presents the results of an analysis of data extracted from marketing textbook glossaries. The findings are then discussed, followed by limitations, with the conclusions presented in the final section.

27.1.1 Development of an Ontology

As a starting point, it was considered appropriate to consider what terminology and knowledge students have on entering a course. By capturing their use of marketing terminology in an assignment setting one could obtain an insight into which concepts are used by students. While there are numerous machine learning tools that are able to provide summaries of terms used in documents, they do not provide a comprehensive set of relevant discipline terms. However, this is not to say they are not able to assist in evaluating terms once identified. The approach taken in this project was to develop a list of terms based on an ontology.

An ontology represents the concepts and the relationship between them for a specialized domain [10]. In computer science, the term ontology refers to all the core concepts, including their terms, attributes, values, and relationships relating to a specified knowledge domain. Introduced in data mining in early 2000 ontologies can be used that are based on domain knowledge, for data mining processes or with metadata [22]. Their popularity is based on them being able to transfer knowledge domains between people and applications.

At the simplest level, an ontology consists of a list of terms that represent those used in discipline. At a more detailed level, they can be grouped into levels and sub-levels. For example, based on the Australian Business Deans Council 2012 report which provided a framework for threshold teaching standards in marketing (ABDC, 2012) five areas could be used for an ontology. The framework includes communication, judgment, knowledge, analysis, and social responsibility.

Yet there are few marketing ontologies that have been developed and those that have are for specific purposes. For example, Yilmaz and Alpetekin [22] developed an ontology-based data mining approach for strategic marketing decisions that were developed for a mobile phone user database. Likewise, Pinto, Gago and Santos [11] developed an ontology for database marketing. The existing public databases, Ontolingua and DAML, were also searched without success. This is despite ontologies being used in other disciplines for teaching purposes including in nursing [3], design [4] and distance learning [2].

For this project, previous research had developed a list of 100 marketing terms based on a survey of marketing academics [20]. This list had been tested as having highly relevant terms relating to marketing. However, while the list was a helpful starting point it was not systematically tested against any authoritative sources nor had a thorough theoretical foundation. This research sought to expand on this prior research and validate the use of terms identified. The following section outlines the methodology for the study.

27.2 Methodology

To establish the ontology of marketing discipline terms, two textbooks were identified for use. The first was an introductory Marketing Principles text [1] which was widely used in marketing courses in Australia. The text was an adaptation of a popular US-based text. The second text was a third-year Strategic Marketing text [12]. It was believed an introductory text was a good basis for developing the list as it was comprehensive and included a considerable glossary. By contrast, the second text was chosen because it had a stronger strategic marketing focus and was used in a final year subject for the course. It is believed the text gave additional depth to the subject content. These two textbooks, while not necessarily being representative of all marketing textbooks were chosen as 'representative' of other marketing textbooks. Future research could explore other textbooks for additional terms.

A list of terms was obtained by extracting the glossary terms from each textbook. The glossary is the authors summary of key terms featured in the text. In the Armstrong text, there were 368 terms. These included single and multi-word concepts. This list was then broken down into the single term words resulting in 418 individual terms. By comparison, the glossary in the second textbook was much shorter—resulting in a total of 200 terms. For the purposes of this analysis only the single term words were used as these form the basis for any multi-word terms. In addition, it was noted that many of the multi-term words were adjective-noun combinations (e.g., 'causal research') or terms that gave context to the concept being presented.

Only 82 words were common in the two textbook lists, so the two lists were combined resulting in a textbook list of 536 marketing words. The textbook list was then compared to an earlier developed list of 100 words [20] obtained from marketing academics (expert list). The expert list revealed an additional 57 unique marketing terms, which were added to the textbook marketing term list. The three lists were combined and resulted in a combined set of 593 unique marketing terms. The list is available from the author on request.

The combined list was then used as the basis for evaluating marketing term usage in 227 first-year marketing assignments. The assignment required students to write a 1500-word report based on the marketing issues facing the food manufacturing industry in Australia. More particularly a hypothetical scenario was presented to the students: A large investor company wishes to examine the current market for biscuit (cookie) manufacture, with a view to investing in this industry. The guidance for the assignment required them to research the industry using online resources such as websites, journal articles, newspaper articles, and databases and present a report that addresses three criteria. The first was to perform a macro-environmental analysis of the industry, second to identify and describe the potential market segments (and segmenting variables) used in the industry and finally to make a recommendation as to the segment of the market they thought the company should look to invest in.

The pre-processing of the assignments after being downloaded from the LMS included removing the reference list, removing tables and images, converting the documents to a text format, and then concatenating them into a single file. For the purpose of this exercise, the assignments were not treated individually but instead as a single large corpus. The file was first sentence tokenized to obtain 'documents' and cleaned by converting text to lower case, removing numbers and punctuation. The sentences formed the basis for documents that were further analyzed as part of the project. The sentences were further tokenized into individual words. See Table 27.1 for a summary of the document statistics.

Table 27.1. Corpus statistics

Assignments	227
Sentence documents	13,375
Total tokens	235,480
Number of unique words extracted	8,989
Most frequent word—the	9,146
Most frequent marketing term—industry	3,188
Average word frequency	26.20

27.3 Determination of TFIDF Scores

For this project stop words were not removed before tokenization as this can change the frequency of terms in the corpus which affects the TFIDF score. Consequently, the objective was to maintain the complete text of the documents [9, 21]. Hyphenated bigrams that were included as part of the textbook lists were maintained.

To determine the TFIDF scores a tool called Gensim was used [13]. Gensim allows for the quick training of a TFIDF model using the whole corpus. Each document was transformed into a bag-of-words (BOW) vector by counting each word (TF), and then multiplying it by the word's inverse document frequency (IDF). TF is the raw count of a word in a document while IDF is a measure of the terms use across all documents. It is calculated as the log of the total number of documents in the corpus divided by the number of documents containing the term.

Hence, TFIDF is calculated as

$$TFIDF = TF\,x\,IDF \tag{27.1}$$

A term that has a high frequency in a document, but with a low overall frequency across all document results in a higher TFIDF score. As a result, common terms, such as stop words, tend to be filtered out and have lower weights. Scores range from zero to one.

27.3.1 Salience of Terms

A total of 8989 terms were extracted from the assignments, with just on 41% being used only once. A further 3191 terms were used up to ten times while only 2,116 (23.5%) were used more than ten times. As the analysis calculates an individual TFIDF score for each term, the maximum TFIDF score was used as a measure of salience [8]. The top 25 terms indicating the highest frequency and highest TFIDF score are presented in Table 27.2.

Table 27.2. Top 25 terms based on frequency and maximum TFIDF

Terms in assignments	Average TFIDF	Max TFIDF	Term Freq	Terms in assignments	Average TFIDF	Max TFIDF	Term Freq
Industry	0.113	0.8159	3188	Marketing	0.175	1.0000	568
Market	0.134	0.9661	3082	Targeting	0.197	1.0000	513
Segmentation	0.188	0.9520	1074	Growth	0.221	1.0000	430
Environment	0.204	0.8585	851	Introduction	0.265	1.0000	293
Factors	0.190	0.8264	823	Online	0.332	1.0000	46
Segment	0.186	0.6270	775	Total	0.383	0.9938	204
Product	0.163	0.5846	761	Share	0.265	0.9851	407
Analysis	0.193	0.8247	691	Sales	0.229	0.9844	360
Marketing	0.175	1.0000	568	Reference	0.319	0.9812	35
Targeting	0.197	1.0000	513	Structure	0.325	0.9800	59
Demographic	0.200	0.7369	502	Distribution	0.288	0.9715	123
Consumer	0.178	0.6567	491	Legal	0.269	0.9706	125
Economic	0.212	0.7377	490	Market	0.134	0.9661	3082
Business	0.199	0.5752	460	Line	0.318	0.9574	84
Growth	0.221	1.0000	430	Segmentation	0.188	0.9520	1074
Target	0.191	0.8484	419	Political	0.219	0.9498	410
Political	0.219	0.9498	410	Control	0.283	0.9494	62
Share	0.265	0.9851	407	Value	0.291	0.9092	343
Income	0.215	0.7938	379	Strategic	0.344	0.8811	26
Technological	0.219	0.7608	378	Environment	0.204	0.8585	851

(continued)

Table 27.2. (*continued*)

Terms in assignments	Average TFIDF	Max TFIDF	Term Freq	Terms in assignments	Average TFIDF	Max TFIDF	Term Freq
Manufacturers	0.191	0.7071	377	Target	0.191	0.8484	419
Production	0.200	0.5276	374	Cookies	0.312	0.8370	300
Demand	0.209	0.5260	366	Concentrated	0.331	0.8307	58
Sales	0.229	0.9844	360	Factors	0.190	0.8264	823
Value	0.291	0.9092	343	Analysis	0.193	0.8247	691

While it would be possible to sum the scores, or take the average score for each term, an examination of the results revealed that taking the maximum TFIDF was more informative in understanding the distribution of terms across the corpus.

27.4 Findings and Discussion

For this project, our interest was on the performance of the key marketing terms that had been extracted from the textbooks and the expert list obtained from academic opinion. This list totaled 593 unique terms. However, of the terms in the list, only 316 had appeared in the corpus, which represented only 6% of the total number of terms in the corpus with a frequency greater than one.

The results show that the most frequently used terms appeared to represent common terms from the marketing discipline—for example ‘segmentation’, ‘product’, ‘marketing’, ‘targeting’, and ‘consumer’ while other terms appeared to be more representative of business in general, such as ‘industry’, ‘environment’, ‘factors’, ‘analysis’. By and large, the 25 most frequent words appeared to adequately represent the type of terms that would be used in the assignment set. Of note, is that the terms with the highest frequency had lower TFIDF scores when compared to those lower in the list which is consistent with the intent of the TFIDF formula.

Turning to the top 25 terms with respect to TFIDF scores a review of the terms appear to have a stronger ‘marketing’ flavor with a less business emphasis as evidenced by the inclusion of ‘online’, ‘share’, ‘political’, ‘distribution’, and ‘strategic’. It is also apparent that using the top scores also includes terms with much lower frequencies indicating they were not used as frequently as other terms.

While it is apparent that common marketing terms were included in the corpus as would be expected, the analysis also reveals insight into the range of terms both being used and not being used. The absence of terms could indicate that the assignment did not capture the concepts being covered by these terms, for example, as might be expected to appear in more advanced marketing subjects. Likewise, the exercise has been informative in terms of the breadth of term usage. There were only 40 terms used from the list that had a frequency greater than 200, or one per assignment. This appears relatively narrow and may provide an opportunity for further investigation among other groups of students.

27.5 Limitations and Future Scope of Research

It is acknowledged the use of textbooks as a source for discipline-specific terms has limitations. For example, textbooks may be outdated because of the production process with texts often having lead times of more than 12 months and only being revised every 3–5 years. While the current textbooks were the most recent editions, they had been in the market for several years. An alternative to using terms in textbooks would be to use more current works, such as lists available on websites and blogs, or from the analysis of recent journal articles. While these approaches may provide additional terms, they may not be what is currently being used for instruction in the class and known to students.

It is also recognized there is considerably more refinement that can be undertaken in reviewing the current list. For example, consideration for other lists, such as the academic word list [5] and the ABDC threshold learning standards referred to earlier could be considered. In addition, refining the key terms based on parts-of-speech tags or grouping terms based on disciplinary area as is often undertaken in the development of ontologies provides opportunities for future research.

The development of employability skills in students is a key challenge for Universities [19]. The use of computers for detailed analysis of assignments and student's work provides teaching staff with the ability to obtain a greater understanding of student performance and concept understanding. The development of such terms lists can inform emerging tools, such as intelligent tutoring systems, and enable more detailed exploration of student performance so that it can be incorporated into evidence-based learning and used to evaluate assessment tasks [6]. While developing discipline-specific knowledge will not necessarily be sufficient alone [7], it is believed building a strong understanding of discipline terminology enables students to develop their confidence and graduate capabilities and provides them with skills that will allow them to transition into employment more readily.

27.6 Conclusions

Machine learning algorithms, such as TFIDF, have a great deal to offer teachers and academics as a means of understanding student learning and discipline understanding. In this study, TFIDF has been a useful tool in highlighting the importance of particular terms being used by students in their assignments for a marketing project. The study has found for this first-year cohort that over half of the identified marketing terms were used. This provides a good baseline for measuring the development and use of marketing terms by students as they progress through their course. More generally, the project has been informative in terms of which terms have and have not been used in the assignments. This information can be used by teaching staff and curriculum designers in the development of assignments and tasks, which can support student learning.

References

1. Armstrong, G., Adam, S., Denize, S., Volkov, M., Kotler, P.: Principles of Marketing. 7th edn. Pearson Education Australia (2017)
2. Boufardea, E., Garofalakis, J.: A predictive system for distance learning based on ontologies and data mining. In: The Fourth International Conference on Advanced Cognitive Technologies and Applications, pp. 151–158 (2012)
3. Clarke, P.: Discipline-specific knowledge: time for clarity. Nurs. Sci. Q. **25**(2), 149–150 (2012)
4. Cliff, A., Woodward, R.: How do academics come to know? the structure and contestation of discipline-specific knowledge in a design school. High. Educ. Int. J. High. Educ. Educ. Plan. **48**(3), 269–290 (2004)
5. Coxhead, A.: A new academic list. TESOL Q. **34**(2), 213–238 (2000)
6. Gutierrez, F., Dou, D., Fickas, S. Griffiths, G.: Providing grades and feedback for student summaries by ontology-based information extraction. In: Chen, X., Lebanon, G., Wang, H., Zaki, M. (eds.) Proceedings of the 21st ACM International Conference on Information and Knowledge Management. ACM, Maui, HI, USA, October 29–November 2, 1722–1726 (2012)
7. Hager, P., Holland, S., Beckett, D.: Enhancing the learning and employability of graduates: the role of generic skills. Business/Higher Education Round Table Position Paper No. 9. Melbourne, Australia (2002)
8. Horn, F., Arras, L., Montavon, G., Muller, K., Samek, W.: Exploring text datasets by visualizing relevant words. arXiv:1707.05261v1 [cs.CL] 17 Jul. Accessed 05 July 2019
9. Manning, C., Raghavan, P., Schutze, H.: Introduction to Information retrieval. Cambridge University Press, New York, NY (2008)
10. Noy, N., McGuinness, D.: Ontology development 101: a guide to creating your first ontology, stanford knowledge systems laboratory technical report KSL-01-05 and stanford medical informatics technical report SMI-2001-0880, March (2001)
11. Pinto, F., Gago, P., Santos, M.: Marketing database knowledge extraction—toward a domain ontology. In: International Conference on Intelligent Engineering Systems. Barbados, 16–18 April. https://doi.org/10.1109/INES.2009.4924761 (2009)
12. Reed, P.: Strategic Marketing: Decision Making and Planning, 4th Ed. Cengage Learning Australia (2014)
13. Rehurek, R. Sojka, P.: Software Framework for Topic Modelling with Large Corpora. In: Proceedings of the LREC 2010 Workshop on New Challenges for NLP Frameworks, pp. 45–50, Valletta, Malta, May. ELRA (2010)
14. Salloum, S., Al-Emran, M., Abdel Monem, A., Shaalan, K.: A survey of text mining in social media: facebook and twitter perspectives. Adv. Sci. Technol. Eng. Syst. J. **2**(1), 127–133 (2017)
15. Shermis, M. D., Burstein, J.: Handbook of Automated Essay Evaluation: Current Applications and New Directions. Routledge (2013)
16. Shu, K., Sliva, A., Wang, S., Tang, J., Liu, H.: Fake news detection on social media: a data mining perspective. ACM SIGKDD Explor. Newsl. **19**(1), 22–36 (2017)
17. Siemens, G.: LAK'11, 1st International Conference on Learning Analytics and Knowledge 2011. https://tekri.athabascau.ca/analytics. Accessed 5 July 2019
18. Slater, S., Joksimovic, S., Kovanovic, V., Baker, R., Gasevic, D.: Tools for educational data mining: a review. J. Educ. Behav. Stat. **42**(1), 85–106 (2017)
19. Treleaven, L., Voola, R.: Integrating the development of graduate attributes through constructive alignment. J. Mark. Educ. **30**, 160–173 (2008)

20. Vitartas, P., Ahmed, T., Alahakoon, D., Midford, S., Nathawitharana, N., Ong, K.L., Sullivan-Mort, G.: Using learning analytics to guide learning: an analysis of marketing assignments. In: Fortin, D., Ozanne, L. (eds.) Marketing in a Post-Disciplinary Era. Proceedings of the Australian and New Zealand Marketing Academy Conference (ANZMAC), Christchurch, 5–7 December, pp. 623–629 (2016)
21. Yang, Y., Pedersen, J.: A comparative study on feature selection in text categorization. In: Proceedings of the Fourteenth International Conference on Machine Learning, ICML '97, San Francisco, CA, pp. 412–420 (1997)
22. Yilmaz, N., Alptekin, G.: An ontology-based data mining approach for strategic marketing. In: Proceedings of the World Multi-Conference on Systemics, Cybernetics and Informatics: WMSCI 2013, Orlando, Florida July 9–12 (2013)

Chapter 28
Developing a Web Application for Recognizing Emotions in Neuromarketing

Filip Filipović, Luka Baljak, Tamara Naumović, Aleksandra Labus, and Zorica Bogdanović(✉)

Faculty of Organizational Sciences, University of Belgrade, 11000 Belgrade, Serbia
{filipfilipovic, lukabaljak, tamara, aleksandra, zorica}@elab.rs

Abstract. The subject of this paper is the development of a web application for recognizing emotions in neuromarketing. The goal is to develop an open system that should enable identification of user emotions using a webcam when accessing web content. Recognition of user emotions will be based on computer vision algorithms and machine learning. The developed web applications will serve as the basis for the development of a system that enables personalization of Internet marketing services based on user emotions in near real time. Evaluation of the developed solution has been done through an experiment. Results show that the users' emotions can be identified using the developed system, with a satisfactory level of precision.

28.1 Introduction

Companies frequently face changes in how consumers make decisions, as well as changes in the buying habits they have. The need to understand the consumer's behavior rises, because companies are trying to ensure that certain products are available at the appropriate price and at the appropriate locations while stimulating purchasing through targeted promotions [1].

Advertising the product in a certain way can cause desirable consumer emotions and increase the likelihood of purchasing a product or service. Several problems have been recognized: the identification of desirable emotions, the way of creating advertising material to stimulate those emotions and evaluation of the success of the advertisement. Neuromarketing tools and techniques are considered to be a possible solution.

The area of neuromarketing is developing fast and neuromarketing research methods gradually become standard tools for research in various areas [2]. The application, described in the paper below, offers support for the research and evaluation in the area of personalized advertising. Personalization allows us to treat the people we're marketing to more like individuals. The goal of personalization isn't to engage with one audience of many, but to engage with infinite audiences of one. Nearly three-

Á. Rocha et al. (eds.), *Marketing and Smart Technologies*, Smart Innovation, Systems and Technologies 167, https://doi.org/10.1007/978-981-15-1564-4_28

fourths (74%) of online consumers get frustrated with websites when the content has nothing to do with their interests [3].

The goal of this paper is to develop a web application that enables detecting user emotions while exposed to certain advertising content, with the final goal to personalize future content based on detected emotions.

28.2 Related Work

28.2.1 Neuromarketing

Neuromarketing was created by the historical merging of science and marketing [4]. David Lewis, a French neuropsychiatrist, is the first author to combine these two sciences [5]. Today, neuromarketing is an area that applies the principles of neuroscience for marketing research. It studies the sensorimotor, cognitive and affective responses of consumers. Due to its ability to understand consumer sentiment, neuromarketing is increasingly used in the field of marketing [1].

In neuromarketing research, neuroscientific methods and techniques are used to analyze and understand consumer behavior, as well as appliances that have been used for medical purposes for years. Techniques such as brain imaging (fMRI, PET, EEG) and physiological measures (heart rate, respiratory rate, and galvanic skin reaction) are frequently used [6], as well as eye motion tracking and emotion detection.

Eye motion monitoring. This technique enables the creation of an eye-movement pattern. The possibilities of applying this technique are in the commercial domain, and frequently for online shopping. It can also be used to research to what extent people pay attention to the product being advertised. The equipment used to monitor the movement of the eyes is very light and portable, it is possible to monitor in real time and register the natural view of the respondents [7]. Using the results of the application of this technique, it is possible to obtain the patterns of the view of the respondents, such as the attention of the respondents to the advertising labels.

Face expression monitoring. The idea of concluding on the feelings of people based on facial expressions was first initiated by Charles Darwin in 1872, and the idea has since been dealt with in detail by a large number of psychologists [7]. Equipment used to monitor facial expressions can measure subtle, often subconscious reactions to stimuli that contain information about what a person feels about something. In this way, the behavior that follows after these reactions can be predicted [7]. Facial expressions can be determined in different ways. The precise mean is facial electromyography (Facial EMG) to measure the activity of muscles. Data is collected by measuring electrical impulses caused by muscle movements [8]. However, this approach cannot easily be implemented to collect user data in real time, so for our application, we have chosen an approach based on machine learning and neural networks. Neural networks can be trained by large amounts of images to recognize specific emotions. Recognition can be done by comparing the learned facial expressions, which we are sure to represent an emotion, with new expressions that the application has collected. Some of the real-world use cases are detection in interviews, testing for video games, market research, and smart cars [9].

28.2.2 Neural Networks

The history of neural networks begins in the early 1940s when Warren McCulloch and Walter Pitts introduced models of neurological networks and showed that even simple networks are able to calculate nearly any logic or arithmetic function [10]. A technical neural network consists of the neurons (simple processing units), and directed, weighted connections between those neurons [11]. A neural network is a set of connected neurons organized in layers:

- **input layer**: brings the initial data into the system. That data is further processed by subsequent layers of artificial neurons.
- **hidden layer**: between input layers and output layers. Their artificial neurons take in a set of weighted inputs and produce an output through an activation function.
- **output layer**: the last layer of neurons that produces given outputs for the program [12].

There are many types of neural networks, but for our application, convolutional neural networks are of key importance. A Convolutional Neural Network, also known as CNN, is specialized in processing data that has a grid-like topology such as an image. Our application should analyze the images of the respondents, captured during the viewing of advertising materials. The benefit of using CNNs is their ability to develop an internal representation of a two-dimensional image. This allows the model to learn position and scale invariant structures in the data, which is important when working with images [13].

Neuron in a CNN processes data only in its receptive field (restricted region of the visual field). The layers are arranged in such a way so that they detect simpler patterns first (lines, curves, etc.) and more complex patterns (faces, objects, etc.) further along [14].

28.3 Designing an Application

The goal of the research is to design an application that will serve as a tool in neuromarket research. The platform should provide features that enable detecting user's reaction to specific content, and reports on the results. The basic functionality of the application is to identify the emotions of the respondents during the review of the content such as image, advertisement or web content. These emotions will be the main factor in evaluating the success of the content. The architecture of the system is shown in Fig. 28.1.

The application has two types of users: a company and a consumer. The consumer has one role, which is the evaluation of advertising content published by companies. The user actively evaluates the content by filling in the poll, and also allows the company to detect emotional reactions to the presented content. Companies place advertising material on the platform and monitor the results of active research.

The evaluation of advertising materials by consumers is carried out as follows:

1. The consumer chooses the advertisement they want to see, independently or based on the recommendation of the system
2. The app brings content and accesses the camera of the consumer's devices
3. The camera begins to record the consumer while advertising content is displayed
4. Depending on the type of content (video or image), the consumer's facial expressions are captured
5. These images are analyzed and the results are sent to the database. The results are used for the personalization of future content.

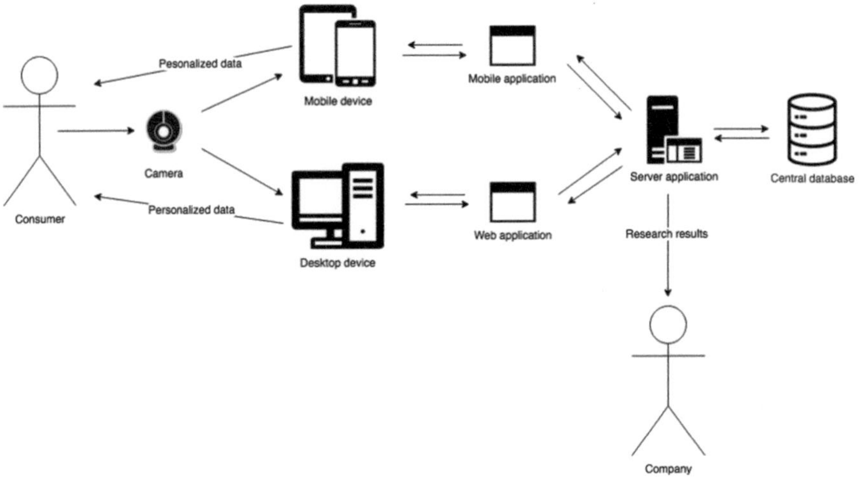

Fig. 28.1. System architecture

The set of technologies chosen for the design of the platform is MEAN stack. The abbreviation represents a package of the following technologies: Angular, Express, Node.js, and MongoDB. All parts of the application have been written in the same language, in this case, JavaScript.

In the database, we store the following data collections: Users, Ads, Surveys, and Feedback.

Face-api.js is a JavaScript API used for recognition of facial expressions. It is made with the tenserflow.js core API. It uses a set of convolutional neural networks (CNN). It implements the following models: SSD Mobilenet V1, Tiny Face Detector, and MTCNN. In this application, for face detection, a Tiny Face Detector model was used.

Tiny Face Detector is a model for real-time face detection. This is the ideal model for our case, due to the small size, speed, and moderate resource consumption [15]. The model is compatible with the web and mobile devices and is, therefore, the most suitable for our application, which is intended to function on both platforms. The model is powered by a 14K image dataset and is a smaller version of the Tiny Yolo V2.

In addition to the Tiny Face Detector model, Face Expression Recognition and Age and Gender Recognition models were used. The Face Expression Recognition model has emerged from publicly available datasets and recognizes the following emotions: sleeplessness, anger, sorrow, excitement, disgust, fear and neutrality. For each of the emotions analyzed on the basis of the image, it prints a percentage. The Age and Gender Recognition model recognizes the gender and age of the respondents with a very high degree of accuracy [16].

Face-api.js returns the results for a particular image in JSON format. The result of the results varies depending on the selection of the models used for the analysis. In our case, the result returns a set of information, each element containing the following:

```
{
        age: number,
        detection: { information about the analyzed image and the de-
tected person's face},
        expressions: {
                angry: number,
                disgusted: number,
                fearful: number,
                happy: number,
                neutral: number,
                sad: number,
                surprised: number (the percentage of the security of the
model that the detectives have a particular emotion)
        }
        gender: string,
        genderProbability: number

}
```

It is necessary to state that the respondent is treated differently depending on whether the advertisement is a picture or a video. In the case of an image, we will analyze the first 5 s of the reaction of the respondents, and in the case of video, the respondent is observed in the duration of the same. For every 500 ms, the application takes a photo of the subject and analyzes the image. Finally, it stores all the data into one object and sends them to the server. In the case of video, it is necessary to record the moment when the emotions are detected.

The server application is separate from the client and has a REST architecture (see Fig. 28.2). REST architecture allows us to easily communicate between server and client applications that are intended for multiple platforms. The basic client application is the SPA (Single-Page Application) web application created in the Angular Framework, which will be described in detail in the next section. The software architecture of the system is shown in Fig. 28.3.

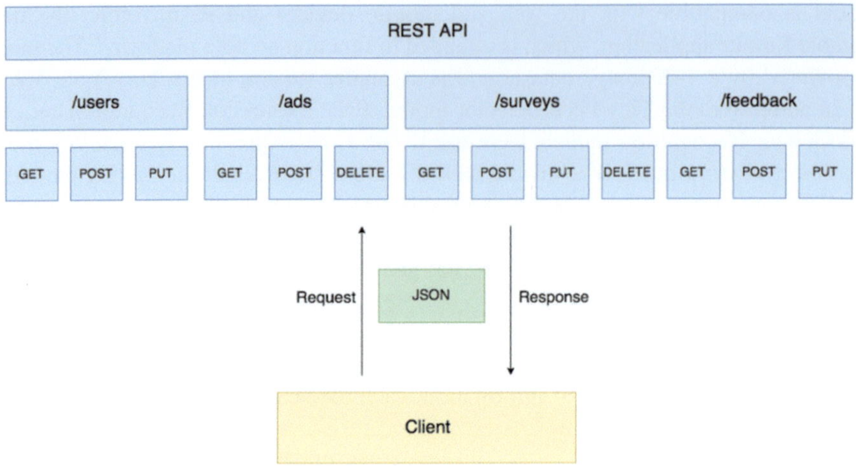

Fig. 28.2. REST architecture of application

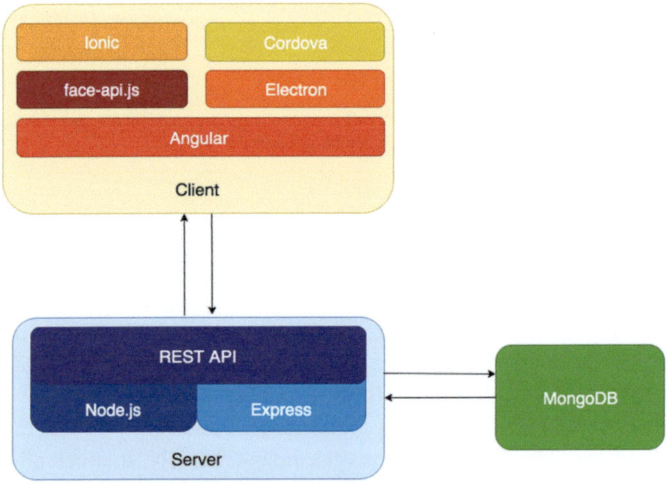

Fig. 28.3. Software architecture of the system

The appearance of the application should be in accordance with the principles of attention-driven design. Attention-Driven Design is a disciplined adherence to visual design principles in order to reduce visual complexity and enhance the ease of comprehension [17]. The goal is to make the respondent feel relaxed when using the application and to focus on the research as much as possible.

List of main application views:

1. Home view—Initial view for authentication of users, companies, and consumers (respondents). Access to other parts of the application is not allowed without authentication.

2. Profile view—Used to enter information about the user.
3. Listing view—Display of advertising materials intended for consumers. It is possible to search and filter advertisements by parameters, which are calculated or entered by companies.
4. Ad view—The display of the selected advertising material, whereby advertising is collected in the advertisement.
5. Administrative view—A section dedicated to companies, within which it is possible to add and manage advertising material. Also, they are graphically displayed feedback ads, based on which the company has an idea of the effect of its advertising materials.

The developed web application for neuromarketing research has a unique design that guides the administrator through three phases of research creation:

1. Creating an ad in the research (Fig. 28.4).
2. Creating a research survey (Fig. 28.5).
3. Publishing the ad and enabling the survey (Fig. 28.6).

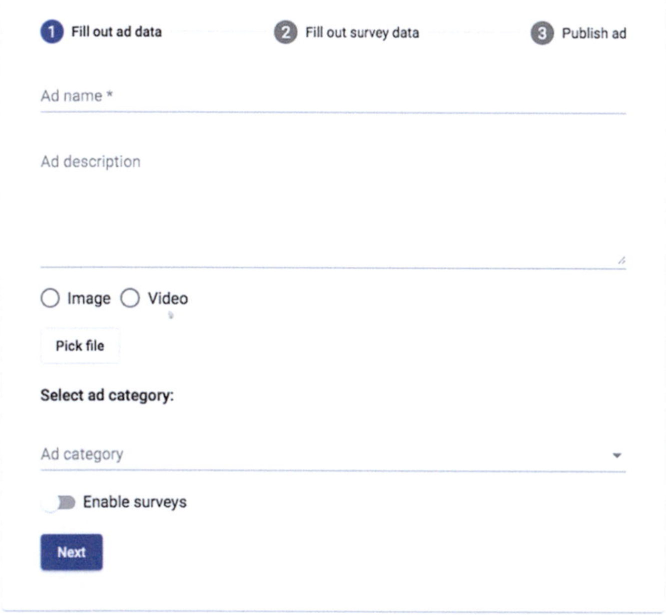

Fig. 28.4. Creating an ad in the research

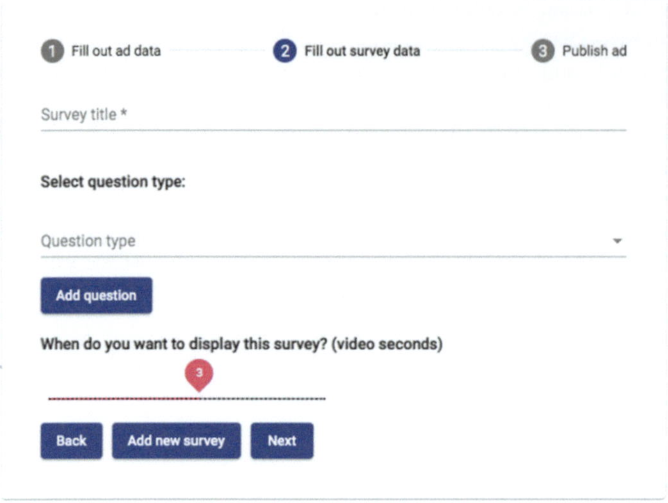

Fig. 28.5. Creating a research survey

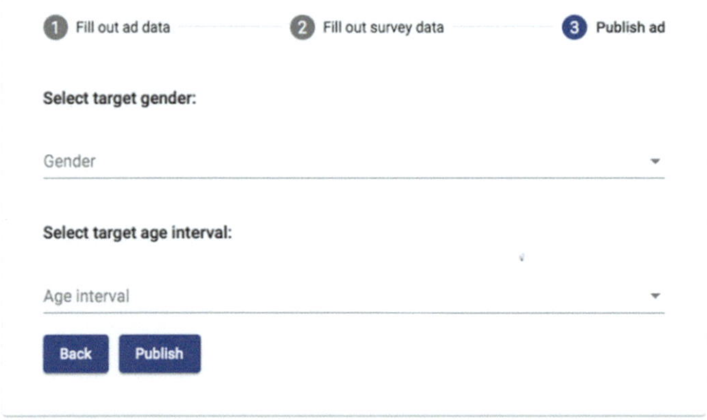

Fig. 28.6. Publishing the ad and enabling the survey

After the publication of the advertisement, respondents who belong to the target group can start evaluating. The interface showing the list of available ads is shown in Fig. 28.6. After the interviewer chooses and evaluates the advertisement, the assessment data is entered into the database and becomes available to the company. The interface that displays the analyzed data will be displayed within the experiment in the next section (Fig. 28.7).

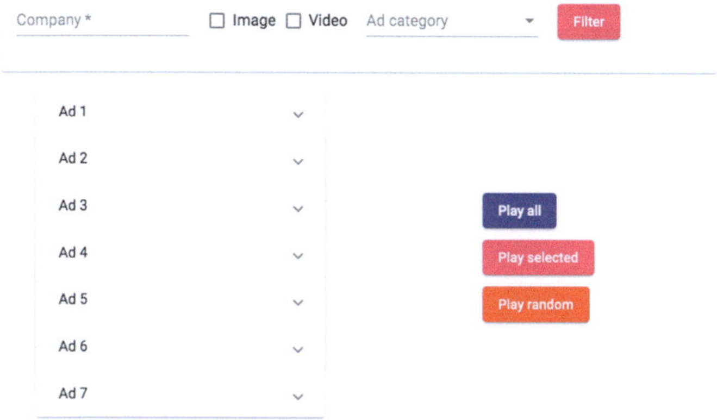

Fig. 28.7. Ads listing view of the application

28.4 Evaluation

We will test the application over one male respondent who will be shown a picture (Fig. 28.8).

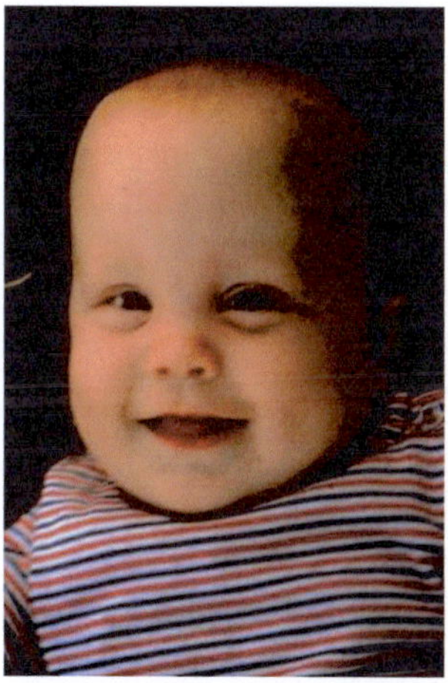

Fig. 28.8. Picture used for application testing

The data retrieved during the research will be evaluated and the emotions recognized. It is expected that the presented figure induces the emotion of happiness. The results are presented in Fig. 28.9.

Male 45-50 years

Second of video	Angry	Disgusted	Fearful	Happy	Neutral	Sad	Surprised
0.5	≈0%	≈0%	≈0%	9.81%	90.17%	≈0%	≈0%
1	≈0%	≈0%	≈0%	40.22%	59.74%	≈0%	≈0%
1.5	≈0%	≈0%	≈0%	97.42%	2.09%	0.47%	≈0%
2	≈0%	≈0%	≈0%	46.39%	52.55%	1.04%	≈0%
2.5	≈0%	≈0%	≈0%	90.79%	8.44%	0.74%	≈0%
3	≈0%	≈0%	≈0%	27.8%	71.91%	0.27%	≈0%
3.5	≈0%	≈0%	≈0%	3.49%	96.46%	0.03%	≈0%
4	≈0%	≈0%	≈0%	9.06%	90.63%	0.33%	≈0%
4.5	≈0%	≈0%	≈0%	11.18%	88.75%	≈0%	≈0%
5	≈0%	≈0%	≈0%	0.02%	99.96%	≈0%	≈0%

Fig. 28.9. Tabular view of result data

On the basis of the obtained data, we see that happiness is the dominant emotion in the respondent. The percentage of certainty of the detected emotion varies while the subject changes the facial expression. Two moments are observed when the program was almost completely sure that the respondent was happy, and that was after 1.5 and 2.5 s. We can also notice the growth of happiness at the beginning of the test and the fall in the end, to the extent that the program is sure that the respondent does not show any emotion.

28.5 Conclusion

Neuromarketing is a developing research field, constantly expanding with new methods and tools [2]. Both neuromarketing and traditional marketing research, give us a perspective of the consumers decision-making process, where we can acknowledge different emotions like consumer's fatigue, stress, and even aggression. Those emotions are a key component in understanding consumers' perception of the product/service, brand, or a company itself.

In the previous section, the experiment showed us that the developed application functionally meets the basic requirement. The implementation of additional functionalities is foreseen in order to increase the benefits for the researcher. Future improvements will include: new visual representations, statistics, and the collection of new types of data such as eye-movement monitoring and speech emotion recognition.

Speech emotion recognition, which aims to recognize emotion states from speech signals, has been drawing increasing attention. Speech emotion recognition is a very challenging task of which extracting effective emotional features is an open question [18, 19]. The platform should be adapted to the respondent, depending on their mood and personality. The core of this requirement would be the implementation of a high-quality recommender system based on biofeedback [20, 21]. Content-based recommendation systems are very important for our application. Content-based recommendation systems try to recommend items similar to those a given user has liked in the past [22].

The main contribution of this paper is the development of the system for emotion recognition for neuromarketing research. This system allows us to gather and analyze data about consumers' emotions toward the shown advertisement image or video. The provided analysis gives companies the insight in consumers' emotional response to their product/service placement which can later lead to the purchase.

Acknowledgements. The authors are thankful to the Ministry of Education, Science and Technological Development, Grant no. 174031.

References

1. Arthmann, C., Li, I.-P.: The Art and Science of Marketing and Neurosciences Enabled by IoT Technologies (2017)
2. Vasiljević, T., Bogdanović, Z., Rodić, B., Naumović, T., Labus, A.: Designing IoT infrastructure for neuromarketing research. In: World Conference on Information Systems and Technologies, pp. 928–935. Springer, Cham (2019)
3. Devaney, E.: How to master personalized marketing. HubSpot (2014)
4. Krajnović, A., Šikirić, D., Jakšić, D.: Neuromarketing and Customers Free Will. University of Zadar (2012)
5. Meola, A.: What is the Internet of Things (IoT) (2016)
6. Jelić, N.: Bihevioralna ekonomija, neuroekonomija, neuromarketing. JAHR-Eur. J. Bioeth. (2014)
7. Reitsma, L.: Five neuromarketing techniques every marketer should know about, newneuromarketing. https://www.newneuromarketing.com/5-neuromarketing-techniques-every-marketershould-know-about (2017)
8. Dooley, R.: Facial EMG: muscles don't lie?—neuromarketing, neuromarketing. https://www.neurosciencemarketing.com/blog/articles/facialemg.html (2011)
9. Facial emotion detection using AI: use-cases, paralleldots. https://blog.paralleldots.com/product/facial-emotion-detection-using-ai/
10. McCulloch, W.S., Pitts, W., A logical calculus of the ideas immanent in nervous activity. Bull. Math. Biol. 115–133 (1943)
11. Kriesel, D.: A Brief Introduction to Neural Networks (2007)
12. Zerium, A.: Artificial neural networks explained, medium. https://blog.goodaudience.com/artificial-neural-networks-explained-436fcf36e75 (2018)
13. When to use MLP, CNN, and RNN neural networks, machine learning mastery. https://machinelearningmastery.com/when-to-use-mlp-cnn-and-rnn-neural-networks/
14. Mishra, M.: Convolutional neural networks, explained. Datascience. https://www.datascience.com/blog/convolutional-neural-network (2019)

15. Face-api.js: JavaScript Face Recognition Leveraging TensorFlow.js. https://www.infoq.com/news/2018/11/faces-api-js/
16. Face-api.js, GitHub. https://github.com/justadudewhohacks/face-api.js?files=1
17. Gardner, O.: Attention-driven design: 23 visual principles for designing more persuasive landing pages. Unbounce (2015)
18. El Ayadi, M., Kamel, M.S., Karray, F.: Survey on speech emotion recogntion: features, classification schemes, and databases. Pattern Recogn. **44**(3), 572–587 (2011)
19. Schuller, B., Batliner, A., Steidl, S., Seppi, D.: Recognising realistic emotions and affect in speech: state of the art and lessons learnt from the first challenge. Speech Commun. **53**(9), 1062–1087 (2011)
20. Burke, R.: Hybrid web recommender systems. The AdaptiveWeb, pp. 377–408. Springer, Berlin, Heidelberg (2007)
21. Mahmood, T., Ricci, F.: Improving recommender systems with adaptive conversational strategies. In: Cattuto, C., Ruffo, G., Menczer, F., (eds.) Hypertext, pp. 73–82. ACM (2009)
22. Lops, P., De Gemmis, M., Semeraro, G.: Content-based recommender systems: state of the art and trends, pp. 73–105 (2010)

Chapter 29
Application of Genetic Algorithm on Multi-objective Email Marketing Delivery Problem

Lei Zhang(✉), Jun He, Zhenyu Yan, Wuyang Dai, and Abhishek Pani

Department of Media & Advertising Solutions, Adobe Systems Incorporated, San Jose, CA 95110, USA
{lzhang, juhe, wyam, dai, apani}@adobe.com

Abstract. Customer journey optimization in email marketing is the art and science of finding the optimal time to deliver a sequence of marketing messages along the consumer's decision journey. It aims to promote individual customer's engagement and experience by optimizing the message delivery schedule over a time horizon, in order to maximize the customer's opens, clicks, and avoid inducing fatigue (cancel subscription) to the customer. This paper proposed a solution by applying the genetic algorithm to optimize the customer journey with different objective functions for different business purposes. The simulation results show that the proposed method can effectively improve different business objectives.

29.1 Introduction

29.1.1 Email Marketing

Email marketing is an important medium of marketing communication especially for companies seeking to build and maintain closer relationships with customers [1]. It has at least double of return on investment than other main forms of online marketing methods, such as webpage banners, search engine marketing, keyword ads [2].

Based on a recent survey from DemandGen [3], the greatest challenges for online marketers are *content* and *timing*. Delivering attractive content in the marketing message to fit individual customer's interest is the most important and challenging point, this requires extensive research on customers' interests and cannot easily be done by an automatic marketing system.

On the other hand, delivering a marketing message at the right timing is also important. There are two major reasons: (1) Regularly sending marketing messages could maintain the buyer's brand awareness. (2) Regularly sending marketing messages could promote customer's engagement when he has an immediate need. Both brand awareness and engagement frequency are crucial to the customer's lifetime value, which leads to business success.

Á. Rocha et al. (eds.), *Marketing and Smart Technologies*, Smart Innovation, Systems and Technologies 167, https://doi.org/10.1007/978-981-15-1564-4_29

Although sending emails is free, it doesn't mean one could send emails unlimitedly. Abusing email marketing could reduce customers' experience, making them unsubscribe from the list. According to the recent research [4], the top reason for email opt-outs is excessive frequency, which counts for 46.4% of the total unsubscription population. Therefore, a successful email marketing must consider each recipient's tolerability, and optimize the email delivery frequency for each recipient base on it.

29.1.2 Existing Applications of Genetic Algorithms in Marketing

Genetic algorithms (GA) are a family of optimization methods inspired by evolution. It has wide applications in banking, engineering, finance, medical, economic, etc. [5–9]. While genetic algorithms are popular and powerful, there are limited works of its application in solving marketing problems. Hao et al. propose a GA based approach to understand customer preferences and overall satisfaction for online travel agencies [10]. Cui et al. apply GA to solve budget-constrained optimization problems for targeting high-value customers in direct marketing [11]. Mahdavi et al. formulate customer-oriented catalog segmentation problems for e-CRM and use GA to solve problems [12]. Meena and Sarmah proposed a buyer supply strategy under multiple suppliers to minimize the risks of supply disruption, and applied GA to solve it [13]. Fung et al. propose a GA based rule-mining approach to discover attributes in customer affective product design [14].

29.1.3 The Scope of This Paper

In this paper, we formulate the email marketing into an optimization problem. The objective is to improve both of the recipients' open rate and click rate by delivering the email at the best time, while prevent recipients' unsubscription by keeping the delivery frequency within their tolerance level. In order to solve this nonconvex optimization problem, we applied the genetic algorithm, which in general, can solve any form of the objective function. In the simulation part, we compare the solutions generated from the genetic algorithm and the greedy algorithm.

29.2 Problem Formulation

In this section, we formulate the email marketing problem into three mathematic optimization problems, from basic to the more complicated form of objective functions.

29.2.1 Open Rate and Click Rate Maximization (Static)

In the first problem, we want to maximize each recipient's utility function, i.e., weighted open rate and click rate over a time horizon. For each recipient, we assume every day's different open rate and click rate. Thus the optimization problem becomes to select delivery days with the highest utility values.

The problem is under two constraints:

(1) Fatigue limitation. Every recipient has his own fatigue rate, i.e. adverse rate per each delivery. For each recipient, the total fatigue rate during this horizon must be controlled within a limit, in order to prevent the recipient from unsubscribing.
(2) Total delivery limitation. This is the maximum of emails that can be sent to one recipient.

Based on the above, the problem can be formulated as

$$\text{objective} : \max \sum_{k=1}^{K} \sum_{i=1}^{N} \left(w_o \text{Open\_Rate}_{k,i}, + w_c \text{Click\_Rate}_{k,i} \right) \cdot x_{k,i} \tag{29.1}$$

s.t.

$$\sum_{i=1}^{N} f_k x_{k,i} \leq T \tag{29.2}$$

$$\sum_{i=1}^{N} x_{k,i} \leq E \tag{29.3}$$

The objective function (29.1) is to maximize weighted open rate and click rate for each recipient over the planning horizon, where i is day index; k is the recipient index; N is number of days in the planning horizon; K is the total number of recipients; $x_{k,i}$: binary decision variable $\in \{0, 1\}$, i.e., send or not send email on day i, for the kth recipient; Open_Rate$_{k,i}$ is predicted the open rate at day i, kth recipient; Click_Rate$_{k,i}$ is predicted click rate at day i, kth recipient. w_o is weighting factor for open rate; w_c is the weighting factor for click rate.

Constraint (29.2) is the recipient's fatigue constraints. The individual recipient's total fatigue probability should be under the total fatigue threshold T. f_k is fatigue probability for each delivery for kth recipient; Constraint (29.3) is the recipient's delivery constraint, where E is the total delivery number limit per recipient.

29.2.2 Maximization with Open Rate Decay

Open rate decay describes the phenomenon that recipients may be temporarily tired of opening new emails when there are too many emails sent in a short period of time. Each recipient has his own ***comfortable delivery frequency***. If the sending frequency is higher than one's comfortable frequency, his open rate would be reduced. In order to quantify this phenomenon, open rate discount function is introduced as follows:

$$y(r_i, s_k) = \begin{cases} \frac{1}{s_k}(s_k - r_i) \; if \; r_i \leq s_k \\ 1 \quad \text{otherwise} \end{cases} \tag{29.4}$$

where: r_i is the last sent time from the day i, i.e. sent recency,

$$r_i = \text{argmax}_{j \in [1,i-1]}(j|X_{k,i-1}, x_j = 1)$$

$X_{k,i-1} = \{x_{k,1}, x_{k,2}, \ldots, x_{k,i-1}\}$ is the past decision sequence from day 1 to day $i-1$; s_k is the comfortable delivery frequency of the recipient k (Fig. 29.1).

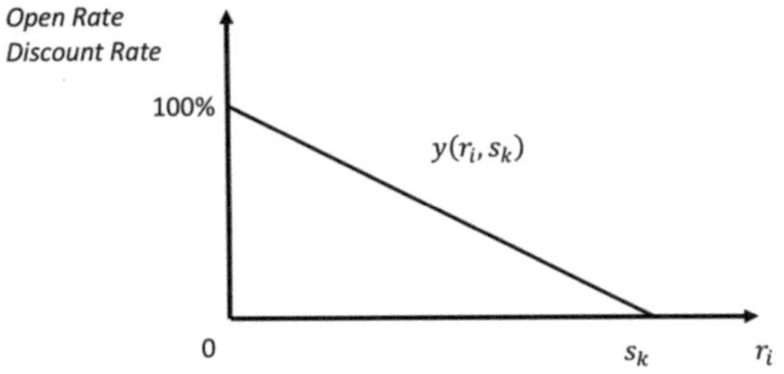

Fig. 29.1. Open rate discount function

With the introduction of open rate decay function on the open rate for each recipient, the modified formulation from (29.1) is changed to

$$\text{obj} : \max \sum_{k=1}^{K} \sum_{i=1}^{N} \left(w_o \cdot \boldsymbol{y}(\boldsymbol{r}_i, \boldsymbol{f}_k) \cdot \text{Open\_Rate}_{k,i} + w_c \text{Click\_Rate}_{k,i} \right) \cdot x_{k,i} \tag{29.5}$$

$$s.t. \sum_{i=1}^{N} f_k x_{k,i} \leq T \tag{29.6}$$

$$\sum_{i=1}^{N} x_{k,i} \leq E \tag{29.7}$$

Note that the open rate decay function was only added to the open rate. This is under the assumption that opening an email and clicking email are two independent actions. The probability of click the email would not be affected by the decay phenomenon.

29.2.3 At Least One Open/Click Probability Maximization

For email marketers, maximizing at least one open/click probability over a time horizon is an important goal. Marketers want recipients open/click emails at least one time during this period, in order to ensure that recipients receive the latest marketing information.

For each recipient, given a sequence of open/click probability $p_1, p_2, \ldots, p_n$, the at least probability can be calculated by

$$p_{\geq 1} = 1 - (1 - p_1)(1 - p_2)\ldots(1 - p_n) = 1 - \prod(1 - p_i) \tag{29.8}$$

For all recipients, the problem of at least one open/click maximization can be formulated as:

$$\max \sum_{k=1}^{K} \left\{ w_o \prod_{i=1}^{N} (1 - \text{Open\_Rate}_{k,i})^{x_{k,i}} + w_c \prod_{i=1}^{N} (1 - \text{Click\_Rate}_{k,i})^{x_{k,i}} \right\} \tag{29.9}$$

$$\sum_{i=1}^{N} f_k x_{k,i} \leq T \tag{29.10}$$

$$\sum_{i=1}^{N} x_{k,i} \leq E \tag{29.11}$$

29.3 Algorithms

29.3.1 Greedy Algorithm

Greedy algorithm is a simple, intuitive algorithm, which is used to solve optimization problems that are simple and does not contain too many local optimal points. They are easy to implement, and they are most time-efficient to solve most of the optimization problems. However, the greedy algorithm is myopic, in the sense that they take decisions based only on the best result for the current state, without worrying about the future consequence of these decisions. Therefore, for many problems, the greedy algorithm may yield a suboptimal solution.

In the email marketing problems introduced in Sect. 29.2, only Problem 1 can yield a global optimal solution. Problem 2 may yield a suboptimal solution. Problem 3 can not be directly solved by the greedy algorithm. The algorithm workflow can be described as

Greedy Algorithm for Open/ Click Rate Maximization

```
INPUT: open_rates // daily open rate, set ,indexed by day
       click_rates // daily click rate, set, indexed by day
       wo // open rate weight
       wc // click rate weight:
       N // maximum delivery number
       f // fatigue per delivery
       T // total fatigue tolerance

MAKE_SCHEDULE (INPUT)
    n = 0  // number of deliveries made
    delivery_days ={}  // planned delivery days
    tf = 0 // total fatigue rate
    score (i) = wo * open_rates (i) + wc * click_rates(i). // calculate objective function
    WHILE  n <N and tf <= T-f:
        di = argmax_i(score) //find the day index which has the highest score
        score(i) = -1  // label this day has been selected
        delivery_days += {di}
        tf +=f
        n = n + 1

    RETURN delivery_days
```

29.3.2 Genetic Algorithm

Genetic algorithm is a stochastic heuristic optimization method for solving both constrained and unconstrained optimization problems. It is based on the idea of the evolution process, "survival of the fitness". Since it is intelligent exploitation of random search provided with historical data to direct the search into the region of better performance in solution space, it is commonly used to generate high-quality solutions for large-scale optimization problems.

Figure 29.2 shows the workflow of the genetic algorithm. There are five phases in a genetic algorithm, *Initial population*, *Fitness assignment*, *Selection*, *Crossover*, and *Mutation*. In the next part, we will describe the implementation of each phase for an email delivery problem.

From Sect. 29.2, in this study, a solution candidate is a vector of binary variables (elements) which indicates delivering email or not delivering email on a day to the recipient. The *initial population* is the group of feasible solution candidates meet the constraints of the problem. In this study, the initial population includes

(1) The solution from the greedy method described in Sect. 29.3.1
(2) Randomly Generated
(3) The solutions with only 1 delivery.

Fitness assignment is the process to evaluate the quality of the solution candidate through fitness function. In this process, an evaluation function is used to measure how well a solution candidate fits the task. In this study, the fitness functions are the objective functions listed in Sect. 29.2.

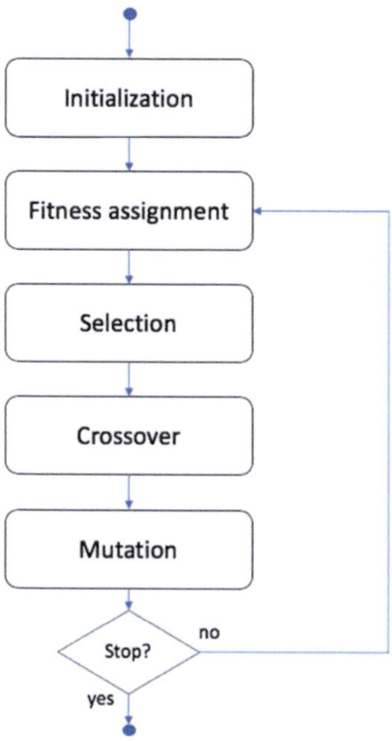

Fig. 29.2. The workflow of genetic algorithm

The idea of *selection* is to select the fittest individuals and let them pass their genes to the next generation. In this process, suboptimal solution candidates may be eliminated from the population, depending on the survival rate, and the selection probability distribution. The selection probability distribution is the distribution to ensure that solution candidates with high fitness have more chance to be selected for reproduction. In this study, the parameters are set as

- Survival Rate: 30%
- Maximum number of solutions survived: 200
- Minimum number of solutions survived:10
- Selection distribution: rejection sampling by candidate's fitness ranking.

Crossover is the process of combining two solution candidates to create new solutions candidates for inclusion in the next generation. It is the prime distinguishing factor of the genetic algorithm from other optimization techniques. In this process, the elements (gene) of the solution candidates are randomly exchanged thus creating a completely new solution candidate (offspring). This process may generate a better solution candidate if the good portions of the gene from parents were passed down.

In this study, the new solution candidates (offspring) are generated by the following procedure:

(1) Randomly select two solution candidates (1, 2)
(2) For each element in the solution 1, crossover with solution 2 at the same element, with half-chance for each solution that its element will be inherited.

(3) Remove the newly generated solution candidate if it is not feasible to the problem.

Mutation process is inspired by the role of mutation of an organism's DNA in natural evolution. With some low probability, a portion of the solution candidate's element will be modified. Mutation process plays an important role in maintaining diversity within the population and inhibits premature convergence.

In this study, each component has an equal chance to be mutated. Besides, only one element will be changed during the mutation process. This is because the number of maximal delivery times is usually small, mutating too many elements may result in the generation of infeasible solution. The mutation procedure can be described as

(1) Randomly select one solution candidates
(2) Randomly mutate one element for this candidate
(3) Remove the newly generated solution candidate if it is not feasible to the problem.

The *stopping criteria* of the genetic algorithm in this study is running 10 iterations without any improvement, or a maximum of 500 iterations reached.

29.4 Simulation and Result

In the simulation, we will solve the email marketing problems listed in Sect. 29.2 using greedy and genetic algorithm. The data used in the simulation is collected from an online flower retailer, where causal customers take a major part of the whole population. The goal of the marketer is to maintain customers' brand awareness.

Each recipient's daily open rate and click rate is predicted through random forest models, which are built based on the past 6 months of the historical email delivery and tracking data of this online retailer. The detail is out of the scope of this paper, so we do not step into this topic.

29.4.1 Open Rate and Click Rate Maximization (Static)

For the open rate and click rate maximization problem, due to the nature of the objective function, the optimization results of the greedy method and genetic algorithm are the same. This result is used as a benchmark for the next two optimization problems. In this test, we set the parameters as follows: $w_o = 1$, $w_c = 1$, $N = 21$, $T = 0.01$, $E = 5$ (Table 29.1).

29.4.2 Open Rate and Click Rate with Open Rate Decay

With the introduction of open rate decay function in the objective function, the problem is no longer suitable to the greedy algorithm, since the next decision is based on the last decision. While the greedy algorithm is able to solve the problem, it may only yield the suboptimal solution. Table 29.2 shows the results from both the greedy algorithm and the genetic algorithm, with the same settings in Sect. 29.4.1.

Table 29.1. Open rate and click rate maximization simulation result

	Greedy algorithm
Total delivery	378,656
Sum of total open rate	103,249
Sum of total click rate	4283
Sum of total fatigue rate	729
Avg. open rate per delivery	0.272
Avg. click rate per delivery	0.0113
Avg. fatigue rate per delivery	0.0019
At least one open probability	0.67
At least one click probability	0.04
Fatigue probability	0.007

From Table 29.2, the genetic algorithm improves more than 9% of the objective function on top of the greedy algorithm. It can be seen that genetic algorithm found a better solution with 9% more total number of opens and 2% less total number of clicks, with the same number of total deliveries. Since the greedy solution is myopic, it always picks the delivery days with the highest open rates, regardless of the decay impact to the rest of the days. The genetic solution, on the other side, seeks the global optimal point by improving objective function, does not have the shortage of the greedy algorithm.

Table 29.2. Optimization with open rate decay simulation result

	Greedy	Genetic	Inc. %
Objective function	140,243.41	152,984.28	9.08
Total delivery	411,354	411,352	0.00
Sum of total open rate	130,537	143,476	9.91
Sum of total click rate	9706	9507	−2.05
Sum of total fatigue rate	892	892	0.00
Avg. open rate per delivery	0.32	0.35	9.91
Avg. click rate per delivery	0.0235	0.0231	−2.05
Avg. fatigue rate per delivery	0.002	0.002	0.00
At least one open probability	0.75	0.79	4.59
At least one click probability	0.093	0.091	−2.59
Fatigue probability	0.0089	0.0089	−0.54

Figure 29.3 shows the value improvement of objective function during the iteration process. As the greedy solution is included in the initial population, and the greedy solution, in general, is not bad, comparing with randomly generated solutions, it can be expected that most of the time the initial value of the objective function starts from the greedy solution.

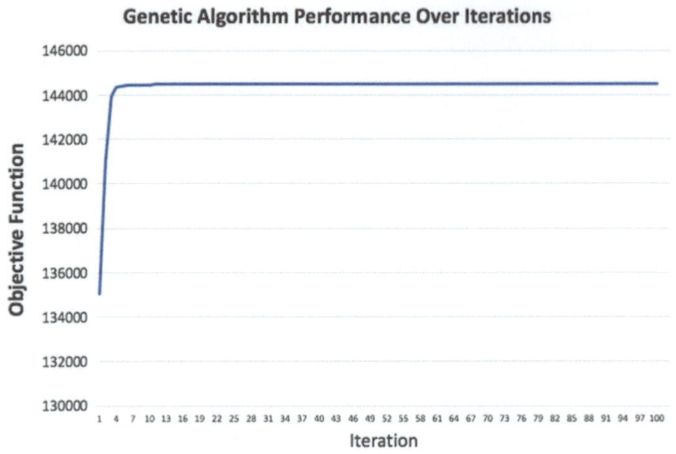

Fig. 29.3. The value improvement of objective function during iterations

29.4.3 At Least One Open/Click Probability Maximization

For this problem, due to the complex of the objective function, it can only be solved by the genetic algorithm. However, since at least one open/click probability maximization is very close to the Problem 1 (open rate and click rate maximization), we will compare the result from greedy solution on Problem 1, and the solution from the genetic algorithm on Problem 3. The results are shown in Table 29.3.

Table 29.3. At least one open/click probability maximization

	Greedy	Genetic	Inc. %
Objective function	956,45.82	974,90.24	1.93
Total delivery	411,354	411,354	0.00
Sum of total open rate	168,593.4	160,252	−4.95
Sum of total click rate	9697.087	131,72.77	35.84
Sum of total fatigue rate	892.193	892.193	0.00
Avg. open rate per delivery	0.41	0.39	−4.95
Avg. click rate per delivery	0.024	0.032	35.84
Avg. fatigue rate per delivery	0.002	0.002	0.00
At least one open probability	0.87	0.856	−1.57
At least one click probability	0.093	0.124	33.43
Fatigue probability	0.009	0.009	0.58

Although Problem 1 and Problem 3 are different, they are quite close, so we can see the potential improvement is limited. From Table 29.3, there is a 1.93% improvement on genetic algorithm over the greedy algorithm. The objective function improvement mostly benefits from the increase in total number of clicks, which largely increases the average at least one click probability.

29.5 Conclusion

In this study, we applied a genetic algorithm to solve email marketing problems with different objective functions. Comparing with the greedy algorithm, genetic algorithm improves the optimization result significantly. Although it naturally time-consuming, the algorithm can be easily parallelized and deployed onto the computing cluster to accelerate its computation speed. With its advantage of generalization to any kind problem, we consider the genetic algorithm as a practical solution to email marketing problems.

References

1. Ellis, C.F., Doherty, N.F.: Web advertising: the role of email marketing. J. Bus. Res. **65**(6), 843–848 (2012)
2. Cho, C.H., Khang, H.K.: The state of internet-related research in communications, marketing, and advertising: 1994–2003. J. Advert. **35**(3), 143–163 (2006)
3. Surveymonkey. https://www.surveymonkey.com/curiosity/map-customer-journey-keep-customers-happy/. Accessed 1 June 2019
4. New research. https://lab.getapp.com/new-research-getdata-why-do-people-unsubscribe-from-email-newsletters/. Accessed 1 June 2019
5. Oreski, S., Oreski, G.: Genetic algorithm-based heuristic for feature selection in credit risk assessment. Expert Syst. Appl. **41**(4), 2052–2064 (2014)
6. Asadi, E., Silva, M.G., Antunes, C.H., Dias, L., Glicksman, L.: Multi-objective optimization for building retrofit: a model using genetic algorithm and artificial neural network and an application. Energy Build. **81**, 444–456 (2014)
7. Soleimani, H., Kannan, G.: A hybrid particle swarm optimization and genetic algorithm for closed-loop supply chain network design in large-scale networks. Appl. Math. Model. **39** (14), 3990–4012 (2015)
8. Deng, S., Yoshiyama, K., Mitsubuchi, T.: Hybrid method of multiple kernel learning and genetic algorithm for forecasting short-term foreign exchange rates. Comput. Econ. 45–49 (2015)
9. Hashem, S., Rafiee, Shahin.: Application of multi-objective genetic algorithms for optimization of energy, economics and environmental life cycle assessment in oilseed production. J. Clean. Prod. **140**, 804–815 (2017)
10. Hao, J., Yu, Y., Law R., Fong, D.K.C.: A genetic algorithm-based learning approach to understand customer satisfaction with OTA websites. Tour. Manag. **48**, 231–241 (2015)
11. Cui, G., Wong, M.L., Wan, X.: Targeting high value customers while under resource constraint: partial order constrained optimization with genetic algorithm. J. Interact. Mark. **29**, 27–37 (2015)
12. Mahdavi, L., Movahednejad, M., Adbesh, F.: Designing customer-oriented catalogs in e-CRM using an effective self-adaptive genetic algorithm. Expert Syst. with Appl. **38**(1), 631–639 (2011)

13. Meena, P.L., Sarmah, S.P.: Multiple sourcing under supplier failure risk and quantity discount: A genetic algorithm approach. Transp. Res. Part E: Logist. Transp. Rev. **50**, 84–97 (2013)
14. Fung, K.Y., Kwong, C.K., Siu, K.W.M., Yu, K.M.: A multi-objective genetic algorithm approach to rule mining for affective product design. Expert Syst. Appl. **39**(8), 7411–7419 (2012)

Chapter 30
End User Involvement in the Big Data Based Service Development Process

Rui Carreira(✉)

Research Centre N2i, Polytechnic Institute of Maia (IPMaia), Maia, Portugal
rcarreira@ipmaia.pt

Abstract. Even though extant research has addressed end user involvement in the process of Information Systems' Development within a Smart City environment, it has not done it in its early or Fuzzy Front End phase. Therefore, this paper promotes and describes a concrete big data based service process in which end users and other individual stakeholders are involved since the early phases of development. Researchers used this end user data to define which part of the big data would be opened for developers and citizens in future stages of the project.

30.1 Introduction

Information and communication technology (ICT) systems are often developed and studied from a systemic organizational viewpoint—how such systems support organizational activities and processes—rather than how they support individual users' preferences, qualities, and actions. However, ICT systems are not used by organizations but by individuals. Those individuals, end users, and other stakeholders, utilize ICT systems to accomplish their professional and family tasks individually and/or in groups. Contemporary ICT systems should be consequently designed not only from the organizational needs' but mostly from individuals' viewpoints, fostering human characteristics.

Even though there is wide agreement that Smart Cities are characterized by pervasive use of ICT systems, which should help cities make better use of their resources [1], there is scarce research on user involvement in ICT requirements' elicitation within a city-scale dimension.

The Future Cities European project aimed at building research capacity for city-scale experiments in Porto (Portugal) by acquiring and installing critical equipment for developing large-scale test-beds and prototypes in the areas of intelligent transportation systems and urban sensor networks (mobile and static), which would leverage Porto as a natural living lab for smart future city and big data technologies. Thus, this research project involved end users (i.e., which includes citizens and other individual stakeholders throughout the paper) in three main stages of the Fuzzy Front End of the New Service Development process [2]:

Á. Rocha et al. (eds.), *Marketing and Smart Technologies*, Smart Innovation, Systems and Technologies 167, https://doi.org/10.1007/978-981-15-1564-4_30

First stage	the main objective was to broadly understand what data (and why) end users would like to have access to;
Second stage	the end users were consulted to define where to locate the static sensors around the city;
Third stage	how do people interact with the city and how would the collected data (and what data specifically) enable the provision of services intended by the city stakeholders.

30.2 Related Work

According to several researchers' (e.g., [1, 3]) view of Smart Cities, the mere deployment of ICT should not imply that a city may be designated as smart since smart initiatives do not only entail technology changes, but also investments in human capital and changes in urban living practices and conditions. Therefore ICT should be rather considered a General Purpose Technology [4], which is complementary to human and organizational capital and whose usage is shaped by political choices and by the urban ecosystem of the citizens, technology vendors, and local authorities, depending on the city stakeholders' needs and habits. There is a clear need to enhance user involvement and user support in experimental research related to the Future Internet within Smart Cities [5]. User participation in information systems development (ISD) is argued (e.g., [6, 7]) as a potential effective practice to achieve various favorable outcomes, including enhanced user satisfaction and increased system quality. Nevertheless, extant research does not usually promote user involvement in the early phases of service development. Therefore, building on positive results from other research areas (e.g., public transportation services) methods for users' input should enhance their proactive participation since the beginning of the discussion in ISD and be adapted to local contexts according to different user needs and conditions [8].

The "open data" movement is pressing urban policy makers to provide their data freely available in a widely used format so that developers can build transit and other applications on top of it. Success has been mixed, as revealed by City-Go-Round, a website that provides access to "useful" transit apps. According to this movement, the website states that as of 2016 only 247 out of 864 transit agencies in the United States had open data, whilst in Estonia, none of 32 listed transport operators provided this service. The goal is clear: city data needs to be provided in an open format, similar to the way in which London Underground transmits its data [9]. Nevertheless, not all of the urban big data collected is equally relevant and so some of it should be selected to be (eventually processed and) opened for developers and end users [10]. Thus, based on a Living Labs approach, the access policy to data should be weighted by project or political deciders and limited to nonsensitive information (e.g., personal or strategic) to guarantee data protection and prevent misuse of the information provided [5].

30.3 Methodology

The research adopted a three-stage approach in order to obtain different kinds of information in distinctive moments of the project from the most relevant stakeholders of the Future Cities European project (i.e., researchers, municipality, companies, and citizens). Each of these stages focused on addressing questions such as Why, What, Where, and How.

30.3.1 First Stage (Why, What)

First, researchers on mobility and urban planning were contacted, while policy makers also demonstrated their strategic and operational (mostly citizen-focused) perspectives. Several meetings took place with six of the most significant research/political deciders in order to understand from a broad perspective of why and what kinds of information would they like to collect from the city. Moreover, in this stage, it was also evaluated the conditions required at the potential locations for the sensors.

Next, a workshop on large-scale city sensing was organized, during which a larger sample of 37 participants brainstormed on possible use cases and pilots. In this early stage of the project, it was important to have an innovative perspective on leading research subjects that could create value both for the project and for the city. Therefore, the participants in this workshop were mostly institutional stakeholders such as researchers, municipalities and startup companies. After an introduction to the project (e.g., essentially to the planned crowdsourcing and urban-sensing test-beds) and to technical subjects (e.g., mobile wireless networks, etc.) the participants used post-it notes to suggest information or services they would like to have access to.

30.3.2 Second Stage (Where)

This stage followed an iterative process with four steps to find the best locations for the sensors:

- Step 1 The high-level locations were selected for each end-use area (e.g., districts, neighborhoods, streets);
- Step 2 Each high-level location was specified in a set of concrete coordinates in order to assess the technical viability such as network coverage and power availability (several locations might be viable in a given street);
- Step 3 The same six stakeholders of the first stage were contacted to rank the suggested concrete locations;
- Step 4 The best viable locations were chosen taking into account several perspectives in order to satisfy most of the stakeholders.

30.3.3 Third Stage (How)

In this final stage of the research study, a qualitative approach was chosen because it can be used to obtain the intricate details about complex phenomena [11] such as cognitive processes that are difficult to extract or learn about through more conventional research methods. The qualitative study was held with the collaboration of the most relevant key decision-makers, who were interviewed in order to broadly understand how do they interact with the city and what do they need to know about it. Interviews were performed to eleven institutional stakeholders, including the municipal police chief, senior academics, startup's CEOs, and heads of municipal departments. In addition, to also include the citizens' perspective, the presidents of an environmental NGO, of a bicycle promoting and of a visual impaired people associations were also interviewed.

The interviews were semi-structured in accordance with a predetermined protocol that aimed at supporting them and at gathering information using open-ended questions, in order to maintain the flexibility for the interviewer to explore other issues that could arise spontaneously during the interviews. The main questions were essentially flexible and exploratory in order not to bias the interviewees' answers. Whenever needed, probing questions were used to obtain further details. The interviews were sound recorded so that the interviewer could dedicate all attention to the interviewees.

30.4 Results

The data collected included information on why and what kind of data would be made available, which would influence the type and location of the sensors required to collect the data, and in turn would influence how to make it openly available for future big data city services. Taking into consideration the iterative process taken to collect information from the end users, extended data results were aggregated into different-level categories using tools such as the Affinity diagram [12].

30.4.1 First Stage (Why, What)

Depending on the type of stakeholder contacted, information on the reasons to collect city data was diverse and focused on their different aims. Even though all stakeholders had the final objective of improving the urban quality of life, the different municipality departments were more focused on the measurement and operational optimization of the environments where people usually are, therefore they were for example interested in counting people in specific areas; traffic planning scientists were mostly concerned with pollution (noise and air) around high traffic areas and with improving mobility; environmental researchers preferred to measure environmental parameters homogeneously across the city despite the density of people around each location.

In this first stage, the results about what kind of data to collect were mostly aggregated into broad categories related to the planned crowdsourcing or/and urban-sensing test-beds of the Future Cities project:

- Crowdsourcing: data essentially related to individual mobility and promoting communication channels between citizens and policy/transport planners;
- Urban-sensing: as this test-bed aimed at involving both mobile (i.e., in vehicles) and static sensors, results varied between trying to use the vehicular network to

exchange static sensor data and complementing static info (e.g., environmental) with mobility updates;
- Both test-beds: mostly associated with merging the data collected from both test-beds.

30.4.2 Second Stage (Where)

Following the results obtained in the first stage, Luis et al. [13] concluded that each data collecting unit might have to include several types of sensors such as mobility, environmental, or video-cameras. Whereas citizens spend more time in certain streets and areas such as parks, sensors should be there located from a people sensing perspective. On the other hand, street-by-street sensing would enable covering the city homogeneously. Taking into consideration the interest of the different stakeholders involved in measuring various sets of variables (e.g., where most people are, high traffic areas or uniform distribution around the city, etc.), decisions on best location had to account for different factors and weights. See Fig. 30.1 for a spatial distribution in Porto of the final location for the 50 data collecting units.

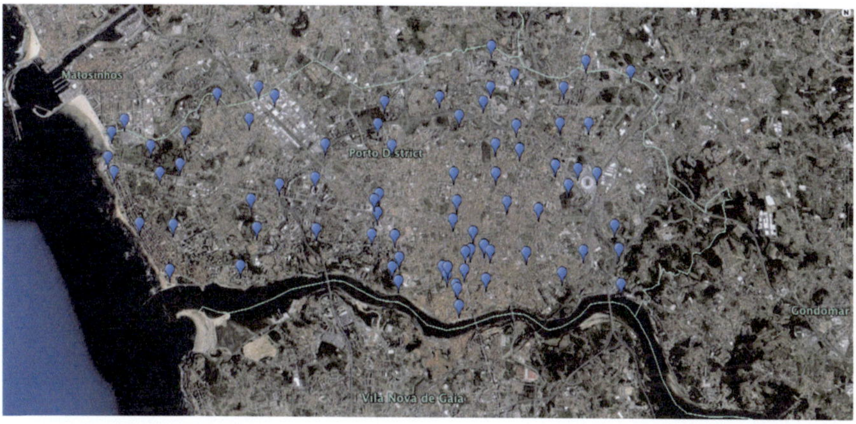

Fig. 30.1. Distribution of the 50 data collecting units' final locations

30.4.3 Third Stage (How)

The data collected in this stage was aggregated into the main categories of information to which the participants would like to have access to. This categorization was revised by Future Cities' researchers during iterative meetings, and the final aggregation of data included the institutional stakeholder and the end user (association) perspectives.

Institutional stakeholder perspective

Main question: What would you like to know about Porto?

Probing question: To what data would you like to have access to?

- Multiplicity of traffic dynamics, Network inference, Multicriteria routes and their optimization, Urban logistics, and accessibility
  - Moving elements
    - Direction,
    - Position (e.g., length of stay),
    - Routes,
    - Average speed,
    - Accidents (e.g., pedestrian runovers),
    - Number of people (e.g., purpose of the trip), and
    - Traffic volume.
  - Transportation services
    - Available modes,
    - Mode demand, and
    - Network and frequencies.
- Citizen learning about the city (and vice versa)
  - Citizens' financial availability,
  - Citizens' household size,
  - Citizens' age,
  - Archeology: historical data,
  - Mood sensor index,
  - Smart city index, and
  - Priority areas based on thermal risk.
- Two-way communication channel with municipality/infrastructure manager
  - Public services,
  - Mobility, and
  - Environment.
- People civility
  - Illegal parking,
  - Training of motorized road users, and
  - Not conflicting with other people's rights.
- Security
  - Thefts,
  - Vandalism.
- Climatology, Health and their assessments before/after particular urban changes
  - Indoor air quality,
  - Bioclimatic comfort,
  - Pollens,
  - Particles,
  - Pollution,
  - Sounds,
  - Shadow areas, and
  - Climatology.
    - Wind (quadrant, speed, or dispersion),
    - Rel. humidity,

Rain,
Fog, and
Temperature.

- Resources
  - Location, Physical accessibility, Point of interest
    Available parks (e.g., times and fees),
    Elderly support,
    Hotels,
    Deliveries, and
    Stores (Status, e.g., open, closed).
  - Energy consumption
    "Automatic light" on in isolated areas.
  - Water consumption
- Data correlations
  - Impact of environment on birds' behavior,
  - Impact of pollution on health,
  - Impact of temperature on happiness,
  - Impact of green areas on health and thermal comfort,
  - Count the number of S. João (municipal holiday) balloons and assess fire risk based on wind direction and temperature, and
  - Impact of environmental indexes on health.

End user (association) perspective

Main question: What is important for an associate in his/her relationship with the city?
Probing question: To what data would like to have access to?

- Geo-referenced mobility information
  - Mobility
    Mapping routes adapted to bicycles (not cars),
    Direction,
    Average speed,
    Fuel use,
    Traffic black spots,
    Volume of traffic,
    Road slope, and
    Type of road floor.
  - Transportation services
    Preferable integration of bicycles in traffic circulation (except for children).
    Multimodal.
      Easier input or output transport in rush hour.
      Positive discrimination (e.g., integrated ticket in app).
- Two-way communication channel with municipality/infrastructure manager
  - Public services,
  - Mobility, and
  - Environment.

- People civility
  - Training of motorized road users,
  - Not conflicting with other people's rights.
- Quality of life
  - Access to employment,
  - Physical accessibility,
  - Culture, leisure,
  - E-accessibility, and
  - Involvement in the community.
- Weather conditions
  - Forecast and estimation of change during travel.
- Resources
  - Pedestrian crossing locations (e.g., vibrating signal traffic lights.).
  - Main points of tourist attraction
    - Occupation of public space,
    - Petrol stations,
    - Bike repair-shops (discounts), and
    - Bike fixation/security/covered parking.
- Better public lighting at night.

30.5 Discussion

Even though extant research has addressed end user involvement in the process of ISD within a Smart City environment [5], it has not done it in the early or Fuzzy Front End phase. Therefore, this paper promotes and describes a concrete big data-based service process in which end users and other stakeholders are involved since the early phases of development. The methodology used enabled gathering information on a broad perspective of end users and stakeholders such as professionals, scientists and user associations. Thus, it was possible to understand the intricate details about how do urban stakeholders interact with the city by adopting an open qualitative approach [11] in important phases of research. Otherwise, traditional ISD methodologies [6] would not enable obtaining complex phenomena such as end user cognitive processes and answering questions such as Why, What, Where, and How should city data be collected and made available for its stakeholders.

30.6 Conclusions and Future Research

ICT systems are often developed and studied from a systemic organizational viewpoint, however they should also be developed from the individuals' perspective. This paper involves several individual perspectives, either professional or family ones, and it contributes to a new breadth of knowledge within the big data based service development process.

Future possible research would benefit from using these data results for researchers to discuss Access Platform Interfaces' requirements in order to define which part of the big data would be opened for developers and citizens. Nevertheless, not all of the urban big data collected would be equally relevant and so some of it would have to be selected to be opened according to different priorities.

Acknowledgements. This work was funded by the European Union under the 7th Framework Programme for Research and Innovation (FP7-REGPOT-2012-2013-1, Future Cities project, grant number 316296).

References

1. Neirotti, P., De Marco, A., Cagliano, A.C., Mangano, G., Scorrano, F.: Current trends in smart city initiatives: some stylised facts. Cities **38**, 25–36 (2014)
2. Khurana, A., Rosenthal, S.R.: Integrating the fuzzy front end of new product development. Sloan Manag. Rev. **38**(2), 103–120 (1997). Winter97
3. Caragliu, A., Bo, C.D., Nijkamp, P.: Smart cities in Europe. In: 3rd Central European Conference in Regional Science—CERS, Košice, Slovak Republic, 7–9 October 2009 (2009)
4. Bresnahan, T.F., Trajtenberg, M.: General purpose technologies 'engines of growth'? J. Econom. **65**(1), 83–108 (1995)
5. Schaffers, H., Sallstrom, A., Pallot, M., Hernandez-Munoz, J.M., Santoro, R., Trousse, B.: Integrating living labs with future internet experimental platforms for co-creating services within smart cities, pp. 1–11
6. He, J., King, W.R.: The role of user participation in information systems development: implications from a meta-analysis. J. Manag. Inf. Syst. **25**(1), 301–331 (2008)
7. Chen, C.C., Liu, J.Y.-C., Chen, H.-G.: Discriminative effect of user influence and user responsibility on information system development processes and project management. Inf. Softw. Technol. **53**(2), 149–158 (2011)
8. Carreira, R., Patrício, L., Natal Jorge, R., Magee, C.: Development of an extended Kansei engineering method to incorporate experience requirements in product-service system design. J. Eng. Des. **24**(10), 738–764 (2013)
9. Goodall, W., Dixon, S., Fishman, T., Perricos, C.: Transport in the Digital Age—Disruptive Trends for Smart Mobility. Deloitte (2015)
10. Janssen, M., Charalabidis, Y., Zuiderwijk, A.: Benefits, adoption barriers and myths of open data and open government. Inf. Syst. Manag. **29**(4), 258–268 (2012)
11. Strauss, A., Corbin, J.: Basics of Qualitative Research, 2nd edn. Sage, Thousand Oaks (1998)
12. Bergman, B., Klefsjo, B.: Quality: From Customer Needs to Customer Satisfaction, Lund. McGraw-Hill, Sweden (1994)
13. Luis, Y., Santos, P.M., Lourenco, T., Pérez-Penichet, C., Calcada, T., Aguiar, A.: UrbanSense: an urban-scale sensing platform for the internet of things. In: 2016 IEEE International Smart Cities Conference (ISC2), pp. 1–6

Chapter 31
Managing Human Resources Resistance to Organizational Change in the Context of Innovation

José-Luis Rodríguez-Sánchez[1(✉)], Nohora Mercado-Caruso[2], and Amelec Viloria[2]

[1] Departamento de Economía de la Empresa (ADO), Economía Aplicada II y Fundamentos del Análisis Económico, Universidad Rey Juan Carlos, Madrid, Spain
joseluis.rodriguez.sanchez@urjc.es
[2] Departamento de Gestiòn Industrial, Agroindustrial y de Operaciones, Universidad de la Costa CUC, Barranquilla, Colombia
{nmercado1,aviloria7}@cuc.edu.co

Abstract. Current global and complex economies demand change and restructuring processes. Besides the usual continuous change in their personal lives, when talking about changes in their working environment, employees offer considerable resistance. Difficulties related to managing this resistance on the part of the company will ultimately lead to organizational failure. Aimed at filling this gap, the present research provides a management model addressing both implicit and explicit resistance. The model comprises six key factors: leadership, communication, valuable HR retention, training, participation, and flexibility. Besides the theoretical model, it has been carried out a case study of a multinational company in the mechanical engineering sector, providing researchers and professionals with a roadmap of actions to manage the identified key factors. Furthermore, results reveal that Human Resources are no longer operational support. Instead, they represent a crucial asset to achieve the expected results in innovation and obtaining a sustainable competitive advantage.

31.1 Introduction

Nowadays organizations have to position and compete in rapidly and constantly changing environments, defined as hypercompetitive and even chaotic in some cases [1]. The demands of the environment force organizations to make changes if they want to continue competing effectively in the market. Change is a complex phenomenon that appears constantly in the current context. According to [2], change is the only permanent thing. As a result, there is an undeniable increase in academic and professional interest regarding the management of organizational change processes. Faced with this scene, organizations need to respond appropriately. For this purpose, it

Á. Rocha et al. (eds.), *Marketing and Smart Technologies*, Smart Innovation, Systems and Technologies 167, https://doi.org/10.1007/978-981-15-1564-4_31

is necessary to understand and analyze when there is a need for change, to create value in order to achieve the expected results.

The attitude and positioning of organizations with regard to change—in an increasingly competitive market—is a determining factor in the evolution and success of a company. It is important to shorten reaction times and proactively implement the change process. Otherwise, ignoring change and preserving traditional ways of doing things and thinking can lead to stagnation, decline, and failure.

Despite the fact that change is a frequent and necessary phenomenon in all organizations, in 70% of cases the expected results are not achieved [3]. The latest studies, such as a 2008 McKinsey survey of 3,199 executives worldwide, reveal that only one out of three transformations is successful [4]. Analyzing the results, the following research question arises: why are HRs so resistant to changes in companies, if it is a practice that develops day by day in their daily lives to adapt to current technological innovation?

The aim of this paper is to delve into the factors that cause employees' resistance to change and to propose a model for managing it. These resistances are often seen as a natural reaction, and employees' resistance to change is unavoidable [5]. People perceive change as a process in which their satisfaction, security, social interaction, prestige, or self-esteem are negatively affected [6].

However, organizations have within their reach human resources (HR) management tools that can help significantly to reduce them, as well as increase the chance of success. Organizations have traditionally focused on managing economic factors (benefits, profitability, ratios, etc.) but it has been demonstrated that the functioning of an organization depends on the attitude of its employees. Aspects such as communication, leadership, or participation are determining factors given that employees' resistance or not becomes a key factor for the implementation and success of the change process, it is essential to understand the determining factors in order to manage the process. For this reason, this paper reviews the literature on the determining causes of employees' resistance during the change processes. After analyzing the causes of employees' resistance, the next step consists of developing a management model. As a result of the work and implications for literature and professionals, we have created a framework of good management practices in order to achieve the expected results during change processes.

The work has been structured in several sections. First, we carry out a review and analysis of the existing literature on organizational change and the reason for personnel resistance. In the second section, the methodological aspects of the case study are analyzed. Third, we analyze the state of affairs regarding organizational change management and its most important aspects then, the generated resistance. We examine HR factors that are crucial to cope with the resistance and to achieve the expected results in change processes, and it is analyzed through a case study of an international change and restructuring process. Finally, conclusions and implications are presented.

31.2 Theory Framework

31.2.1 Organizational Change and Restructuring Process

Change is recurrent in organizations, so it must be managed and planned. Organizational change refers to a series of changes and readjustments that are made to adapt the new strategy to the organization. According to [7] change is an observation of differences in form, quality, or long-term style of an organizational entity. Puertas [8] explains that planned organizational change is the design and establishment of a structural innovation, a new action plan or a change in operating philosophy, climate or style.

Tsouka and Chia [9] argues that change must not be thought of as a property of the organization. Rather, the organization must be understood as an emerging property of change, thus formulating the paradox that the organization drives change, but at the same time, it regenerates itself in this process. Any change that occurs in a part of the company affects [10].

31.2.2 Steps of the Organizational Change and Restructuring Process

The success of an organizational change and restructuring process depends on the planning of its implementation. At this point, many companies resist change and end up failing during the process. The model by [3] consists of three steps: (1) unfreezing the existing status quo. The main management issues arise at this first stage. It involves preventing employees from entering into conflict or seeing their interests and objectives affected. At this stage, employees are motivated to break down the existing way of doing things. A stage is completed, and the old state closed [11]. (2) Introduction of the changes. It is at this stage that changes are made. The necessary modifications are introduced to achieve new objectives. Employees are in a transitioning stage. Getting through this stage is essential for them to adapt to the new situation. Bridges and Mitchell [12] calls his second phase "the neutral zone". And, (3) refreezing. The aim is to ensure that people do not revert back to their old habits prior to the implementation of the change. The introduction of any change alters the stability of the organization and this can lead to resistance. At this stage, employees need to understand the new situation. Referred to by [13] as the "new beginning", changes become permanent, the indications are applied.

31.2.3 Personnel Resistance

There are different classifications that help us know better what these resistances are and thus be able to manage them more effectively. Traditionally, personnel resistance has been divided into three levels according to the resistance pyramid developed by [14] not willing, not able and not knowing.

Resistance to change can manifest in different ways, explicitly or implicitly [15]. According to [16], it can be expressed: explicit or overt resistances refer to situations that are recognized quickly and directly, for example, strikes, labor disputes, sabotage or defective work. Implicit or covert resistances are more difficult to perceive and therefore more dangerous. They usually consist of loss of confidence in the company, disloyalty, delays, absenteeism, higher rates of accidents or decreased motivation.

These resistances decrease employees' motivation and productivity, lead to dissatisfaction toward HR management and discrimination through restricted communication. As we can see, there is a cycle in which resistance to change varies from individual to collective. A contagious effect can occur, leading to consequences such as abandonment and loss of valuable employees. Even so, the organization is affected because losing valuable employees is one of the most important issues in organizational value, leading to poor implementation of change. The aim of the work is to build a management model of the most relevant HR factors during the change process to avoid personnel resistance. It is essential to involve them in the process.

31.3 Methodology

Perry [17] states that within qualitative research, case studies contribute to the comprehensive understanding of a specific social phenomenon. They are relevant when the aim is to understand, solve or improve a procedure performed in the professional field. The multinational has conducted several operations as a means of growth in emerging countries or consolidation in mature markets. Several operations in mature markets, in which the multinational was not yet consolidated and needed further development, such as in China and the USA (1980 and 1989, respectively), stand out. Recently, in 2009, large acquisition of a local company within the emerging Saudi market represented the opportunity to penetrate that growing market.

For the collection of evidence, 14 semi-structured interviews with various informants were performed with the aim of promoting an open dialogue about their role in the change and restructuring process. This process involved working with two groups of staff. The first group was composed of four members of the management committee (HR manager, operations manager, communication and marketing manager, and training and development manager). Second, open interviews were performed, and informal contacts established with more than 45 staff members from different intermediate organizational levels from both the parent and target companies. The diversity of the origins of the informants provides different interesting perspectives to obtain conclusions about the research.

The number of interviews and informants was not established at the beginning of the research. Therefore, it was determined by "theoretical saturation," namely, the interviews were conducted to the point at which the informants reported similar and coherent information to support the validity and quality of the theoretical model. To complement the interviews, we participated in meetings, communication plans, or events organized by the company, in which the values of the organization were promoted in terms of HR management, in addition to the commitment and development of integrity and trust. In addition, different internal documents were consulted through the corporate intranet, highlighting the manual for the execution of the HR strategy, developed and signed by the Chief Human Resources Officer of the multinational worldwide. The motto "support our business by supporting our employees" stands out.

During the transcription stage, several filters were used to extract relevant information for the outcome of the case study, given its influence on HR management. Due to the expansion and size of the multinational, it publishes multiple standards, guidelines and procedures on its corporate intranet, which were analyzed to extract the most determinant points. During the transcription of data from the interviews, the informants were contacted via email, and the second round of interviews was held to clarify doubts and verify that the field notes were accurate. Once generated, the results were presented to the informants for their approval or to provide critical feedback that allowed performing any modifications that were necessary.

31.4 Case Study Results

This section proposes and analyzes in depth the determining Key factors for managing personnel resistance in the change and restructuring process.

31.4.1 Leadership

Personnel resistance can be reduced by using a proper and appropriate leadership style [6]. The relationship between leaders and employees is direct as they provide a first insight into the change HR is about to undertake. This essential even crucial factor can set a lead for change [18]. We can distinguish two sets of processes that explain why leaders' personal attributes would be related to employees' resistance to change. Leaders' personal attributes influence the choices and the decisions they make in the organization [19]. These decisions influence followers' attitudes and beliefs [20].

The operations manager explained: "strategic changes and modifications are usually initiated and formulated by senior management. The leader is the link with the lower echelons, to convey this change and to be understood by employees as a benefit of the whole company and for the purpose of improving and maintaining a strong position."

"Top managers need to identify the positive aspects of change, both for the organization and for employees. On the one hand, they see changes as an opportunity to renew the organization in order to be more successful. They also see organizational changes as possibilities to advance," the communication and marketing manager considered.

The HR manager stressed the important role of Leaders' personal attributes values and traits predispose leaders to certain types of decisions in the organization: "on the one hand, values influence leaders' behavior and choices of leaders through their influence on the organization and the types of behaviors and outcomes they view as desirable in order to reward employees. On the other hand, traits influence the types of behaviors they engage in and the decisions they make. For example, leaders who are risk averse would be less likely to make organizational decisions that involve risk taking."

It is necessary to manage the influence and behaviors of employees' beliefs. The training and development manager suggests: "leaders, particularly senior leaders, shape shared beliefs. How leaders respond to organizational processes influences the type of beliefs and attitudes that should be held by organization members. When deciding who to attract, select and retain in the organization, leaders create organizational climates in which people are similar to one another in their beliefs and attitudes."

P1: Highly qualified leaders can reduce HR resistance to organizational change.

31.4.2 Communication Process

The communication process plays a key role in the conflicts that arise in the change processes. When a change process is to be undertaken, communication becomes an essential tool in order to implement successfully the change. But in practice lack of communication is one of the most important causes for employees' resistance to change [13] Employees have doubts and need the information to mitigate their uncertainty [21]. Employees resist any type of change because they do not have enough information about the change, why it is implemented, and its impact on personal life [22]. Likewise, members of the organization perceive change negatively because they are given a biased view of change and may not really understand the real purpose.

The communication and marketing manager stated the following: "an effective internal communication system is useful for disseminating and making information transparent. Thus, employees can take up the general functioning of the organization, assume the strategy, have objective elements to propose changes and have guidelines for their individual activity. Also, it promotes the interaction and exchange between employees in aspects related to professional life."

Employees have doubts and need the information to mitigate their uncertainty. According to the training and development manager: "a good communication process must contain information on: (1) Importance of the change. Make employees believe in the usefulness of change and reduce the burden of responsibility after its implementation. (2) The adequate interpretation of the project. Change managers should ensure that employees have sufficient information in order to understand the change. And, (3) risk for employees."

P2: An efficient internal communication process can reduce HR resistance to organizational change.

31.4.3 Valuable Employees

Another important factor in analyzing employees' resistance is their own personality. In general, all employees resist change by nature [23], but beyond that, there are some personal characteristics that sharpen or mitigate resistance. Resistance is based on emotional reactions that derive from low tolerance for change, the fear of losing power or status [24].

The HR manager stated: "each employee has different personal characteristics, different professional characteristics and different strategies for coping with changes." The operations manager expressed the characteristics that can make a difference are: (1) active methods for dealing with change. Points out two possible coping strategies: active and avoidance coping. The first group sees opportunities in change, makes action plans and shares concerns to mitigate the stress. In contrast, the second group tries to avoid confrontation with the new status quo. It is characterized by forgetting everything related to change until it becomes effective. (2) Tolerance for ambiguity. Employees with high ambiguity tolerance perceive ambiguous situations as desirable. At the other extreme, people with low ambiguity tolerance interpret ambiguous situations as threats. Employees with high ambiguity tolerance will be less resistant to change. (3) Self-efficacy. It refers to employees believe in their capability to achieve the expected results. Employees who have high levels of self-efficacy adapt better to changes than those who have low levels. And, (4) a flexible character. Flexible people are those who can tolerate words like "probably" or "maybe". They do not need everything to be exactly in the right place and they are willing to make changes even after decisions are made. In his study, we cannot say that the more flexible employees are, the less they are likely to resist change. This feature moderately affects the generation of resistances.

Some employees have positive emotions of excitement. But the training and development manager stressed: "in almost all significant changes, negative feelings appear, therefore some valuable employees have some positive feelings, such as enthusiasm with the possibility of a better future, the release of the old problems and the expectations of personal development."

Retaining valued employees is critical to achieving the expected results, but in practice, many companies fail to do it. Top achieve this objective the communication and marketing leader stated the following: "this involves assessing the characteristics and capabilities of employees and placing them in the most appropriate position. In order to produce a sustainable competitive advantage, the company must retain employees who have better skills and capabilities than their competitors."

P3: Efficient company retention policies can reduce HR resistance to organizational change.

31.4.4 Training

Training plays a key role in the change process. In times of change, those who are open to learning will take over the future, while those who find themselves beautifully equipped will deal with a world that no longer exists [25]. Employees may resist change for reasons related to training such as low confidence about their skills or lack of knowledge on how to deal with change. Therefore, it is important for organizations to offer a range of training that can improve employees' skills and competences [26].

"To successfully implement organizational change, trained leaders have to manage that challenge efficiently and with the voluntary support of employees. The best way is to provide leaders with training that gives them a better insight into the options change can bring," the operations manager referred. It is no wonder that employees place more credibility in those leaders who are well prepared and offer more confidence. In turn, it will be easier for leaders to provide support and information to their staff if they have been trained to do so.

Another way to reduce resistance is to ensure that employees have acquired the necessary skills required by change. To manage this factor the communication and marketing manager explained: "this can be achieved by providing the appropriate instruction or training, devoting time and energy, so that employees meet the new requirements. In this way, they feel that the organization invests time and resources in their adaptation and training."

Creating a work environment in which employees feel like they are active agents of change reduces stress and tension among them. The training and development manager expressed: "the training offered by the organization is a sign of its commitment, as it involves a cost (economic, time, etc.). Training also helps improve the effectiveness of knowledge integration and absorbs or acquires new knowledge by reliance on manuals, databases, processes and routines."

Training may not only be understood as a "professional training", but also as a means which facilitates the proposed reforms and new working practices. In this way, the HR manager considered the following to be essential: "it becomes especially important because it is a privileged tool to achieve a behavioral change. Organizations should provide training courses to employees on how to behave during the change process and encourage them to manage stress rather than trying to avoid it. In order to help managers and employees handle stress, low morale, and low productivity, practices such as educational seminars must be developed and used to reduce uncertainty".

P4: Efficient training can reduce HR resistance to organizational change.

31.4.5 Participation

Employees' involvement in the change process increases the chances of success. It is difficult for employees to resist change in which they have been involved since its inception. Therefore, it is essential to achieve its commitment through participation [27]. This participation means a reduction in the psychological trauma (stress, anxiety, confusion or feelings of incompetence) that can be experienced during the change. Because of this positive impact on the process, it is interesting to encourage employees' involvement [10].

The communication and marketing manager commented: "organizations should promote the open information interchange and the development of face-to-face personal relationships among employees. Seminars, training activities, guided tours can help reduce tension. It is necessary to consult employees, report on small achievements and problems in the change process, and to involve staff in providing solutions. In addition, organizations may give employees control and management capacity over their activities, show their influence and benefits for change."

It is crucial that the staff must recognize people who help promote the change process, support and reward them. The operations manager stated the following: "their actions help prevent bad behaviors from the rest of the employees. These relationships create a degree of friendship that condition positive attitudes towards participation."

"Another way of encouraging participation is the empowerment, which consists of making jobs more meaningful and interesting by giving employees the opportunity for personal growth and fulfillment. It means employees feel that they influence the decision-making," explained the training and development manager.

P5: Employee participation in the change process can reduce HR resistance to organizational change.

31.5 Conclusions

Two types of contributions derive from our work. For the scientific community, the work offers a model of HR management in the change and restructuring process, integrating the previous theoretical contributions of experts in the field and corroborating their research with primary information from a leading multinational company in the sector. In addition, the work offers a series of practical implications for how to manage HR management in the change and restructuring process. More specifically, a roadmap (Table 31.1) that can serve as a reference to the different directors involved in the process. In this manner, each manager will know the most important HR factors to be managed, and the actions that can be taken to influence success.

Table 31.1. Propositions, practices, and results

Why are HR resistant to organizational changes, even if these practices are crucial to adapt to the current context of technological innovation?		
Propositions	HR practices	Results
P_1: leadership	• Empathy • Active listening • Influence	(1) Career development (2) Rewards (3) Pride of belonging
P_2: communication	• Transparency • Interaction • Intranet	(1) Certainty (2) Proper interpretation (3) Feedback
P_3: valuable HR	• Characteristics • Competences • Selection and promotion	(1) Valuable HR retention (2) Motivation (3) Examples to follow
P_4: training	• Investment • Databases • Processes and routines	(1) Psychology (2) Knowledge integration (3) Dissemination
P_5: participation	• Socialization period • Open information • Focus groups	(1) Positive feedback (2) Bilateral solutions (3) Delegation capability

Change is a natural, constant, and unavoidable phenomenon that affects all organizations, and therefore all members of such organizations. Change is a complex process that must be well managed and led in order to be successful. It aims at making the necessary changes in the organization in order to achieve the expected purpose, abandoning the old ways of doing things in the implementation of change: the success of change depends on the human factor [28, 29].

References

1. Maheshwari, S., Vohra, V.: Identifying critical HR practice impacting employee perception and commitment during organization change. J. Organ. Change Manag. **28**(5), 872–894 (2015)
2. Casado, J.M.: El Valor de la Persona: Nuevos Principios para la Gestión del Capital Humano. Prentice Hall (2003)
3. Kotter, J.P.: Leading change: why transformation efforts fail. Harvard Bus. Rev. Change 1–21 (1998). A Harvard Business Review Paperback
4. Aiken, C., Keller, S.: The Irrational Side of Change Management. The Quarterly McKinsey & Company, Toronto, EEUU (2009)
5. Amorós, E.: Comportamiento organizacional: En busca del desarrollo de ventajas competitivas. USAT Escuela de Economía Chiclayo, Lambayeque, Perú (2007)
6. Kreitner, R., Kinicki, A.: Organizational Behavior, 5th edn. McGrawHill, New York (2001)
7. Ruona, W.E., Gibson, S.K.: The making of twenty-first century HR: an analysis of the convergence of HRM, HRD and OD. Hum. Resour. Manag. **43**(1), 49–66 (2004)
8. Puertas, F.: Romper los límites actuales de los recursos humanos. Harvard Desusto Bus. Rev. **245**, 63–66 (2015)
9. Tsouka, H., Chia, R.: On organizational becoming: rethinking organizational change. Organ. Sci. **13**(5), 567–582 (2002)
10. Quirant Espinosa, A., Ortega Giménez, A.: El cambio organizacional: la importancia del factor humano para lograr el éxito del proceso del cambio. Revista de empresa **18**, 50–63 (2006)
11. Rees, C., Johari, H.: Senior managers' perceptions of the HRM function during times of strategic organizational change, case study evidence from a public sector banking institution in Malaysia. J. Organ. Change Manag. **23**(5), 517–536 (2010)
12. Bridges, W., Mitchell, S.: Leading transition: a new model for change. Lead. Lead. **16**(3), 30–36 (2000)
13. Alas, R., Sun, W.: Organizational changes in Chinese companies: a resource-based view. Chin. Manag. Stud. **1**(4), 225–242 (2007)
14. Ngoh, T.: Maximising human resource potential in the midst of organizational change. Singap. Manag. Rev. **27**(2), 25–35 (2005)
15. Guerras Martín, L.A., Navas López, J.A.: Dirección Estratégica de la Empresa. Teoría y Aplicaciones, 5ª edn. Civitas Ediciones, Madrid (2015)
16. Robbins, S., Sanghi, S., Judge, T.: Organizational behaviour: concept, features and importance. Pearson Education, London (2009)
17. Perry, C.: Processes of a case de study methodology for postgraduate research in marketing. Eur. J. Mark. **32**, 785–802 (1998)
18. Pater, R.: Leveraging leadership: making the most from the least. Prof. Saf. **56**(4), 55–57 (2011)
19. Carpenter, M.A., Geletkanycz, M.A., Sanders, W.G.: Upper echelons research revisited: antecedents, elements and consequences of top management team composition. J. Manag. **30**, 749–778 (2004)
20. Berson, Y., Oreg, S., Dvir, T.: CEO values, organizational culture and firm outcomes. J. Organ. Behav. **29**, 615–633 (2008)
21. Fiedler, S.: Managing resistance in an organizational transformation: a case study from a mobile operator company. Int. J. Project Manag. **28**, 370–383 (2010)
22. Kalyani, M., Prakashan Sahoo, M.: Human resource strategy: a tool of managing change for organizational excellence. Int. J. Bus. Manag. **8**(6), 280–286 (2011)

23. Schillin, A., Werr, A., Grand, S., Sardas, J.C.: Understanding professionals′ reactions to strategic change: the role of threatened professional identities. Serv. Ind. J. **32**(8), 1229–1245 (2012)
24. Armenakis, A.A., Bedeian, A.G.: Organizational change: a review of theory and research in the 1990s. J. Manag. **25**, 293–315 (1999)
25. Chaffey, D.: E-Business and E-Commerce Management: Strategy, Implementation and Practice, 2nd edn. Harlow, Prentice Hall, England (2004)
26. Viloria, A.: Commercial strategies providers pharmaceutical chains for logistics cost reduction. Indian J. Sci. Technol. **8**(1), Q16 (2016)
27. Goulet, P.K., Schwieger, D.M.: Managing culture and human resources in mergers and acquisitions. Handbook of Resarch in International Human Resource Management, Edward Elgar Publishing, Cheltien, UK and Northampton, MA, USA, pp. 405–429 (2006)
28. Amelec, V.: Increased efficiency in a company of development of technological solutions in the areas commercial and of consultancy. Adv. Sci. Lett. **21**(5), 1406–1408 (2015)
29. Amelec, V., Carmen, V.: Design of a model of evaluation of productivity for microfinance institutions. Adv. Sci. Lett. **21**(5), 1529–1533 (2015)

Chapter 32
Fournier (BRQ) Versus Rusbult (IM): Is Love Really Important in Consumer-Brand Relationship?

Ricardo Cayolla[1,4], Sandra Loureiro[3], and José Luís Reis[2](✉)

[1] ISMAI, Maia University Institute, Maia, Portugal
rcayolla@ismai.pt
[2] ISMAI, Maia University Institute, Research Unit CEDTUR/CETRAD, Maia, Portugal
jreis@ismai.pt
[3] ISCTE - IUL, University Institute of Lisbon, Lisbon, Portugal
sandra.loureiro@iscte-iul.pt
[4] Universidade de Aveiro, GOVCOPP, DEGEIT, UA, Aveiro, Portugal

Abstract. Assuming the meaning and value that brands have for consumers, this paper developed a structured viewpoint in the consumer-brand relationship. This paper explores the fact that there are deep investments by consumers in their relationships with brands. The objectives of the article are, first, to present the two main consumer-brand relationship models (Brand Relationship Quality—BRQ, Investment Model—IM) and, second, to challenge the importance of Love in the relationships established between consumers and brands.

32.1 Introduction

Early discussions on branding emerge in the United States at the beginning of the twentieth century. For instance, Copeland (1923) states that buyers will be reluctant to purchase a product if there is no recognizable brand name from a reputable manufacturer [1]. Later, Gardner and Levy (1955) claim that a brand is embedded within a complex symbolism that represents a variety of attributes and ideas, and brand names are thus more than labels distinguishing products and allowing consumers to tell them apart in the context of complex buying situations. Gardner and Levy (1955) also mention that a brand comprises sets of ideas, feelings, and attitudes that the consumer has about an organization or a product [2]. Based on this, consumers make choices through one or several elements of these sets that they find most appropriate [1, 2].

Consumer–brand relationship (CBR) "itself suggests that there are relationships between consumers and brands" [3]. CBR is a field of research in which several authors —Fournier's (1998) model called Brand Relationship Quality (BRQ) is the most applied have developed a set of theories trying to explain how consumers and brands are related [4]. The central question is twofold: why consumers choose one brand or

Á. Rocha et al. (eds.), *Marketing and Smart Technologies*, Smart Innovation, Systems and Technologies 167, https://doi.org/10.1007/978-981-15-1564-4_32

another and why consumers tend to repeat behavior and purchase a brand and avoid others. CBR studies the process by which a bidirectional relationship is established and maintained between the consumer and the brand, which may involve multiple relationships [5], without forgetting that CBR largely depends on consumer perceptions [3]. Rusbult Investment Model (IM) wants to know how people persist in their interpersonal relationships [6].

The article is organized as follows: First, the article describes the brand concept concerning is meaning and value as it is defined in the literature. Second, the focus it is on CBR foundations and main constructs. Next, the article refers to the theoretical background of IM. Finally, the article presents the discussion BRQ versus IM.

32.2 Brand Meaning and Value

The twentieth century ended with two main approaches to the brand concept: product plus and holistic [7]. The first approach views a brand as an extra component added to a product (it serves to identify the product). The holistic approach refers to the group of attributes (tangible/intangible, rational or emotional) that act to enhance consumer satisfaction. In the twenty-first century, brands have been regarded as a highly contextualized entity open to diverse contemporary approaches and understandings and consequently to a never-ending process of theoretical development [8]. Therefore, it is highly unlikely that we can congregate all a brand's characteristics in a single, consensual, and widely accepted definition. Gabbott and Jevons (2009) state there will never be a unifying definition, only a constantly evolving series of contexts, features, and circumstances.

Whether from a managerial or consumer perspective, a brand must be able to create relationships, generate expectations, and stand out among the competition. The brand being the result of a process where the brand and the consumer interact, the relationship is formed through the experiences and emotions of both: a consumer and a brand [9]. When the consumer does not see more than the visible part of the brand—meaning the advertising and the final price in comparison to other similar products or brands [10], the brand is no more than an organization's promise to try to give the goods or service a predictable performance (Kotler and Keller 2006). Marketers have the job of creating contexts that help the development of the brand in consumers' minds [9].

It is not insignificant to consider that it is the marketing effect that makes a customer compare a product or service with another branded or unbranded one, and it is this awareness that makes him understand and behave in favor of the brand [11]. It is through a set of behaviors by various players (customers, distributors and suppliers) that it is possible to achieve a higher volume of sales or margins through the brand name as well as a stronger position and greater differential advantage, considering the important role played by memory in the centrality and primacy of the consumer and the brand value of the asset itself [12].

If the builders of the brand are the customers [13], consumer behavior is a good indicator of consumers' perceptions of the brand and of the meaning they give to the brand in their life, a space where memory plays an important role [11]. Brands can provide value for consumers by facilitating decision-making [14], reducing search costs [15], reducing the risks inherent to product acquisition and by providing emotional, hedonic, and symbolic benefits [16].

This mutual relationship represents a dichotomy between functional and emotional aspects because when choosing a brand, the consumer makes an evolution from both a functional and emotional point of view [17]. For Barnes (2003), it is in the interplay of these two parts—functional and emotional—that the consumer expresses an individual level either with the company or with the brand [18]. Rosenbaum-Elliott et al. (2011) distinguish between functional and symbolic brands considering two dimensions: the level of involvement and cognitive-emotional nature [19]. At the core of Fournier's (1998) work, we find relationships between consumers/customers and brands [4]. The relationship goes far beyond a functional utilitarian question, or a merely symbolic one as a generator of emotions and feelings. The reasons why consumers establish relationships with brands (or their image) are related to the meaning they have acquired in their lives and what brands add to it.

32.3 Consumer-Brand Relationship (CBR)

Since the 1990s, researchers have given CBR greater interest and importance. Due to organizations' interest in knowing more and more about consumer behavior, marketing base structuring and exchange become not only transactional but also relational [20]. Although there was already major concern with measuring a range of factors related to the parties involved consumers, suppliers, manufacturers, and employees, as well as customer satisfaction and loyalty [21, 22], the existence of two major schools of thought regarding the evaluation of performance or results were crucial in developing "customer relationship management" and opened the door to a new paradigm of relationship marketing [5]. These two schools are: (a) the group of Parasuraman et al. (1985, 1988), the United States school, with its SERVQUAL and Gap's model to measure the quality perceived by consumers [23, 24]; and (b) the group of Grönroos (1990) and Gummesson (1987), the Nordic school, with the Image Model (quality dimension) and the four dimensions of quality (conception, production, delivery and relationship) [9, 25]. The emergence of new constructs (from both the academic and managerial side) and new consumer behavior led to a quest for knowledge about this relationship, intending to deepen understanding of the relationship between brands and consumers, and vice versa.

Various constructs are used in CBR. Brand Love is a concept gaining prominence in marketing and consumer behavior, although few empirical studies deal with the manifestation of love for a brand. Other constructs are also recurrently quoted. In marketing, brand love, despite little consensus on what it means to love the brand, is an important topic, being the basic construct on which the whole CBR theory rests [26]. Several studies focus on different constructs in CBR (see Fig. 32.1).

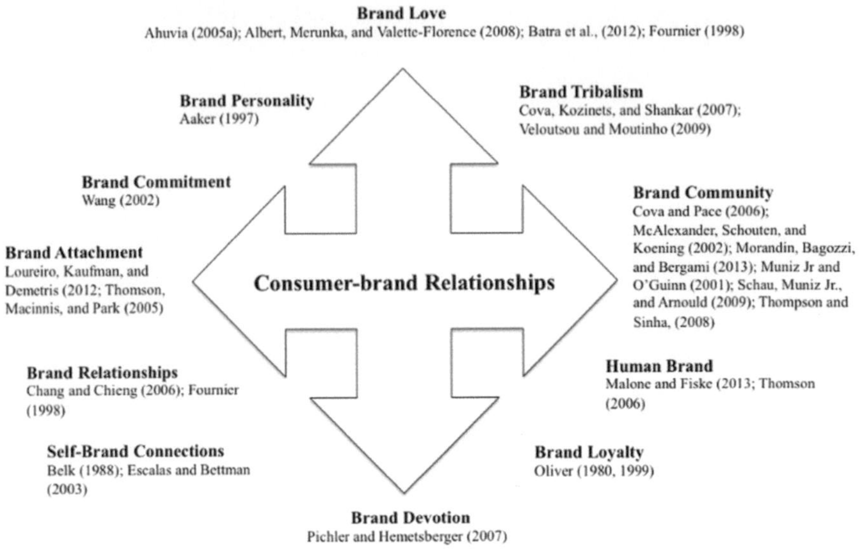

Fig. 32.1. Main constructs employed in CBR. Adapted from authors cited in the figure

32.4 Deep Relationship Investment

Rusbult (1980) proposed the Investment Model (IM), a model widely applied in various fields of knowledge, in both interpersonal relationships and non-relational contexts [27]. The IM is theoretically grounded in Interdependence Theory (IT), to examine the processes by which people persist within interpersonal relationships [6]. With its roots in the IT, which theorizes how behavior and motivation are structured and have a bidirectional relationship [28, 29], the concept of dependence is a major component of the theory [30].

In fact, IT was developed to explain how behavior and motivation are structured and have a bidirectional relationship [28, 29]. Therefore, the concept of dependence is a major component in the theory [30]. In the context of this theory, the rewards or costs could be physical or only psychological and may exist objectively or merely in the individual's subjective perception [27]. So, the IM aims to examine the processes by which people persist within interpersonal relationships [6].

The IM emerged in the 1980s as a new approach to interpersonal relationships, introducing the construct commitment, but its basic concepts are firmly rooted in existing psychological and sociological literature [27]. The outcome value and alternative value are parameters brought directly from the IT [28, 29]. To understand the concept of commitment, there are the studies of Becker (1960), Schelling (1956) and extrinsic investments as "side bets" in understanding the importance of commitment [31, 32], or Blau (2009, p. 79) that in the original 1964 edition shows "how valuable a woman is as a love object to a man depends to a considerable extent on her apparent popularity with other men" [33].

The Investment Model (see Fig. 32.2) extends interdependence propositions in two ways: the model identifies the third base of dependence (investment size) and suggests that dependence produces the psychological experience of commitment [30]. Investment size relates to "those concrete or intangible resources attached to the partnership that would be lost or seriously diminished upon relationship dissolution" [6], that could be "extrinsic (for example, home) or intrinsic (emotional involvement") (Rusbult 1980). Commitment is "the specific mechanism of constraint behavior through previous side bets" [31] (p. 40).

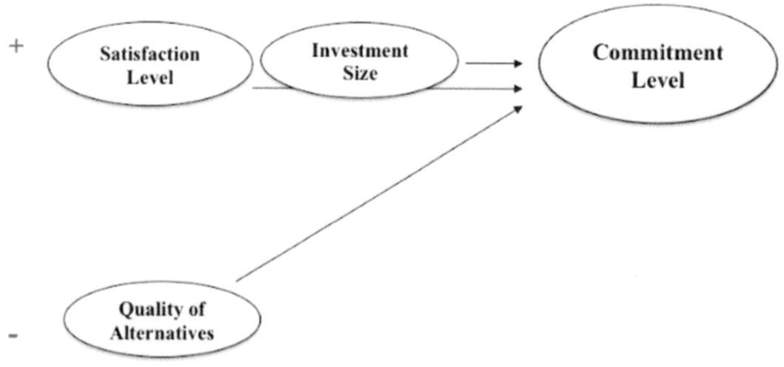

Fig. 32.2. The investment model—Rusbult (1980)

The first three components (satisfaction, quality of alternatives and investment size) are dependences (structural state), which lead to the fourth, commitment, representing the outcome of the model. Dependence is the descriptive, structural state of a relationship, whereas commitment is the psychological experience of that state [6]. In fact, it is the psychological experience of commitment, rather than the structural state of dependence, that is claimed to influence everyday behavior in relationships [30].

Commitment varies not only according to the outcomes and alternatives of the relationship but also the amount that someone invests in it [27]. The greater the satisfaction (some important needs are fulfilled) and the investment size (direct—time, money, or indirect—children, shared physical things) and the lower the quality of available alternatives (some important needs might not be satisfied without the relationship), the stronger the commitment to the relationship (Rusbult et al. 1998; Wieselquist et al. 1999). The importance of commitment is not only as a mediator in relationships, but it also acts as a "kind of glue" positively influencing the consistency of relationships [27, 34, 35].

Numerous studies (see [35]) have shown the use and robustness of the IM in predicting the end of the relationship or otherwise, partners' perspectives, willingness to sacrifice and the chimera of the superiority of one relationship over others. In addition, the IM has been used in a range of studies on diverse ethnicities [36], heterosexual and homosexual relationships [37], abusive relationships [38] and friendships [36]. Although the majority of results come from the field of interpersonal relationships, the model has also been applied in non-relational contexts, such as exercise behaviour [39], sport [40, 41], musical activities [42], clinical regime [43], schools [44], job commitment [34], constructs [45] and similarly in a residential community [46].

The primary goal of the IM is to predict the degree of commitment and satisfaction with a variety of forms of ongoing association (e.g., romantic, friendship, business) with wide ranges of duration and involvement [27], but it is by moving beyond the positivity effect that we can start to understand how and why some relationships persist, extending the interdependence orientation to better understand a variety of central processes in ongoing close relationships [35].

32.5 Discussion and Future Research

Rooted in social psychology regarding interpersonal relationships, Fournier (1998) introduced consumer-brand relationship theory which, from the perspective of consumer behaviour, proves to be a valid theory to help us understand the dynamics of the link between consumption and brand as well as determining the importance of the brand in the consumer's life [47]. Fournier (1998) proposed a theory of the relationship with brands through personalization of human traits in the brand, work that was the driving force in creating a new field of research: the consumer-brand relationship. While a brand cannot act and think like a human being, it can have distinctive features created by designers and exists through a bidirectional relationship with the consumer. Supported by the work of Batra et al. (2012), we believe that, for our study, the most important CBR models are those that incorporate love for the brand in the actual construction of consumer identity such as Ahuvia (2005); Albert et al. (2008); Escalas and Bettman (2005) and Fournier (1998), not taking into consideration those that do not use how consumers relate to brands to build or design their identity, that is, all theories derived from Sternberg (1986) [4, 48–50].

Considering the models presented and the work of Breivik and Thorbjørnsen (2008), Fournier's (1998) model called Brand Relationship Quality (BRQ) is the most representative [3, 4]: first, it is one of the most frequently applied models in empirical research into relationship brand management; secondly, it was developed specifically to assess the strength of consumer-brand ties; thirdly, it contains several relationship dimensions that influence relationship stability and durability; fourthly the relationship dimensions are founded on various interpersonal theories originating in Kelley and Thibaut (1978) [28]; and finally, love is a central construct.

Based on theories of close relationship but without the concept of love, another important and cited model emerges in CBR, with Relationship Investment (RI) at the core and called the Investment Model (IM). The Investment Model is proposed by Rusbult (1980) and found in social psychology [27]. According to Breivik and Thorbjørnsen (2008), who tested BRQ and RI and compare them "on the basis of the empirical model fit and interpretation" [3] (p. 443) the two models (BRQ and RI) are probably the relationship models most frequently applied in the re-concept search study of brand management being both global models, trying to provide an integrated view of the structure of consumer-brand relationships.

BRQ model was developed specifically to assess the strength of consumer-brand ties (Fournier 1998) and the RI model was originally developed to understand satisfaction and commitment in romantic relationships (Rusbult 1980), meaning that they were developed with different aims [3, 27]. That way both models vary either in the way they approach the structural model or in how the metaphorical relationship is transferred to the configuration of the consumer with the brand. However, they have one thing in common: both originate in theories of interpersonal relationships, i.e., have similar bases, in that they are derived from social psychology where theories of close relationships are found.

In our opinion, two aspects giving RI an advantage over BRQ: 1st the appearance of its conceptual framework, being clearly aligned with existing models and theories in marketing, where commitment is seen as a construct mediating factors that subsequently influence the final consumer behavior; 2nd with respect to the metaphorical transfer, it is structurally based on the same domain and most concepts are well established in the field of marketing.

The findings suggest that the RI model, even modified, offers a straightforward interpretation of consumer-brand relationships varying in intensity [3], which agrees with the work of other authors [47] by stating that the results of their study provide strong support for the RI model in predicting consumer-brand relationship ties. Therefore, the IM model is "one of the most frequently cited and applied relationship models in interpersonal relationship research" and a "new, fresh perspective on how consumers relate to brands" [3] (p. 447). Future research is needed in order to understand the implications of the IM not only in other fields besides the positive ones but also on brands.

References

1. Copeland, M.T.: Relation of consumers' buying habits to marketing methods. Harv. Bus. Rev. 1(3), 282–289 (1923)
2. Gardner, B.B., Levy, S.J.: The product and the brand. Harv. Bus. Rev. 33–39 (1955)
3. Breivik, E., Thorbjørnsen, H.: Consumer brand relationships: an investigation of two alternative models. J. Acad. Mark. Sci. **36**(4), 443–472 (2008). https://doi.org/10.1007/s11747-008-0115-z
4. Fournier, S.: Consumers and their brands: developing relationship theory in consumer research. J. Consum. Res. **24**(4), 343–353 (1998). https://doi.org/10.1086/209515
5. Loureiro, S.M.C.: Consumer-brand relationship relationship: Consumer foundation and state-of-art. In: Kaufmann H., Panni M. (eds.) Customer-Centric Marketing Strategies: Tools for Building Organizational Performance, pp. 414–434. Business Science Reference, Hershey, PA (2012). https://doi.org/10.4018/978-1-4666-2524-2.ch020
6. Le, B., Agnew, C.R.: Commitment and its theorized determinants: a meta-analysis of the investment model 10 (2003)
7. Ambler, T., Styles, C.: Brand development versus new product development: towards a process model of extension decisions. J. Prod. Brand Manag. **6**(1), 13–26 (1997). https://doi.org/10.1108/10610429710160002
8. Gabbott, M., Jevons, C.: Brand community in search of theory: an endless spiral of ambiguity. Mark. Theory **9**(1), 119–122 (2009). https://doi.org/10.1177/1470593108100066

9. Grönroos, C.: Relationship approach to marketing in service contexts: the marketing and organizational behavior interface. J. Bus. Res. **20**(1), 3–11 (1990)
10. Kapferer, J.-N., Laurent, G.: La sensibilité aux marques. In: Kapferer J.N., Thoenig J.C. (eds.) La Marque: Moteur de la Competitivité des Entreprises et de la Croissance de l'Économie, pp. 93–124. McGraw-Hill, Paris (1998)
11. Keller, K.L.: Strategic Brand Management: Building, Measuring, and Managing Brand Equity, 2nd edn. (2003)
12. Srivastava, R.K., Shocker, A.D.: Brand equity: a perspective on its meaning and measurement. Marketing Science Institute (1991)
13. Brito, C.: Uma abordagem relacional ao valor da marca. Revista Portuguesa e Brasileira de Gestão 9(1–2), 49–63 (2010)
14. Jacoby, J., Szybillo, G.J., Busato-Schach, J.: Information acquisition behavior in brand choice situations. J. Consum. Res. **3**(4), 209 (1977). https://doi.org/10.1086/208669
15. Murphy, J.: What is branding? In: Hart S., Murphy, J. (eds.) Brands, The New Wealth Creators, pp. 1–12. Macmillan, London (1998)
16. Srinivasan, T.C.: An integrative approach to consumer choice. In: Wallendorf M., Anderson P. (eds.) Advances in Consumer Research, pp. 96–100. Association for Consumer Research, Provo, Ut (1987)
17. Chernatony, L.: From brand vision to brand evaluation. J. Commun. Manag. Lond. **6**, 295–296 (2002)
18. Barnes, J.G.: Establishing meaningful customer relationships: why some companies and brands mean more to their customers. Manag. Serv. Qual. Int. J. **13**(3), 178–186 (2003). https://doi.org/10.1108/09604520310476445
19. Rosenbaum-Elliott, R., Percy, L., Pervan, S.: Strategic brand management (2011)
20. Morgan, R.M., Hunt, S.D.: The commitment-trust theory of relationship marketing. J. Mark. **58**(July), 20–38 (1994)
21. Oliver, R.L.: A cognitive model of the antecedents and consequences of satisfaction decisions. J. Mark. Res. **17**, 460–469 (1980)
22. Oliver, R.L.: Whence consumer loyalty? J. Mark. 63, 33–44; Bax, M.: Design science: filosofia da pesquisa em ciência da informação e tecnologia. Ciência da Informação 42(2), 298–312 (1999)
23. Parasuraman, A., Zeithaml, V.A., Berry, L.L.: A conceptual model of service quality and its implications for future research. J. Mark. **49**, 41–50 (1985)
24. Parasuraman, A., Zeithaml, V.A., Berry, L.L.: SERVQUAL: A multiple-item scale for measuring consumer perceptions of service quality. J. Retail. **64**(1), 12–40 (1988)
25. Gummesson, E.: The new marketing-developing long-term interactive relationships. Long Range Plan. **20**(4), 10–20 (1987)
26. Batra, R., Ahuvia, A., Bagozzi, R.P.: Brand love. J. Mark. **76**(March), 1–16 (2012)
27. Rusbult, C.E.: Commitment and Satisfaction in Romantic Associations: A Test of the investment Model **186**, 172–186 (1980)
28. Kelley, H.H., Thibaut, J.W.: Interpersonal relations: a theory of interdependence (1978)
29. Thibaut, J.W., Kelley, H.H.: The social psychology of groups (1959)
30. Agnew, C.R., Van Lange, P.A.M., Rusbult, C. E., Langston, C.: Cognitive interdependence: commitment and the mental representation of close relationships. J. Personal. Soc. Psychol. 74(4), 939–954 (1998)
31. Becker, H.S.: Notes on the concept of commitment. Am. J. Sociol. **66**(1), 32–40 (1960)
32. Schelling, T.C.: An essay on bargaining. Am. Econ. Rev. **46**(3), 281–306 (1956)
33. Blau, P.M.: Exchange and power in social life (2009)

34. Rusbult, C.E.: A longitudinal test of the investment model: the impact on job satisfaction, job commitment, and turnover of variations in rewards, costs, alternatives, and investments 68(3), 429–438 (1983)
35. Rusbult, C.E., Martz, O.J.M., Agnew, C.R.: The investment model scale: measuring commitment level, satisfaction level, quality of alternatives, and investment size 5, 357–391 (1998)
36. Lin, Y.-H.W., Rusbult, C.E.: Commitment to dating relationships and cross-sex friendships in America and China. J. Soc.Pers. Relationsh. **12**(1), 7–26 (1995). https://doi.org/10.1177/0265407595121002
37. Duffy, S.M., Rusbult, C.E.: Satisfaction and commitment in homosexual and heterosexual relationships. J. Homosex. **12**(2), 1–23 (1986)
38. Rusbult, C.E., Martz, J.M.: Remaining in an abusive relationship: an investment model analysis of nonvoluntary dependence. Pers. Soc. Psychol. Bull. **21**, 558–571 (1995)
39. Wilson, P.M., Rodgers, W.M., Carpenter, P.J., Hall, C., Hardy, J., Fraser, S.N.: The relationship between commitment and exercise behavior **5**, 405–421 (2004). https://doi.org/10.1016/S1469-0292(03)00035-9
40. Carpenter, P.J., Scanlan, T.K.: Changes over time in the determinants of sport commitment. Pediatr. Exerc. Sci. 10(see 20), 356–365 (1998)
41. Raedeke, T.D.: Is athlete burnout more than just stress? A sport commitment perspective. J. Sport Exerc. Psychol. **19**, 396–417 (1997)
42. Koslowsky, M., Kluger, A.: Commitment to participation in musical activities: an extension and application of the investment model. J. Appl. Soc. Psychol. **16**(9), 831–844 (1986)
43. Putnam, D.E., Finney, J.W., Barkley, P.L., Bonner, M.J.: Enhancing commitment improves adherence to a medical regimen. J. Consult. Clin. Psychol. **62**(1), 191–194 (1994)
44. Geyer, P.D., Brannon, Y.S., Shearon, R.W.: The prediction of students' satisfaction with community college vocational training. J. Psychol. **121**(6), 591–597 (1987)
45. Ping, R.A.: Voice in business-to-business relationships: cost-of-exit and demographic antecedents **73**, 261–281 (1997)
46. Lyons, W.E., Lowery, D.: Citizens responses to dissatisfaction in urban communities: a partial test of a general model. J. Polit. **51**(4), 841–868 (1989)
47. Sung, Y., Choi, S.M.: "I won't leave you although you disappoint me": the interplay between satisfaction, investment, and alternatives in determining consumer-brand relationship commitment. Psychol. Mark. 27(11), 1050–1073 (2010). https://doi.org/10.1002/mar
48. Ahuvia, A.: Beyond the extended self: loved objects and consumers' identity narratives. J. Consum. Res. **32**(June), 171–184 (2005)
49. Albert, N., Merunka, D., Valette-Florence, P.: When consumers love their brands: exploring the concept and its dimensions. J. Bus. Res. **61**(10), 1062–1075 (2008). https://doi.org/10.1016/j.jbusres.2007.09.014
50. Escalas, J.E., Bettman, J.R.: Self-construal, reference groups, and brand meaning. J. Consum. Res. **32**(3), 378–389 (2005). https://doi.org/10.1086/497549

Chapter 33
Information Society: Access to Public Information

Susana Molina, Gina Parrales, Soraya Linzán, and Arturo Clery(✉)

Universidad Estatal Península de Santa Elena, La Libertad, Ecuador
clery@upse.edu.ec

Abstract. From the various international declarations and the rules on the right of access to information, its specific content can be better understood. In this scenario, this paper will explore the framework of the principles that must be supported by the new Ecuadorian legislation on the right to access public information. The general guarantee of the right to access public information under international law establishes a general opening in favor of the dissemination of information that is held by public entities. This implies not only that states must guarantee the right to information, but also that effective systems must be established to enforce it. Although the regimes on the right to information vary in different countries of the world, there are also several notable similarities (Mendel, El derecho a la información en América Latina: comparación jurídica. Organización de las Naciones Unidas para la Educación, la Ciencia y la Cultura, Quito (2009)) [3]. These similarities will be the basis of the principles discussed below.

33.1 Obligation of Active Publication and Access Facilitation

Public entities must have, in the first instance, the active obligation to publish the most relevant information. In order to give effect in practice to the right to information, it is not enough to require public entities to access information requests; Effective access depends on these entities actively publishing and disseminating key categories of information even in the absence of any request.

Freedom of information implies that State entities publish and disseminate documents of significant public interest, for example, on operational information on how the entity works and the content of any decision or policy that affects the public.

Public entities must be required, even in the absence of a request, to actively publish important information of significant public interest. Every public entity, at its own initiative and when appropriate, must disseminate information with a view to promote public administration transparency, administrative efficiency, and well-informed public participation.

Proactive dissemination of a broad spectrum of information, including policy information, consultation opportunities, budget, subsidies, benefits, contracts, all activities that affect community members is important.

Á. Rocha et al. (eds.), *Marketing and Smart Technologies*, Smart Innovation, Systems and Technologies 167, https://doi.org/10.1007/978-981-15-1564-4_33

Laws on the right to information must require public entities to produce and reproduce lists of information. The scope of this obligation depends to some extent on resource limitations, but the amount of information covered should increase over time, particularly as new technologies facilitate the publication and dissemination of information.

Systems should be established to increase, over time, the quantity and quality of information subject to routine disclosure. The long-term objective should be to make the information proactively available, to minimize the need for individuals to have to resort to requests to access it.

Requests for information should be processed in a timely and fair manner, and an independent review of any rejection should be available. Ensuring the right to information effectively in practice requires not only proactive disclosure by public entities but also that anyone can request and receive any information they have, subject to exceptions.

This, in turn, requires that clear procedures be established, according to which public entities will process requests for information. In addition, it requires a system for independent review of internal decisions by these public entities. The processes for accessing information are complex and these normally occupy a large part of the laws on the right to information.

Public entities must establish open and accessible internal systems to ensure the public's right to receive information, specifically referring, in this regard, to the need for strict time limits for the processing of information requests and to give notice of any refusal to give access, which will include substantive written reasons for the denial.

The procedures must be fast, with clear, fair, non-discriminatory, and simple rules to process applications, must include clear and reasonable time limits, forecasts to assist those who request information, free or low-cost access, that do not exceed the cost of copy and send the information, and a requirement that when access is denied, reasons are provided that include the specific basis for the denial immediately [2].

The following are some recommendations on the processes:

- Applications must be processed by any public entity that has the information, in terms of equality and with minimal formality.
- Applicants should not provide reasons for their requests.
- Applications must be addressed in a timely manner and within the established deadlines.
- Assistance should be provided as far as possible.
- The reasons for any refusal to grant access must be given.
- Applicants should be given access in the way they prefer, by inspecting the registry or providing them with a copy.

It is also well established that any refusal by the public entity to disclose the information, or the failure to respond to the requests in the prescribed manner, must be subject to appeal. Many laws have an internal appeal to a higher authority within the same public entity to which the request was made. This is a useful approach, which can help amend errors immediately and increase the confidence of officials to disseminate information.

However, it is crucial that the right of appeal to an independent entity be available to review decisions made by public entities. In the absence of this resource, it cannot be affirmed that individuals really have a right to access information that is in the hands of public entities, but merely the right to have their requests for information considered. Without an independent review, there is a lot of information that will never be disclosed.

While the various international declarations on the right to information clearly require an independent review, they are somewhat less clear as to the nature of such review and in particular, if the review by an independent oversight entity, such as an Information Commission or the Ombudsman is required, or if a review by ordinary justice would suffice.

In practice, the most progressive laws on the right to information do have an appeal before an independent monitoring entity. This is much more accessible than ordinary courts of justice, for ordinary people seeking information and is an effective way to secure this right. It does not matter if it is necessary to establish a new entity for this purpose or if the task is assigned to an existing entity, what matters is that the entity has adequate protection against political interference.

Ultimately, it is also favorable to provide that the law allows the appeal to ordinary courts of justice. Only these courts have the real authority to set the disclosure standards in the controversial areas and ensure the possibility of a full and well-reasoned approach to difficult disclosure issues.

An important aspect of the appeals that is also widely respected in the norms analyzed and that constitutes best procedural practices in the right of access to information is the fact that public entities that seek to deny access to information have the obligation to verify that your refusal is legitimate. This results from the idea—and is even central to the concept—that access to information is a right and that the obligation of proof must correspond to the party that intends to deny access, which has already been mentioned in a previous section.

Respect the costs of access, individuals should not be prevented from requesting information through the implementation of excessive costs. Charging fees for access to information is a very difficult matter. On the one hand, if the rates are excessive, they will constitute a barrier to access, which will harm the right. On the other hand, it does have its costs for public entities to provide the services of allowing access and these Administrations should have some way to recover these costs.

The cost of access should not be so high that it discourages potential applicants, denying the intention of the same law. Access must be free or low-cost. Collections for copies should not exceed the costs actually incurred.

The public authorities that intend to impose the payment of a right for the information they provide, will inform the applicants of the information the fees of the rights to be paid, indicating the cases in which the authorities may waive the perception of those rights and the cases in which the communication of information is subject to an advance payment.

33.2 The Need to Change the Administrative Culture

Public entities must actively promote government openness. In most countries during the last century, there was a deep-rooted culture of government secrecy, based on long-standing practices and attitudes. The right to public information depends on changing this culture since it is virtually impossible to force public servants to be generous, even with more progressive legislation.

Long-term success depends on convincing public officials that openness is not only an obligation but also a fundamental and central right to effective governance. A whole range of promotional measures may be necessary to overcome the culture of secrecy and ensure that the people know the right to information and its implications.

States must take measures to promote, implement, and enforce the right to information, including, specifically, aspects of record management, training of officials, programs to raise public awareness, and reports on implementation.

The specific promotional measures required will vary from one country to another. The assignment of central responsibility for various measures, for example to a dedicated oversight body, such as an Information Commissioner, the Ombudsman or a central government department, will provide a point of responsibility to ensure adequate resources are allocated and attention to this important issue, which was already analyzed in a previous section.

A useful tool, provided in many laws on the right to information, is the publication of a simple and accessible guide on how to request information; in this task, social media play a key role and electronic media plays a particularly important role.

Another instrument that will help to overcome the culture of secrecy is to establish penalties for those who intentionally obstruct access to information in some way, including the destruction of records or interference with the work of the watchdog.

Any person who intentionally denies or obstructs access to information, in violation of the rules, must be subject to sanctions. Such penalties may be administrative, civil or criminal, or some combination of the three classes. For example, claim damages when losses are suffered as a result of the violation of the law; likewise, criminal sanctions will offer an important and radical message to officials, preventing any obstruction of access to information that will not be tolerated.

Other means have been tried to address the culture of secrecy, including incentives for those who perform well and discredit for those who do not, and also ensure the monitoring of informational progress through annual reports on the performance of public entities in the application of the law at the information.

In many countries, one of the biggest obstacles to access information is the poor conditions in which records are kept. Officials often do not know what information they have or, if they do, they cannot locate the records they are looking for. Adequate management of official documents is central to the effective application of the right to information. Information management is one of the basic functions of modern government and doing it well is crucial for the effective provision of all public services.

Public authorities must be required to comply with the minimum standards in records management. Systems must be established to promote more demanding standards over time; governments must maintain and preserve records; as well as the need for clear and established rules for the preservation and destruction of documents.

33.3 Hierarchy of the Rules of Access to Information and Protection for Whistleblowers

It is up to the States to determine how to address the issue of exceptions to the right of access to information produced by norms prior (or subsequent) to the law on access to information. States often have a range of laws on confidentiality in their laws, but many do not comply with several of the standards analyzed, particularly those related to exceptions.

Over time, different states must commit to the revision of their laws that restrict the disclosure of information and limit access to public information, with a view to align them with the law of the right to access information.

The rules of confidentiality of the different laws should be amended as necessary, to comply with the principles of freedom of information. However, in most cases, this is a medium-term solution. A short-term solution, which allows an immediate effect on the right to information, is to provide that the law that establishes the right to information has a higher hierarchy over the laws of confidentiality.

This should be achieved by applying a restrictive interpretation of the laws on confidentiality; in this way, when a conflict arises, the law on the right to information may precede the law of confidentiality that is in conflict.

The law of access to information must, to the extent that there is any inconsistency, prevail over other laws. Most laws on the right to information include a complete set of exceptions that protect all legitimate interests of confidentiality, so there should be no need for these exceptions to be extended through confidentiality laws.

Some systems to resolve conflicts between legal regulations will be necessary to prevent public servants from being placed in the position that a confidentiality law prohibits them from disclosing information, which, on the contrary, the law of the right to information requires them to do.

On the protection of whistleblowers, individuals who provide information on incorrect actions should receive protection. If officials can expose themselves to penalties, for erroneously delivering information in accordance with the right to information, they will probably tend to be conservative, preferring the secret that is more usual anyway.

As a result, many laws on the right to information provide liability protection for officials who, in good faith, disclose information under the law of the right to information. This protection is important to change the culture of secrecy within the government and foster a climate of openness [4].

Similar protection is provided in many countries to individuals who provide information about wrongdoing. It is often unclear whether the disclosure of information about improper actions under the law is justified, even if that law includes taking into account the best public interest and individuals seeking to disclose information for the benefit of the public interest cannot be expected., undertake a complex calculation, weighing the different interests that could come into play. By providing this protection, the flow of information to the public about various types of wrong actions is encouraged.

No one should submit to any sanction for delivering in good faith information about improper actions, or facts or actions that reveal a serious threat to health, safety or the environment, except when the imposition of sanctions meets a legitimate interest and is necessary for a society democratic. It is necessary to protect the official from any legal, administrative or labor sanction for delivering information about undue actions.

33.4 Limitations to the Exercise of the Right of Access

Legislation on the right of access to information must be guided by the principle of maximum openness, which is derived from the primary guarantees of the right to public information, which represents its central meaning. The principle of maximum openness implies that the scope of the right to public information must be as wide as the range of information and public and sometimes private entities, as well as individuals who can claim the right.

Based on this principle, the assumption is made that all information is accessible, subject to just a brief system of exceptions, while limitations are necessary, "the general rule is that for any restriction of rights, they must be directed to satisfy an imperative public interest and among several options to achieve that objective, one must choose the one that restricts the protected right on a smaller scale" [1].

In practice, the laws, regulations, and decisions discussed in previous sections did define information and its access in a fairly broad way; however, these same rules also proposed a list of access limits, necessary to safeguard other higher rights and interests.

The right of access to public information claims to comply with the obligation to favor, at all times, the principle of maximum openness and publicity of the information, making effective the possibility of delivering documents containing confidential and reserved information, provided that it is protected [5].

Access to information is not an absolute right. Despite being governed by characteristics that promote maximum openness of information, it has restrictions and exceptions, which must be based on principles of reasonableness and proportionality. Like administrative practices, the interpretation of these restrictions and compliance with the principle of publicity of information have a direct impact on the satisfaction or not of the exercise of the right of access to information.

To affirm that public information is of general interest, free access and property of citizens, with reasonable exceptions, implicitly implies the existence of a catalog of restrictions that imply the obligation to protect certain information. This information, in addition to personal data, also reaches that which compromises national security or puts public safety or even public health at risk.

- The reasoning for which it is not possible to know what was requested is duly justified, informing not only of the existence of a legal cause of restriction, but also verifying that in effect the request falls within that hypothesis.

- The time of reservation of the information is indicated.
- The obligation to report the existence of a resource is fulfilled.

Deficiencies in complying with the imperative to establish and motivate the classification of information (whether reserved or confidential) immediately result in a violation of the right of access to information, since the right only admits justified exceptions [6].

Therefore, the answers in which a denial of access is notified must incorporate the logical reasoning that proves that the particular case is the one provided in the legal assumption and demonstrate, with reasonable arguments, that the disclosure of the information implies present damage, material and direct.

The exceptions must be clearly and demandingly defined, and subject to strict evidence of damage and public interest, although in practice it is complex to determine the legitimate scope of the exceptions for the exercise of the right to access public information. On the one hand, a system of exceptions that is too broad can seriously harm the right. In some cases, the laws on the right to information that are very effective are greatly weakened by an excessively broad or open system of exceptions.

On the other hand, it is obviously important that all legitimate confidentiality interests be adequately addressed because otherwise it will be legally required for public entities to disclose the information even if it causes them exaggerated damages.

The right of access must be subject to a narrow and carefully crafted system of exceptions to protect the superior public and private interests. The exceptions should be applicable only when there is a risk of substantial damage to the protected interest and when that damage is greater than the global public interest of being able to access the information.

The complexity and also the importance of this issue were reflected in the international standards analyzed in the previous sections. Exceptions to the right of access to public information must be limited and formulated in precise terms. The limits to the right of access must be exceptional, previously established by legal means and in response to a real and imminent danger, which threatens national security in a democratic society.

The refusal to disclose information cannot be based on the purpose of protecting governments from an embarrassing situation or on the disclosure of their wrongdoing; A complete list of legitimate purposes that could justify the non-disclosure of public information should be provided in the law and exceptions should be formulated in precise terms, to avoid the inclusion of exaggerated restrictions that do not affect the legitimate interest.

Different laws on the right to access information recognize several legitimate purposes that may be subject to exceptions to the right to access, a subject that never ceases to be controversial. Some of the possible limitations for access to official documents can be:

- National security, national defense, and the protection of international relations.
- Public safety.
- The prevention, investigation, and prosecution of criminal activities.
- Privacy and other legitimate private interests.
- Commercial and other economic interests, whether private or public.

- The equality of the parties in legal proceedings.
- Inspection, control, and supervision by public authorities.
- The economic, monetary, and exchange rate policies of the State.
- Confidentiality of deliberations within or between public authorities during the internal elaboration of a matter, among others.

The reasons for refusal mentioned above and others that may arise from the reality of each company must be interpreted restrictively, taking into account the interest that the disclosure of the information requested would have for the public.

It is clear from the analysis of the general principles on the right to information, that it is not legitimate to deny access to information simply because it relates to any of the listed interests, but that the justification of the denial must be proven. The burden of this proof of denial of access to public information must fall on the public body that denies the request and must be duly based on three pillars:

- The information must relate to a legitimate purpose listed in the law.
- The disclosure must pose the threat of substantial harm to that purpose.
- The damage to the purpose must be greater than the public interest in having the information.

In a small number of cases, the damage is implicit in the nature of the exception. This is the case, for example, of legally privileged information or when the disclosure represents a breach of trust. However, the vast majority of exceptions that do not include a specific reference to the damage do not have this automatic damage.

For example, the public force has a lot of information that has a tangential relationship with its operations, administrative information related to its purchases of food or stationery. It is clearly not legitimate to deny access to this information because they are not direct expenses for the national defense since its disclosure would not harm any interest of the defense, however, the same principle could not be applied if it were weapons or espionage activities legitimate.

In some countries, the exceptions themselves are subject to limits (exceptions to the exceptions) to take into account cases in which there is no harm to the legitimate purpose. An example of this is presented when the information is already publicly available, in which case any damage has already been done, or when an affected third party has given consent for disclosure, effectively waiving any claim for the damage.

As carefully as a system of exceptions is drawn up, there will always be cases in which the higher public interest requires the disclosure of the information, even if this causes some damage to the protected interest. This happens partly because it is not possible to draw up exceptions that take into account or encompass all cases of higher public interest and partly because the particular circumstances of the given moment may symbolize that the general public interest requires the disclosure of information.

An example would be sensitive military information that exposed corruption in the armed forces. Although the disclosure might seem, at first glance, a weakening of national defense, on the other hand, to eliminate corruption in the armed forces, it may eventually strengthen it.

It is also very established in practice that when only part of a record is confidential, if possible the rest should be disclosed.

On the same issue, the general time limits for retaining information can help ensure that the damages of yesteryear do not continue to keep the information secret indefinitely. In many cases, the risk of harm that originally justified the limitation disappears or is substantially reduced over time.

For example, most laws on the right to information provide some protection for internal deliberative processes or advice within the government. Although this is justified in the short-term, the risk of disclosure in 15 or 20 years can hardly scare those who freely and frankly provide advice today, which is a key interest that is protected by this exception.

The time limits for the dissemination of historical information allow us to presume that the original damages are no longer important after several years, after which, if those interested in limiting access to such information persist, it would be necessary to justify specifically a continuation of the retention of that information.

Another aspect that has generated the greatest concern in the administration of data banks and public records is the possible irresponsible dissemination of personal data, also called sensitive information, that has been entrusted to public authorities.

It is important to point out that the principle of maximum publicity or as open as possible, can and should harmonize perfectly with the protection and guarantee of safeguarding information about an identified or identifiable person, information that is held by an institution, such as that refer to age, religion, ethnicity, sex, political affiliation, domicile, marital status, heritage, among others.

In this sense, there are criteria that allow, in certain particular cases, to overcome the dilemma of delivering information that contains personal data. In the first instance, those data that, being effectively personal, justify the suitability of the hiring of a certain person for a public service (for example, work and professional background), or those related to their salary perception, should not be considered confidential, including bonuses, compensation, benefits, and other income [5].

This information is closely related to government transparency and adequate public management, so they are not related to the private sphere of officials, but to the public aspect of their actions.

Another criterion is that which allows the reproduction of documents containing information considered as reserved or confidential, provided that the elimination of the parts classified as confidential or reserved is allowed. In this way, the development of public versions is the alternative that harmonizes two rights, allowing access to documents and safeguarding that information that according to the law is restricted.

33.5 Main Procedures for the Satisfactory Attention of Requests for Access to Information

The right of access to information cannot be understood except through its exercise. Reason why administrative practices and procedures have a direct impact on achieving or not your satisfaction. This section presents some proposals that can contribute to achieving them.

In the context of carrying out a procedure, it is important to keep a record of the presentation of a trade before an authority, public body or government agency. In the case of requests for access to information, this procedure is directly linked to the formal actions that have to be initiated to address it and the certainty that this procedure has actually begun.

Therefore, it is considered that the first aspect to be addressed in the attention of requests for information is the immediate notification of acknowledgment of receipt of the request submitted. The acknowledgment must indicate, at a minimum, the date on which the application is presented and the signature or reference of who is taking the process as initiated. It can be complemented with the deadlines for responding, the transcript of the request being received, the confirmation of the applicant's data and the method in which he prefers the information to be delivered.

Issuing an acknowledgment of receipt with these characteristics is relatively simple and the benefits are immense: it grants certainty about the receipt of the request and the text of the petition, commits the public servant as responsible for the processing, and sets the deadlines for the completion of the Procedure. Therefore, it is a practice that must be implemented in the action to facilitate and facilitate access to information.

Regarding the response time to a request for access to public information, access to information is sometimes a precondition for the satisfactory exercise of other rights. Obtaining it in a timely manner can determine the legal sphere of the person who requires the information, such as in the case of an adequate legal defense. In these cases, in order for the information to be useful, it is required, in addition to its simple delivery, that it be done expeditiously.

In terms of access to information, it is essential, to be satisfied with the right, to obtain a response in time. This implies the existence of reasonable deadlines that are in effect observed. Regardless of the days indicated by each legislation, it is a fact that must be complied with, ensuring that the citizens have access to the required information. For these purposes, response times must be considered based on the time elapsed between the date on which the request is registered and the time when the administrative unit effectively makes the response available to the applicant.

It is clear that responding to a request may require more time due to logistical or material complications. This is to be assumed, that the greater the diversity of the information requested, administrative areas involved, seniority, volume and classification requirements, the information will be made available to the applicant over a longer period of time; to the extent that the above aspects or other conditions that involve difficulties decrease, the delivery time should also decrease [5].

Therefore, the maximum terms that legislation contemplates to deliver the information should be exhausted only exceptionally and for sufficiently justified reasons. Efforts should focus on getting information to the public in the shortest possible time.

33.6 Attention as Much as Possible to the Request and Types of Responses

One measure to increase the quality of the responses is that the request of the interested party is fulfilled as much as possible. For this it is important, when dealing with various requirements, to individually address each of the approaches made by the applicant.

This implies systematically and logically addressing each of the questions posed, stating in the answers explicitly the item that you intend to satisfy. For example, in cases where file or document reproductions are attached or accompanied, it must also be indicated to which point each one corresponds to what is requested. The foregoing, despite being strictly considered in form, has a broad impact on the qualitative and substantive aspects of the responses.

In accordance with the criterion of attending to the request as much as possible, in cases where it is not possible to deliver the information, it is appropriate to clarify the reasons: for nonexistence; because what is requested is not understood; or because some restriction provided by law was updated.

The practice of simply ignoring the request must be considered unacceptable. It is true that the authority before which the request is presented is obliged to deliver only the information that is in its possession, but this does not imply omitting this precision in the response to the petitioner.

Otherwise, as the reason why a specifically requested aspect has not been addressed is not clarified, the interpretation could be that this information has been unduly denied.

Another important aspect of assessment is the way in which a file is concluded, that is, the classification by which it is terminated: information submitted, nonexistent information, information to be delivered, classification of confidential or confidential, etc.

This becomes relevant for two reasons: first, it is the way in which the institution will consider this matter in its internal indicators and with them it will try to correct its performance; second, through the statistical grouping of this information, the aim is to show the citizens the degree of compliance with the obligation to give access to the information. If the information is not disseminated correctly and truthfully, the public perception will be wrong.

On this matter, it should be noted that there are considerable differences between a properly answered request and the information submitted. For example, if a citizen is notified in a timely manner that what was requested was considered confidential by the respective administrative unit, said request will be considered as properly attended, but not as information provided. Similar exercises can be done regarding nonexistent information or information to integrate.

On this point in particular, it is relevant that in the place where the classification or status of information requests is reported to the public, the reasons and arguments that accompany such classifications are also made public. In this way, one would have notion not only of the type of information that the public is interested in but also of the criteria under which access is granted or denied; even, if information that should exist has not been integrated and, therefore, it was not possible to deliver it.

33.7 Notification of the Existence of Resources and Other Requirements

Most of the regulatory frameworks regarding access to information provided that the applicant has the right to file an appeal, not only because of the denial of access to information but also because they consider that the response time, cost or modality delivery were not adequate.

The Resources Can also Be Presented When the Information Offered Is Considered Incomplete or Does not Correspond to that Required

It is suggested that all resolutions and notifications issued by the administrative units regarding the processing of an application, state that an appeal may be filed, indicating the reasons on which it proceeds and the legal basis for it. But in a special way, the notification of the existence of a resource must be made known to the applicant when access is denied for having classified the information as reserved or confidential.

The above is relevant since most applicants are not experts in the area of access to information. By making its knowledge the possibility of challenging those cases in which the response does not satisfy them, the possibilities of exercising this right in all the dimension that has been conceived are extended and, if necessary, the lack of expertise in the topic.

In order to contribute to the prevailing principle of greater openness, it is desirable that in practice the administrative units carry out actions that promote, strengthen, and facilitate the use of the resources of nonconformity provided. This would promote a review of the classifications that imply restrictions on access to information, while at the same time correcting administrative errors or wrong classification criteria. Small proactive actions in this regard would have a relevant impact. The notification of the right to file an appeal is one of them.

In addition to the previous points, it is decisive to achieve the objectives of access to information that the responses meet other minimum quality standards, among which are: the presentation of information in simple formats (to facilitate greater understanding by the citizen), privilege the delivery method indicated by the applicant and avoid unnecessary transfers or expenses. It is enough to deliver incomprehensible, incomplete or insufficient information that does not correspond to what was requested or establish complex procedures to affect the object and purpose of access to information.

Access to a document or even a file, can be a failed action, when meaningless data is obtained or when accessing the file it is unknown how to obtain the desired information. Therefore, when the right of access to public information is exercised, it is done with the conviction that the person has the right to access documents and files, and in general, the information held by the Public Administration. In this sense, access to public information will be applied to that recorded in the records and files held by public administrations, to documents and to information that may be inferred from their computer systems [7].

For example, if access to a list of files is requested with the intention of knowing which authorities were involved in the complaints, what was the conflict resolution process, the region or place where the events occurred and the classification of the act allegedly Violation of rights, the information provided must be clear from what was requested. It should not be presented, for example, as a set of numerical keys in vertical columns, which do not allow to know what each one corresponds to; and in any case, the use of codes and nomenclatures must be accompanied by their respective explanations [5].

Regarding the delivery method that the requesting party prefers, that is, the direct access, the simple copy, the certified copy, the electronic version or any other recognized by the applicable regulatory framework, this is important, since for the access the information is considered complete, it is not enough that the documents are made available to the applicant through any of the means indicated, but it is also essential that the medium selected by him be privileged. In other words, except for fully justified exceptions, it must comply with the requested modality.

References

1. Covarrubias, I.: La vida privada de los funcionarios públicos frente a dos derechos: el acceso a la información pública y la libertad de expresión. Revista Lus et Praxis **21**(1), 217–270 (2015)
2. Martínez, E.: Transparencia en la publicidad institucional de la administración pública. El papel del derecho de la información en la sociedad del conocimiento. Actas del Seminario Internacional Complutense, pp. 285–295 (2008)
3. Mendel, T: El derecho a la información en América Latina: comparación jurídica. Organización de las Naciones Unidas para la Educación, la Ciencia y la Cultura, Quito (2009)
4. Mendel, T.: Libertad de información: comparación jurídica (segunda edición). Organización de las Naciones Unidas para la Educación, la Ciencia y la Cultura, Quito (2008)
5. Pulido, M.: El acceso a la información es un derecho humano: propuesta para un estándar de acceso a la información de organismos públicos de derechos humanos, vol. 2. Mono Comunicación, México (2006)
6. Ramos, F.: El acceso a la información pública y su control político. En El Papel del Derecho de la Información en la Sociedad del Conocimiento, pp. 249–266 (2008)
7. Sánchez-de-Diego, M.: Un derecho fundamental a acceder a la información pública. El papel del derecho de la información En La sociedad del conocimiento: el derecho de acceso a la información pública, pp. 7–41 (2008)

Chapter 34
Facebook and Polícia de Segurança Pública: An Exploratory Study of Follower's Engagement

Sónia M. A. Morgado[1(✉)], Tiago Moniz[2], and Sérgio Felgueiras[1]

[1] Major Events Laboratory, Research Center (ICPOL), Instituto Superior de Ciências Policiais e Segurança Interna, Lisbon, Portugal
smmorgado@psp.pt
[2] P.S.P, Largo da Penha de França 1, Lisbon, Portugal

Abstract. The engagement with social networks promotes a flow of information for users and owners of the social media pages. Even in law enforcement agencies, it's necessary to evaluate the interaction of social media users. As a first approach to this subject, this study describes the engagement of the followers with the Facebook page of Polícia de Segurança Pública. An exploratory study was done, sustained by an online survey, resulting in 2.311 operational responses from the followers. Our results are displayed according to the different types of engagement: cognitive, behavioural and emotional and the global one. The cognitive and emotional engagement stands out in the analysis. Furthermore, the analysis of the result shows that global engagement is positive. The sample is mainly composed of feminine elements, with a university degree, and that spends more or less on to four hours a day on social media. The findings can be useful for the communication and public relations department of PSP to evaluate the social media strategy that has been followed.

34.1 Introduction

Social media provides the spurs and the limits that form this paper so that it has character development and pattern. The story behind the schemes of social media starts in the '90s, of the twentieth century with the development of the World Wide Web, alongside the development of the networked society [1]. This development led to the search for new ways of communication, accessibility to contents, transmuting processes [2] and inducing pragmatism in those who use the strategy and tactics to make the best use of social media.

Since web 2.0, defined as participative web, each one delivers information, contributes to development of applications and resources [3], therefore it is undeniable the dynamism of the information society [4] and the knowledge contained by the users and followers, that turn on or off from the pages in consideration to interaction and depth of the dialogue [5].

Á. Rocha et al. (eds.), *Marketing and Smart Technologies*, Smart Innovation, Systems and Technologies 167, https://doi.org/10.1007/978-981-15-1564-4_34

As national, transnational and global phenomena are spatially unlimited, every organisation, institution and brands have a social page in multiple social media networks, promoting engagement between stakeholders. The followers and the population had an inherent factor within, the factor of power. Thus, the importance of enhancing the bonds, to overcome psychological barriers concerning law enforcement agencies and control the systematic outburst of cognitive, emotional and behavioural vies.

As a result of an integrative process and cultural homogenisation [6], globalisation overcomes all the barriers, social, geographic, economic and cultural ones [7]. Its core is sustained by the virtual environment, mainly in social media. Following the path of previous investigations on social media in law enforcement [8–10], this study is another contribution.

Among other social networks, Facebook is the most common one for enterprises, brands, political parties, religious organisations, non-governmental organisations and law enforcement agencies [11].

Arising from the urgency of presenting a theoretical corpus that allows the comprehension of the engagement, in which permanency would be like going to the fictional universe of Nose Dive, from Black Mirror of Joe Wright, or even in the neo-depressive and techno-futuristic societies in which life is lived in a dystopian, post-apocalyptic country of The Hunger Games of Collins [12], is necessary.

The contact with the basic and theoretical concepts of engagement and social media is required, to ensure a diachronic platform that results from the cross theoretical and empirical approach.

In the current environment, strong relations between law enforcement agencies, and in this case, Polícia de Segurança Pública (PSP), with the followers or his citizens, provides a measure of success of the proximity strategy, and provides a competitive advantage for being able to perform the tasks with more efficiency and positive public recognition. Police officers and other policymakers recognise the importance of the dimensions of technology, quality and performance. Therefore, the linkage between knowledge allows the incessant search for knowledge to determine the engagement between followers and the PSP Facebook page. This study incorporates the levels of engagement and the capacity of PSP to promote synergy, in a constant action–reaction frame, establishing diachronic dynamics. The dynamics might explain the coherence and the shift of PSP paradigm towards the social media environment.

The paper is organised in a traditional form starting with the introduction; state of the art (Information and Communication Technologies [ICT], Facebook; engagement); method (conceptual model, data and analysis); results and discussion and conclusions.

34.2 State of the Art

This section starts with a list of four elements and preconditions for the establishment of engagement and social media: ICT, social media, Facebook and engagement. These features are not independent of each other but conjointly underpinning. Increasing use of ICT means more ability to be an effective user of social media. Using social media implies a choice, or multiple uses of different platforms that go from Facebook to Twitter, Snapchat, or others. In our case, the focus is on Facebook. As social media interaction advanced, relations between organisations increased in complexity and a strategy for keeping up to date with the new paradigm is necessary. The levels of engagement were also a result of an interactive process, requiring an open approach to the context in a, what can be considered, somewhat closed organisation.

Gaining terrain in knowledge and among followers or the millennials reduces the disruption in the relation between the parties involved.

34.2.1 Information and Communication Technologies and Networked Society

The change from traditional media and fear of the virtual environment was the defining characteristic of the previous centuries. The milestone for this change was the dissemination of technology that involved a transformational turn in the '60 and '70s of the twentieth century. The drive for society is technology, which improves productivity and sustains power [1, 13, 14].

The new era required an adjustment to the globalisation of economic activities, the omnipresence of a virtual media system, which is a sudden upsurge in technological disease with a profound effect on the interrelation of human civilization.

From the read-only web [15], or web of information [16], or simply web 1.0, that configures a repository of information for the organisations, to the future web 4.0, involving artificial intelligence, the mankind has gone on an evolutional path.

With web 2.0 a new phase began, because a platform of creative intelligence was in place, with the www, increasing the quality and the number of users [17, 18]. Even though there is no clear distinction between web 1.0 and 2.0 [19–21] the fact is that technology provides the necessary tools for a more interactive and bidirectional interrelation between the stakeholders. The transition for a read and write web allowed the propagation of new languages, such as Asynchronous JavaScript and XML (AJAX) [15, 18] and of applications like Myspace, Flickr, Facebook, Youtube and others [22].

Web 3.0, or semantic web, reveals its functionality in being projected for human consumption [23]. As an extension of the existing web, the cooperation between hardware and individuals, allows them to adjust the consumption to their needs [24]. Ultimately the question is the possibility of interconnecting and analysing information for a better flow of information and to be accessible in mobile devices. Experience, and client-oriented strategy has shown that client satisfaction is the main core of the strategy [25].

Addiction to technologies reveals the surroundings in which humankind and millennials are now conducting themselves. In Portugal, the phenomenon is well reported in the last census. 67% of the population with more than 15 years uses the Internet [26].

The widespread use of applications led to a new conception of communication sustained in a virtual environment in which the proliferation of social media was and is the main effect.

34.2.2 Social Networks and Facebook

There is no way to fully separate ICT issues from social media ones. The social media radically changed the way how people communicate. From the development of web

1.0, that allowed the connection between personal computers, to connecting people [27], established new standards. Standards were set up to communicate, publish and collaborate among each and everyone who uses social media [28]. As perfect examples of this connection that incentivizes sharing there are the blogs, the wikis, social media, such as Facebook, LinkedIn, Twitter (for example) [29].

Social media is defined as a group of people or entities connected by a network (friendship, work or information sharing) [30] upheld in face-to-face interaction or ICT context. In the last case, the adjustment is fast because the capacity to exchange information overcoming space and time continuum [31] is a fact.

The question remains: What are the consequences of fast adjustment and lagging adjustment? During this process, the unprecedented ability to interact tend to continue so that users are more likely to claim and gain power [32].

Since the interaction, and empowering continues to rise, a new paradigm between organisation and client/user is established [33]. In fact, from the unidirectional communication paradigm to the bidirectional one [18] the dispute between production and consumption of information is becoming less apparent, which is understandable, in face of merging and amalgamating of both [15].

The boom feels so good that consumers are now prosumer's [34, 35]. Someone that intervenes actively in the production of information, contents, identities, involving themselves in a participative democratic culture from everyone and to all [36].

During the upturn, the power of the prosumer is not a temporary phenomenon. Truly, new strategies must be considered [37]. This has some favourable consequences. In practice, such a journey involves reinforcement of the relations, trust and meeting expectations and to act collectively, as to have the prosumers stand up for the organisation and the brand [38], and to establish brand loyalty [39].

Even though changes have been occurring in the social networks share [40], the fact is that Facebook has 1.65 billion users [41]. One of the consequences can be realised quickly. Eventually, we will have to engage the prosumers by social networks, coordinating the traditional marketing approach to a social one [42]. Therefore, most organisations aim for interactive and co-creative partnership marketing [43], centred in the personality of the brand, and reducing the costs in investment on their platforms [44].

The research on the social network about consumer behaviour has been developed [40, 45], however, the use by brands/organisations is recent [46, 47].

To 'give power to share and make the world more open and connected' [48], barriers were broken, and the limits of interaction were broadened. It is the largest social network, considering the global reach, the number of active users and shared culture [49]. As so, its main ingredient, access easiness [37], and users' attitudes towards a particular page or brand enable a metric for engagement [50]. The externalities arising from the users affect the philosophy of customer relation [51], because it's a worldwide scale of interaction, it's cheaper to advertise products and services [49] and it makes the prosumer as an active partner [52].

The development of Facebook tools has been the subject of some investigation in law enforcement, namely in PSP, so, this paper is one more contribution to its development [8–10, 53, 54].

34.2.3 Engagement

Apart from it being timeless [55], the paradigm of relational marketing has been studied since 1987 [56–58]. The study of the long-term relationship oriented is indeed a positive focus for satisfaction, trust and loyalty [39, 59–62].

Now the challenge is the engagement that has been of interest to many authors [63–66], and has conquered a place of honour in recent years [63, 66].

Through the need for a metric for recognising the relations and its deepness between stakeholders [5], engagement is a powerful tool in postmodern consumer behaviour and the decision-making process [67]. As the focal point is the intangible aspects of the relation (abilities, information, knowledge and interaction connectivity) [68] the tightening is strategic.

As a result, the literature supports a plethora of concepts varying from consumer engagement [68], online user engagement [67, 69] and community engagement [70]. On the other hand, there is a relational nature of the concept in which contact, experience, interaction and the connection is valued [71].

The third approach is made considering a motivational angle, as it's a mental state that allows a co-creation between partners [63] and also coveys in the involvement and immersion in the process [72].

Besides, the fourth element involves the behavioural manifestations developed by the users towards the organisation or brand, even without being associated with a product or service buy [66–68].

The last one is due to be analysed. Engagement is a multidisciplinary construct [63, 66, 73] with cognitive, emotional and behavioural dimensions, which must be determined [65, 74, 75]. Thus, as customer relationship continues to evolve, the concept that determines the orientation of the paper is this one.

The amazing growth of Facebook was followed by almost equally sensational augmentations of new channels of communication, requiring an optimal approach to the engagement, intensity [46] and the predictors associated. Some of the predictors are: (i) interesting and relevant content; (ii) simple and unique contents [76]; (iii) consumer identification; (iv) nostalgic publications; (v) intrinsic humour; (vi) associated with lifestyles and, (vii) demonstration of the backstage and frontstage of the organisation or brand activities [77]. Promotions, events, games and competitions are also an element for rising engagement [78]. To growth, the use of influencers or anonymous people, and the consequent share of images or information is a trend and a factor of engagement [79].

With the advent of Facebook, organisations began to focus on it. The law of keeping up to date with technology, behaviour patterns, increases the need for transforming the way organisations interact. Within this spectrum, PSP has also turned to Facebook and other social networks. Gathering information on what kind of engagement do followers have is an indispensable requirement.

34.3 Method

The necessary steps are taken for obtaining a valid response to the matter of engagement with the PSP Facebook page, resulted from the appliance of the scientific method, centred in the literature review, and quantitative method.

The choice is the foundation of all social sciences, economics, law, and philosophy. The dynamic reality of a society dominated by the media and social media oriented, knowing the projection of the PSP Facebook page, and the engagement of the followers is crucial. This might provide the pillar for adjusting the behavioural patterns towards PSP, as also the compliance with the legal regulations, and civic behaviour, promoting the regular functioning of a democratic state, by ensuring safety and protection of life and property before dangers [80].

34.3.1 Conceptual Model and Hypothesis

Sometimes a descriptive knowledge of a situation is not enough. It requires a formal structure adjusted to the subject. For perceiving a logical relation between propositions, we formulate hypothesis [81] based on the conceptual model (Fig. 34.1).

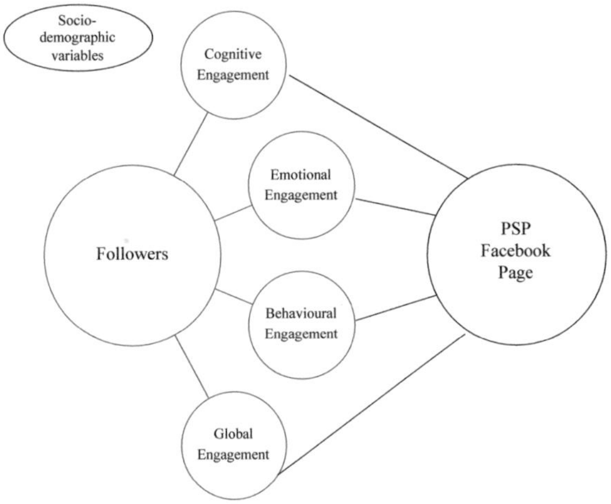

Fig. 34.1. Conceptual model

The cognitive, emotional and behavioural investment is a prerequisite for the engagement [75, 76] enabling the relation on social networks between interested parties [67, 69]. Thus, the following hypothesis is the subject of our attention:

H_1: The followers do not express any kind of engagement with the PSP Facebook page.

On the other hand, as the engagement is related to cognitive, emotional and behavioural dimensions [67, 68, 72, 82], tests for assessing which one is the most prominent. Considering this another hypothesis is formulated:

H_2: The engagement does not present a unique dimension as a fundamental one.

Even though, the literature review lacks from information associating socio-demographic variables (age, gender, marital status and educational qualifications) to engagement, the potential existence of relations between leads to figure out the following hypothesis:

H_3: The socio-demographic variables do not influence a different kind of engagement.

As to perform this research an exploratory, a descriptive, correlational and inferential study was made. This conveys the empirical settings that provide the necessary results to test the hypothesis and are in line with the literature [63, 73, 83].

From the universe of 600.111 followers, a sample of 2.311 subjects was collected. The margin error for this study was 1% with 99% credibility.

The participants of the study were mainly female (65.9%). The mode age is between 30 and in this context, the leadership of spending more hours connected is the female gender, with an astonishing 44% against 22.5% for the male. As for the marital status, 42.6% are single, and a university degree is the main academic qualification (40.8%).

Based on the literature [46] a survey was applied for gathering data of the followers that completed the questionnaire responses in direct mode [84]. They answer online after the link has been made available on the official page of PSP. Consolidating the Portuguese version of the survey internal consistency was performed by measuring the number of items and each dimension [85]. Considering 0.70 of the alpha de Cronbach as acceptable [86] our results comply with this scenario, because each dimension of engagement displays high values (cognitive $\alpha = 0.89$; emotional $\alpha = 0.85$; behavioural $\alpha = 0.77$. and global $\alpha = 0.92$).

As for the method of analysis descriptive statistics of means (M), standard deviation (DP), maximum and minimum were calculated for the variables [87]. A Spearman correlation was also made. Complementing a bivariate analysis was performed by applying the one-way ANOVA and *t*-student test. Every time the *p*-values presented values higher than 0.05, post hoc analysis was performed. The first step was considering the homogeneity of the variances with Levene and as a consequence the Scheffe test to find out which pairs of means are fully significant.

All data were analysed using Statistical Package for the Social Sciences, version 25.0(IBM SPSS Inc., USA). The confidence level was $p \leq 0.05$.

34.4 Results and Discussion

The responses about every question concerning each dimension were attained, resulting in Table 34.1 that gives a summary of the variables:

Table 34.1. Descriptive statistics for engagement dimensions

Engagement	Items	M	SD	Min	Max
Cognitive	4	3.78	0.79	1	5
Emotional	4	3.79	0.83	1	5
Behavioural	3	3	0.93	1	5
Global	15	3.52	0.7	1	5

Table 34.1 shows that all respondents have positive levels of engagement. In fact, the emotional link between followers and the PSP Facebook page is the highest. This establishes a pattern of a positive attitude when browsing the page (M = 3.90; DP = 0.92), promoting happiness (M = 3.61; DP = 0.95) and well-being of the followers (M = 3.85; DP = 0.88). The rational linkage to the page the tendency is also positively high. In fact, considering the items underlying the analysis, followers tend to reflect (M = 3.96; DP = 0.88), think about PSP (M = 3.52; DP = 1.04), and the interest in knowing more is stimulated (M = 3.88; DP = 0.92). Publishing pedagogical and preventive content to fortify emotional experience more often to increase at which followers see content.

Even if when comparing with other pages it's not where followers spend most of the time (M = 2.44; DP = 1.03), the fact is that when searching for internal security information the page is highly browsed (M = 3.13; DP = 1.17), which makes the consultation regular (M = 2.45; DP = 1.03).

The existing parity between cognitive and emotional may bear in mind what previous studies might have encountered [53] because the interest is stimulated by rational and informational content, namely data and facts, warning campaigns and prevention. Sharing messages concerning social causes, in which PSP is actively involved ('earth day', 'energy day', and others), can be useful for retaining followers on the page.

Alongside this scenario, the conclusions can also be supported by other authors [46, 77] because identification with the content, sharing and commenting are the main elements of engagement.

Even in the borderline of positive engagement for the behavioural one, the null hypothesis is rejected, prevailing the approaches that bring to light the importance of the engagement as a fact in the whole strategy of an organisation [67, 69, 75, 76], in which PSP is not an exception. One strategy that can be of added-value concerns the use of games with embedded (educational, preventive, etc.) messages to promote higher levels of engagement.

When comparing which dimension is a fundamental part of the engagement, the results have shown that the cognitive and emotional ones are the mainframe of engagement (p-value $\leq$ 0.05), which are in line with what previous research has revealed [67, 68, 72, 82]. The existence of statistically significant differences in behaviour and engagement has not been proven (p-value > 0.05). Several reasons could be stated. They might spend more time on other pages, it is possible that the high academic qualifications may improve the readiness to assimilate and perceive the information, and the message.

In the face of the results, established in Table 34.2 to test the third and last hypothesis statistical, differences between genders can be observed.

Table 34.2. Groups and engagement analysis (p-values)

Engagement	Gender[1]	Age[2]	Marital status[2]	Academic qualifications[2]
Cognitive	0.000*	0.400	0.000*	0.000*
Emotional	0.000*	0.890	0.000*	0.000*
Behavioural	0.000*	0.560	0.000*	0.000*
Global	0.000*	0.000*	0.610	0.000*

Note: [1]T-student test; [2]One-way ANOVA; *significant at a level of 0.05

Following this line of thought female gender displays higher values in each dimension of engagement (e.g. Emotional—*F*: $M = 3.91$, DP = 1.54; *M*: $M = 3.54$, DP = 0.89). Not surprisingly emotional and cognitive also display the highest mean values of engagement (3.80 and 3.94, respectively).

Along with gender, academic qualifications also have statistical differences between the groups. The difference is only sustained between the university degrees and high school graduates (Levene: $p = 0.000 \leq 0.005$; Scheffe: $p = 0.001 \leq 0.005$). Arguably the more educated show means values higher in emotional and cognitive, and lower in the behavioural engagement in comparison with the less educated ones.

One can point out that one of the reasons for this scenario is the level of awareness and understanding of the matter by the educated ones. The less educated may be led to pursue search more vigorously to enhance their skills and knowledge.

Differences in the engagement between the groups of 30–39 and 50–59 years old (Levene: $p = 0.000 \leq 0.005$; Scheffe: $p = 0.000 \leq 0.005$). This might be a result of a higher level of engagement in the youngsters ($M = 3.84$, SD = 0.69 against $M = 3.20$, SD = 0.72).

As for the time spent in social networks, the levels of association between engagement are all positive, which means, as more time is spent on social networks the level of engagement grows. It's a direct association between the variables, that in case of the emotional engagement and spending more than 8 h in social networks is moderate (rho = 0.37, p-value ≤ 0.005).

The third hypothesis found multiple degrees of considerations. In the absence of literature review, further research is advisable to investigate to what extent the intra-group differences (age, academic qualifications and other pair of variables) are solid. PSP should do a more profound study for understanding the demographic characteristics of followers. This way the sharing of images, experiences and information might be more efficient and increase knowledge and emotional engagement.

34.5 Conclusion

The results of this exploratory study corroborate many of the earlier research findings, shedding new light and providing some new understandings of the follower's engagement towards the PSP Facebook page.

In the past Facebook was seen as an unreal tool of communication, however, derived from a rapid evolution of ICT, the reality of social networks has caught up with society.

In light of these findings, the engagement is an optimal predictor of positive behaviour and attitudinal results towards the PSP organisation. New attitudes and techniques of synergic relations might guarantee rationality, emotion and behaviour. It constitutes a potential for achievement of police and social goals, in an information society.

The modest predisposition in the behavioural dimension of engagement might suggest that the page competes with similar ones. Thus, it is possible for the PSP to strengthen the ties, by making the search for internal security more dependent on the PSP page.

Admittedly the relation between stakeholders is not easy. The freedom of followers to interact with the page requires network social indoctrination. The predictably of the responses, likes, shares and reactions to the page it incorporates constant and repetitive cooperation of the organisation.

A scientific study is inextricably linked to limitations. The limitations listed in this study have been determined by the sampling, and by responders. Even though we have a high response rate, it involves an opt-out sample. Secondly, the survey is based on self-reported information. Finally, given the clear characterisation of the sample, a bias might be valued. The answers might incorporate followers that are not regular users of the social network but were willing to participate in the study.

References

1. Castells, M.: A Sociedade em Rede, 4th edn. Gulbenkian, Lisboa (2012)
2. Morgado, S.M.A., Ferraz, R.: Social media: facebook e a gestão policial no planeamento de grandes eventos. Int. J. Mark. Commun. New Media (Special Number 1—QRMCNM) 27–47 (2016)
3. Antonius, N., Rich, L.: Discovering collection and analysis techniques for social media to improve public safety. Int. Technol. Manage. Rev. **3**(1), 42–53 (2013). https://doi.org/10.2991/itmr.2013.3.1.4
4. Storey, D.J.: Understanding the small business sector. https://papers.ssrn.com/sol3/papers.cfm?abstract_id=1496214. Last Accessed 29 July 2019 (1994)
5. Ryan, L., Leong, M.: Engaging the new consumer. In: Esomar Asia Pacific Conference 2007, pp. 1–5. ESOMAR, Singapura (2007)
6. Morgado, S.: Going global: health organizations and networking—information society and social media. In: Proceedings in Scientific Conference 2013, vol. 1, pp. 47–51. EDIS—Publishing Institution of the University of Sikina, Slovakia (2013)
7. Waters, M.: Globalização. Celta Editora, Oeiras (1999)
8. Morgado, S.M.A., Ferraz, R.: Análise de conteúdo e pesquisa em ciências policiais: Contextualização em ambiente virtual, Facebook, e o planeamento de grandes eventos. Atas CIAIQ/2016—Investigação Qualitativa em Ciências Sociais **3**, 600–608 (2016)
9. Morgado, S.M.A., Ferraz, R.: Social media: facebook and police major planning events. Int. J. Mark. Commun. New Media **1**(Special number—QRMCNM), 27–47 (2016)
10. Ferraz, R.: Facebook e o planeamento de grandes eventos: Um contributo operacional na área das informações (Unpublished Master's thesis). ISCPSI, Lisboa (2015)
11. Bullock, L.: 6 Predictions for content marketing in 2017. https://www.socialmediatoday.com/marketing/6-predictions-content-marketing-2017 (2016). Last Accessed 08 July 2019
12. Cleveland, J., Holder, K., O'Roark, B.: The economics of hunger games. Int. J. Pluralism Econ. Educ. **7**(2), 152–169 (2016). https://doi.org/10.1504/IJPEE.2016.078860
13. Castells, M.: A sociedade em rede: Do conhecimento à política. In: Castells, M., Cardoso, G. (orgs.), A sociedade em rede: do conhecimento à acção política. Debates–Presidência da República, pp. 17–30. Imprensa Nacional, Lisboa (2005)
14. Castells, M.: Communication, power and counter-power in the network society. Int. J. Commun. **1**(1), 238–266 (2007)
15. Hiremath, B., Kenchakkanavar, A.: An alteration of the Web 1.0, Web 2.0 and Web 3.0: a comparative study. Imperial J. Interdisc. Res. **2**, 705–710 (2016)

16. Choudhury, N.: World Wide Web and its journey from Web 1.0 to Web 4.0. Int. J. Comput. Sci. Inf. Technol. **5**(6), 8096–8100 (2014)
17. Harrison, T., Barthel, B.: Wielding new media in Web 2.0: exploring the history of engagement with the collaborative construction of media products. New Media Soc. **11**(1–2), 155–178. https://doi.org/10.1177/1461444808099580 (2009)
18. O'Reilly, T.: What is Web 2.0: design patterns and business models for the next generation of software. http://www.oreilly.com/pub/a//web2/archive/what-is-web-20.html (2005). Last Accessed 25 June 2019
19. Dwivedi, Y., Williams, M., Mitra, A., Niranjan, S., Weerakkody, V.: Understanding advances in web technologies: evolution from Web 2.0 to Web 3.0. In: 19th European Conference on Information Systems. ECIS 2011, pp. 1–13. ECIS, Helsinki (2011)
20. Fuchs, C., Hofkirchner, W., Schafranek, M., Raffl, C., Sandoval, M., Bichler, R.: Theoretical foundations of the web: cognition, communication, and co-operation. towards an understanding of Web 1.0, 2.0, 3.0. Future Internet **2**(1), 41–59. https://doi.org/10.3390/fi2010041 (2010)
21. Song, F.W.: Theorizing Web 2.0: a cultural perspective. Inf. Commun. Soc. **13**(2), 249–275. https://doi.org/10.1080/13691180902914610 (2010)
22. Hendler, J.: Web 3.0 emerging. Computer **42**(1), 88–90. https://doi.org/10.1109/mc.2009.30 (2009)
23. Floridi, L.: Web 2.0 vs. the semantic web: a philosophical assessment. Episteme **6**(1), 25–37. https://doi.org/10.3366/e174236000800052x (2009)
24. Berners-Lee, T., Cailliau, R., Groff, R., Pollermann, B.: World Wide Web: the information universe. Electron. Netw. Res. Appl. Policy **2**(1), 52–58 (1992)
25. Aghaei, S., Nematbakhsh, M.A., Farsani, H.K.: Evolution of the World Wide Web: from Web 1.0 to Web 4.0. Int. J. Web Semant. Technol. **3**(1), 1–10. https://doi.org/10.5121/ijwest.2012.3101 (2012)
26. INE: Censos 2011, https://censos.ine.pt/ (2015). Last Accessed 01 October 2019
27. Weinberg, T.: The new community rules: marketing on the social web. O'Reilly Media, Inc. (2009)
28. Cooke, M., Buckley, N.: Web 2.0, social networks and the future of market research. Int. J. Mark. Res. **50**(2), 267–292 (2008)
29. Smith, T.: The social media revolution. Int. J. Mark. Res. **51**(4), 559–561 (2009). https://doi.org/10.2501/S1470785309200773
30. Garton, L., Haythornthwaite, C., Wellman, B.: Studying online social networks. J. Comput. Med. Commun. **3**(1), 1–32. https://doi.org/10.1111/j.1083-6101.1997.tb00062.x (1997)
31. Kozinets, R.V., de Valck, K., Wojnicki, A.C., Wilner, S.J.: Networked narratives: understanding word-of-mouth marketing in online communities. J. Mark. **74**(2), 71–89 (2010). https://doi.org/10.1509/jmkg.74.2.71
32. Kucuk, U.: Consumer empowerment model: from unspeakable to undeniable. Dir. Mark. Int. J. **3**(4), 327–342 (2009). https://doi.org/10.1108/17505930911000892
33. Rodrigues, S.: O impacto das redes sociais no relacionamento entre marcas e consumidores (Unpublished Master's thesis). ISCTE-UL, Lisboa (2012)
34. Shcherbina, I.: Reading in the age of Web 2.0. Russ. Soc. Sci. Rev. **58**(1), 86–108. https://doi.org/10.1080/10611428.2017.1275204 (2017)
35. Toffler, A.: The Third Wave. Bantan Books, New York (1980)
36. Lévy, P.: O futuro da internet: Em direção a uma ciberdemocracia planetária. Paulus, São (1999)
37. Pires, G.D., Stanton, J., Rita, P.: The internet, consumer empowerment and marketing strategies. Eur. J. Mark. **40**(9/10), 936–949 (2006). https://doi.org/10.1108/03090560610680943

38. Kietzmann, J.H., Hermkens, K., McCarthy, I.P., Silvestre, B.S.: Social media? Get serious! Understanding the functional building blocks of social media. Bus. Horiz. **54**(3), 241–251 (2011). https://doi.org/10.1016/j.bushor.2011.01.005
39. Reynolds, K.E., Beatty, S.E.: Customer benefits and company consequences of customer-salesperson relationships in retailing. J. Retail. **75**(1), 11–32 (1999). https://doi.org/10.1016/S0022-4359(99)80002-5
40. Global Web Index: Global Web Index's Quarterly Report on the Latest Trends in Social Networking https://www.globalwebindex.net/hs-fs/hub/304927/file-2812772150-pdf/Reports/GWI_Social_Summary_Report_Q1_2015.pdf (2015). Last Accessed 15 Dec 2017
41. Statista: Number of Facebook Users Worldwide 2008–2017 https://www.statista.com/statistics/264810/number-of-monthly-active-Facebook-users-worldwide/. Last Accessed 15 Dec 2017
42. Drury, G.: Social media: should marketers engage and how can it be done effectively? J. Dir. Data Digit. Mark. Pract. **9**(3), 274–277 (2008). https://doi.org/10.1057/palgrave.dddmp.4350096
43. Heinonen, K.: Consumer activity in social media: managerial approaches to consumers' social media behavior: consumer activity in social media. J. Consum. Behav. **10**(6), 356–364 (2011). https://doi.org/10.1002/cb.376
44. Simmons, K.S.: Intergenerational communication in the workplace. Online J. Certif. Manag. (2008)
45. Castilho, A.S., Antunes, M.J.: Modelo de gestão de conteúdos nos social media: os estudos de casos das startups. Revista Comunicando **12**, 102–123 (2013)
46. De Vries, L., Gensler, S., Leeflang, P.S.H.: Popularity of brand posts on brand fan pages: an investigation of the effects of social media marketing. J. Interact. Mark. **26**(2), 83–91 (2012). https://doi.org/10.1016/j.intmar.2012.01.003
47. Malhotra, A., Malhotra, C. K., See, A.: How to Create Brand Engagement on Facebook. https://sloanreview.mit.edu/article/how-to-create-brand-engagement-on-Facebook/. Last Accessed 25 May 2019
48. Facebook: Política de Dados. https://www.Facebook.com/policy.php. Last Accessed 25 May 2019
49. Kirkpatrick, D.: The Facebook Effect: the Inside Story of the Company that is Connecting the World, 1st edn. Simon & Schuster Paperbacks, New York (2011)
50. Chan, T.K.H., Zheng, X., Cheung, C.M.K., Lee, M.K.O., Lee, Z.W.Y.: Antecedents and consequences of customer engagement in online brand communities. J. Mark. Anal. **2**(2), 81–97 (2014). https://doi.org/10.1057/jma.2014.9
51. Berry, L.L.: Relationship marketing of services perspectives from 1983 and 2000. J. Relat. Mark. **1**(1), 59–77 (2002). https://doi.org/10.1300/J366v01n01_05
52. Duncan, T.: IMC: using advertising and promotion to build brands, 1st edn. McGraw-Hill, Boston (2002)
53. Silva, R.: Relações públicas policiais em ambiente virtual (Unpublished Master's thesis). ISCPSI, Lisboa (2014)
54. Moniz, T.: O engagement dos utilizadores do Facebook com a página da Polícia de Segurança Pública (Unpublished Master's thesis). ISCPSI, Lisboa (2018)
55. Wakabayashi, J.L.: Research on relational marketing: an analysis on literature content 2007–2008. Bi-Annu. Acad. Publ. Univ. ESAN **15**(29), 119–131 (2010)
56. Gummesson, E.: The new marketing: developing long-term interactive relationships. Long Range Plan. **20**(4), 10–20 (1987). https://doi.org/10.1016/0024-6301(87)90151-8
57. Gummesson, E.: Total relationship marketing: experimenting with a synthesis of research frontiers. Aust. Mark. J. (AMJ) **7**(1), 72–85 (1999). https://doi.org/10.1016/S1441-3582(99)70204-1

58. Grönroos, C.: Relationship marketing: strategic and tactical implications. Manag. Decis. **34** (3), 5–14 (1996)
59. Bowden, J.L.H.: The process of customer engagement: a conceptual framework. J. Mark. Theor. Pract. **17**(1), 63–74 (2009)
60. Grégoire, Y., Tripp, T.M., Legoux, R.: When customer love turns into lasting hate: the effects of relationship strength and time on customer revenge and avoidance. J. Mark. **73**(6), 18–32 (2009). https://doi.org/10.1509/jmkg.73.6.18
61. Patterson, P., Yu, T., De Ruyter, K: Understanding customer engagement in services. In: Proceedings of ANZMAC 2006 Conference Advancing Theory, Maintaining Relevance, pp. 4–6. ANZMAC, Brisbane (2006)
62. Pham, M.T., Avnet, T.: Rethinking regulatory engagement theory. J. Consum. Psychol. **19** (2), 115–123 (2009). https://doi.org/10.1016/j.jcps.2009.02.003
63. Brodie, R.J., Hollebeek, L.D., Smith, S.D.: Engagement: an important bridging concept for the emerging SD logic lexicon. In: 2011 Naples Forum on Service, University of Auckland Business School (2011)
64. Erat, P., Desouza, K.C., Schäfer-Jugel, A., Kurzawa, M.: Business customer communities and knowledge sharing: exploratory study of critical issues. Eur. J. Inf. Syst. **15**(5), 511–524 (2006). https://doi.org/10.1057/palgrave.ejis.3000643
65. Hollebeek, L.D.: Demystifying customer brand engagement: exploring the loyalty nexus. J. Mark. Manage. **27**(7–8), 785–807 (2011). https://doi.org/10.1080/0267257X.2010.500132
66. Vivek, S.D., Beatty, S.E., Dalela, V., Morgan, R.M.: A generalized multidimensional scale for measuring customer engagement. J. Mark. Theor. Pract. **22**(4), 401–420 (2014). https://doi.org/10.2753/MTP1069-6679220404
67. Gambetti, R.C., Graffigna, G.: The concept of engagement: a systematic analysis of the ongoing marketing debate. Int. J. Mark. Res. **52**(6), 801–826 (2010). https://doi.org/10.2501/S1470785310201661
68. Vivek, S.D., Beatty, S.E., Morgan, R.M.: Customer engagement: exploring customer relationships beyond purchase. J. Mark. Theor. Pract. **20**(2), 122–146 (2012). https://doi.org/10.2753/MTP1069-6679200201
69. Mollen, A., Wilson, H.: Engagement, telepresence and interactivity in online consumer experience: reconciling scholastic and managerial perspectives. J. Bus. Res. **63**(9–10), 919–925 (2010). https://doi.org/10.1016/j.jbusres.2009.05.014
70. Algesheimer, R., Dholakia, U.M., Herrmann, A.: The social influence of brand community: evidence from european car clubs. J. Mark. **69**(3), 19–34 (2005). https://doi.org/10.1509/jmkg.69.3.19.66363
71. Sashi, C.M.: Customer engagement, buyer-seller relationships, and social media. Manag. Decis. **50**(2), 253–272 (2012). https://doi.org/10.1108/00251741211203551
72. Higgins, E.T., Scholer, A.A.: Engaging the consumer: the science and art of the value creation process. J. Consum. Psychol. **19**(2), 100–114 (2009). https://doi.org/10.1016/j.jcps.2009.02.002
73. Calder, B.J., Malthouse, E.C., Schaedel, U.: An experimental study of the relationship between online engagement and advertising effectiveness. J. Interact. Mark. **23**(4), 321–331 (2009). https://doi.org/10.1016/j.intmar.2009.07.002
74. Hollebeek, L.D., Glynn, M.S., Brodie, R.J.: Consumer brand engagement in social media: conceptualization, scale development and validation. J. Interact. Mark. **28**(2), 149–165 (2014). https://doi.org/10.1016/j.intmar.2013.12.00
75. Hollebeek, L., Chen, T.: Exploring positively-versus negatively-valenced brand engagement: a conceptual model. J. Prod. Brand Manage. **23**(1), 62–74 (2014). https://doi.org/10.1108/JPBM-06-2013-0332

76. Long, K.E.: How Does it Make You Feel?: a Visual Content Analysis of Emotions Portrayed in Consumer Brands' Images on Social Media. Lunds University, Lunds (2014)
77. Palmer, A., Koenig-Lewis, N.: An experiential, social network-based approach to direct marketing. Dir. Mark. Int. J. **3**(3), 162–176 (2009). https://doi.org/10.1108/17505930910985116
78. Mangold, W.G., Faulds, D.J.: Social media: the new hybrid element of the promotion mix. Bus. Horiz. **52**(4), 357–365 (2009). https://doi.org/10.1016/j.bushor.2009.03.002
79. Gensler, S., Völckner, F., Liu-Thompkins, Y., Wiertz, C.: Managing brands in the social media environment. J. Interact. Mark. **27**(4), 242–256 (2013). https://doi.org/10.1016/j.intmar.2013.09.004
80. Mendes, S., Morgado, S.: Intelligence services intervention: constraints in Portuguese democratic state. In: Teixeira, N.S., Oliveira, C.S., Lopes, M., Sardinha, B., Santos, A., Macedo, M. (eds.), International Conference on Risks, Security and Citizens: Proceedings/Atas, pp. 285–297. Município de Setúbal, Setúbal (2017)
81. Morgado, S.M.A., Anjos, O.: Qualitative methodology helping police sciences: building a model for prevention of road fatalities in São Tomé and Principe. In: Costa, A., Reis, L.A., Moreira, A. (eds), Computer Supported Qualitative Research, WCQR 2018. Advances in Intelligent Systems and Computing, vol. 861, pp. 291–304. Springer, Cham (2019)
82. Van Doorn, J., Lemon, K.N., Mittal, V., Nass, S., Pick, D., Pirner, P., Verhoef, P.C.: Customer engagement behavior: theoretical foundations and research directions. J. Serv. Res. **13**(3), 253–266 (2010). https://doi.org/10.1177/1094670510375599
83. Leckie, C., Nyadzayo, M.W., Johnson, L.W.: Antecedents of consumer brand engagement and brand loyalty. J. Mark. Manage. **32**(5–6), 558–578 (2016). https://doi.org/10.1080/0267257X.2015.1131735
84. Quivy, R., van Campenhoudt, L.: Manual de investigação em ciências sociais. 2nd edn. Gradiva, Lisboa (2008)
85. Tavakol, M., Dennick, R.: Making sense of Cronbach's alpha. Int. J. Med. Educ. **2**, 53–55 (2011). https://doi.org/10.5116/ijme.4dfb.8dfd
86. Nunnally, J.C.: Psychometric Theory, 2nd edn. McGraw-Hill, New York
87. Morgado, S.M.A.: Does health promote economic growth? Portuguese case study: from dictatorship to full democracy. Eur. J. Health Econ. **15**(6), 591–598 (2014). https://doi.org/10.1007/s10198-013-0497-9

Chapter 35
The LexDoBusiness Collaborative Platform

Helder Azeredo[1], José Luís Reis[1,2(✉)], and Agostinho Sousa Pinto[1]

[1] ISCAP, Porto Superior Institute of Accounting and Administration, Research Unit CEOS.PP/ISCAP/P.PORTO, Matosinhos, Portugal
{2183554,apinto}@iscap.ipp.pt, jreis@ismai.pt

[2] ISMAI, Maia University Institute, Research Unit CEDTUR/CETRAD, Maia, Portugal

Abstract. The objective of this paper is to present a proposal for the creation of an online business consulting business plan, through a collaborative platform called LexDoBusiness. The literature review concluded that a business project must address several issues, namely: digital transformation, business plan and model, E-Commerce, E-Business and M-Business, marketing mix, digital marketing, Customer Relationship Management, and Content Management Systems. The value proposition must be the most appropriate for each of the customer segments, thus distinguishing itself from the competition. The research work is intended to assess the acceptability of a business model of the LexDoBusiness web platform, which brings together several services essential to its success, such as legal, accounting, accounting, fiscal, and technology services. Regarding the research methodology, Design Science Research was chosen because it is considered the most appropriate for this project, which will have external secondary data collection present in the INE and PORDATA platforms, as well as primary data collection through a semi-structured survey addressed to clients of a network of lawyers.

35.1 Introduction

The LexDoBusiness aims to be an integrated online collaborative platform solution, with various services in order to help entrepreneurs develop their business as well as entrepreneurs who wish to turn their business idea into a company. This paper aims to propose to present the key issues to a business model plan so that in future they can guide the development of the online platform. In this paper it will be presented a systematization of the literature review with a concept matrix, and some relevant aspects associated to a business plan, like digital transformation, digital marketing, customer relationship management, etc. that allow the analysis, develop and implement of LexDoBusiness project, considering the genesis of the idea of the project, the customers target, and the ethical issues inherent in the provision of electronic platform law services.

Á. Rocha et al. (eds.), *Marketing and Smart Technologies*, Smart Innovation, Systems and Technologies 167, https://doi.org/10.1007/978-981-15-1564-4_35

35.2 Systematization of the Literature Review

Searches in the databases, based on the search criteria, returned twenty publications that were considered more relevant for this work, which are presented in Table 35.1 in an adaptation of the concept matrix developed by Webster and Watson (2011). The concepts identified in Table 35.1 provided the basis for developing the present work and will be developed in the following points for a better understanding.

Table 35.1. Concepts matrix.

Adaptation from [1]

Authors	Concepts							
	E-Commerce, E-Business e M-Business	Business model	Business plan	Digital law	Digital transformation	Marketing mix	CMS	CRM	Digital marketing
Adewojo e Bass (2018)							✓		
Baltes (2015)					✓				✓
Borden (1964)						✓			
Cabot (2018)							✓		
Chaffey e Smith (2013)					✓				✓
Conrado (2011)					✓				✓
Dominici e Seaf (2009)						✓			
Freitas (2018)		✓							
Kalakota e Robinson (2001)	✓							✓	
Kibasi (2018)			✓	✓	✓				
IAPMEI (2011)				✓					
Laudon e Traver (2017)	✓	✓							
Massa, Tucci e Afuah (2017)		✓							
Osterwalder e Pigneur (2011)		✓							
Putra, Hasibuan, e Sunarya (2017)			✓		✓				
Torrubia (2013)			✓	✓	✓				
Turban, King, Lee, Liang, e Turban (2015)	✓	✓							
Wang e Fesenmaier (2006)								✓	
Werthein (2000)					✓				
Kotler (2003)	✓				✓				

35.2.1 Digital Transformation

Traditional media are slowly being overtaken by a new digital age. According to Werthein (2000), society is constantly changing, with new technologies being mainly responsible for saying that we currently live in an information society [2]. For this reason, the law firm must grasp the opportunities that technology offers and thus adapt to its clients' new ways of communication, their needs, lifestyle, and preferences.

Not infrequently, when they have a legal problem, they no longer go to a law firm at first, preferring to search the Internet for solutions to their problems or share their difficulties on the Internet, waiting for someone to have been through similar situations.

According to the Resilience through Change PwC Law Firms Survey 2018 27th Annual Law Firms Survey [3] (p. 26), the speed of technological change is a major concern for law firms. Thus, it is easily concluded that adaptation will have to be rapid. In fact, the report on the same theme, released in 2017, already reported the urgent need for change by law firms and their adaptation to customer needs. This report mentions that one of the reasons for clients' discontent and frustration is that the cost of providing full legal services and, as a result, they prefer to hire lawyers as appropriate situations arise [3].

In 2013, during the first virtual advocacy day at the Universitat Oberta de Catalunya, there was a reflection and debate on "How internet and technology are transforming the profession" [4] (p. 97) and how new technologies affect legal services. This journey also aimed to help implement a virtual business and show that the internet is a professional opportunity.

35.2.2 The Business Plan

People who have their own business, when it succeeds, provide more results than having a career subordinate to a company. The Internet is a fertile area of great opportunity, as knowledge about e-commerce is growing ever wider and more exploited.

The Internet makes it possible to leverage the small and medium-sized enterprises (SMEs) that are the engines of economic development in many countries. About 95% of all companies in the world are classified as SMEs and employ up to 60% of the global workforce [5]. Online business does not require physical store costs, being the cost of maintaining a lower ".com" business and offering opportunities to sell products and services anywhere and designing an online business plan critical to success.

According to IAPMEI (2016) the business plan has three major functions [6]:

1. "During its preparation, help the entrepreneur understand the various aspects of his business (market, product, financing model, business model, growth drivers and business risks);"
2. "Guide the development of the company in the short, medium and long term, through a set of previously established and agreed objectives;"
3. "Be a support for introducing the company to other stakeholders (investors, partners, etc.)."

The following points present some concepts necessary for the implementation of an online business plan.

35.2.3 The Business Model

The term business model can have several meanings. According to Freitas (2018), "the term is often used generally in very different areas of research" [7] (p. 1), and adds that the result "of the widespread use of the concept, the term has been used to describe everything from the way the company makes revenues to how it organizes itself" [7] (p. 1).

Massa, Tucci, and Afuah (2017) explain that the business model can designate a real company attribute, cognitive scheme, or conceptual representation of an activity, as well as a structure or new perspective on organizations and their performance [8].

According to Osterwalder and Pigneur (2011) "A business model describes the logic of value creation, delivery, and capture by an organization and consists of nine elements that demonstrate the logic of how an organization intends to generate value" [9] (p. 14). These elements (Customer Segments, Value Propositions, Channels, Customer Relationships, Revenue Streams, Key Resources, Key Activities, Key Partnerships, and Cost Structure), form the basis for a useful tool, called the Business Model Frame or Model Canvas [9].

35.2.4 E-Commerce, E-Business, and M-Business

With the increasing computerization, new forms of business transactions have also been created through e-commerce. According to Kotler (2003), "The Internet offers an excellent platform for communication, buying and selling. Over time, your benefits will increase. Business leaders extol their potential" [10].

Laudon and Traver (2017) define e-commerce as "The use of the Internet, the Web and mobile applications and browsers to conduct business transactions. More formally, digitally supported business transactions between individuals and organizations and organizations among themselves" [11]. While other authors define it as "using the Internet and other networks to buy, sell, transport or market data, goods or services" [12].

The increasing pace of new application emergence, development of new distribution channels and competitive dynamics has been one of the effects on the business revolution brought about by e-commerce innovation [13]. The same authors explain that e-business encompasses all electronically supported internal and external activities, including e-commerce.

35.2.5 Marketing Mix

The concept and expression Marketing Mix was coined by Borden (1964) by reading a paper on marketing costs and introduced in his speech to the American Marketing Association [14]. Jerome McCarthy, in 1964, defined the 4 P's as a combination of all factors that managers can use to meet market needs. In McCarthy's version, the marketing mix consists of Product, Price, Place, and Promotion. Each of these P's comprises a series of management policies described by Borden in 1964, thus including other sub-policies within each P [15].

Since then there have been several studies and researches that derive from the need to find a new marketing paradigm that, surpassing the 4 P's, aims to define more specifically the marketing levers especially in the digital world we live in. One of these studies led Chen in 2006 to highlight the importance attached to the environment and interaction in the model proposed by Chinese researchers at the National University of Taiwan (1999), which considers 8 P's, four of which are traditional P's (cited by [15], p. 24), as we can see in Fig. 35.1.

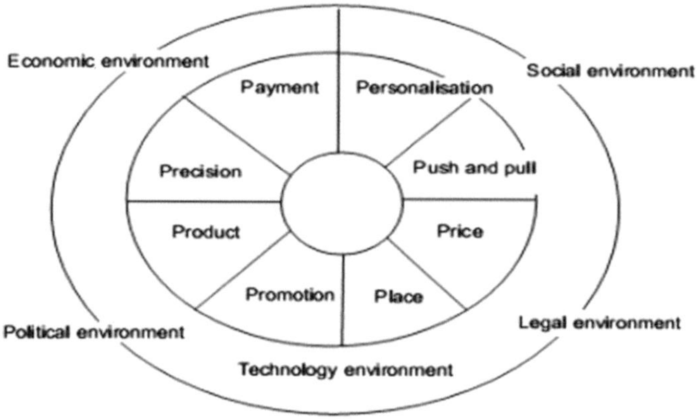

Fig. 35.1. The Internet Marketing Mix 8P's model (cited by [15], p. 24)

35.2.6 Digital Marketing

Organizations have to know to avail the benefits of the Internet, which open them to new perspectives of business, unable to expose and advertise their products and services, reaching a number of potential customers much higher than they would by traditional means, thanks to the ease of access to obtain information about prospects and customers.

The digital reality revolutionized the market in such a way that any organization that wants to remain competitive is forced to rethink their marketing strategies [16].

There are several approaches to building a Digital Marketing Plan. According to the bibliography consulted, it is possible to highlight two strands: the 8P Methodology proposed by Conrado (2011) and the SOSTAC® model proposed by Chaffey and Smith (2013) [17, 18].

The 8Ps' methodology, according to Conrado (2011) "treats the whole process of digital marketing from research to precision, which is the measurement of the results obtained, in a systemic and cyclical way. The 8P's refer to the Research (Pesquisa), Planning, Production, Publishing, Promotion, Propagation, Personalization and Precision" [17].

The SOSTAC® model is one of the most widely used and simplest methodologies to use in developing a Digital Marketing strategy. The acronym SOSTAC refers to the six main points of this model, S-Situation Analysis, O-Objectives, S-Strategy, T-Tactics, A-Action, and C-Control [18].

After analyzing the two approaches presented, it was found that the methodology of 8P's, proposed by Conrado (2011), advocates a viral communication to propagate the services of organizations. If we consider the statutory limitations imposed by the Portuguese Bar Association, this methodology does not fit the business plan that is intended to be implemented. Thus, it is believed that the plan most appropriate to the case of the LexDoBusiness project is that of Chaffey and Smith (2013), as it is the most comprehensive, detailed, and complete.

35.2.7 Customer Relationship Management—CRM

A CRM system is based on a relational strategy between the company and its customer, so it is a business strategy defined in order to know the customer as well as possible to understand him and to know what he wants, when he wants, in the right amount to create value and better define your consumption needs in terms of goods or services.

According to Wang and Fesenmaier (2006), effective CRM programs are needed to retain customers on the site and convert visitors into loyal customers. Getting new visitors becomes increasingly challenging, and keeping existing users becomes even more valuable, because loyalty to a site dramatically reduces the cost of maintaining site traffic [19].

CRM is a combination of business processes and technology, its goal is to understand a company's customers from a multifaceted perspective: who they are?, what they do?, and what they like? Is CRM critical to business survival in the age of a never satisfied customer? Industry leaders believe so. Increased competition, globalization, rising cost of customer acquisition, and high customer turnover are important issues in industries as disparate as financial services, telecommunications, and retail [13].

35.2.8 Content Management System—CMS

Armbrust et al., Rimal and Maier, Walraven et al. (cited by Adewojo and Bass, 2018) report that cloud computing provides scaling, resource provisioning, IT infrastructure, development platforms, data warehousing, and software applications on the Internet. Cloud vendors have recently witnessed an increase in the number of users using cloud software applications commonly known as Software as a Service (SaaS). SaaS is a cloud computing model for accessing software applications over the Internet that eliminates customer infrastructure installation, centralizes maintenance, and lowers total costs [20].

There is Content Management System (CMS), which is one of the SaaS models that allows you to design and implement a website or e-commerce platform. According to Cabot (2018), a CMS is any system that facilitates the creation and publication of digital content. This category includes platforms for creating static websites, blogs, forums, and online stores [21].

The most popular CMS is Wordpress, however, there are other competing CMS's with similar features, such as Joomla, Drupal, Shopify, and Squarespace, all with lower market share. Sixty percent of all sites using a content management system use WordPress, the only one with a double-digit market share, Joomla is second with a 6.4% market share [21].

Adewojo and Bass (2018) in their research project used WordPress as a benchmark application because it is an open-source distributed content management system that is widely used and used by internet users to test the performance of the three multi-tenancy standards, getting great results and offers free plug-ins and extensions [20].

Both Wordpress and Joomla, through their well-known plug-ins and e-commerce extensions, WooCommerce and VirtueMart respectively, allow you to deploy a custom-made platform with the desired features, with eye-catching design and intuitive navigation.

35.3 Project Methodology

After reviewing the literature, which systematized the various important concepts that serve as the knowledge base to turn the idea into business, it is important to define the methodology to be used in the research project.

The methodology used in the research project to be developed will be Hevner's Design Science Research (DSR) (2004). DSR is a results-based information technology research methodology that provides specific guidelines for evaluation and iteration in research projects [22].

DSR involves building, investigating, validating, and evaluating artifacts such as constructs, frameworks, models, methods, and instances of the information system to solve new practical problems [23].

According to Hevner (2004, p. 83) DSR is inherently a problem-solving process based on seven guidelines: "Guideline 1: Design as an Artifact; Guideline 2: Relevance of the Problem; Guideline 3: Project Evaluation; Guideline 4: Research Contributions; Guideline 5: Research Rigor; Guideline 6: Design as a Survey; Process; Guideline 7: Research Communication" [22].

35.3.1 The Idea for the LexDoBusiness

LexDoBusiness aims to be a B2B electronic consulting platform focused on entrepreneurship development. The main objective of LexDoBusiness will be to work for all those interested in starting a business in Portugal.

The vision will be to allow anyone to have a real opportunity to pursue their entrepreneurial dream. The mission will be the transformation of ideas into successful businesses, taking advantage of each other's entrepreneurial potential.

The provision of LexDoBusiness services ranges from training entrepreneurs through online and in-person training, structuring project ideas, and monitoring the company throughout its life cycle.

With a multidisciplinary team that will provide the necessary support in the legal area, conflict mediation, accounting, tax, and informatics through the development of the company's website and its hosting. The existing partner network will help in Portugal and abroad, namely, in Italy and Brazil. The B2B project will operate in the consulting area, covering the areas of computer science, accounting, tax, and legal.

It is now common for companies to hire services from other companies. LexDoBusiness aims to be a solution that avoids the dispersion of services by reducing costs and time, presenting a wide range of services that can be found in one place, as shown in Fig. 35.2.

Fig. 35.2. LexDoBusiness core services

35.3.2 Model and Working Plan

The main objective of this project is the realization of the idea, presented in the previous point, in a successful company. The purpose of this proposal is to transform the idea into a business model, as well as to elaborate the appropriate business plan that will allow the implementation of a company in accordance with the initial idea and expectations.

The business model will have as its starting point in Portugal, because LexDoBusiness's target audience is national SMEs, however, and given that about 50% of the attorneys'[1] clients who will join LexDoBusiness contact them from abroad, ensuring the provision of services to these markets as crucial.

According to IAPMEI the internationalization of an online home business is almost immediate, "When the home business was created on an online platform, its internationalization can be almost immediately guaranteed. In the limited case of the

[1] Lawyer Cristiane de Souza Reis and Lawyer Mauro Rossoni.

development and distribution of computer applications, accessibility to foreign markets can be guaranteed without major adaptations" [24] (p. 82). In this way and in accordance with the above, the communication, access, and maintenance of foreign customers are thus ensured.

The research methodology that will be used in the project is Design Science Research, which will have external secondary data collection present in the INE and PORDATA platforms, for the primary data, in order to test the acceptance of the provision of legal services fully online, an online questionnaire will be conducted. The type of sampling to be used in this research will be qualitative, through the non-probability quota sampling technique. The scope of the sample will be the geographic area of Portugal and Brazil, in the sector of law activity, during the month of December 2019. The target population will be the universe of clients and contacts of lawyers, as well as their partners, all over 18 years old.

The various components of the LexDoBusiness business model will be analyzed, as well as the extent to which the platform users consider the most relevant aspects and services, as well as the degree of acceptance of legal services exclusively online. This analysis is justified considering that the law firm is traditionalist in contrast to what LexDoBusiness intends to be. For this reason, the intent is to understand whether clients are willing to hire legal services via the web and to what extent digital transformation has influenced clients seeking, for example, a lawyer.

35.4 Ethical Issues

There are statutory limitations imposed by the Portuguese Bar Association. As an example, it is not allowed to do direct marketing with offers for hiring services, as it is considered customer acquisition, prohibited by the Bar Association.[2]

All communication made by LexDoBusiness must reconcile the statutory requirements, established by the Bar, with the opportunities that the digital age confers in the relationship with clients. In the light of the above, it is necessary to develop a business plan that creates a strong digital presence for LexDoBusiness, taking advantage of current trends, new forms of customer-business communication, while maintaining an intimate relationship level with which LexDoBusiness sector requires, because it involves the disclosure of personal information, without ever neglecting, the legal precepts, principles, and deontological rules of the profession.

The most suitable solution to reconcile the above-mentioned factors is Inbound Marketing which will allow you to promote LexDoBusiness through blogs, free eBooks, free webinar format legal advice, courses, and videos without infringing any legal imposition.

[2] Al.e) of article 94 of the Statute of the Bar, approved by Law No. 145/2015 of 9 September.

35.5 Conclusions

This paper aims to establish some guides for the future implementation of a business plan for the project LexDoBusiness. The project aims to create an innovative B2B collaborative electronic consulting platform focused on entrepreneurship development.

The paper presents a literature review, with the state of the art about topics that are somehow related to the project, in some way, namely those that justify the investments that need to be made in the study of the plan and the business model, which is supported by content available on a platform that allows the relationship of the partners involved in the project to start a business, and somehow support the platform's customers in whatever aspects they deem necessary.

The paper mentions the ethical aspects to be taken into consideration, as such platforms that provide legal support services to their clients must comply with the rules imposed by current legislation.

The paper also presents the project methodology that will be used, and the guidelines that will guide the development of the LexDoBusiness collaborative platform project.

References

1. Webster, J., Watson, R.T.: Analyzing the past to prepare for the future: writing a review. MIS Q. **26**(2), 12 (2011). www.jstor.org/stable/4132319
2. Werthein, J.: A sociedade da informação e seus desafios. Ciência da Informação **29**(2), 71–77 (2017). https://doi.org/10.1590/S0100-19652000000200009
3. Kibasi, T.: Time for change. IPPR Prog. Rev. **25**(2), 133–136 (2018). https://doi.org/10.1111/newe.12104
4. Torrubia, B.: Primera Jornada de Abogacía Virtual, 97–100 (2013)
5. Putra, P.O.H., Hasibuan, Z.A., Sunarya, P.O.A.: Depicting E-business application adoption for small and medium enterprises (SMEs): the e-business triangle model. In 2017 International Conference on Computing, Engineering, and Design (ICCED), pp. 1–6 (2017). https://doi.org/10.1109/CED.2017.8308135
6. IAPMEI: Guia Prático do Empreendedor (2016). Retrieved from https://www.iapmei.pt/getattachment/PRODUTOS-E-SERVICOS/Empreendedorismo-Inovacao/Empreendedorismo/Guias-e-Manuais-deApoio/GuiaEmpreendedor.pdf.aspx?lang = pt-PT
7. Freitas, S.: Invovação e Formulação de Novas Ideias de Negócio, Unpublished text, ISCAP - Porto Superior Institute of Accounting and Administration (2018)
8. Massa, L., Tucci, C., Afuah, A.: A Critical Assessment of Business Model Research (2017). Retrieved from https://aom.org/uploadedFiles/Publications/Annals/ANNALS-2014-0072.R4.pdf
9. Osterwalder, A., Pigneur, Y.: Business Model Generation - Inovação em Modelos de Negócios, vol. 1. Alta Books, Rio de Janeiro (2011)
10. Kotler, P.: Marketing de A a Z, 80 Conceitos que Todo Profissional Precisa Saber, 3º edn. Campus/ Elsevier, Rio de Janeiro (2003)
11. Laudon, K.C., Traver, C.G.: E-commerce: Business. Technology. Society (2018). Retrieved from www.pearsonhighered.com
12. Turban, E., King, D.R., Lee, J.K., Liang, T.-P., Turban, D.C.: Electronic commerce: a managerial and social networks perspective. Springer texts in business and economics, 8th

edn. Springer, Cham (2015). Retrieved from https://sabraz.files.wordpress.com/2017/08/e-commerce-turban.pdf
13. Kalakota, R., Robinson, M.: e-business 2.0: Roadmap for success (2001). Retrieved from http://dinus.ac.id/repository/docs/ajar/e-business_roadmap_for_success_full.pdf
14. Borden, N.: The concept of the marketing mix (1964). Retrieved from https://manchester.rl.talis.com/items/1EB5E50E-2CDF-D475-E086-53E913CAA4A2.html
15. Dominici, G., Seaf, D.: From marketing mix to E-marketing mix: a literature overview and classification. Int. J. Bus. Manag., 17–24 (2009). Retrieved from file:///Users/karentng/Downloads/2606-11252-1-PB.pdf
16. Baltes, L.: Content marketing—the fundamental tool of digital marketing. Bull. Transilv. Univ. Braşov **8**(2), 8 (2015)
17. Conrado, A.: Os 8 Ps Do Marketing Digital: O guia estratégico do marketing digital, 2nd edn. Texto Editores, Lda, Alfragide (2011)
18. Chaffey, D., Smith, P.: Emarketing Excellence: Planning and Optimizing Your Digital Marketing, 4th edn. Routledge, Oxon (2013)
19. Wang, Y., Fesenmaier, D.R.: Identifying the success factors of web-based marketing strategy: an investigation of convention and visitors bureaus in the United States. J. Travel. Res. **44**(3), 239–249 (2006). https://doi.org/10.1177/0047287505279007
20. Adewojo, A.A., Bass, J.M.: Evaluating the effect of multi-tenancy patterns in containerized cloud-hosted content management system. In: 2018 26th Euromicro International Conference on Parallel, Distributed and Network-based Processing (PDP), pp. 278–282. IEEE (2018). https://doi.org/10.1109/PDP2018.2018.00047
21. Cabot, J.: Word press: a content management system to democratize publishing. IEEE Softw. **35**(3), 89–92 (2018). https://doi.org/10.1109/MS.2018.2141016
22. Hevner, A.R.: Design science in information systems research. J. Med. Chem., 75–105 (2004). https://doi.org/10.1021/jm991076c
23. Bax, M.: Design science: filosofia da pesquisa em ciência da informação e tecnologia. Ciência da Informação **42**(2), 298–312 (2015). https://doi.org/10.18225/ci.inf..v42i2.1388
24. IAPMEI: Manual do Empreendedor (2016). Retrieved from https://www.iapmei.pt/getattachment/PRODUTOS-E-SERVICOS/Empreendedorismo-Inovacao/Empreendedorismo/Guias-e-Manuais-de-Apoio/ManualdoEmpreendedor-(4).pdf.aspx?lang=pt-PT

Chapter 36
Machine Learning Applied to the H Index of Colombian Authors with Publications in Scopus

Amelec Viloria[1](✉), Jenny Paola Lis-Gutiérrez[2,3], Mercedes Gaitán-Angulo[4], Carmen Luisa Vásquez Stanescu[5], and Tito Crissien[1]

[1] Universidad de la Costa, Barranquilla, Colombia
{aviloria, rectoria}@cuc.edu.co
[2] Corporación Universitaria del Meta, Villavicencio, Colombia
jenny.lis@unimeta.edu.co
[3] Universidad Nacional de Colombia, Bogota, Colombia
[4] Corporación Universitaria de Salamanca, Barranquilla, Colombia
m_gaitan689@cues.edu.co
[5] Universidad Nacional Experimental Politécnica "Antonio José de Sucre", Barquisimeto, Venezuela
cvasquez@unexpo.edu.ve

Abstract. Our research aims to establish how to predict the H index of Colombian authors with publications in Scopus until 2016. The selection of the date was because, as mentioned earlier, the number of documents indexed per year exceeded 10,000 and they obtained the highest number of documents cited. To accomplish this purpose, a quantitative, nonexperimental, cross-sectional, descriptive, explanatory, and predictive research was designed using supervised learning algorithms. These were applied to information from 8,840 Colombian authors. Among the findings we can highlight that: (i) Colombia is in the fifth position in the scope of countries of South America and the Caribbean, in terms of the number of products and citations; (ii) the largest number of Colombian authors with products in Scopus until 2016, belonged mainly to the area of natural sciences, followed by medical sciences and health; (iii) most of the Colombian authors were men (64.2%, or 5,442) and they have higher H index rates than women; (iv) using random cross validation for 10 iterations, the methods with the best predictive value using R2 and the minimization of mean absolute error (MAE) correspond to: AdaBoost (96.6% and 0.397, respectively); Random Forest (96.8% and 0.431, respectively); KNN (94.4% and 0.525, respectively); Tree (94.9% and 0.53, respectively); and Neural Network (93.3% and 0.7, respectively); and (v) the variables that help predict the H index in the case of the Colombian authors, in addition to the citations, correspond to: the quantity of products, number of products in Q1, and international collaboration.

Á. Rocha et al. (eds.), *Marketing and Smart Technologies*, Smart Innovation, Systems and Technologies , https://doi.org/10.1007/978-981-15-1564-4_36

36.1 Introduction

According to [1] Colombia went from 587 documents in Scopus in 1996 to 12,651 in 2018, surpassing for the first time the 10,000 documents in 2016 (10,377). In this same period, the cited documents went from 481 to 2,582 and again 2016 is a year of inflection, since the number of works cited in the period was the maximum (6,002).

In addition, the number of citations, in 1996 the works of Colombian authors received 12,468, filled to a maximum value of 67,601 in 2015. International collaboration in 1996 was 54.86%, the maximum percentage was 58.2% in 1997 and 48.93% in 2018. Colombia's H index in 2018 is 261 and the number of citations per document corresponds to 0.53 [1].

In terms of the cumulative number of documents, citable documents and citations, between 1996 and 2018, Colombia ranks fifth, after Brazil, Mexico, Argentina, and Chile. This is important because authors such as [2] found that the production of scientific knowledge is a social process, and that the social, economic, and political context plays a fundamental role in high-quality scientific production, since it can promote or discourage research. This was corroborated by Csomós [3] who analyzed the scientific production of cities worldwide in terms of the number of journal articles indexed by the Scopus database between 1986 and 2015. The results showed that the cities with the highest scientific output are found mainly in developed countries and China; the United States is also the most relevant contributor to more than 75% of all cities.

Authors such as [4], highlight the relationship between academic production and human development, considering that human beings use capacities, abilities, and skills as instruments to solve difficulties, raise the quality of life or accumulate wealth.

In this context, our research aims to establish how to predict the H index of Colombian authors with publications in Scopus until 2016. The selection of the date was because, as mentioned earlier, the number of documents indexed per year exceeded 10,000 and they obtained the highest number of documents cited.

To accomplish this purpose, a quantitative, nonexperimental, cross-sectional, descriptive, explanatory, and predictive research was designed using supervised learning algorithms.

36.2 Recent Application of Machine Learning

Below, we mention some of the most recent works on machine learning application. Authors like [5], propose new directions for research in the age of artificial intelligence in relation to power and decision making. De Sousa et al. [6] maintains that research on artificial intelligence in the public sector is scarce, but countries such as India and the United States stand out for being the most active in this field, economic affairs and environmental protection are the areas most developed with these techniques. It emphasizes that artificial neural networks as the technique most used in research studies for Artificial Intelligence solutions in the public sector.

Likewise, [7–9], and [10] study some applications of artificial intelligence for museums, documentation, and data archiving. Owens [11] showed the need to articulate research centers and researchers in information science, informatics, and computer engineering to make digital cultural records more useful and usable.

Among the specific works associated with the application of machine learning to the H index is that of [12], who applied prediction algorithms for the H index of neuroscience journals. Abrishami and Aliakbary [13] proposed a neural network-based method for predicting long-term citations of an article, considering the number of citations received by the work in the first few years after publication. The authors identified the advantages of applying this algorithm, as opposed to other methods, with respect to accuracy in predicting annual and total citations. Kilicoglu et al. [14] applied the support vector machines and two neural network variants: (i) convolutional neural network, (ii) bidirectional long short-term memory, to 4,182 citations in clinical publications. This is done in order to perform an analysis of feelings and identify whether they were positive, negative, or neutral citations. The authors indicate that this may be an input for future predictions of citations and the H index.

36.3 Method

In this section, we present the variables and method used for the analysis.

36.3.1 Data

The study includes information from 8,840 authors with Colombian affiliation in Scopus until 2016 (Table 36.1).

Table 36.1. Variables

Variables	Typology
Number of papers	Numerical
Sex	Categorical
Main publications area	Categorical
Secondary publications area	Categorical
Discipline	Categorical
Number of citations	Categorical
Citations by document	Numerical
Publications in international journals	Numerical
Publications in national journals	Numerical
Publications in mixed journals	Numerical
Products without collaboration	Numerical
Products with national collaboration	Numerical
Products with international collaboration	Numerical
Quantity of products Q1	Numerical
Products of excellence	Numerical
Main author products	Numerical
Products of excellence in which he is the principal author	Numerical
H index	Numerical
Normalized impact	Numerical
Gross force (normalized impact * number of items)	Numerical

36.3.2 Algorithms Used

For information processing, several methods (algorithms) of supervised learning were applied, which are summarized in Table 36.2. The selection of automatic learning for data analysis is due to the following reasons: (i) it allows the application of algorithms that do not require normal data, i.e., they are nonparametric; (ii) algorithms are calibrated with few data and even work with missing data or complex interactions; and (iii) cross validation is used for prediction [15].

Table 36.2. Supervised learning algorithms

Method
AdaBoost
Random Forest
Support Vector Machines (SVM)
Neural network
KNN
Naive Bayes
Learning algorithm of the decision tree

36.4 Results

36.4.1 Descriptive Results

We analyze the variables indicated in Table 36.1. Some of the most relevant results are presented below. Figure 36.1 shows the relationship between the cumulative H index between 1996 and 2018, and the citations per document of the countries of South America and the Caribbean. We can see that the leading country in the region is Brazil and that Colombia is in the fifth position.

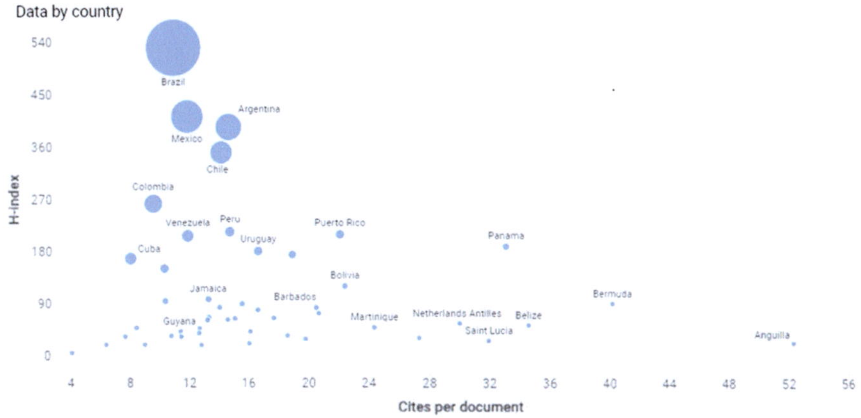

Fig. 36.1. H index and cites per Document (2018) [1]

In Fig. 36.2, we can see the distribution of citations by document and the H index of citable documents, according to the area of knowledge, for the case of Colombia in 2016. The most important areas were medicine, engineering, agricultural and biological science, social science, physics and astronomy, and computer science. It is worth mentioning that the distribution of the H index is asymmetrically positive (Fig. 36.3). This indicates that we are not working with a variable with normal distribution.

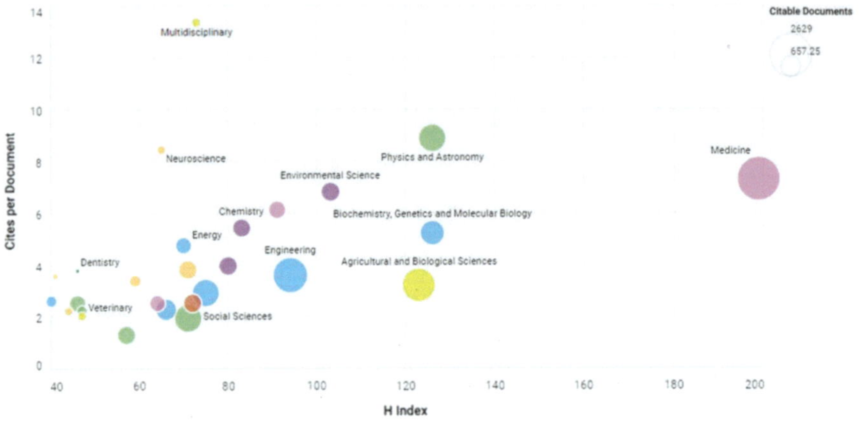

Fig. 36.2. Subject bubble chart [1]

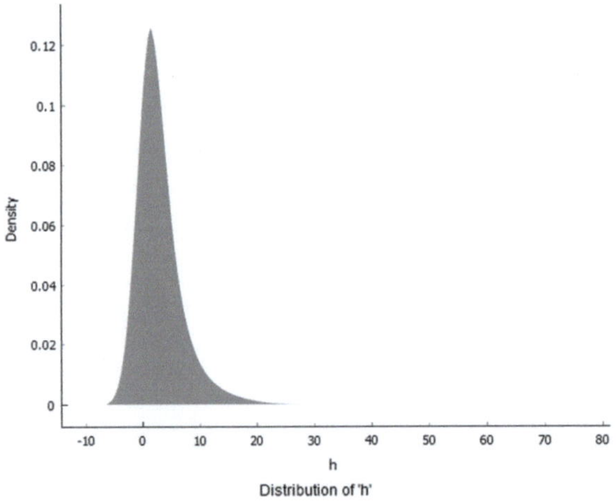

Fig. 36.3. Distribution of the H index using Orange [16]

Figure 36.4 shows the behavior of the values of the H index segmented by area of knowledge. In this graph we can see that the largest number of Colombian authors with products in Scopus until 2016, belonged mainly to the area of natural sciences, followed by medical sciences and health. These two areas also had the highest H indices. The area with the lowest number of authors and the lowest H index was humanities.

In Fig. 36.5, the number 1 represents women and the number 2 represents men. This figure shows that most of the authors were men (64.2%, or 5,442), with women accounting for 35.8% (3,038). Likewise, we can see that men have higher H index rates than women. On the other hand, Fig. 36.6 shows that the areas where Colombian authors published in journals classified in quartile q1 were mainly: medical sciences, natural mathematical sciences, and engineering,

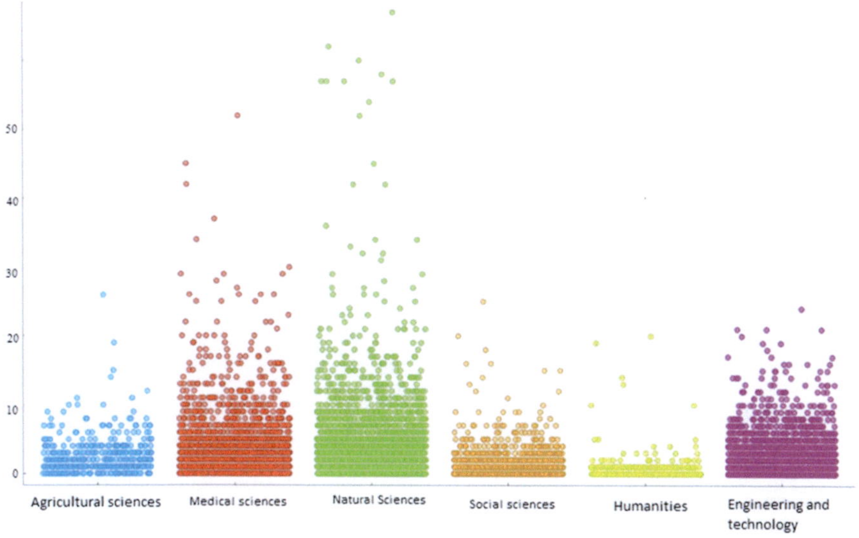

Fig. 36.4. Distribution of the H index using Orange [16]

Fig. 36.5. Distribution of the H index using Orange [16]

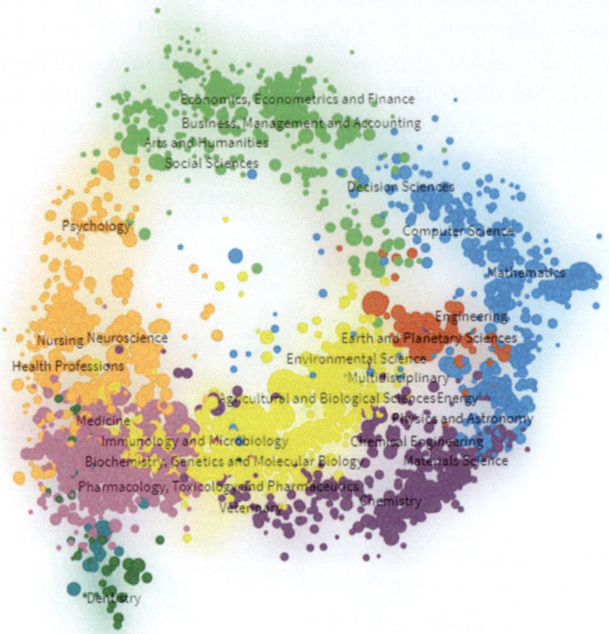

Fig. 36.6. SJR of journals by area of knowledge with Colombian authors [1]

36.4.2 Algorithms Used

This section shows the application of automatic learning to the information mentioned in Sect. 36.3.1. In this case, we use like "target" variable the H index. Different methods (algorithms) of supervised learning were applied, which would make it possible to replicate what has been learned in the past to new data. Figure 36.7 shows the different learning and prediction algorithms used and Fig. 36.8 shows the precision calculations of the learning and prediction algorithms.

Except the Support Vector Machines (SVM), all algorithms have a performance higher than 93.3%. We use a mechanism to calculate the sample for the evaluation, it is called cross validation "which splits the data into k folds and uses k – 1 folds for training and the remaining fold for testing. This procedure is repeated, so that each fold has been used for testing exactly once" [17].

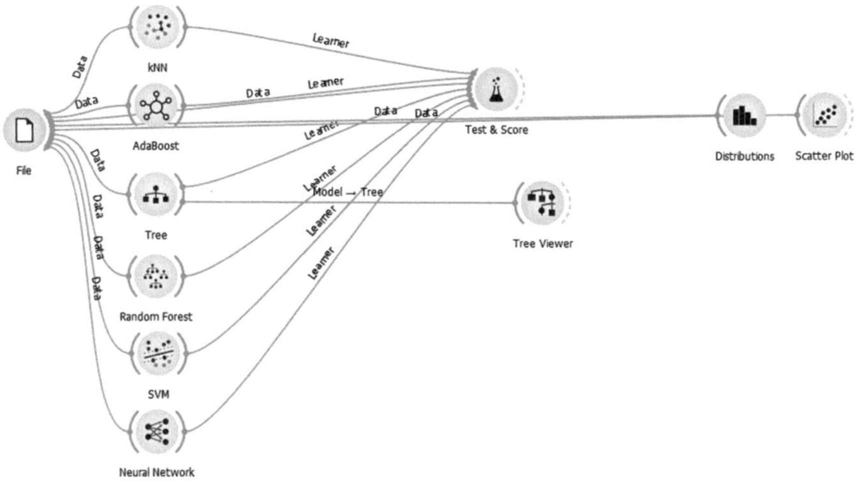

Fig. 36.7. Image of the model representation using Orange [16]

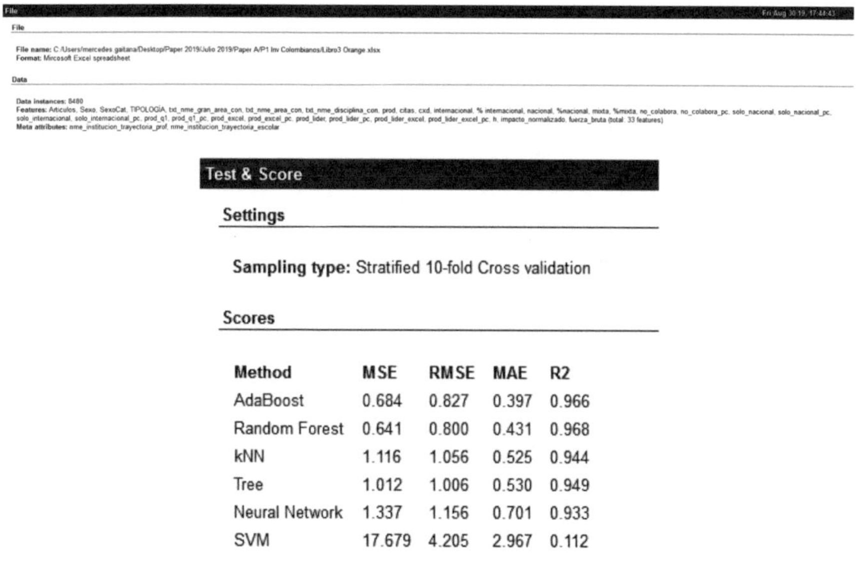

File — Fri Aug 30 19, 17:44:43

File

File name: C:/Users/mercedes.gaitana/Desktop/Paper 2019/Julio 2019/Paper A/P1 Inv Colombianos/Libro3 Orange.xlsx
Format: Mircosoft Excel spreadsheet

Data

Data instances: 8480
Features: Articulos, Sexo, SexoCat, TIPOLOGÍA, txt_nme_gran_area_con, txt_nme_area_con, txt_nme_disciplina_con, prod, citas, cxd, internacional, % internacional, nacional, %nacional, mixta, %mixta, no_colabora, no_colabora_pc, solo_nacional, solo_nacional_pc, solo_internacional, solo_internacional_pc, prod_q1, prod_q1_pc, prod_excel, prod_excel_pc, prod_lider, prod_lider_pc, prod_lider_excel, prod_lider_excel_pc, h, impacto_normalizado, fuerza_bruta (total: 33 features)
Meta attributes: nme_institucion_trayectoria_prof, nme_institucion_trayectoria_escolar

Test & Score

Settings

Sampling type: Stratified 10-fold Cross validation

Scores

Method	MSE	RMSE	MAE	R2
AdaBoost	0.684	0.827	0.397	0.966
Random Forest	0.641	0.800	0.431	0.968
kNN	1.116	1.056	0.525	0.944
Tree	1.012	1.006	0.530	0.949
Neural Network	1.337	1.156	0.701	0.933
SVM	17.679	4.205	2.967	0.112

Fig. 36.8. Calculation of learning and prediction algorithms using Orange [16]

Using random cross validation for 10 iterations, the methods with the best predictive value using R^2[1] and the minimization of mean absolute error (MAE) correspond to: AdaBoost (96.6% and 0.397, respectively); Random Forest (96.8% and 0.431, respectively); KNN (94.4% and 0.525, respectively); Tree (94.9% and 0.53,

[1] The R^2 "is interpreted as the proportion of the variance in the dependent variable that is predictable from the independent variable" [18].

respectively); and Neural Network (93.3% and 0.7, respectively). In the latter case, when changing the parameter of neurons per hidden layer in neural networks from 100 to 1000, the results changed to: 92.8% and 0.73. This implies that the most parsimonious model is more efficient.

In the case of the learning algorithm of the decision tree, we found that the value of 178 citations is a determining point for the prediction of the H index. For authors with 178 citations or less, the relevant variables for the prediction of the H index are the number of citations and products. While for authors with more than 178 citations, other variables like number of products Q1 and collaboration, are significant for the prediction.

36.5 Conclusions

The analysis of the information on the Colombian academic production indexed in Scopus, until 2016 allowed to identify that: (i) Colombia is in the fifth position in the scope of countries of South America and the Caribbean, in terms of the number of products and citations; (ii) the largest number of Colombian authors with products in Scopus until 2016, belonged mainly to the area of natural sciences, followed by medical sciences and health; (iii) most of Colombian authors were men (64.2%, or 5,442) and they have a higher H index rates than women; and (iv) using random cross validation for 10 iterations, the methods with the best predictive value using R2 and the minimization of mean absolute error (MAE) correspond to: AdaBoost (96.6% and 0.397, respectively); Random Forest (96.8% and 0.431, respectively); KNN (94.4% and 0.525, respectively); Tree (94.9% and 0.53, respectively) and Neural Network (93.3% and 0.7, respectively); and (v) the variables that help predict the H index in the case of the Colombian authors, in addition to the citations, correspond to: the quantity of products, number of products in Q1, and international collaboration [19].

Now, within the future works derived from the present investigation it is suggested, (i) carry out an analysis for 2017 and 2018, seeking to establish the changes in the variables of Colombian academic production indexed in Scopus; (ii) prepare a similar analysis for other countries of the region; and (iii) contrast the findings with countries of other continents, in order to validate the results obtained in this study.

References

1. Scimago Lab. Viz Tools. Scimago, Granada (2019)
2. Wuestman, M., Hoekman, J., Frenken, K.: The geography of scientific citations. Res. Policy (7), 1771–1780 (2019)
3. Csomós, G.: A spatial scientometric analysis of the publication output of cities worldwide. J. Inf. **12**(2), 547–566 (2018)
4. Quinapanta, M.A., Lescano, L.R.F., Barral, O.P., Jiménez, R.A.F., Rivera, D.N.: Medición del rendimiento del talento humano en instituciones de educación superior: producción científica. Ingeniería Industrial **40**(1), 24–36 (2019)
5. Fast, N.J., Schroeder, J.: Power and decision making: new directions for research in the age of artificial intelligence. Curr. Opin. Psychol. **33**, 172–176 (2020)

6. De Sousa, W.G., de Melo, E.R.P., Bermejo, P.H.D.S., Farias, R.A.S., Gomes, A.O.: How and where is artificial intelligence in the public sector going? A literature review and research agenda. Government Information Quarterly, 101392 (2019). [In Press]
7. Rosales, R., Castañón-Puga, M., Lara-Rosano, F., Flores-Parra, J.M., Evans, R., Osuna-Millan, N., Gaxiola-Pacheco, C.: Modelling the interaction levels in HCI using an intelligent hybrid system with interactive agents: a case study of an interactive museum exhibition module in Mexico. Appl. Sci. **8**(3), 446 (2018)
8. Belhi, A., Bouras, A.: Towards a multimodal classification of cultural heritage. In: Qatar Foundation Annual Research Conference Proceedings, vol. 2018, No. 3, p. ICTPD1010. HBKU Press, Qatar (2018)
9. Charalampos, G., Dimitrios, C., Georgios, P., Tzoumanika, A.: Enhanced digital cultural experience. In: 2017 8th International Conference on Information, Intelligence, Systems & Applications (IISA), pp. 1–5. IEEE (2017)
10. Yasser, A.M., Clawson, K., Bowerman, C.: Saving cultural heritage with digital make-believe: machine learning and digital techniques to the rescue. In: Proceedings of the 31st British Computer Society Human Computer Interaction Conference, p. 97. BCS Learning & Development Ltd. (2017)
11. Owens, T.: We have interesting problems: some applied grand challenges from digital libraries, archives and museums. In: Proceedings of the 18th ACM/IEEE on Joint Conference on Digital Libraries, pp. 1–1. ACM (2018)
12. Ibáñez, A., Bielza, C., Larrañaga, P.: Cost-sensitive selective naive Bayes classifiers for predicting the increase of the h-index for scientific journals. Neurocomputing **135**, 42–52 (2014)
13. Abrishami, A., Aliakbary, S.: Predicting citation counts based on deep neural network learning techniques. J. Inf. **13**(2), 485–499 (2019)
14. Kilicoglu, H., Peng, Z., Tafreshi, S., Tran, T., Rosemblat, G., Schneider, J.: Confirm or refute?: A comparative study on citation sentiment classification in clinical research publications. J. Biomed. Inform. **91**, 103123 (2019)
15. Torres-Samuel M., Vásquez C.L., Viloria A., Varela N., Hernández-Fernandez L., Portillo-Medina R.: Analysis of patterns in the University World Rankings Webometrics, Shanghai, QS and SIR-SCimago: Case Latin America. In: Tan, Y., Shi, Y., Tang, Q. (eds.) Data Mining and Big Data. DMBD 2018. Lecture Notes in Computer Science, vol. 10943. Springer, Cham (2018)
16. Demsar, J., Curk, T., Erjavec, A., Gorup, C,, Hocevar, T., Milutinovic, M., Mozina, M., Polajnar, M., Toplak, M., Staric, A., Stajdohar, M., Umek, L., Zagar, L., Zbontar, J., Zitnik, M., Zupan, B.: Orange: data mining toolbox in python. J. Mach. Learn. Res. **14**(Aug), 2349–2353 (2013)
17. Pretnar, A.: The Mystery of Test & Score. University of Ljubljana, Ljubljana (2019). Retrieved from https://orange.biolab.si/blog/2019/1/28/the-mystery-of-test-and-score/
18. Demšar, J., Zupan, B.: Orange: data mining fruitful and fun-a historical perspective. Informatica **37**(1), 55–60 (2013)
19. Torres-Samuel M., et al.: Efficiency analysis of the visibility of Latin American Universities and their impact on the ranking web. In: Tan, Y., Shi, Y., Tang, Q. (eds.) Data Mining and Big Data. DMBD 2018. Lecture Notes in Computer Science, vol. 10943. Springer, Cham (2018)

Chapter 37
Management Model for the Logistics and Competitiveness of SMEs in the City of Barranquilla

Juan de la Hoz Hernandez[1], Hugo Martinez Caraballo[2], Jairo Martinez Ventura[1], Hugo Hernandez Palma[1], Julio Cesar Mojica Herazo[3], Sandra Milena Acosta Orozco[3], Andres Rodriguez Toscano[3], and Jesus Silva[4](✉)

[1] Corporacion Universitaria Latinoamericana, Barranquilla, Colombia
{jdelahoz,academico,hhernandez}@ul.edu.co
[2] Universidad Simon Bolivar, Barranquilla, Colombia
hugo.martinez@unisimonbolivar.edu.co
[3] Universidad de La Costa, Barranquilla, Colombia
{jmojica5,sacostall,arodrigu83}@cuc.edu.co
[4] Universidad Peruana de Ciencias Aplicadas, Lima, Peru
jesussilvaUPC@gmail.com

Abstract. In Colombia, small and medium-sized enterprises (SMEs) are the most powerful engines of economic development, since they generate a high volume of jobs in the national territory, while diversifying productive, commercial, and service activities. In order to propose a model for logistics management as a component for the competitiveness of SMEs, a qualitative study is proposed that will allow us to collect from a structured review of recent literature, identifying the elements to be taken into account for logistics management in order to provide timely support and open space for continuous improvement. The proposed model is structured in six main blocks: characterization of processes according to the activity of the SME, external factors of influence, internal factors of influence, feeding of information, monitoring and control of operations, and feedback between support areas. The above elements have been analyzed and suggested taking into account the specific aspects of SMEs in the country, and taking into account the peculiarities of these small productive cells.

37.1 Introduction

Currently, small business is established as the support of the development of a country [1]; In Colombia, this reality is not far, because in recent years, a significant economic growth has emerged from SMEs, the former is due to the characteristics of its constitution [2], its flexible structure as the barriers of entry to the markets are not excessive. However, it must be borne in mind that although SMEs enjoy great success, it is important for their survival and competitiveness that they have operational activities that facilitate positioning in the economic environment [3].

Á. Rocha et al. (eds.), *Marketing and Smart Technologies*, Smart Innovation, Systems and Technologies 167, https://doi.org/10.1007/978-981-15-1564-4_37

In this sense, competitiveness is becoming increasingly important in the business sector and in the development of the private sector, particularly SMEs, since, according to Listerri et al. [4], the skills of these economic agents to assimilate knowledge and technology to produce, to cooperate, and to compete in markets are crucial for nations to make efficient use of their resources and to achieve high levels of productivity and competitiveness [5]. Due to the above, SMEs are forced to assume new strategies, through constant innovation of products and production processes, to respond quickly to changes in the environment, placing greater emphasis on cost reduction and increased productivity, the permanent search for market niches and the promotion of skilled labor, all with the aim of entering the global market [6].

37.2 Methodology

Based on the qualitative paradigm, the theoretical/descriptive research method was specified, using the bibliographic review as a tool for extracting information. This guidance is based on rigorous scientific research, as an option to explore theoretical inputs on a specific topic and thus establish its most essential characteristics [7]. As part of the methodological design, some criteria were established for the inclusion of the publications to be selected, such as: scientific articles published in journals indexed in current recognized databases in English or Spanish and with at least 10 years of validity, documents and books by authors aligned with the topic and consultation of reports from national, and foreign organizations with progress in the topic proposed in academic conferences or events.

The flexibility of this type of method makes it easier for the researcher to consider the theoretical relevance of the topic in question, since it will be possible to compare, consider and compile the different observations of the authors, thus supporting the process of critical analysis and allowing the construction of a contribution from reflection [8]. For the organization of the information, a matrix of analysis of the documents was designed, where they were classified as shown in Table 37.1. After compiling the information, the most important contributions were collected and presented in the document.

Table 37.1. Control Matrix for document analysis

No. of publications	Publication period	Type of publication
22	2005–2018	Scientific article
6	2011–2017	Books
2	2013–2015	Conferences

37.3 Importance of Logistics for the Development of SMEs

Properly managing supply chains is one of the most significant factors in enabling organizations to gain an advantage in the business sector [9]. Logistics management begins with the identification of needs until the satisfaction of the end customer is achieved, which demonstrates the importance of adequate administration of the processes involved in it [10]; hence the role played by each of the links in this chain, which must have the capacity for interaction that allows them to both speed up and ensure success during the incorporation of new products or services in the market.

For SMEs, supply chain management has a special connotation, which means that they coordinate raw materials, intermediate goods, finished products, and material handling in the best possible way within the organization, specifically within the warehouse [11]; for this reason, it should not be forgotten that it is important to give added value to each activity in the handling of materials and therefore it is necessary to know in detail the processes in the handling of materials, the needs of the production processes and the immediate need for cost control, specifically in elements related to transport, raw materials, acquisition expenses, manufacturing, distribution and inventory costs [12].

The above shows that the efficiency of supply chains in SMEs depends on a number of factors that make it possible to avoid demand uncertainty and cooperation among the organizations involved in a supply activity as much as possible [11]; however, among the most important elements to be coordinated in supply chain management with SMEs is the determination of which suppliers will be involved in supply management, which internal factors may affect them and also the recognition of the internal components that exert influence [13].

All of the above has led to supply chain management focusing on relations with suppliers, as this dynamic is the central axis of the process. To this end, it is considered vital to establish an information system between the parties of interest in order to guarantee close operational integration among them [14]. For SMEs, this implies that supply chain management has a high influence on the internal processes of organizations, and with it the activities related to the management of processes that must be coordinated in order to have competitive advantages in the business environment, which is carried out in parallel through information systems [15]. The cycle described is shown in Fig. 37.1.

In relation to what has been said, authors such as Ding and Peters [16] indicate that, in order to achieve competitiveness, companies must create collaborations with other companies, which must be characterized by being inter-organizational, because this facilitates the flow of information through them. These authors review the existing literature on information systems, concluding that practices in this field vary from company to company. This, they conclude, facilitates knowledge flows and, at the same time, allows for innovation, provided that these collaborative networks are created and properly managed.

On the other hand, researchers like Grant [17] believe that the key to information management in the supply chain is more about accessing knowledge from its partners than about acquiring it. This access allows an increase in the use of knowledge-based resources, which becomes an advantage for the company when there is uncertainty about technological changes, for example; for this purpose, the success of the SME networks in northern Italy stands out, pointing out that this is due to the fact that companies are usually created from informal relationships, based on common interests and a willingness to share experiences, which is much more effective when it comes to integrating and transferring knowledge.

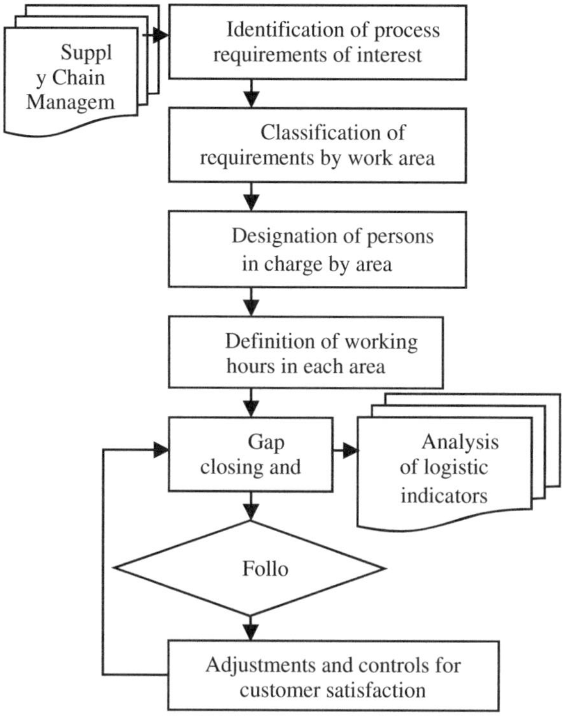

Fig. 37.1. Supply chain processes

Finally, it is necessary to rescue contributions from experts such as Capó et al. [18], who have identified in the particular case of SMEs, in order to produce a positive interaction between the different parties, so that an adequate information system can be generated, it is necessary to meet a series of requirements, which require a similarity between management systems, culture, language, objectives, etc., which are not always aligned in the first instance [19]. In particular, full involvement of all components of each value chain must be achieved, creating a climate of collaboration and mutual trust [20]. This is only possible through more stable and long-lasting interactions, in which a relationship of equals between these components is established, so that links are developed and knowledge is exchanged, creating what are called knowledge networks [21].

37.4 Logistics Model for Competitiveness in SMEs

Achieving a logistical flow in accordance with the requirements of SMEs requires not only the articulation of the components, but also improvement mechanisms that allow for a permanent review and adjustment of the tasks developed [22]. For this reason, and taking into account the aspects mentioned above, a standard model is proposed that can be applied by any SME and that, as progress is made after its adoption, can be improved and modified as the parties of interest deem appropriate. This model has been divided into two large blocks which are illustrated in Figs. 37.2 and 37.3, respectively.

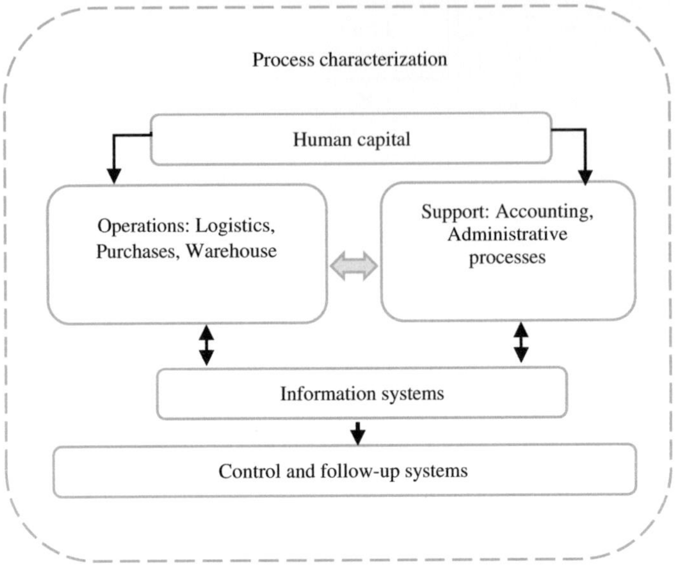

Fig. 37.2. Components of internal influence in the supply chain

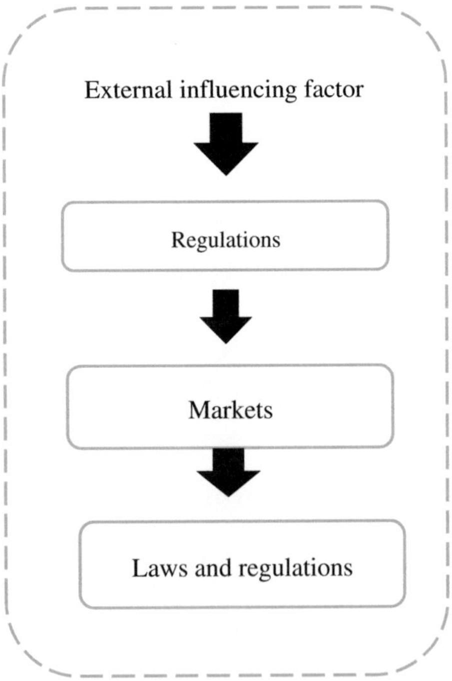

Fig. 37.3. Components of external influence in the supply chain

The proposed model is made up of six fundamental blocks: firstly, the characterization of the processes according to the SME's activity is visualized, followed by the factors of influence internal to the organization, related to the operations, support processes and human capital, the information systems that in turn feed the control and monitoring systems of the model, and finally, the external factors of influence, represented in regulations, laws and regulations, and the market.

Initially, the characterization of processes in accordance with the SME's activity allows defining, based on the diagnosis, which are the most important particularities of the processes, taking into account the activity that the company develops, this aspect stands out, since the object of the organization defines very specific characteristics depending on the sector [23]. On the other hand, internal factors of influence constitute an element that provides information on the variables that intervene in the model from the point of view of the organization, that is, those aspects where leaders can exert influence to change or improve them, from the strategic point of view, correspond to strengths and weaknesses [24].

Within the internal context, the operations, commonly formed by the logistics, purchasing and warehouse areas, among others, maintain direct and coordinated communication with the support areas, such as accounting, administration, and finance. For this reason, mechanisms for the analysis of information, supported by structured systems, allow for rapid, efficient, and fluid decision-making [25, 26].

This information, as support for decision-making, becomes the main input for the management and control systems of operations, facilitating the identification of improvement opportunities and other critical points of attention that may affect the normal development of operations [27]. In addition, the information systems and data analyzed there, as well as management control and monitoring, provide senior management with knowledge for the strategic direction of the organization [28].

Finally, external influencing factors, represented on the one hand by laws, regulations, and rules, are elements that the organization cannot change or influence, in which case it can only adapt. In addition, the market, with all its complexity and dynamics, is the one that sets the standard for the company's logistics and other internal processes [29–32].

37.5 Conclusions

Once the bibliographic review has been completed and the suggested model has been presented, it can be observed that in order to achieve high levels of competitiveness, SMEs must adopt operational strategies that facilitate and give value to internal processes. In this sense, the supply chain plays a transcendental role, which requires constant supervision and renewal. That is why it is of vital importance, the management of information and communication with all the integrated figures in the management of the supply chain, which provides significant benefits in the operation, with the management of services and stakeholders which can generate a constant flow of information and knowledge between suppliers, distributors, and customers. In this way, the most important aspects of the organization can be strengthened, the failures that can impact customer satisfaction can be minimized and the brand positioning can be increased to reach levels of market participation and increase essential aspects such as sustainability, planning, innovation, and continuous improvement.

References

1. Salcedo Miranda, R., Crespo, F., Ramirez, D., Neira, D., Castro Bolaño, L.J., Puello, N.: Estrategias de cooperación como fuente de competitividad: Caso de estudio del clúster de muebles del departamento del Atlántico (Colombia) (2017)
2. Cardona, D., Rada, A.M., Palma, H.G.H.: Creación de empresa como pilar para el desarrollo social e integral de la región caribe: Apuntes críticos. Saber, ciencia y libertad **12**(1), 15 (2017)
3. Hernández, H.G., Cardona, D.A., Del Rio, J.L.: Direccionamiento Estratégico: Proyección de la Innovación Tecnológica y Gestión Administrativa en las Pequeñas Empresas. Información tecnológica **28**(5), 15–22 (2017)
4. Listerri, J., Angelelli, P., Painter, F., Chrisney, M., Nieder, F., Mico, A., Wilson, S.: Guía Operativa para Programas de Competitividad para la Pequeña y Mediana Empresa. Informe de trabajo. BID, Washington (2012)
5. Kaur, S.P., Kumar, J., Kumar, R.: The relationship between flexibility of manufacturing system components, competitiveness of SMEs and business performance: a study of manufacturing SMEs in Northern India. Glob. J. Flex. Syst. Manag. **18**(2), 123–137 (2017)
6. Castorena, O.H., Cerros, J.A.R., Valenzo-Jiménez, M.A.: La Relación entre el uso de las TIC y la Gestión de la Cadena de Suministro en la Pyme manufacturera de Florencia, Caquetá, Colombia. Revista de la Facultad de Contaduría y Ciencias Administrativas **3**(5), 30–37 (2018)
7. Taylor, S.J., Bogdan, R., DeVault, M.: Introduction to Qualitative Research Methods: A Guidebook and Resource. Wiley (2015)
8. Alvesson, M., Sköldberg, K.: Reflexive Methodology: New Vistas for Qualitative Research. Sage (2017)
9. Tatoglu, E., Bayraktar, E., Golgeci, I., Koh, S.L., Demirbag, M., Zaim, S.: How do supply chain management and information systems practices influence operational performance? evidence from emerging country SMEs. Int. J. Logist. Res. Appl. **19**(3), 181–199 (2016)
10. Domínguez, J.D.M., Domínguez, A.B., Acosta, L.R., Montenegro, M.F., Palma, H.H., Cardona, S.A.: La función administrativa en la era de las TIC. Revista Lasallista de Investigación **12**(2), 139–151 (2015)
11. Kamalahmadi, M., Parast, M.M.: A review of the literature on the principles of enterprise and supply chain resilience: major findings and directions for future research. Int. J. Prod. Econ. **171**, 116–133 (2016)
12. Kumar, R., Singh, R.K., Shankar, R.: Critical success factors for implementation of supply chain management in Indian small and medium enterprises and their impact on performance. IIMB Manag. Rev. **27**(2), 92–104 (2015)
13. Jayaram, J., Dixit, M., Motwani, J.: Supply chain management capability of small and medium sized family businesses in India: a multiple case study approach. Int. J. Prod. Econ. **147**, 472–485 (2014)
14. Song, H., Yu, K., Ganguly, A., Turson, R.: Supply chain network, information sharing and SME credit quality. Ind. Manag. Data Syst. **116**(4), 740–758 (2016)

15. Qrunfleh, S., Tarafdar, M.: Supply chain information systems strategy: impacts on supply chain performance and firm performance. Int. J. Prod. Econ. **147**, 340–350 (2014)
16. Ding, H.B., Peters, L.S.: Inter-firm knowledge management practices for technology and new product development in discontinuous innovation. Int. J. Technol. Manag. **20**(5–8), 588–600 (2010)
17. Grant, R.M.: Knowledge and organization. In: Nonaka, I., Teece, D.J. (eds.) Managing Industrial Knowledge: Creation, Transfer and Utilization. Sage, London (2011)
18. Capó, J., Expósito, M., Masiá, E.: La Gestión del Conocimiento en las redes de PYMES. El caso del cluster textil valenciano. Revista de Economía industrial **355**, 305–315 (2005)
19. Li, M., Gan, L., He, X.: The Research on integrated strategy of supply chain information systems in the automobile industry based on order-to-delivery mode. In: Proceedings of the 2015 International Conference on Applied Mechanics, Mechatronics and Intelligent Systems (AMMIS2015), pp. 519–526 (2016)
20. Pang, Z., Chen, Q., Han, W., Zheng, L.: Value-centric design of the internet-of-things solution for food supply chain: value creation, sensor portfolio and information fusion. Inf. Syst. Front. **17**(2), 289–319 (2015)
21. Laudon, K.C., Laudon, J.P.: Management Information Systems, vol. 8. Prentice Hall (2015)
22. Giannakis, M., Louis, M.: A multi-agent based system with big data processing for enhanced supply chain agility. J. Enterp. Inf. Manag. **29**(5), 706–727 (2016)
23. Prajogo, D., Oke, A., Olhager, J.: Supply chain processes: linking supply logistics integration, supply performance, lean processes and competitive performance. Int. J. Oper. Prod. Manag. **36**(2), 220–238 (2016)
24. Caporale, T., Citak, M., Lehner, J., Schoknecht, A., Ullrich, M., Social BPM lab–characterization of a collaborative approach for business process management education. In: 2013 IEEE 15th Conference on Business Informatics (CBI), pp. 367–373. IEEE (2013)
25. Rushton, A., Croucher, P., Baker, P.: The Handbook of Logistics and Distribution Management: Understanding the Supply Chain. Kogan Page Publishers (2014)
26. Diaz-Martinez, J.J., Ruiz-Ariza, J.D., Contreras-Salinas, J., Hernández-Palma, H.G.: Technology management to increase the efficiency of the supply chain. J. Theor. Appl. Inf. Technol. **95**(19) (2017)
27. Van-Der Meer, M., Kurth-Nelson, Z., Redish, A.D.: Information processing in decision-making systems. Neurosci. **18**(4), 342–359 (2012)
28. Vidal-Ledo, M.J., Pérez, A.B.A.: Gestión de la información y el conocimiento. Revista Cubana de educación médica superior **26**(3), 474–484 (2012)
29. Yap, J.: Adapting porter's strategy formulation process and generic strategies model in today's competitive business environment. Singap. Manag. J. **1**(1), 38–99 (2012)
30. Viloria, A.: Commercial strategies providers pharmaceutical chains for logistics cost reduction. Indian J. Sci. Technol. **8**(1), Q16 (2016)
31. Amelec, V.: Increased efficiency in a company of development of technological solutions in the areas commercial and of consultancy. Adv. Sci. Lett. **21**(5), 1406–1408 (2015)
32. Amelec, V., Carmen, V.: Design of a model of evaluation of productivity for microfinance institutions. Adv. Sci. Lett. **21**(5), 1529–1533 (2015)

Chapter 38
Business Intelligence and Data Mining to Support Sales in Retail

Francisca Castelo-Branco[1], José Luís Reis[2(✉)], José Carvalho Vieira[3], and Ricardo Cayolla[1,4]

[1] ISMAI, Maia University Institute, Maia, Portugal
fmacbrancos@gmail.com, rcayolla@ismai.pt
[2] Research Unit CEDTUR/CETRAD, ISMAI, Maia University Institute, Maia, Portugal
jreis@ismai.pt
[3] Research Unit UNICES, ISMAI, Maia University Institute, Maia, Portugal
jvieira@ismai.pt
[4] GOVCOPP, DEGEIT, Universidade de Aveiro, UA, Aveiro, Portugal

Abstract. Markets are increasingly competitive, and organizations are constantly finding ways to improve their data processing in order to be able to find their customers' behavioral patterns when they buy their products. The aim of this study is to create a body of knowledge, so that a project can use the tools and techniques associated with data mining in retail sales in a proper way, presented concepts and key techniques like market basket analysis, association rules and cross-selling and up-selling. Companies make expert use of statistics and modeling to improve a wide variety of functions. In the paper are presented some common applications of business intelligence. As well as successful examples of data mining applied to retail sales, some information is also presented in this work to explain the prospects for the future of retail.

38.1 Introduction

Markets are increasingly competitive in almost all areas of economic activity. The retail industry encompasses the journey of a good or service. This typically starts with the manufacturing of a product and ends with the said product being purchased by a consumer from a retailer. Global retail sales in 2016 were around 22 trillion U.S. dollars and projected to amount around 28 trillion U.S. dollars by 2020 [1]. For example, the sports market is a global market and is growing driven by an increasing portion of the population living healthier lifestyles and stronger participation in sports and fitness activities [2]. In 2019, the revenue in the e-commerce market segment of Sports & Outdoor amounts to €213 million and is expected to show an annual growth rate of 6.6% [3]. The objective of this work is the study of solutions to be used in a project that applies business intelligence solutions that improve business competitiveness and efficiency in the retail sector, relating information using data mining techniques, like market basket analysis.

Á. Rocha et al. (eds.), *Marketing and Smart Technologies*, Smart Innovation, Systems and Technologies 167, https://doi.org/10.1007/978-981-15-1564-4_38

38.2 Business Intelligence

In modern retail business, increasing standards, automation, and technologies have led to vast amounts of data becoming available, however, business decision-making process has become complicated. Business Intelligence (BI) technologies emerge as the times require. Business Intelligence is a very wide-ranging set of the collection, consolidation, analysis and information access capabilities for a solution, data warehouse, data query and reporting, multidimensional data analysis, data mining, and other technologies [4].

Business Intelligence and Analytics (BI&A) and the related field of big data analytics have become increasingly important in both the academic and the business communities over the past two decades. BI became a popular term in the business and IT communities only in the 1990s. In the late 2000s, business analytics was introduced to represent the key analytical component in BI [5].

The main features of BI are [6, 7]:

1. Fast data processing: BI can access data in any time or department segment, realize the function of data array, data selection, and data format conversion. Data processing speed and capacity are fully protected.
2. Intelligent correlation analysis: BI makes full use of the mathematical model, profoundly reveals enterprises operating rules, and provides enough scientific decision-making material.
3. Multidimensional analysis target: BI gives a combined analysis in the object of merchandise, brands, manufacturers, suppliers, category, price, and so on, which creates the material for multidimensional data structure.
4. Progressive analysis result: In practical applications, BI can get the most specific decision-making data through continuous analysis and progressive selection. BI often uses Key Performance Indicators (KPI) to assess the present state of business and to prescribe a course of action.

There are key technologies of business intelligence like Data Warehouse, Online Analytical Processing (OLAP), Data Mining, and Release and Express technology. Some of them will be explained in point 3, OLAP and the Release and Express Technology will be explained in the following points.

38.2.1 Online Analytical Processing (OLAP)

Online Analytical Processing (OLAP) is an approach to quickly provide answers to analytical queries that are multidimensional in nature. The typical applications of OLAP are in business reporting for sales, marketing, management reporting, business process management, budgeting and forecasting, financial reporting, and similar areas. In the core of any OLAP system is a concept of an OLAP cube (also called a multidimensional cube or a hypercube) [7, 9]. It consists of numeric facts called measures which are categorized by dimensions. The cube metadata is typically created from a star schema or snowflake schema of tables in a relational database. Measures are derived from the records in the fact table and dimensions are derived from the dimension tables [8].

38.2.2 Release and Express Technology

Release and Express Technology is a solution that BI provides with the popularity of web applications, like web-based application services, which expand the BI release of the information. The query and reporting tools, especially the Multidimensional Data Display tool can provide analysis result for users in a visualization way [4]. Analytics competitors make expert use of statistics and modeling to improve a wide variety of functions. In Table 38.1 are some common applications.

Table 38.1 Applications of business intelligence—adapted from [9]

Function	Description	Examples
Supply Chain	Simulate and optimize supple chain flows; reduce inventory and stock-outs	Dell, Walmart, Amazon
Customer selection, loyalty, and service	Identify customers with the greatest profit potential; increase likelihood that they will want the product or service offering; and retain loyalty	Harrah's, Capital One, Barclays
Pricing	Identify the price that will maximize yield, or profit	Progressive, Marriott
Human Capital	Select the best employees for tasks or jobs, at compensation levels	New England Patriots, Oakland A's, Boston Red Sox
Product and service quality	Detect quality problems early and minimize them	Honda, Intel
Financial Performance	Better understand the drivers of financial performance and the effects of nonfinancial factors	MCI, Verizon
Research and development	Improve quality, efficacy, and, where applicable, safety of products and services	Novartis, Amazon, Yahoo

The UK Consumer Cards and Loans business within Barclays Bank, for example, spent five years executing its plan to apply analytics to the marketing of credit cards and other financial products. The company had to make process changes in virtually every aspect of its consumer business: underwriting risk, setting credit limits, servicing accounts, controlling fraud, cross-selling, and so on. On the technical side, it had to integrate data on 10 million Barclaycard customers, improve the quality of the data, and build systems to step up data collection and analysis. In addition, the company embarked on a long series of small tests to begin learning how to attract and retain the best customers at the lowest price. And it had to hire new people with top-drawer quantitative [9].

38.3 Data Mining

Data mining, as the term is used, is the exploration and analysis of large quantities of data in order to discover meaningful patterns and rules [10]. Data mining supports knowledge discovery by finding hidden patterns and associations, constructing analytical models, performing classification and prediction, and presenting the mining results using visualization tools [11].

The retail industry is a well-fit application area for data mining since it collects huge amounts of data on sales, customer shopping history, goods transportation, consumption, and service. The quantity of data collected continues to expand rapidly, especially due to the increasing availability, ease, and popularity of business conducted on the web, or e-commerce. Retail data mining can help identify customer buying behaviors, discover customer shopping patterns and trends, improve the quality of customer service, achieve better customer retention and satisfaction, enhance goods consumption ratios, design more touching goods transportation and distribution policies, and reduce the cost of business [11].

38.3.1 Data Mining Forms and Cycle

Data Mining comes in two forms: directed and undirected. Directed data mining attempts to explain or categorize some target fields such as income or response classification, estimation, and prediction. Undirected data mining attempts to find patterns or similarities among groups of records without the use of a target field or collection of predefined classes, as examples of undirected tasks are affinity grouping and clustering [10].

The virtuous cycle of data mining in practice has four stages [10]:

1. Identifying the business problem.
2. Mining data to transform the data into actionable information.
3. Acting on the information.
4. Measuring the results.

Data Mining provides many different techniques to extract knowledge from data. It is a multidisciplinary field of research that has many extremely useful applications. At the present time the techniques are becoming more commonly used but have not been applied adequately in the store layout. Store layout problem is motivated by applications known as market basket analysis to find relationships between items purchased by customers [12]. The store layout is an important retailing decision that can help or harm sales and store profitability. Store layouts are extremely important because they strongly influence in-store traffic patterns, shopping atmosphere, shopping behavior, and operational efficiency. Store layout design can play a key role not only in satisfying buyers' requirements but also in influencing their wants and preferences. Store layout affects consumers' price acceptability, which is positively related to purchase intentions [13]. They also report that superstores are currently revolutionizing the nature of retail service, mainly by creating more effective self-service arrangements as a result of improvements in store layout design [12].

38.3.2 Data Warehouse

Data warehouses collect and coalesce data from across an enterprise, often from multiple transaction-processing systems, each with its own database. Data is gathered because it is needed for some operational purpose, such as inventory control or the billing and, once it has served that purpose, it languishes on disk or tape or is discarded [10]. Analytical systems can access data warehouses. Data warehousing may be a facilitating technology of data mining. It is not always necessary, as most data mining does not access a data warehouse, but firms that decide to invest in data warehouses often can apply data mining more broadly and more deeply in the organization [14]. For example, if a data warehouse integrates records from sales and billing as well as from human resources, it can be used to find characteristic patterns of effective salespeople.

The goals of an initial data warehouse implementation should be specific, achievable, and measurable [11]. Building and using a data warehouse is a complex task because it requires business skills to understand how systems store and manage their data, technology skills to understand how to make assessments from quantitative information and derive facts based on conclusions from historical information in the data warehouse and program management skills to answer the need to interface with many technologies, vendors, and end-users in order to deliver results in a timely and cost-effective manner [11].

In general, the warehouse design process consists of the following steps [11]:

1. Choose a business process to model (e.g., orders, invoices, shipments, inventory, account administration, sales, or the general ledger).
2. Choose the business process grain, which is the fundamental, atomic level of data to be represented in the fact table for this process (e.g., individual transactions, individual daily snapshots, and so on).
3. Choose the dimensions that will apply to each fact table record. Typical dimensions are time, item, customer, supplier, warehouse, transaction type, and status.
4. Choose the measures that will populate each fact table record. Typical measures are numeric additive quantities like dollars-sold and units-sold.

Data warehouse administration includes data refreshment, data source synchronization, planning for disaster recovery, managing access control and security, managing data growth, managing database performance, and data warehouse enhancement and extension [11].

Firstly, having a data warehouse may provide a competitive advantage by presenting relevant information from which to measure performance and make critical adjustments to help win over competitors. Secondly, a data warehouse can enhance business productivity because it is able to quickly and efficiently gather the information that accurately describes the organization. Thirdly, a data warehouse facilitates customer relationship management because it provides a consistent view of customers and items across all lines of business. Finally, a data warehouse may bring about cost reduction by tracking trends, patterns, and exceptions over long periods in a consistent and reliable manner [11].

38.3.3 Market Basket Analysis

Market basket analysis (MBA) is used to determine which products sell together. The input data to a market basket analysis is normally a list of sales transactions, where each has two dimensions, one represents a product and the other represents a customer, depending on whether the goal of the analysis is to find which items sell together to the same person [13]. The main aim of MBA in marketing is to provide the information to the retailer to understand the purchase behavior of the buyer, which can help the retailer in correct decision-making [15]. MBA is transaction data that describes three fundamentally different entities: customers, orders, and items [10].

In a relational database, the data structure for market basket data often looks like Fig. 38.1. This data structure includes three important entities [10]:

- The order is the fundamental data structure for market basket data. An order represents a single purchase event by a customer. This includes the total amount of the purchase, the total amount, additional shipping charges, payment type, and whatever other data is relevant about the transaction.
- The item table also typically has a link to a product reference table, which provides more descriptive information about each product. This descriptive information should include the product hierarchy and other information that might prove valuable for analysis.
- The customer table is an optional table and should be available when a customer can be identified, for example, on a website that requires registration or when the customer uses an affinity card during the transaction.

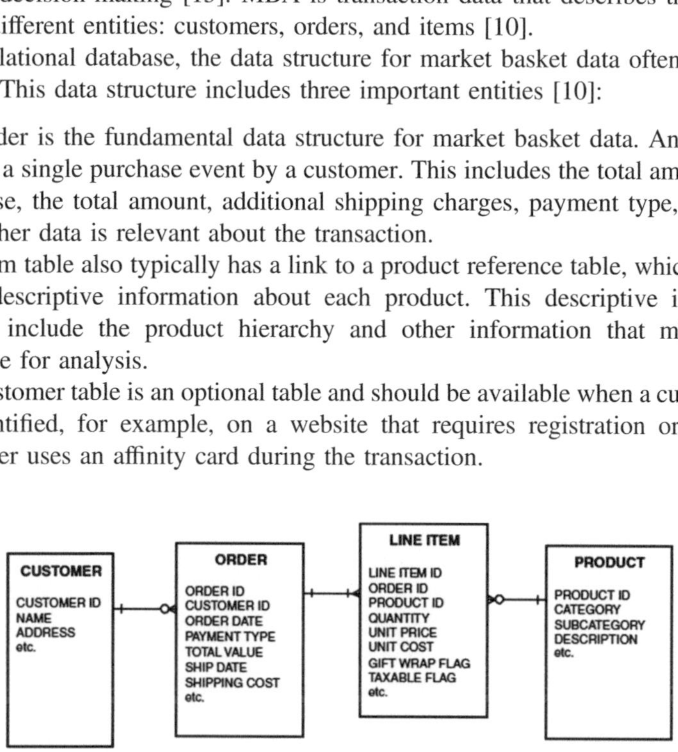

Fig. 38.1. Data model for the transaction-level market basket [10]

All three levels of market basket data are important. For instance, to understand orders, there are some basic measures:

- What is the average number of orders per customer?
- What is the average number of unique items per order?
- What is the average number of items per order?
- For a given product, what is the proportion of customers who have ever purchased the product?

MBA is a technique that assists in determining which sports items are purchased together according to the association rules, primarily with the aim of identifying cross-selling opportunities. Its primary objective is to improve the effectiveness of marketing and sales tactics using the customer's data [16]. Supposing a manager of a sports store wants to learn about the buying habit of the customers so that it can be determined which groups or sets of items customers are more likely to buy from the store. One plausible option to determine this is by performing market basket analysis on the retail history, i.e., the retail data collected from previous transactions. The main application of this analysis is in optimizing the store layout, where the items that are likely to be sold together can be placed closer to each other [16].

The algorithm that will be used is the FP-Growth algorithm. FP-Growth algorithm is an efficient algorithm for calculating frequently co-occurring items in a transaction database. To understand how it works, let's start with some terminology, using a customer transaction as an example [17]:

- item—any object that is sold on the market;
- basket—a container for one or more items selected by the customer;
- itemset—any subset of items that are sold together, in the same shopping basket;
- transaction—the complete set of items in an individual shopping basket, in the moment of purchase;
- transaction database—the complete set of shopping baskets/transactions recorded by the merchant.

38.3.4 Association Rules

Given a set of records (i.e., transactions), each of which containing several items from a given collection, the goal of association-rule discovery is to produce dependency rules that can predict the occurrence of an item based on occurrences of other items [18]. For example: A store manager may wish to know how likely a customer would buy a tennis ball if he buys tennis rackets. Such knowledge can help the store manager in product arrangement, shelf space planning, and effective implementation of product promotion strategies. The items that co-occur frequently can be discovered from a customer transactions database using association rules [16].

The applications of data mining in retail trade enterprises are mainly concentrated in association rules mining. Association rule mining is an initial data exploration approach that is often applied to an extremely large data set. An example is grocery store market basket data. Association rules mining provides valuable information in assessing significant correlations. By mining association rules, marketing analysts try to find sets of products that are frequently bought together [12].

Associations rules are helpful for the organization to improve themselves [16]:

- Product recommendations and promotion;
- Product placement;
- Next best offers strategies.

38.3.5 Cross-Selling and Up-Selling

Cross-selling is an old and valuable technique used by salespeople to increase order size and to transform single-product buyers into multiproduct ones. Cross-selling involves the sales of additional items related (or sometimes unrelated) to a previously purchased item, while up-selling involves the increase of order volume either by the sales of more units of the same purchased item, or the upgrading into a more expensive version of the purchased item. Rather than relying on a sales or services representative to decide whether to cross-sell and which item to offer, modern cross-selling utilizes analytical tools to study the customer's past behavior, correlate this information with similar customers, and then identify potential cross-selling opportunities at each contact with the customer [19].

Within the context of customer relationship management, cross-selling has become a valuable strategy for customer development, for several reasons:

1. There is a belief that it costs five times less to serve an existing customer than to acquire a new one [20].
2. Cross-selling leads to a broader scope for the customer relationship, increasing not only share of wallet but also the firm's "share of mind" with the customer [19].
3. As the customer buys more products and services from the firm and broadens the scope of the relationship, the firm learns more about the customer's needs and preferences, improving their ability to target marketing efforts and to cross-sell [20].

This information advantage, added with the higher costs of switching, produces a virtual local monopoly for the firm, which is then better able to compete for its customers than other firms that do not have an established relationship or access to the same information about their needs and preferences. For these reasons above, many customer-focused enterprises are taking advantage of cross-selling as a tool for customer development [19]. While implementing these cross-selling strategies, firms realized that cross-selling is more effective in inbound than outbound customer contacts. In other words, it is better to cross-sell when the customer calls the firm than to call the customer for the purpose of cross-selling. Some of the reasons for this important finding are intuitive. First, costs are lower, as the contact is initiated by the customer, and there is no waste in reaching the customer. Second, since the customer initiated the contact, the mindset is already centered on the firm and its services, simplifying the sales task. Third, if the customer called with a problem and it is solved to their satisfaction, the customer is more receptive to the cross-selling suggestion, particularly when this suggestion meets their needs. The analytical tools for cross-selling reviewed so far focused on identifying the "next-to-buy" or "next-to-offer" service, based on "natural" acquisition sequences observed across customers and over time [19].

38.4 The Future of Retail

Many retail marketing organizations still operate much as they did a decade ago focused on pushing products and offers to mass segments in the hopes of driving traffic that may result in a sale. Yet, that is when a shift should have occurred when the retail epicenter shifted from products to consumers—see Fig. 38.2. It's a shift most large retailers are still trying to master, while their smaller digital competitors build loyal followings and lifetime value by creating more relevant and engaging experiences that enable shopping in the moment of desire anywhere the customer works, lives, or plays [21].

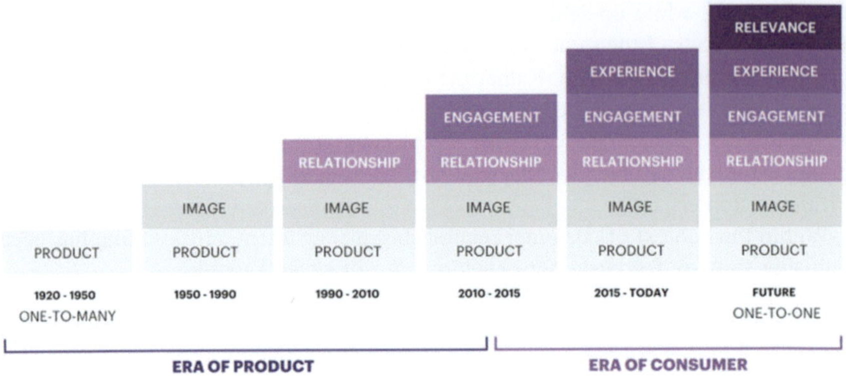

Fig. 38.2. The shift from products to consumers [21]

In the past, offering the right product at the right price might have been enough to win over consumers, but no more. To win over consumers who can buy from whomever, whenever, retailers need to have a clearly defined purpose in order to build trust. In this new scenario, consumers expect to shop how, when, and where they want across multiple points of engagement [22]. The Accenture Consulting see customer-centered growth as the future of retail, and the big picture involves focusing on six key areas [21]:

- Extended Ecosystem: capacity to manage agencies, partners, and internal functions that are constantly evolving.
- Data and Insights: capacity to best leverage technologies and data to prioritize and continually improve CX.
- Evolving marketing for the future: capacity to do differently and better to win, centered on customer experience.
- Customer Experience: capacity to build brands with purpose and deliver experiences consistently at scale.
- Customer Value proposition: capacity to identify the biggest growth opportunities, those will require new services, products or business models.
- Organization Capability: capacity to engage and empower the organization's people and ensure that there is the right talent to be successful.

For instance, shoppers might want to search and browse products on an indoor visualization of a store to obtain information such as an item's price, color, size, and approximate location in the store. Radio frequency identification (RFID), has become promising technology in the retail environment for inventory management, supply chain management, and retail marketing, because it can be configured to detect an object's presence, absence, or approximate location [23].

38.5 CRISP-DM Model

The Cross Industry Standard Process for Data Mining (CRISP-DM) is a freely available model that has become the leading methodology in data mining. Because of its industry and tool independence, CRISP-DM can provide guidelines for organized and transparent execution of any project [24].

The CRISP-DM methodology is described in terms of a hierarchical process model, consisting of sets of tasks described at four levels of abstraction (from general to specific): phase, generic task, specialized task, and process instant. At the top level, the data mining process is organized into several phases and each phase consists of several second-level generic tasks. The second level is called generic because it is intended to be general enough to cover all possible data mining situations. The third level, the specialized task level, is the place to describe how actions in the generic tasks should be carried out in certain specific situations. For example, at the second level there might be a generic task called clean data. The third level describes how this task differs in different situations, such as cleaning numeric values versus cleaning categorical values [25]. The description of phases and tasks as discrete steps performed in a specific order represents an idealized sequence of events. In practice, many of the tasks can be performed in a different order, and it will often be necessary to repeatedly backtrack to previous tasks and repeat certain actions.

Horizontally, the CRISP-DM methodology can be represented by the reference model, which presents a quick overview of phases, tasks and their outputs. It is composed by six phases that are explained in detail on Appendix A, but typically it groups all scheduled tasks into six consecutive phases [24]:

1. Business Understanding (understanding business objectives and converting them into data mining problems, developing project plan),
2. Data Understanding (collection of data, verification of data quality, data exploration),
3. Data Preparation (data selection, cleaning, data formatting and integration, constructing the final dataset for modeling),
4. Modeling (selection of modeling techniques, building and assessment of a model),
5. Evaluation (evaluation of results in the context of business objectives, reviewing the quality of the process, further decisions on deployment),
6. Deployment (deployment plan, deployment execution and monitoring, final report and review of the project).

Despite clear distinction between the groups of tasks in CRISP-DM, it has been acknowledged that there is no single universal practice in information discovery. The variety of data and related issues as well as business objectives may require various degrees of flexibility in applying the CRISP-DM reference model [25].

By analyzing historical and ongoing data, companies could generate valuable knowledge, which could contribute to the optimization of processes. However, the process of data mining is a process with high complexity and a structured process model is necessary to successfully execute such projects. The most commonly preferred process model, CRISP-DM, already has proven to be a successful guideline in a variety of cases. These case studies identified relevant patterns and created predictive models from the data sources, which could be used to optimize processes [26].

38.6 Case Studies of Data Mining in Retail

There are several case studies and good examples of the application of data mining solutions in different areas, for example:

1. The Analytics team at Hewlett-Packard recently executed a manually driven cross-sell/up-sell pilot in the Small and Medium Business online store and call center. The pilot, for which management dictated a 1-month development timeframe, utilized sales transaction, product configuration, and product availability data. Leveraging market basket analysis techniques among a small subset of available product SKUs, the pilot yielded an ROI of more than $300 K/month and more importantly, gave birth to greater opportunities to further showcase the power of analytics and data-driven decision-making at HP [27].
2. The Swedish interior giant IKEA featured image recognition and augmented reality for the first time when it showcased its 2013 catalog. Customers could scan through the catalog with their mobile devices to highlight products they were interested in, and from this, the brand offered personalized digital content and reviews to inform their purchase. The brand also used image-recognition technology, with which customers can scan catalog items and virtually place them in their own homes to see what they would look like. They can then select the colors and sizes that work best in the space, without having to go to the store and purchase the product. This allowed the catalog readers to make informed purchases, resulting in higher customer satisfaction and fewer returned items [28].
3. Macy's upmarket department store has a long history of providing excellent customer service and has become a household name. Despite the heritage established since the first store opened in 1858, the brand has taken to the digital age like a fish to water. Macy's uses big data to offer a smarter customer's experience. The brand analyzes multiple data points, such as stock levels and price promotions, and combines these findings with stock keeping unit data from a product at a particular location—as well as customer data—to ascertain which products are on sale in each store. This ensures that its chosen products suit the buying habits of customers in each location. On top of this, Macy's collects customer data ranging from visit frequency to style preference. This data is used to personalize the customer experience, offering incentives at the point of sale with loyalty rewards and promotions. This data also enables it to send targeted direct mail to its customers to boost conversions [28].
4. Amazon Go is the latest wave of technology in retail that is expected to lead the way to the future of AI in retail. The basic concept behind Amazon Go is that it is a new kind of store that flourishes on the concept of no checkout requirements. Consumers who walk into a store can take whatever they want without having to go

through the hassle of lines and waiting for checkout. The checkout-free shopping experience in Amazon Go is only made possible using the same technology that is currently in place behind computer vision, sensor fusion, and self-driving cars. The technology automatically detects all that is being taken and keeps track of them in a virtual cart. Shortly after the consumer leaves, they will be sent a receipt and charged through their Amazon account [29].

5. As Gerri Martin-Flickinger, the Chief Technology Officer and EVP at Starbucks, told CIO.com: "With about 90 million transactions a week, we know a lot about what people are buying, where they're buying, how they're buying. And if we combine this information with other data, like weather, promotions, inventory, rearview mirror perspective insights into local events, we can deliver better-personalized service to forward-looking, agile optimization on campaign performance after other customers." For example, by combining predictive analytics with location fueled by predictive analytics it's too late to make an impact data, Starbucks can send an offer to customers who are near a Starbucks based and experimentation on what the data predicts they will want on a certain day and time [21].

Retailers must redefine customer experience, engagement, and relationship—all more complex than ever because of an integrated marketplace. To be a retailer of choice, companies must embed themselves into the very fabric of a consumer's everyday life—providing an adaptable, beneficial experience that moves beyond a simple product sale [21].

There are some interesting facts that pertain to the use of Artificial Intelligence (AI) in retail, to build the insight into what to expect during the feature;

- It is expected that customers will manage 85% of their relationship with the enterprise without interacting with a human.
- According to a report by Business Insider, it is said that customers who interact in online opinions and reviews with retailers are 97% more likely to convert along with the retailer during this phase of change [29].

38.7 Conclusions

This paper presented some of the relevant aspects associated with the application of data mining to optimize retail store sales. In addition to the aspects associated with business intelligence, concepts and key techniques, and tools (market basket analysis, association rules and cross-selling and up-selling) that can be applied in a data mining project were discussed. In addition to a brief presentation of the CRISP-DM model, the future of retail and data mining application success stories were also presented, creating the conditions for this body of knowledge can be used in a future work related to data mining to support retail sales can be used.

References

1. O'Connell, L.: Total retail sales worldwide from 2015 to 2020 (2019). https://www.statista.com/statistics/443522/global-retail-sales/
2. Markets Insider: The growth of sales in sportswear (2017). https://markets.businessinsider.com/news/stocks/the-growth-of-sales-in-sportswear-1002249734. Accessed 15 Mar 2019
3. Statista Market Forecast: Sports & outdoor—Portugal (2019). https://www.statista.com/outlook/259/147/sports-outdoor/portugal?currency=eur
4. Gang, T., Kai, C., Bei, S.: The research & application of business intelligence system in retail industry. In: Proceedings of the IEEE International Conference on Automation and Logistics, ICAL 2008, September, pp. 87–91 (2008). https://doi.org/10.1109/ICAL.2008.4636125
5. Chen, H., Storey, V.C.: Business intelligence and analytics: from big data to big impact. MIS Q. **36**(4), 1165–1188 (2012)
6. Acito, F., Khatri, V.: Business analytics: why now and what next? Bus. Horiz. **57**(5), 565–570 (2014). https://doi.org/10.1016/j.bushor.2014.06.001
7. Krishnamoorthi, S., Mathew, S.K.: Business analytics and business value: a comparative case study. Inf. Manag. **55**(5), 643–666 (2018). https://doi.org/10.1016/j.im.2018.01.005
8. Chaudhuri, S., Dayal, U.: An overview of data warehousing and OLAP technology. ACM SIGMOD Rec. **26**(1), 65–74 (1997)
9. Davenport, T.H.: Competing on Analytics. Harvard Business Review (2006)
10. Berry, M.J.A., Linoff, G.S.: Data mining techniques: for marketing, sales, and customers. SIGMOD Rec. **25** (1996). https://doi.org/10.1145/235968.280351
11. Han, J., Kamber, M., Pei, J.: Data Mining. Elsevier Inc. (2012)
12. Cil, I., Ay, D., S Turkan, Y.: Data driven decision support to supermarket layout. In 8th WSEAS International Conference on Artificial Intelligence, Knowledge Engineering & Data Bases (AIKED '09) (2009). www.internal-pdf://237.233.27.24/Cil-2009-Data driven decision support to super.pdf
13. Bäckström, K., Johansson, U.: Creating and consuming experiences in retail store environments: comparing retailer and consumer perspectives. J. Retail. Consum. Serv. **13** (6), 417–430 (2006). https://doi.org/10.1016/j.jretconser.2006.02.005
14. Provost, F., Fawcett, T.: Data science for business. Mach. Learn. (2011). https://doi.org/10.1007/s13398-014-0173-7.2
15. Kaur, M., Kang, S.: Market basket analysis: identify the changing trends of market data using association rule mining. Procedia Comput. Sci. **85**(2016), 78–85 (2016)
16. Kaur, H., Singh, K.: Market basket analysis of sports store using association. Int. J. Recent. Trends Electr. Electron. Eng. **3**(1), 81–85 (2013)
17. RapidMiner.: FP-growth (2019). https://docs.rapidminer.com/latest/studio/operators/modeling/associations/fp_growth.html. Accessed 15 Apr 2019
18. Gullo, F.: From patterns in data to knowledge discovery: what data mining can do. Phys. Procedia **62**, 18–22 (2015). https://doi.org/10.1016/j.phpro.2015.02.005
19. Kamakura, W.A.: Cross-selling: offering the right product to the right customer at the right time. J. Relat. Mark. **6**(3–4), 41–58 (2008)
20. Rothfeder, J.: Trend: cross-selling (2003). https://www.cioinsight.com/c/a/Trends/Trend-CrossSelling. Accessed 15 Apr 2019
21. Accenture Consulting: The new retail(er) (2018). https://www.accenture.com/t20181024T092627Z__w__/mz-en/_acnmedia/PDF-87/Accenture-Retail-Living-Marketing-Updated.pdf

22. PWC.: Rethinking retail: The role of the physical store (2018). https://www.pwc.be/en/documents/20180627-rethinking-retail.pdf
23. Nur, K., Carreras, A., Pous, R., Morenza-Cinos, M.: Projection of RFID—obtained product information on a retail store's indoor panoramas. IEEE Intell. Syst. (2015)
24. Wowczko, I.: A case study of evaluating job readiness with data mining tools and CRISP-DM methodology. Int. J. Infonomics **8**(3), 1066–1070 (2015). https://doi.org/10.20533/iji.1742.4712.2015.0126
25. Smart Vision Europe.: CRISP-DM (2015). http://crisp-dm.eu/. Accessed 3 May 2019
26. Bloothoofd, M.P., Francken, A., Graas, R.: CRISP-DM Methodology Fact Sheet (2018)
27. Kitts, B., Melli, G., Rexer, K.: Data mining case studies proceedings. In: IEEE International Conference on Data Mining (2005)
28. Greene, V.: Top 7 examples of big data retail personalization (2018). https://bigdata-madesimple.com/7-examples-of-big-data-retail-personalization/. Accessed 12 Apr 2019
29. van Loon, R.: AI's impact on retail—examples of Walmart and Amazon (2018). https://bigdata-madesimple.com/ais-impact-on-retail-examples-of-walmart-and-amazon/

Chapter 39
The Residents' Perception of the Impacts of Tourism in Porto

João Carvalho[1,2(✉)], Sandrina Ribeiro[3], and Marc K. Peter[4]

[1] Oporto Global University, Porto, Portugal
joaomscarvalho@gmail.com
[2] CICS.NOVA.UMinho, REMIT, Covilhã, Portugal
[3] ISMAI, Maia University Institute, Maia, Portugal
sandrinarb.ribeiro@gmail.com
[4] FHNW School of Business, Olten, Switzerland
mark.peter@fhnw.ch

Abstract. The tourism phenomenon interferes directly in the life of individuals, so the study of the impacts of tourism has become an area of interest, namely associated with the concept of societal sustainability. This paper analyses not only the residents' perceptions about the economic, social and ecological impacts of tourism in Porto, but also their attitudes towards tourism and tourists. In addition, it was considered a new variable recently introduced in the literature regarding the concept of support to tourism in a behavioural approach. Thus, this study adopts a more comprehensive plan by developing the sequence of 'perceptions—attitudes—behaviours' of residents. Considering the nature of the research, the methodology used consisted of the application of a questionnaire survey to a sample of 62 residents in the municipality of Porto. The results of this study indicate that the residents value more the economic component, followed by the social dimension, and have a less formed opinion about the environmental impacts of tourism. Although the residents show positive attitudes towards tourism and tourists, they do not provide behavioural support to tourism development. This exploratory analysis of the residents' perceptions, attitudes and behaviours to support the tourist development in Porto makes an important contribution to the development of sustainable tourism planning and management in this region.

39.1 Introduction

Tourism is one of the sectors that has developed most in Portugal in recent years. With the growth of tourism, the Portuguese economy has gained a new breath, with this sector accounting for 7.5% of the national economy in 2017 [1]. The city of Porto is no exception, proving to be a destination that is fashionable, with many businesses targeted to tourists. It is the second largest city in Portugal, situated on the north bank of River Douro. Its Historic Centre has been classified as a World Heritage Site by UNESCO since 1996. Porto is a rising city in terms of tourism activity, having been mentioned and recognized around the world. Proof of this is the awards it has won,

Á. Rocha et al. (eds.), *Marketing and Smart Technologies*, Smart Innovation, Systems and Technologies 167, https://doi.org/10.1007/978-981-15-1564-4_39

standing out in 2017 as the best European destination for the third year after having already won that title in 2012 and 2014 [2].

Although tourism is generally seen as an effective way to revitalize a region's economy, it still depends heavily on the local population's support for its development, which in turn is influenced by residents' perceptions of tourism impacts on their communities. In this sense, several studies in recent years have analysed the residents' perceptions about the impacts of tourism and their attitudes towards the development of this activity [3–8]. Consequently, and based on the tetra-value theory approach [9, 10], the objectives of this exploratory research are to analyse the perceptions of residents in Porto about the positive and negative economic, social and ecological impacts of tourism, as well as the psychological impact that is measured by the residents' attitudes and behaviours towards tourism and tourists.

39.2 Tetra-Value Theory and Societal Sustainability

Tetrad-value theory widened Elkington's concept of the triple bottom line, which was considered a new approach for enterprises to develop sustainable competitive advantages that considers the economic, social and environmental dimension of sustainable development [11]. In the context of sustainable entrepreneurship, Carvalho and Sousa proposed the tetra-value theory [9, 10], which considers economic, social, ecological and psychological dimensions. Carvalho relates them to the concept of societal sustainability that includes economic, social, ecological and psychological sustainability [12–14].

In general, a sustainable entrepreneur must ensure the economic sustainability of the organization, which relies on the organization's ability to develop and grow with the efficient use of resources; to contribute to social sustainability, maintaining levels of social inclusion, security, health, education, justice and quality of life that promote social equity; to preserve natural resources and the environment; and, as psychological sustainability, to promote mental well-being in order to increase the satisfaction of individuals [13, 14]. In this line of reasoning, it can be affirmed that all concepts are interconnected, contributing to societal sustainability.

39.3 The Impacts of Tourism

The term tourism impact has been gaining increasing attention in the literature, given the growing evidence that tourism development not only generates positive impacts but also negative impacts on people and places that host tourists [15]. Several studies on the impacts of tourism have subdivided these impacts into three categories: economic, socio-cultural and environmental [4, 16, 17]. The concept of sustainability recognizes the importance of the involvement of residents in tourism planning and management, which allows the development of tourism to be socially responsible [18].

Tourism economic impacts in the residents' perception highlight in the first-place economic interest [4–6, 19–21], namely creating employment opportunities, developing the local economy, increasing local investment and businesses, diversifying economic activity and raising tax revenues at local and national levels, thus making tourism an important source of income for residents [5, 6, 22–24]. In addition, residents also benefit from tourism as it contributes to a set of improvements in community infrastructure and public facilities [16, 22, 25, 26], which contributes to improving the quality of life of residents [5, 6, 24]. However, these impacts can also be evaluated negatively as a result of residents' perception of an increase in the cost of living, raising the price of goods and services, or even a disparity in the distribution of economic benefits and increase of the local rates [5, 16, 22, 27–30]. Moreover, according to Johnson et al. and Teye et al., residents see tourism as an industry that offers low wages and lamentable working conditions [31, 32].

With regard to social impacts, it should be noted that tourism influences local socio-cultural characteristics, affecting mainly the social life of the residents [33]. The development of tourism promotes the creation of new attractions, leisure activities [4, 22, 34] and local cultural events [22]. The sharing of experiences and traditions increases the pride of the local population and reaffirms cultural identity [16, 35, 36]. However, it is true that, in the long term, tourism can generate feelings of anguish and pressure in the lives of residents, threatening their social reality [33]. In this respect, residents point to some negative impacts, including traffic congestion and parking problems [16, 24, 34]; increasing vandalism [5] and criminality [5, 16, 19, 22, 24]; the use of drugs [5, 19] and alcohol [24], as well as prostitution [6].

In relation to environmental impacts, Andereck et al. point out that, on the one hand, tourism allows for greater awareness of the preservation of natural resources and, on the other hand, leads to increased pollution of air and water, destruction of natural life and excess waste caused by tourism [16].

Finally, we focus on the psychological impacts of tourism, which integrate the concepts of perceptions, attitudes and behaviours. The attitude is a psychological variable that refers to the residents' lasting predisposition in relation to the tourist phenomenon in their communities [37]. In addition, according to Sharpley, one of the main limitations of previous research is that the influence of perceptions and attitudes on residents' behaviour has not been analysed [38].

To provide a better explanation of residents' perceptions and their support for tourism, several theories have been presented, of which the most cited is the Social Exchange Theory [39]. If the residents perceive that there are more positive than negative impacts, then they form more positive attitudes towards tourism and tend to support their development [16, 17, 36, 39–41]. However, some studies have not found significant relationships between residents' perceptions and attitudes [16, 41]. This led to the adoption of an alternative approach focused on social interactions between tourists and residents, which is considered by some to be the essence of tourism [38, 42]. Teye et al. and Woosnam et al. found a relationship between the feelings expressed by residents about tourists and their attitudes, which means that residents develop an affective connection with tourists based in their interactions [32, 43]. Thus, the Social Representation Theory [23, 44] has come to explain that residents perceive the impacts of tourism through the social representations of tourism in their communities based on their experiences and interactions with tourists, but also on information published in the media and other external sources [23, 45]. The study by Martín et al. shows that positive perceptions about social impacts are the factor that most influences residents' attitudes towards tourists, which confirms the Social Representation Theory, since these impacts are the most likely to be conditioned by interactions between tourists and residents [46]. However, Carmichael examined the extent to which positive or negative attitudes are reflected in actions by residents and, surprisingly, the author reveals that neither positive attitudes nor negative attitudes lead to further action. In fact, most residents who have positive attitudes demonstrate passive behaviour and therefore support tourism 'silently'. On the other hand, most residents who express negative attitudes demonstrate 'resigned acceptance'. This suggests that the residents can in fact express negative attitudes towards tourism and tourists, without this implying oppositional actions or behaviour [47].

39.4 Methods

The present study encompasses residents of Porto as they share the same space with tourists daily. Considering the objectives of the study, we opted for the elaboration of an exploratory study, with a descriptive and inferential data analysis, using SPSS 23.

The method of data collection consisted in the application of a questionnaire survey to a convenience sample, using a Google online form, preceded by a pre-test to a sample of five residents in the municipality of Porto. The questionnaire was disseminated through social networks (Facebook, Instagram, LinkedIn and Reddit) through the linking to the form during the period from 3 February to 15 March 2019. To obtain a greater number of responses to the questionnaire, the invitation to participate was published in Facebook groups for the dissemination of online academic questionnaires and sharing pages related to the theme 'travel and tourism' (e.g. 'Online Surveys' and 'Travel Lovers').

After a brief introductory note on the purpose and nature of the study, and a statement ensuring the anonymity and confidentiality of the responses, the questionnaire was divided into two main sections: (1) sociodemographic characterization of the respondents (age, sex, marital status, educational qualifications and local of residence); and (2) some statements about tourism, presented in order to identify the level of agreement/disagreement, regarding residents' perceptions of the positives and negatives impacts related to tourism activity, their attitudes towards tourism and tourists, and their behaviour in support of tourism development. It should be noted that the questionnaire was developed with reference to existing literature [13], although some statements regarding the current and specific situation regarding tourism in the city of Porto are included. The questions concerning perceptions about the impacts of tourism and attitudes towards tourism and tourists were presented following the same pattern of response, that is, through a five-point Likert scale.

The population of this study is that of 'residents in the county of Porto' over 18 years of age. Sampling is non-probabilistic for convenience, and the presented test results should be evaluated with this limitation. Considering the time available for data collection, a total of 131 questionnaires were received, 62 of which were residents and 69 of people living in the vicinity of Porto. Keeping the criterion of inclusion in the sample, we only use the information obtained with residents in Porto, who 72.6% were female. The mean age was 32.05 years (SD = 14.14; Range 18–74); 67.7% of the participants were single, 8.1% divorced or separated, and 24.2% married or living together. They lived in Porto, on average, for 14.6 years (SD = 11.77). As far as education level is concerned, 4.8% had basic studies, 24.2% high school, 4.8% technical studies, 48.4% bachelor's degree, and 17.7% a master's degree. Their professional situation was: 37.1% employees, 12.9% independent workers, 1.6% domestic service, 19.4% students, 17.7% students and workers, 3.2% looking for the first job, 3.2% unemployed, and 4.8% retired.

39.5 Analysis and Discussion of Results

The Kolmogorov-Smirnov test was used in order to ascertain whether the distribution of responses is uniform or not. In relation to residents' perceptions about the economic, social and ecological impacts of tourism in Porto, the description of the results obtained is shown in Table 39.1. It is also worth noting that the score on items with negative statements was inverted.

Table 39.1. Results of the questions regarding the residents' perceptions on the economic, social and ecological impacts of tourism in Porto

Item		M (SD)	$Z_{K\text{-}S}$
Economic impacts	1. Tourism is an important sector for the development of the local economy	4.63 (0.79)	5.969***
	2. The number of local businesses has increased considerably in Porto since the tourism expansion	4.32 (0.97)	4.763***
	3. Tourism attracts more investment to the community and promotes local businesses in Porto	4.47 (0.82)	5.398***
	4. The development of tourism creates more job opportunities	4.42 (0.80)	4.487***
	5. Tourism offers more infrastructures and public facilities (e.g. roads) to the centre of Porto	3.61 (1.23)	2.604***
	6. Tourism has increased the cost of living (i.e. the price of goods and services)	4.52 (0.78)	5.398***
	7. The tourist activity is the cause of the increase in housing prices in the centre of Porto	4.56 (0.880	5.715***
	8. The tourist activity creates unstable jobs under unpleasant conditions	3.48 (1.23)	2.413***
	9. Porto lacks more tourists to develop economically	2.89 (1.26)	1.397*
Social impacts	10. My everyday life has been changed with the development of tourism in the city (e.g. due to lack of tranquillity)	2.87 (1.50)	2.413***
	11. The presence of tourists in Porto removes the presence of the residents	2.50 (1.41)	2.667***
	12. Tourism greatly increases the quality of life of residents	2.58 (1.25)	2.223***

(continued)

Table 39.1. (*continued*)

Item		M (SD)	$Z_{K\text{-}S}$
	13. Tourism leads to a higher quality of local public services (e.g. health centres and hospitals, public transport service)	2.68 (1.30)	1.905**
	14. Tourism values local culture and encourages local festivals and traditions	3.79 (1.15)	3.747***
	15. The development of tourism encourages the restoration and protection of the city's historical heritage	4.10 (0.94)	3.725***
	16. The development of tourism has led to the creation of a greater number of cultural activities and events (e.g. theatres, exhibitions) in Porto	3.97 (0.94)	4.128***
	17. Tourism development generates social problems (e.g. increased theft, crime rate, vandalism, drug use, alcohol consumption) in Porto	3.18 (1.30)	1.524*
	18. Tourism leads to increased traffic congestion, accidents and parking problems in the city centre	3.47 (1.30)	2.286***
Ecological impacts	19. The development of tourism has increased the environmental awareness of local residents	2.63 (1.24)	2.159***
	20. Due to tourism, there is a greater protection of the public spaces of the city of Porto	3.32 (1.25)	2.477***
	21. The development of tourism leads to the destruction of public resources and the natural spaces (public gardens and parks) of the city	2.85 (1.27)	1.397*
	22. The development of tourism (e.g. construction of hotels and other tourist facilities) is a cause of increased pollution, noise and waste	3.37 (1.24)	2.223***
	23. The tourist activity consumes a great part of the natural resources (water, energy, etc.) and compromises the availability of these resources	2.97 (1.20)	1.207
	24. Tourism negatively affects the cleanliness of the streets of Porto and public health	2.79 (1.32)	2.096***

***$p < 0.001$; **$p < 0.01$; *$p < 0.05$

For all impact assessment items, there is statistically significant evidence that the residents' perception responses do not follow a uniform distribution, which means that the respondents present an asymmetric tendency to agree or disagree in each statement, except for the item that refers to the impact of tourism on consumption and availability of natural resources.

The results show that Porto residents highlight the fact that tourism is an important sector for development of the local economy, in addition to attracting more investment and promoting local businesses in Porto. Porto owes its economic recovery largely to tourism [48], which has emerged as a source of revenue and has assumed an important role in attracting investment [49]. It seems then that residents can recognize the economic benefits of tourism after a more troubled period that led to population abandonment and desertification in the centre of Porto [50]. The new dynamics induced by tourism made it possible to recover the commercial fabric after a period of disappearance of the local population and a decline in competitiveness [51] and, in the opinion of the residents, even increased the number of local businesses in Porto. In addition, residents consider that tourism development creates more employment opportunities [49, 52]. All these positive results are in line with those presented in the literature [5, 6, 22].

It should be noted that the increase in employment is not synonymous with better employment, since residents attribute a score closer to the neutral value (3) in the item referring to the fact that tourism creates unstable and unpleasant jobs. These results agree with the studies of Johnson et al. and Teye et al. [31, 32]. The wages remain low and unmotivating, while revenues generated by the sector continue to grow [52]. The average result obtained in the item referring to the fact that tourism offers more infrastructures and public facilities to the centre of Porto is a little larger, which is in line with the existing literature [16, 25, 26].

The increase in prices of goods and services pointed out by the residents is also corroborated by the literature [16, 27, 28] and has been leading, for example, to increased protests and pressure on the political powers to take measures of easier access to housing. Porto, a few years ago, was a decrepit, deserted, dirty city full of buildings falling apart, but now the real estate market has found a response in a rehabilitation of the park built by tourism [51], which set the basis for the city to renew itself, to welcome those who want to take advantage of spaces where, once, nobody wanted to live. It is more profitable for landlords to rent temporary to tourists, than to families who are looking for a place to live [48, 49]. However, the residents in Porto do not feel that the city needs more tourists to develop economically. This is where the economic and social impacts merge, as the quality of life of residents is at stake.

Regarding the social dimension, the residents agree that tourism development encourages the restoration and protection of the city's historical heritage, which meets the study by Ramires et al. [53]. If, on the one hand, excessive tourism is a factor of destruction for historical sites [48], on the other hand, it seems that tourism in Porto is far from calling into question its heritage built over time, which is, above all, a factor of attraction of tourists to the city. The development of tourism is the cause of the urban rehabilitation that has been carried out in Porto and which gave a new life to the city in an attempt to promote urban sustainability [51]. It appears the fact that the tourism activity has led to the creation of a greater number of cultural activities and events in Porto, which is in line with Marques et al., since the city never had a cultural offering as rich as now [54]. In addition, the residents believe that tourism values local culture and stimulates local festivals and traditions. On the other hand, Porto residents do not feel that tourism has increased their quality of life, nor did it lead to a higher quality of local public services, which contradicts results from other studies [5, 6, 24]: García et al. point to the feelings of pressure resulting from the tourist activity in the life of the residents [33]. In Porto, the first signs of discontent and some contestation generated by the tourist pressure can already be seen [49]. In addition, tourism seems to lead to greater traffic congestion, accidents and parking problems in the city centre, as some authors mentioned in their studies [16, 24, 34]. Although many tourists go by foot and enjoy the public transport network to move around the city, new tourist transport services have gained in momentum, as well as car rental [48]. Moreover, residents tend to consider that the development of tourism generates social problems in the centre of Porto, which is supported in the literature [5, 6, 24]. It is also worth noting that the residents do not agree that the presence of tourists in the city removes their presence, nor that tourist development has changed their daily life. Not discarding the possible risk of the city approaching a situation of excessive tourism, what is now happening in the centre of Porto, it is not the expulsion of residents but rather the urban occupation after decades of neglect [50, 51].

Regarding the ecological impacts, it is never too much to remember that the disorderly growth of tourism can lead to the depletion of natural resources, but this is not yet a concern of the residents, who attribute a neutral value to the statement, which means that they have no opinion formed on this subject. This evaluation contradicts the existing literature, which states that tourism allows a greater awareness of the preservation of natural resources [16]. Moreover, residents do not consider tourism development to have increased their environmental awareness. This position may be due to the lack of information and environmental awareness of the local population. The residents attribute a value close to the neutral value regarding the fact that tourism leads to a greater protection of the city's public spaces and, at the same time, do not consider that tourist development leads to the destruction of public resources and the natural spaces of the city, which suggests that tourism has not yet jeopardized the quality of urban life as a reflection of quality of the city's public spaces. The residents also tend to do agree that there is an increase in pollution, noise and litter caused by tourism development, which is supported by the existing literature [16]. Apart from that, the residents are not of the opinion that tourism negatively affects the cleanliness of the streets of Porto and public health.

Regarding the attitudes of the residents towards tourism and tourists in Porto, Table 39.2 provides the results of the Kolmogorov-Smirnov test. The residents support the development of tourism in Porto and believe that the development of tourism is beneficial to the local community, even though they consider that it is important to develop plans to manage tourism growth. Moreover, the residents tend to defend that Porto is 'becoming increasingly global and less traditional due to tourism development'. Finally, the creation of the Municipal Tourism Tax is not totally discouraged by the resident population, since it is known that the funds collected with the tourist tax are intended to improve the conditions of the city for tourists and residents [55]. Nonetheless, the residents feel that they suffer the effects of tourism more strongly because they live in a tourist area. Lastly, the residents feel a general dissatisfaction with the installation of parking metres as a parking regulation solution, because they also have to pay, particularly in residential areas.

Table 39.2. Results of the questions related to the attitudes of residents in relation to tourism and tourists in Porto

Item		M (DP)	$Z_{K\text{-}S}$
Attitudes related to tourism	1. I support the development of tourism in Porto	4.16 (0.961)	4.763***
	2. Porto is becoming increasingly global and less traditional due to tourism development	3.79 (1.133)	3.620***
	3. The Porto City Council should do more to promote tourism in the city	3.19 (1.212)	1.842**
	4. I consider important to develop plans to manage tourism growth in Porto	4.32 (0,919)	4.890***
	5. I, as a resident of Porto, suffer the consequences of tourism more strongly because I live in a tourist area	3.34 (1.378)	2.159***
	6. I am in favour of the creation of the Municipal Tourist Tax (€2 per night)	3.58 (1.386)	2.731***
	7. I support the installation of parking meters as a way to regulate parking in the city centre	2.48 (1.388)	2.667***
Attitudes related to tourists	8. I like to see tourists in Porto	4.27 (0.908)	5.017***
	9. I would like the number of tourists to visit Porto to increase	3.34 (1.254)	2.032**
	10. I feel that the excess of tourists is a threat to the residents and to the city	3.31 (1.374)	2.223***
	11. Overall, I believe that the development of tourism in Porto is beneficial to the local community	4.02 (1.032)	4.255***

***$p < 0.001$; **$p < 0.01$; *$p < 0.05$

Therefore, the residents perceive more positive than negative impacts and demonstrate more positive attitudes towards tourism and support the development of tourism activity, thus corroborating the existing literature [16, 41, 56]. Moreover, the economic benefits of tourism are the aspects that Porto residents perceive more positively, as some authors indicated in their studies [19–21]. The positive perceptions about economic impacts are the ones that most influence the attitudes towards tourism [21], which may explain the positive attitudes of Porto residents towards tourism development.

In relation to the residents' attitudes towards tourists, the data shows that most of them like to see tourists in Porto, which is in line with the study by Marques et al. [54]. Although they are pleased to see the number of tourists increasing, residents feel that the excess of tourists is a threat to them and to the city. Porto residents continue to welcome tourists with open arms [48, 49], and although they already feel somehow threatened by their presence, they support the increase in tourists visiting the city. These results agree with several studies [16, 32, 46] that demonstrate that positive perceptions of residents about the impacts of tourism lead to positive attitudes towards tourists.

Finally, considering the behaviour of the residents in relation to tourism in Porto, we analyse the results presented in Table 39.3.

Table 39.3. Results of the questions related to the behaviour of residents in relation to tourism in Porto

Item		M (DP)	$Z_{K\text{-}S}$
Behaviours in relation to tourists	1. I participate in activities relevant to the promotion of tourism in Porto	1.16 (1.34)	2.985***
	2. I work/volunteer at tourism events taking place in Porto	0.42 (0.691)	5.334***
	3. I buy local products in the city of Porto	2.40 (1.093)	2.413***
	4. I share news, views, photos or videos of Porto tourism in social networks	1.84 (1.270)	1.397*
	5. I would like to be involved in the process of developing strategies for tourism in Porto	1.66 (1.588)	2.921***

***$p < 0.001$; **$p < 0.01$; *$p < 0.05$

It is possible to see that, in general, the behavioural support of residents in relation to tourism is quite negative. The one that obtained a less negative score is the purchase of local products in the city of Porto. One possible explanation may be that local products are expensive, which is an obstacle to their consumption, since not all residents are willing to make a financial effort or have the financial capacity to acquire these products [57]. In addition, the sample of this study consists, in large parts, of young adults who tend to prefer international brands to local products [58]. Moreover, the residents tend not to share news, opinions, photos or videos of Porto tourism in social networks. They would not like to be involved in the process of developing strategies for Porto tourism, nor participate in activities relevant to the promotion of Porto tourism, nor participate at ongoing tourism events in the city. These results contradict Martín et al. and Nunkoo and Ramkissoon [46, 59]. Although Porto residents show positive attitudes towards tourism and tourists, they adopt behaviours that do not support tourism. These results are in line with Carmichael, who suggests that the majority of the residents who express positive attitudes give their behavioural support to tourism in a 'silent' way and, therefore, these attitudes are not reflected in actions [47].

39.6 Conclusion

This study shows that there exists a great congruence in the way that residents perceive the impacts of tourism, both in positive and negative aspects.

The economic benefits are the aspects that gather more positive perceptions by the residents in relation to tourism. They value the role of tourism in local economic development, investment attraction, commercial revitalization, increased employment, and construction of public infrastructure and facilities. However, the residents' perception is that the tourism sector creates unstable and low-wage jobs. In addition, they are concerned about aspects such as rising cost of living and real estate speculation in the city centre, and they do not consider that the city needs more tourists to develop economically.

Social impacts are those that residents consider to be the aspects that most challenge their quality of life. Although the development of tourist activity is seen as an incentive for the restoration and protection of the city's historical heritage and leading to the creation of more cultural activities and events that value local culture, local festivals and traditions, the residents do not fail to show some fear of losing the 'soul' of the city. Moreover, the residents do not realize that tourism has led to a higher quality of local public services, or contributed to improving their quality of life. On the contrary, the residents point to increased traffic, accidents, parking problems and even social problems due to tourism. Still, the residents do not consider that the presence of tourists drives them away from the city, nor that tourism development has changed their daily lives.

The environmental dimension of tourism impacts is not significant enough for the residents to have an opinion on the subject. In fact, the resident population does not consider that tourism has increased their environmental awareness. Despite this, the residents defend that tourism development leads to greater protection of public spaces and is not the cause of destruction of public resources and natural spaces of the city. While on the one hand, residents refer to the increase in pollution, noise and litter caused by tourist activity, on the other hand, they consider that tourism impact is not strong enough to negatively affect the cleanliness of the streets of Porto, as well as public health.

With regard to the attitudes of the residents in relation to tourism, a favourable general attitude is observed, which indicates support for the development of the tourist activity in Porto. Although residents point to the importance of drawing up plans to manage tourism growth, they also consider that the Porto City Council should do more to promote tourism in the city. Still, the residents believe that Porto is becoming increasingly global and less traditional due to tourism development and feel that they suffer from the impact of tourism more strongly because they live in a tourist area. The residents are pleased with the presence of tourists in Porto, and even like to see the number of tourists increase, although they are aware that the excess of tourists is a threat to them and to the city. In general, residents feel that tourism development is beneficial to the local population. Although the residents show support for tourism development in an attitudinal approach, it seems that they do not adopt behaviours to support the development of tourism and, therefore, these positive attitudes towards tourism and tourists are not reflected in action.

This study has some limitations related to its capacity of generalization such as the nature of the tourism phenomenon because the research was conducted in places that vary not only geographically, economically and socioculturally, but also in terms of tourist development [38]. Secondly, the sample is not representative of the study population because it is based only on people available to respond to the online survey.

In summary, the perceptions of the participants in this study seem to show that tourist activities in Porto contribute to increased economic sustainability, but with less impact at the other levels, such as social, psychological and ecological sustainability. Thus, it will be important to analyse these perceptions, and to promote a better planning and managing of these activities within the city, namely in coordination with local authorities.

References

1. Dinheiro Vivo: Turismo representa 7,5% da economia nacional em 2017. https://www.dinheirovivo.pt/economia/turismo-representa-75-da-economia-nacional-2017/ (2017). Accessed 17 Dec 2018
2. Público: Porto eleito melhor destino europeu de 2017. https://www.publico.pt/2017/02/10/local/noticia/porto-eleito-melhor-destino-eurupeu-de-2017-1761590 (2017). Accessed 10 Feb 2017
3. Doxey, G.V.: A causation theory of visitor-resident irritants, methodology and research inferences: the impact of tourism. In: Sixth annual conference proceedings of the travel research association, San Diego, California, 8–11 September 1975 (pp. 195–198). Travel and Tourism Research Association, San Diego, CA (1975)
4. Belisle, F.J., Hoy, D.R.: The perceived impact of tourism by residents: a case study in Santa Marta, Colombia. Ann. Tour. Res. **7**(1), 83–101 (1980). https://doi.org/10.1016/S0160-7383(80)80008-9
5. Gursoy, D., Jurowski, C.A., Uysal, M.: Resident attitudes: a structural modeling approach. Ann. Tour. Res. **29**(1), 79–105 (2002). https://doi.org/10.1016/S0160-7383(01)00028-7
6. Haralambopoulos, N., Pizam, A.: Perceived impacts of tourism: the case of Samos. Ann. Tour. Res. **23**(3), 503–526 (1996). https://doi.org/10.1016/0160-7383(95)00075-5
7. Liu, J.C., Var, T.: Resident attitudes towards tourism impacts in Hawaii. Ann. Tour. Res. **13** (2), 193–214 (1986). https://doi.org/10.1016/0160-7383(86)90037-X
8. Pizam, A.: Tourism's impacts: the social costs to the destination community as perceived by its residents. J. Travel Res. **16**(4), 8–12 (1978)
9. Carvalho, J. M.S., Sousa, C.A.A.: Social entrepreneurship, sustainability, and tetrad-value theory. In: Proceedings of the Global Cleaner Production and Sustainable Consumption Conference. Sitges, Spain, 1–4 Nov 2015
10. Carvalho, J.M.S., Sousa, C.A.A.: Is psychological value a missing building block in societal sustainability? Sustainability **10**(12), 4550 (2018). https://doi.org/10.3390/su10124550
11. Elkington, J.: Cannibals with Forks: The Triple Bottom Line of 21st Century Business. Capstone Publishing, Oxford, UK (1997)
12. Carvalho, J.M.S.: Inovação & Empreendedorismo. Ideia, Informação, Implementação, Impacto. Grupo Editorial Vida Económica, Porto (2016)
13. Carvalho, J.M.S.: Social innovation, entrepreneurship, and sustainability. In Carvalho, L.C. (Coord.) Handbook of Research on Entrepreneurial Ecosystems and Social Dynamics in a Globalized World, pp. 147–180. IGI Global, Hershey, USA (2018)
14. Carvalho, J.M.S.: The Ties of Business: A Humanistic Perspective of Innovation, Entrepreneurship and Sustainability. LAP Lambert Academic Publishing, Mauritius (2018)
15. Lankford, S.V., Howard, D.R.: Developing a tourism impact attitude scale. Ann. Tour. Res. **21**(1), 121–139 (1994). https://doi.org/10.1016/0160-7383(94)90008-6
16. Andereck, K.L., Valentine, K.M., Knopf, R.C., Vogt, C.A.: Resident's perceptions of community tourism impacts. Ann. Tour. Res. **32**(4), 1056–1076 (2005). https://doi.org/10.1016/j.annals.2005.03.001
17. Gursoy, D., Rutherford, D.G.: Host attitudes toward tourism: an improved structural model. Ann. Tour. Res. **31**(3), 495–516 (2004). https://doi.org/10.1016/j.annals.2003.08.008
18. Robson, J., Robson, I.: From shareholders to stakeholders: critical issues for tourism marketers. Tour. Manag. **17**(7), 533–540 (1996). https://doi.org/10.1016/S0261-5177(96)00070-2

19. Diedrich, A., García-Buades, E.: Local perceptions of tourism as indicators of destination decline. Tour. Manag. **30**(4), 512–521 (2009). https://doi.org/10.1016/j.tourman.2008.10.009
20. Huh, C., Vogt, C.A.: Changes in residents' attitudes toward tourism over time: a cohort analytical approach. J. Travel Res. **46**(4), 446–455 (2008). https://doi.org/10.1177/0047287507308327
21. Vargas-Sánchez, A., Plaza-Mejía, M., Porras-Bueno, N.: Understanding residents' attitudes toward the development of industrial tourism in a former mining community. J. Travel Res. **47**(3), 373–387 (2009). https://doi.org/10.1177/0047287508322783
22. Andereck, K.L., Vogt, C.A.: The relationship between residents' attitudes toward tourism and tourism development options. J. Travel Res. **39**(1), 27–36 (2000). https://doi.org/10.1177/004728750003900104
23. Fredline, E., Faulkner, B.: Host community reactions: a cluster analysis. Ann. Tour. Res. **27**(3), 763–784 (2000). https://doi.org/10.1016/S0160-7383(99)00103-6
24. Milman, A., Pizam, A.: Social impacts of tourism on Central Florida. Ann. Tour. Res. **15**(2), 191–204 (1988). https://doi.org/10.1016/0160-7383(88)90082-5
25. Kim, K., Uysal, M., Sirgy, M.J.: How does tourism in a community impact the quality of life of community residents? Tour. Manag. **36**, 527–540 (2013). https://doi.org/10.1016/j.tourman.2012.09.005
26. Nunkoo, R., Ramkissoon, H.: Developing a community support model for tourism. Ann. Tour. Res. **38**(3), 964–988 (2011). https://doi.org/10.1016/j.annals.2011.01.017
27. Andriotis, K.: Community groups' perceptions of and preferences for tourism development: evidence from Crete. J. Hosp. Tour. Res. **29**(1), 67–90 (2005). https://doi.org/10.1177/1096348004268196
28. Dyer, P., Gursoy, D., Sharma, B., Carter, J.: Structural modeling of resident perceptions of tourism and associated development on the Sunshine Coast, Australia. Tour. Manag. **28**(2), 409–422 (2007). https://doi.org/10.1016/j.tourman.2006.04.002
29. Wang, Y., Pfister, R.E., Morais, D.B.: Residents' attitudes toward tourism development: a case study of Washington, NC. In: Proceedings of the 2006 North-Eastern Recreation Research Symposium, Bolton Landing, NY, USA, pp. 411–418. U.S. Department of Agriculture, Forest Service, Northern Research Station, Newtown Square, PA, 9–11 Apr 2006
30. Weaver, D.B., Lawton, L.J.: Resident perceptions in the urban-rural fringe. Ann. Tour. Res. **28**(2), 439–458 (2001). https://doi.org/10.1016/S0160-7383(00)00052-9
31. Johnson, J.D., Snepenger, D.J., Akis, S.: Residents' perceptions of tourism development. Ann. Tour. Res. **21**(3), 629–642 (1994). https://doi.org/10.1016/0160-7383(94)90124-4
32. Teye, V., Sönmez, S.F., Sirakaya, E.: Residents' attitudes toward tourism development. Ann. Tour. Res. **29**(3), 668–688 (2002). https://doi.org/10.1016/S0160-7383(01)00074-3
33. García, F.A., Vázquez, A.B., Macías, R.C.: Resident's attitudes towards the impacts of tourism. Tour. Manag. Perspect. **13**(1), 33–40 (2015). https://doi.org/10.1016/j.tmp.2014.11.002
34. Bujosa, A., Rosselló, J.: Modelling environmental attitudes toward tourism. Tour. Manag. **28**(3), 688–695 (2007). https://doi.org/10.1016/j.tourman.2006.04.004
35. Besculides, A., Lee, M.E., Maccormick, P.J.: Resident's perceptions of the cultural benefits of tourism. Ann. Tour. Res. **29**(2), 303–319 (2002). https://doi.org/10.1016/S0160-7383(01)00066-4
36. Yoon, Y., Gursoy, D., Chen, J.S.: Validating a tourism development theory with structural equation modeling. Tour. Manag. **22**(4), 363–372 (2001). https://doi.org/10.1016/S0261-5177(00)00062-5

37. Ap, J.: Residents' perceptions on tourism impacts. Ann. Tour. Res. **19**(4), 665–690. https://doi.org/10.1016/ 0160-7383(92)90060-3 (1992)
38. Sharpley, R.: Host perceptions of tourism: a review of the research. Tour. Manag. **42**(1), 37–49 (2014). https://doi.org/10.1016/j.tourman.2013.10.007
39. Perdue, R.R., Long, P.T., Allen, L.: Resident support for tourism development. Ann. Tour. Res. **17**(4), 586–599 (1990). https://doi.org/10.1016/0160-7383(90)90029-Q
40. Ko, D.W., Stewart, W.P.: A structural equation model of residents' attitudes for tourism development. Tour. Manag. **23**(5), 521–530 (2002). https://doi.org/10.1016/S0261-5177(02)00006-7
41. McGehee, N.G., Andereck, K.: Factors predicting rural residents' support of tourism. J. Travel Res. **43**(2), 131–140 (2004). https://doi.org/10.1177/0047287504268234
42. Yu, J., Lee, T.J.: Impact of tourists' intercultural interactions. J. Travel Res. **53**(2), 225–238 (2014). https://doi.org/10.1177/0047287513496467
43. Woosnam, K.M., Norman, W.C., Ying, T.: Exploring the theoretical framework of emotional solidarity between residents and tourists. J. Travel Res. **48**(2), 245–258 (2009). https://doi.org/10.1177/0047287509332334
44. Andriotis, K., Vaughan, R.D.: Urban residents' attitudes toward tourism development: the case of Crete. J. Travel Res. **42**(2), 172–185 (2003). https://doi.org/10.1177/0047287503257488
45. Zhang, J., Inbakaran, R.J., Jackson, M.S.: Understanding community attitudes towards tourism and host-guest interaction in the urban-rural border region. Tour. Geogr. Int. J. Tour. Sp. Place Environ. **8**(2), 182–204 (2006). https://doi.org/10.1080/14616680600585455
46. Martín, H.S., Sánchez, M., Herrero, Á.: Residents' attitudes and behavioural support for tourism in host communities. J. Travel Tour. Mark. **35**(2), 231–243 (2018). https://doi.org/10.1080/10548408.2017.1357518
47. Carmichael, B.: A matrix model for resident attitudes and behaviours in a rapidly changing tourist area. Tour. Manag. **21**(6), 601–611 (2000). https://doi.org/10.1016/S0261-5177(00)00007-8
48. Der Spiegel: Paradise Lost: How Tourists are Destroying the Places They Love. https://www.spiegel.de/international/paradise-lost-tourists-are-destroying-the-places-they-love-a-1223502.html. Accessed 21 Aug 2018
49. Santos, F.S.: Turistas e residentes: evitar o demasiado cheio. Sapo. https://24.sapo.pt/opiniao/artigos/turistas-e-residentes-evitar-o-demasiado-cheio. Accessed 8 Aug 2017
50. Portal de Notícias do Porto: Números históricos—Porto voltou em 2017 a ganhar residentes após 40 anos de perdas que levaram 100 mil a deixar a cidade. http://www.porto.pt/noticias/numeros-historicos-porto-voltou-em-2017-a-ganhar-residentes-apos-40-anos-de-perdas-que-levaram-100-mil-a-deixar-a-cidade (2018)
51. Boavida-Portugal, L.: O turismo nas áreas históricas de Lisboa e Porto. Ameaça ou oportunidade? Observador. Disponível em: https://observador.pt/opiniao/o-turismo-nas-areas-historicas-de-lisboa-e-porto-ameaca-ou-oportunidade/. Accessed 28 May 2016
52. Pinheiro, A.M., Laranjeiro, A.: Turismo. Escassez de mão-de-obra não sobe salários. Diário de Notícias. https://www.dn.pt/edicao-do-dia/14-fev-2019/interior/turismo-escassez-de-mao-de-obra-nao-sobe-salarios–10574069.html. Accessed 14 Feb 2019
53. Ramires, A., Brandão, F., Sousa, A.C.: Motivation-based cluster analysis of international tourists visiting a World Heritage City: the case of Porto, Portugal. J. Dest. Mark. Manag. **8**, 49–60 (2018). https://doi.org/10.1016/j.jdmm.2016.12.001
54. Marques, M.I.S., Candeias, M.T., Magalhães, C.M., Santos, A.M.: Perceção dos residentes e profissionais do Centro do Porto sobre o desenvolvimento do turismo na cidade. População e Sociedade **26**, 45–61 (2016)

55. Nunes, E.: O turismo está a estragar cidades? Jornal de Notícias **32**. https://tag.jn.pt/o-turismo-esta-a-estragar-as-cidades/. Accessed 9 June 2019
56. Gursoy, D., Kendall, K.W.: Hosting mega events: modelling locals' support. Ann. Tour. Res. **33**(3), 603–623 (2006). https://doi.org/10.1016/j.annals.2006.01.005
57. Gazeta Rural: Preço é o principal entrave ao consumo de produtos locais. http://gazetarural.com/2019/04/30/preco-e-o-principal-entrave-ao-consumo-de-produtos-locais/. Accessed 30 Apr 2019
58. Costa, A.R.: Preço continua a ser barreira ao 'consumo consciente'. Distribuição Hoje. https://www.distribuicaohoje.com/insights/preco-continua-a-ser-barreira-ao-consumo-consciente/. Accessed 24 Oct 2018
59. Nunkoo, R., Ramkissoon, H.: Gendered theory of planned behavior and residents' support for tourism. Curr. Issues Tour. **13**(6), 525–540 (2010). https://doi.org/10.1080/13683500903173967

Chapter 40
Good Practices in Logistics for SMEs: A Strategy for the Global Marketplace

Juan de la Hoz Hernandez[1], Hugo Martinez Caraballo[2], Jairo Martinez Ventura[1], Hugo Hernandez Palma[1], Julio Cesar Mojica Herazo[3], Sandra Milena Acosta Orozco[3], Andres Rodriguez Toscano[3], and Jesus Silva[4(✉)]

[1] Corporacion Universitaria Latinoamericana, Barranquilla, Colombia
{jdelahoz, academico, hhernandez}@ul.edu.co
[2] Universidad Simon Bolivar, Barranquilla, Colombia
hugo.martinez@unisimonbolivar.edu.co
[3] Universidad de La Costa, Barranquilla, Colombia
{jmojica5, sacostall, arodrigu83}@cuc.edu.co
[4] Universidad Peruana de Ciencias Aplicadas, Lima, Peru
jesussilvaUPC@gmail.com

Abstract. Thanks to globalization, the environment in which companies operate today has changed. Whereas a few decades ago, the same government generated protectionist policies, today with the opening of markets, competition is not only national but also international, so companies, regardless of their size, must make changes in their business model, aiming at the internationalization of markets both for the acquisition of materials and goods and for the marketing of their products, being more agile in responding to their customers in terms of product quality, expected delivery time and cost. It is in this scenario where logistics plays a fundamental role as a tool for competitiveness and, therefore, the following work aims to highlight the impact of logistics on SMEs and the need to aim for continuous improvement of their logistics processes. By means of a qualitative/descriptive methodology, relevant aspects are reviewed, finding as a final result essential components that SMEs must incorporate to attend a world-class market.

40.1 Introduction

In the field of international business, an issue that has become increasingly relevant and debated is how to make companies international in a highly competitive environment, generated by the ever-increasing level of demand in world trade [1]. Thus, the internationalization of a company, from a strategic point of view, requires the implementation of strategies that take into account both the company's capacities and resources and the threats of the environment, orienting its activities, either totally or partially, toward an international environment, generating flows of different types (Commercial, Financial, and Knowledge) between different countries [2].

Á. Rocha et al. (eds.), *Marketing and Smart Technologies*, Smart Innovation, Systems and Technologies 167, https://doi.org/10.1007/978-981-15-1564-4_40

Based on the above, for SMEs in Colombia, it has been more complex to implement the internationalization process, due to the constant concerns, gaps, and doubts that arise throughout the process to achieve that their products penetrate the international market and obtain a differential position within the countries with which they relate and compete. For the companies of the so-called first world, the transformation has been more dynamic, so their adaptation to this new stage has been much more fluid, thanks to the technological and cultural tools they have at their disposal, a very different case with Colombian SMEs, where the subject is practically under development and in some cases in the exploratory stage [3], in addition to this, the companies must remain in force against both national and international competitors that come to the domestic market to interact in this scenario.

In this context, logistics emerges as a key tool for companies, and especially SMEs, to successfully compete under the framework of globalization, by generating value through efficient and accessible logistics [4], a logistics that helps the organization to differentiate itself from its competitors, by having processes that allow for customer satisfaction at the right cost, in the right time and with the right quality, through sound supply chain management and continuous improvement of the supply chain, since by developing and maintaining skills, managers will be able to sustain the differential advantages of their companies [5, 6]. We are facing a changing and flexible world; whereas in the traditional model of closed and protected markets, the strategy to follow was to raise the selling price, today, due to globalization, the price is set by the consumer, not the producer [7]. The current market demands efficiency, effectiveness, creativity, and agility from organizations in order to be competitive and define their strategies aimed at business sustainability. Due to this, the concept and importance of logistics has evolved over the last 20 years [8].

40.2 Methodology

For the development of the established objective, a qualitative research of descriptive orientation is designed that will look for from the recent bibliography review, to establish the most relevant aspects that must be considered at the moment of observing, reviewing, and improving the logistic processes in the small and medium company. The descriptive methodology allows us to take the aspects or attributes of greater representation in the face of a problem and from the reflections supported by accepted theories or guidelines, to determine, at the researcher's discretion, those components that can facilitate the approach of the event or situation under investigation [9]. For this reason, and with the aim of generating contributions for improving competitiveness in SMEs, an analysis matrix was designed, in which recent research that has offered progress in this area was reviewed.

For the selection of the publications, the following were taken as inclusion criteria: scientific articles and research of the last five years, documents that handled key words such as logistics, SMEs, competitiveness, continuous improvement, and processes. Also, it was established to carry out the search in Scopus, Scielo Springer, Emerald, and Latindex, with the purpose of selecting material validated and received by the scientific community for its contributions in recent years.

40.3 Results

In Colombia, companies are classified according to Law 905 of August 2, 2004, and are presented in Fig. 40.1.

SMEs:

- They represent 96.4% of the economic establishments [9]

COMPANY	NUMBER OF EMPLOYEES	TOTAL ASSETS IN MINIMUM WAGES
• Microenterprise	• Less than or equal to 10	• Less than 500
• Small	• 11 to 50	• Between 501 and 5,000
• Medium	• 51 to 200	• Between 5.001 and 30.000
• Large	• Greater than 200	• Greater than 30,000

Fig. 40.1 Classification of companies in Colombia

- They generate approximately 81% of employment [9]
- It is estimated that they contribute 40% of GDP [9].

Only about 13% of SMEs export [9] and based on the Business Dynamics Report in Colombia, from the first quarter of 2018 by Confecámaras, it can be observed that of the total number of companies in the country, a very important group of 99.4% is made up of microenterprises from all sectors of the economy, with the commercial, service and industrial sectors being the most important, and the agricultural and livestock sector lagging behind [10]. The impact that SMEs have on Colombia's economy is very high, the same is true in the rest of Latin America and the Caribbean, where small and medium-sized enterprises make up the majority of companies [11].

The dynamics of globalization are driving SMEs to seek the optimum way to carry out their logistics operations, so that they can place their products on the market more efficiently; the vision of logistics for small and medium-sized enterprises must mature from being seen as a procedure that generates obligation, to using it as a tool to obtain competitive advantages, through the proper administration of the same [12]. In Latin America and the Caribbean, logistical costs vary between 18 and 35% of the value of the product, and are even higher for SMEs, at around 45%, which is quite a high cost compared to approximately 8% in the countries of the Organization for Economic Co-operation and Development (OECD).

Although in recent years most countries have understood the impact of logistics and have implemented actions to improve their processes, the region is still lagging behind [11], with logistics being an interdisciplinary task that involves multiple areas of the company, including from the systematization of purchase orders to customer service in the after-sales process, as well as intervening in the supply of raw materials, planning and management of production stages; logistics also participates in the storage, handling, and management of inventories, packaging, physical distribution, and finally in the flow of information [7], all this oriented to meet customer requirements [13]. An equally complete, but simpler definition is the one known as the 7 Cs of Business

Logistics, which seeks to ensure the availability of the right product, in the right quantity and under the right conditions, in the right place, at the right time, for the right consumer, at the right cost [14].

To ensure that logistics is a tool for companies (especially SMEs) that adds value, it is necessary to review each of the links in the supply chain in order to identify the critical success factors that may directly affect competitiveness [15] (Fig. 40.2).

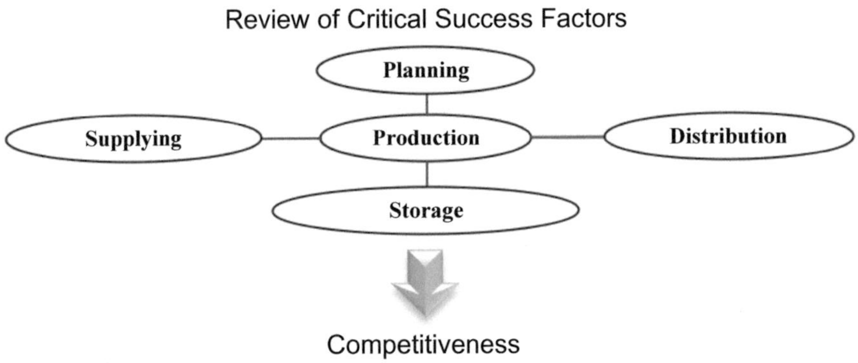

Fig. 40.2 Logistics scheme as a value tool

Likewise, it is necessary to establish indicators that allow a consistent evaluation of the effectiveness of the organization's logistics management, review the fulfillment of goals and the objectives set out [16], detect failures, be able to execute corrective action plans based on the results of these indicators, as well as plans for continuous improvement, for which the PHVA cycle: Planning, Doing, Verifying, Acting, can be very useful, observing that under this model a planning of the logistic chain is carried out, through a process of strategic development, from which the production plan and the supply plan are structured, to then execute based on the planned and consecutively evaluated execution of the previously structured plans, through the follow-up and measurement of each process for its subsequent improvement, using tools such as corrective or preventive actions, according to the needs presented [17]. A system of indicators for logistical processes should be a source that facilitates those responsible for them to monitor compliance with the objectives and identify opportunities for improvement [18]. The above arguments are illustrated in Fig. 40.3.

40.3.1 World Class Logistics

Nowadays, logistic, and storage support is a competitive factor that promotes or affects the functionality of any economic unit in any area of productive development (services, industries, and commerce). Currently, what used to be seen as a secondary process within the organization, and that was destined to the movement of inventory or merchandise, is now a key structure that provides not only support but also practical elements that generate added value [19]. At present, the concept of logistics has been

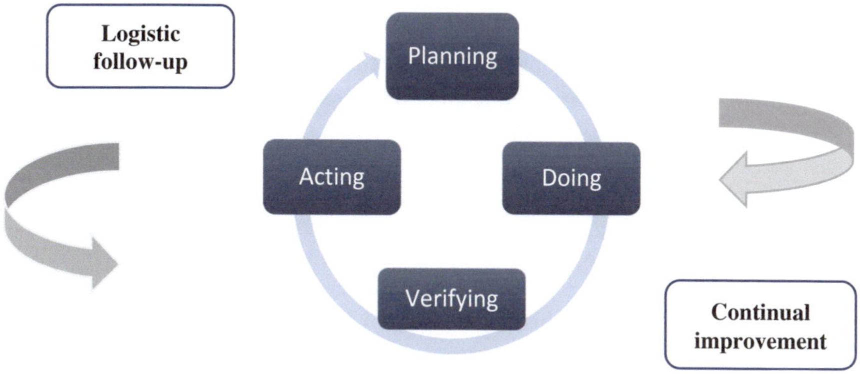

Fig. 40.3 PHVA applied to logistics

transformed and its projection covers other dimensions, taking into account that it is not only about the reception, organization, custody, and movement of goods, but also the integration of all the information generated in this area and that feeds parallel processes to achieve the satisfaction of stakeholders in highly competitive environments [20].

When defining the primary objective of a high-level logistic process, it is found that it is to guarantee the continuous and timely supply of materials, merchandise, means of production, and services so that an organization can successfully carry out its corporate purpose, this to ensure the uninterrupted operation of the entire organization and promote a constant improvement that facilitates competitiveness [21]. Maintaining the uninterrupted provision and support of other processes implies some common principles or practices, which is why a common scheme must be established and applicable to almost any environment that allows compliance with these good practices [22]. Next, we will analyze general practices that have taken on great importance worldwide and that have been defined as feasible in logistics for SMEs, regardless of the corporate purpose of each organization and are illustrated in Fig. 40.4.

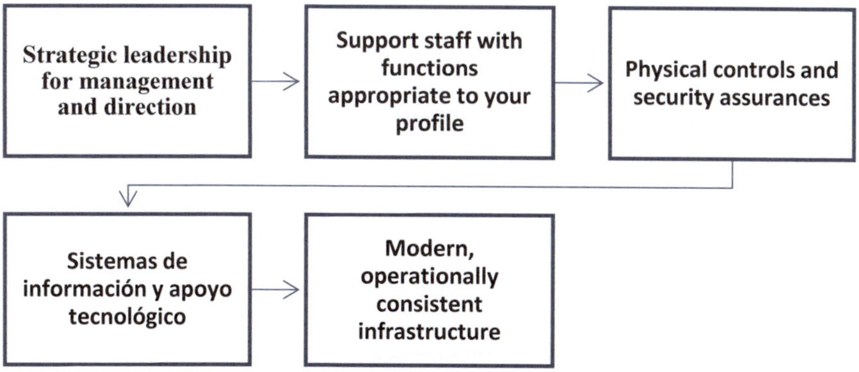

Fig. 40.4 Good practices for logistics in SMEs

The above practices can be described as follows:

- Strategic leadership for management and direction. This first practice involves changing the profile of the person leading the logistics process. Previously, a person with qualities such as responsibility, training in logistics, and numerical reasoning was considered suitable [23]. Currently, we are looking for area leaders to have competencies that cover a more complete level that allows them to lead the process in such a way that they can anticipate the different effects that the processes may have in parallel with the good or bad decisions made in each movement. This in turn demands a commitment from organizations to the continuous training and updating of this leader and his collaborators [24].
- Support staff with functions appropriate to your profile. Until a few years ago the logistics staff was chosen without taking into account factors such as: organizational capacity, revision, attention to detail, numerical agility, and commitment. These competencies make the real difference in the processes; because by properly following the guidelines given by the visible head, the processes are developed in an orderly and competent manner, projecting the fulfillment of the results [25]. This creates an environment of trust and security for the other departments of the organization. This in turn, feeds the satisfaction of the different users or customers by meeting their requirements in a more accurate manner in a higher percentage [26].
- Physical controls and security guarantees. The concept of control and physical security has also evolved, as has human resource factors. It is not only important to keep the products organized according to previously defined standards, but also to have an integral management plan that allows for the eradication of possible risks to which any organizational element may be exposed, which includes everything from goods and services to the information of any company [27].
- Information systems and technological support. This is undoubtedly one of the aspects in which most progress has been made. Currently, information support systems play a priority role in reducing losses in operations and processes. In addition, it will positively impact outside the organization, giving confidence to potential customers or partners. At the same time, accounting, fiscal, and operational identification aspects can be handled simultaneously, complying on time with all the commercial, legal, and financial implications of the organization and its control entities [28].
- Modern infrastructure is consistent with the operation. The logistics areas must have all the necessary equipment, tools, and resources to guarantee the correct implementation of the processes and the phases inherent to them. A correct planning is definitive, because the logistics becomes the platform that facilitates all the processes and their correct interaction according to the demands of the parts of interest [29].

After analyzing all the previous practices, we can affirm that transforming the concept of logistics to take it to a plane of worldwide competition, implies making a transition in the conceptions of this process understanding that a change of vision is essential for the adoption of strategic plans that allows to ponder the needs of the globalized logistic market and to attend opportunely each one of its demands [30, 31]. In addition, it is essential to achieve a level of competence that facilitates the adoption of new information and standards, since changes are taking place in an accelerated and structured manner in all markets. It is clear then that the current productive and economic apparatuses [32], look for in the logistic support, everything required to operate at a much more efficient and dynamic market level, that allows to obtain joint and continuous results offering satisfaction to all the users of this platform or process [33–36].

40.4 Conclusions

Each company is born with the goal of being profitable and maintaining itself over time, and it is indisputable that the globalized environment in which it must operate in order to reach its objective, makes it increasingly more complicated, an environment where customers are responsible for setting the business course, they set the path by which companies must set their objectives, and for SMEs, given their working capital and infrastructure conditions, the challenge becomes even greater.

Logistics evolve along with this movement, to become a process that integrates each of the links, from the supplier to the end customer, in order to carry out a strategic management that generates significant savings in product costs and customer satisfaction, which in the end translates into a competitive advantage; logistics is an area with an important growth projection and its contribution to administrative, commercial, and environmental management makes it a structural support for the business world. Therefore, SMEs must direct their attention and efforts to logistics, constantly updating their knowledge in this field in order to achieve a level of innovation and continuous improvement that allows them to stand out from their national and international competitors in the local and foreign markets.

References

1. Griffith, D., Tamer, S., Shichun, X.: Emerging themes in international business research. J. Int. Bus. Stud. **39**, 1220–1235 (2008)
2. Leandro, A.A.: El proceso de internacionalización de empresas. Tec empresarial. **3**(3), 18–25 (2013)
3. Tabares-Arroyave, S.: Revisión analítica de los procesos de Internacionalización de las pymes. Pensamiento & Gestión. **33**, 67–92 (2012)
4. Olivos, P.C., Carrasco, F.O., Flores, J.L.M., Moreno, Y.M., Nava, G.L.: Modelo de gestión logística para pequeñas y medianas empresas en México. Contaduría y administración. **60** (1), 181–203 (2015)
5. Wagner III, J.A., Hollenbeck, J.R.: Organizational Behavior: Securing Competitive Advantage. Routledge (2014)
6. Meihami, B., Meihami, H.: Knowledge management a way to gain a competitive advantage in firms (evidence of manufacturing companies). Int. Lett. Soc. Hum. Sci. **3**, 80–91 (2014)
7. García, L.A.M.: Gestión Logística Integral: Las mejores prácticas en la cadena de abastecimiento. Ecoe Ediciones (2016)
8. Pinheiro-De Lima, O., Breval-Santiago, S., Rodríguez-Taboada, C., Follmann, N.: Una nueva definición de la logística interna y forma de evaluar la misma. Ingeniare. Revista Chilena de Ingeniería **25**(2), 264–276 (2017)

9. Ministerio de Comercio, Industria y Turismo de Colombia: Dirección de Micro, Pequeña y Mediana empresa. Reporte de Mipymes (2014)
10. Confecámaras, Red de Cámaras de Comercio de Colombia.: Informe de Dinámica Empresarial en Colombia (2018)
11. Guasch, J.L.: La logística como motor de la competitividad en América Latina y el Caribe. Banco Interamericano de Desarrollo (2011)
12. Villarreal-Segoviano, F.: Logística Integral: Una alternativa para crear valor y ventajas competitivas en las pequeñas y medianas empresas (pymes) del Sector Calzado. Nova Scientia. **4**(8), 165–202 (2012)
13. Jim-Wu, Y.C.: Contemporary logistics education: an international perspective. Int. J. Phys. Distrib. Logist. Manag. **37**, 504–528 (2007)
14. Martín, C.: Logística (Aspectos Estratégicos). Editorial Limusa. (2006)
15. Ortiz-Torres, M., Valdés, P., Arias-Castillo, E.: Desempeño logístico y rentabilidad económica. Fundamentos teóricos y resultados prácticos. Economía y Desarrollo. **149**(1), 182–193 (2013)
16. Arellano, A., Ríos, N., Carballo, B., Félix, F.: Arquitectura de procesos de la cadena logística para empresas a integrarse al distrito internacional de agronegocios Pyme. Scientia etTechnica. **3**(43), 141–146 (2009)
17. Mazo, A.Z., Montoya, R.A.G., Henao, S.A.F.: Indicadores logísticos en la cadena de suministro como apoyo al modelo scor. Clío América **8**(15), 90–110 (2014)
18. Kirby, C., Brosa, N.: La logística como factor de competitividad de las Pymes en las Américas. Banco Interamericano de Desarrollo (2011)
19. Denisa, M., Zdenka, M.: Perception of implementation processes of green logistics in SMEs in Slovakia. Procedia Econ. Financ. **26**, 139–143 (2015)
20. Payaro, A., Papa, A.R.: Logistics outsourcing: why do not some Italian SMEs adopt the externalization? Asian Bus. Res. **2**(2), 46 (2017)
21. Hernández, H.G., Cardona, D.A., Del Rio, J.: L: Direccionamiento Estratégico: Proyección de la Innovación Tecnológica y Gestión Administrativa en las Pequeñas Empresas. Información tecnológica **28**(5), 15–22 (2017)
22. Kianto, A., Hussinki, H., Vanhala, M., Nisula, A.M.: The state of knowledge management in logistics SMEs: evidence from two Finnish regions. Knowl. Manag. Res. Pract. 1–11 (2018)
23. Abdullah, N.A.H.N., Yaakub, S., Subhan, M.: Logistics and supply chain related issues faced by Malaysian SMEs: a case study. Int. Rev. Manag. Mark. **6**(3), 432–435 (2016)
24. Kherbach, O., Mocan, M.L.: The importance of logistics and supply chain management in the enhancement of Romanian SMEs. Procedia-Soc. Behav. Sci. **221**, 405–413 (2016)
25. Wang, X.L., Wang, L., Bi, Z., Li, Y.Y., Xu, Y.: Cloud computing in human resource management (HRM) system for small and medium enterprises (SMEs). Int. J. Adv. Manuf. Technol. **84**(1–4), 485–496 (2016)
26. Machuve, D., Mjema, E., Trojer, L., Zlotnikova, I.: Conceptual model of information logistics in value chain analysis of food processing SMEs in Tanzania. J. Math. Comput. Sci. **5**(4), 517–537 (2015)
27. Villegas, S.M.N., Haasis, H.D.: A system dynamics approach for SMEs internationalization networking process. In: International Conference on Dynamics in Logistics, pp. 21–25. Springer, Cham (2018, February)
28. Pan, S., Ballot, E., Fontane, F., Hakimi, D.: Environmental and economic issues arising from the pooling of SMEs' supply chains: case study of the food industry in western France. Flex. Serv. Manuf. J. **26**(1–2), 92–118 (2014)
29. Alon, I., Chen, S., Mandolfo, M.: Supply chain–marketing integration: how do European SMEs go to China via the New Silk Road. Bus. Process. Manag. J. (2018)

30. Palma, H.G.H., Arbeláez, D.A.C., Pardo, A.R.: La innovación empresarial como impulsora de las Pymes en la Región Caribe. Revista Panorama Económico (24), 9 (2016)
31. Salcedo Miranda, R., Crespo, F., Ramirez, D., Neira, D., Castro Bolaño, L.J., Puello, N.: Estrategias de cooperación como fuente de competitividad: Caso de estudio del clúster de muebles del departamento del Atlántico (Colombia) (2017). https://repositorio.cuc.edu.co/xmlui/handle/11323/1182
32. La Palma, H.G.H.: gestión empresarial, un enfoque del siglo XX, desde las teorías administrativas científica, funcional, burocrática y de relaciónes humanas. Escenarios **9**(1), 38–51 (2011)
33. Palma, H.H., Sierra, D.M., Arbeláez, D.C.: Enfoque basado en procesos como estrategia de dirección para las empresas de transformación. Saber, Ciencia y Libertad **11**(1), 141–150 (2016)
34. Viloria, A.: Commercial strategies providers pharmaceutical chains for logistics cost reduction. Indian J. Sci. Technol. **8**(1), Q16 (2016)
35. Amelec, V.: Increased efficiency in a company of development of technological solutions in the areas commercial and of consultancy. Adv. Sci. Lett. **21**(5), 1406–1408 (2015)
36. Amelec, V., Carmen, V.: Design of a model of evaluation of productivity for microfinance institutions. Adv. Sci. Lett. **21**(5), 1529–1533 (2015)

Chapter 41
How Can Marketing Managers Thrive in the Age of Artificial Intelligence?

Gioia V. Volkmar(✉)

University of St. Gallen, 9000 St. Gallen, Switzerland
Gioia.volkmar@unisg.ch

Abstract. Artificial intelligence (AI) is becoming increasingly important in solving many tasks, which were performed before by humans, that is, translating languages as well as in the hiring process and customer service. The AI technology is improving in a fast manner. This has led to the current debate whether AI is entirely beneficial or should be seen as a threat for the job security. In any way, we are now at the point where nearly everyone is talking and thinking about the application of AI within their work structures. Therefore, we will examine whether AI has the potential to substitute managerial tasks in the way that marketing managers exclusively rely on the respective recommendations and results for a certain action. Additionally, we will discover which tasks lead to positive or negative feelings regarding each manager's preference. These results will be used to explain the tendency to replace certain activities with AI.

41.1 Introduction

The rise of powerful AI will be either the best or the worst thing ever to happen to humanity. We do not yet know which.

Stephen Hawking, 2016 [1].

The deployment of artificial intelligence (AI) shows tremendous advances over the past years [2]. Their scope includes various consumer applications as well as different industrial applications, including healthcare, commerce, financial services and marketing [3]. This technological partnership will make consumers as well as companies to be part in the advancement of this defining technology [4]. AI will revolutionize the current state of work and their processes within companies. However, the larger impact will be in enhancing and complementing humans regarding their capabilities, in fact not replacing them [5]. This development will promote collaboration between humans and machines to solve many different kinds of problems that will ensure a trustworthy and beneficial way [6].

Á. Rocha et al. (eds.), *Marketing and Smart Technologies*, Smart Innovation, Systems and Technologies 167, https://doi.org/10.1007/978-981-15-1564-4_41

41.2 Theoretical Background

Many alarms have sounded on the potential for AI technologies to transform business [7] and change the workforce, especially for jobs that are easy to automate [6]. This led to the current debate whether AI is entirely beneficial or should be seen as a threat for the workplace [4], even though AI has the potential to improve efficiency as well as productivity across an organization [8].

In addition, many companies plan to introduce forms of AI applications in the nearest future, though lack a holistic implementation strategy for AI in their business [9]. An identified barrier for the adaptation of AI is the absence of skills regarding managerial capabilities as well as the specific skills on how to deal with this technology [8]. In this context and with the driving digitalization, the role of a manager is being challenged.

Looking at prior research, two main streams can be outlined in order to define the role and responsibilities of a manager. Drucker (1954) dedicated his work among other things to the practice of management and introduced the concept of management by objective [10]. On this basis, he categorized the work of manager into five areas: set objectives, organize, motivate and communicate, measurement and develop people [11]. However, his studies did not specify all the tasks of a manager. It seems crucial to assess the impact of different activities and their workload for a manager.

Yet another research stream is dedicated to define all the activities of a manager. These publications are based on Mintzberg's work: "The nature of managerial work" (1973) [12] and his following publications (e.g. [13–15]). They summarize managerial activities into different categories (e.g. [16–18]). As a result, he defines three areas and their corresponding management roles: interpersonal roles (figurehead, leader, liaison), informational roles (monitor, disseminator, spokesperson) and decisional roles (entrepreneur, disturbance handler, resource allocator, negotiator). While this research stream is more precise and has an explicit focus on the work of a manager, not all external factors were included in these studies. The impact of new and current technology advancements was not taken into consideration as they have not been as relevant as they are today; Mintzberg is missing common business practices as digital communication as well as all forms of working remotely (home or elsewhere). In addition to Mintzberg's findings, other researcher included cultural aspects which show slightly different results when compared to Mintzberg [19].

While there is enough evidence based on this research stream that managers use most of their time for administrative desk work instead of using their time for strategic tasks (e.g. [12, 20]), there is no differentiation on managers working in different fields such as marketing. Especially in marketing, the AI revolution is driven by the influence of more affordable and accessible advanced data sets in the business world. The use of different AI applications has provided marketers with new and more reliable information regarding process automation, market forecasting and decision-making [21].

While both practitioners and researchers in the field of marketing are mostly interested in how to transform unstructured and structured data into smart data in order to gain valuable insights (e.g. [22]), or on how smart technologies can interact with consumers (e.g. [4, 23, 24]), research on how managers can be supported with smart technologies is scarce.

41.3 Challenges and Research Question

We are now at a point where almost everyone is talking and thinking about the application of AI within their daily life as well as in their enterprises and working structures [20]. A rising area of application within enterprises is marketing organizations, which lack support of marketing managers with AI tools.

The key question remains: How can marketing managers thrive in the age of AI?

One of the most challenging issues is trusting machines in their decision-making process. A manager who is working on highly confidential matters and is handling strategic decisions is unlikely to use AI without his supervision [20]. The problem is that machine-learning systems will not explain their decisions, which could mean an algorithm could be performing well for the wrong reasons. All computers have their knowledge according to the data they are fed to. They might pick up a biased view or might be only good in narrow tasks that are very similar to the existing data systems [9]. It does not have common sense and the ability to adapt the given data to unknown situations [25]. Therefore, transparency in the decision-making process is very important in order to use AI accordingly [5]. Managers need enough knowledge to gain a certain amount of trust [26].

The other important aspect is the understanding of AI technologies and how it can be implemented in the daily business, especially in the more strategic positions within the company [26]. In the past years, machines were able to meet and surpass human performance in many different ways [27]. They were able to perform tasks better than a human could possibly do, while using a large amount of data [28]. However, when thinking outside the box or any emotions, AI is lacking specific capabilities of a human being [29]. According to Boden (2016) humans won't be replaced by a machine any time soon when it comes to functions that require a certain amount of empathy and emotional intelligence [30].

There should be a rethinking and a shift of roles and responsibilities between machines and humans. Machines could replace tasks and roles that humans are either not pleased to do or take a lot of time. Instead, humans could focus on more creative and strategic tasks as well as all interpersonal [31].

41.4 Current Study

The current study presented in this paper aims to address this research gap by investigating whether, when and why managers can be supported in their daily work by using different AI tools. More specifically, we will conduct three sequential studies.

The first study examines tasks, activities and challenges of managers during their daily work. We will use an explorative approach in order to determine all potential activities carried out during a day. The questionnaire used will provide the basis to conduct the following studies. A second study analyzes the subjective feelings of marketing managers towards certain activities. We will analyze which tasks under certain activities are associated with positive or negative emotions, such as feelings of joy or frustration. These results could then be used to explain the willingness of a manager to have the tasks substituted or supported by AI tools. The last study will be conducted with technology experts as well as marketing managers to identify the state-of-the-art technologies and future recommendations for marketing. Findings from all three studies will be combined to answer the question whether AI has the potential to support or substitute managerial tasks in a way that marketing managers can rely on AI tools for recommendations as well as results for a certain action.

41.5 Conclusion

This paper aims to explore the potential of AI tools in the field of marketing management. Therefore, the author wants to contribute to the existing literature on managerial tasks, especially in the field of marketing and artificial intelligence. Until now, there has been no comparable study that successfully links these two topics with additional focus to the individual perception of a manager. It is of great value to understand what tasks and challenges marketing managers are currently facing to further determine the suitable use of AI.

These results have thus important implications for marketing practitioners and researchers. Managers at all levels will need to be willing to adapt and include AI technologies and tools in their daily work life to embrace and work alongside AI rather than be replaced by it.

References

1. Saffell, N.: The best or worst thing to happen to humanity—Stephen Hawking launches Centre for the Future of Intelligence. University of Cambridge. https://www.cam.ac.uk/research/news/the-best-or-worst-thing-to-happen-to-humanity-stephen-hawking-launches-centre-for-the-future-of. Accessed 4 July 2019
2. Torra, V., et al: Artificial intelligence, In: Said A., Torra V. (eds.) Data Science in Practice, pp. 9–26. Springer, Berlin (2019)
3. Huang, M.-H., Roland, T., Rust, R.: Artificial intelligence in service. J. Serv. Res. **21**(2), 155–172 (2018)
4. Daugherty, P., Wilson, J.: Collaborative intelligence: humans and AI are joining forces. Harv. Bus. Rev. **96**(4), 114–123 (2018)
5. Jarrahi, H.: Artificial intelligence and the future of work: Human-AI symbiosis in organizational decision making. Bus. Horiz. **61**, 577–586 (2018)
6. Altman, R., et al.: Artificial intelligence and life in 20130. One hundred year study on artificial intelligence. Stanford University, https://ai100.stanford.edu/sites/g/files/sbiybj9861/f/ai_100_report_0831fnl.pdf. Accessed 3 July 2019
7. Daugherty, P., Wilson, J.: Human + Machine: Reimagining Work in the Age of AI. Harvard Business Press, Boston (2018)

8. Bughin, J., et al.: Notes from the AI frontier. Tackling Europe's Gap in Digital and AI. McKinsey Global Institute (2019)
9. Ransbotham, S., et al.: Reshaping business with artificial intelligence. MITSloan Manag. Rev. **59**(1), 1–16 (2017)
10. Drucker, P.: The Practice of Management, 1st edn. Harper & Row, New York (1954)
11. Zahra, S.: The practice of management: reflection on Peter F. Drucker's landmark book. Acad. Manag. 17(3), 16–23 (2003)
12. Mintzberg, H.: The Nature of Managerial Work, 1st edn. Longman, New York (1973)
13. Mintzberg, H.: The manager's job: folklore and fact. Harv. Bus. Rev. 163–176 (1990)
14. Mintzberg, H.: Rounding out the manager's job. Sloan Manag. Rev. **36**(1), 11–26 (1994)
15. Mintzberg, H.: Managing. Berrett-Koehler Publisher, San Francisco (2009)
16. Kurke, L., Aldrich, H.: Note—Mintzberg was right!: a replication and extension of the nature of managerial work. Manag. Sci. **29**(8), 975–984 (1983)
17. Shapira, Z., Dunbar, R.: Testing Mintzberg's managerial roles classification using an in-basket simulation. J. Appl. Psychol. **65**(1), 87–95 (1980)
18. Pavett, C., Lau, A.: Managerial Work: The Influence of hierarchical level and functional speciality. Acad. Manag. J. **26**(1), 170–177 (1983)
19. Pearson, C., Chatterjee, S.: Managerial work roles in Asia: an empirical study of Mintzberg's role formulation in four Asian countries. J. Manag. Dev. **22**(8), 694–707 (2003)
20. Kolbjørnsrud, V., Richard Amico, R., T. Robert J.: How artificial intelligence will redefine management. Harv. Bus. Rev. 1–6 (2016)
21. Conick, H.: The past, present and future of AI in marketing. marketing news. https://www.ama.org/marketing-news/the-past-present-and-future-of-ai-in-marketing/. Accessed 5 July 2019
22. Wedel, M., Kannan, P.: Marketing analytics for data-rich environments. J. Mark. **80**(6), 97–121 (2016)
23. Novak, T., Hoffman, D.: Relationship journeys in the internet of things: a new framework for understanding interactions between consumers and smart objects. J. Acad. Mark. Sci. **47**, 216–237 (2019)
24. Kietzmann, J., Paschen, J., Treen, E.: Artificial intelligence in advertising. how marketers can leverage artificial intelligence along the consumer journey. J. Advert. Res. 58(3), 263–267 (2018)
25. Duan, Y., Edwards, J., Dwivedi, Y.: Artificial intelligence for decision making in the era of BigData–evolution, challenges and research agenda. J. Inf. Manag. **48**, 63–71 (2019)
26. Davenport, Thomas H., Kirby, J.: Just how smart are smart machines? MITSloan Manag Rev **57**(3), 20–25 (2016)
27. Davenport, T., Ronanki, R.: Artificial intelligence for the real world. Don't start with moon shots. Harv. Bus. Rev. 96(1), 108–116 (2018)
28. Kelly, K.: Better than human: why robots will–and must–take our jobs. Wired. https://www.wired.com/2012/12/ff-robots-will-take-our-jobs/. Accessed 2 July 2019
29. Vishal, M.: The future of artificial intelligence in the workplace. Forbes. https://www.forbes.com/sites/vishalmarria/2019/01/11/the-future-of-artificial-intelligence-in-the-workplace/#19f0383173d4. Accessed 1 July 2019
30. Boden, M.: On the singularity, emotions, and computer consciousness. Oxford University Press's Academic Insights for the Thinking World. https://blog.oup.com/2016/06/singularity-artificial-intelligence/. Accessed 4 July 2019
31. Davenport, T., Kirby, J.: (2015), Beyond automation. Harv. Bus. Rev. 59–65 (2015)

Chapter 42
Customer's Acceptance of Humanoid Robots in Services: The Moderating Role of Risk Aversion

Daniel Belanche[1(✉)], Luis V. Casaló[2], and Carlos Flavián[1]

[1] Faculty of Economy and Business, University of Zaragoza (Spain), Gran Vía 2, 50.005, Saragossa, Spain
{belan, cflavian}@unizar.es
[2] Faculty of Business and Public Administration, University of Zaragoza (Spain), Plaza Constitución s/n, 22.001 Huesca, Spain
lcasalo@unizar.es

Abstract. The emerging introduction of humanoid robots in service encounters is becoming a reality in the present and the short-term. Owing to this unstoppable advance, there is a need to better understand customers' perceptions and reactions toward humanoid agents in service encounters. To shed some light on this underexplored phenomenon, this research investigates how the interaction between robot and customer's features may contribute to a successful introduction of this disruptive innovation. Results of an empirical study with a sample of 168 US customers reveal that customer's perceptions of robot's human-likeness increase the intentions to use humanoid service robots. Interestingly, customers' risk aversion moderates this relationship. Specifically, the study found that highly risk-averse customers tend to avoid using humanoids when they are perceived as highly mechanical-like. The discussion highlights the main contributions of the research, which combine previous knowledge on human–robot interaction and risk aversion from a marketing approach. Managerial implications derived from the research findings and the avenues opened for further research are described at the end.

42.1 Introduction

42.1.1 The Rise of Robots

Humanoid service robots are replacing human service providers in many sectors [1]. Among other benefits, automation increases productivity and reduces labor costs [2]. Indeed, robot prices are decreasing continually at the rate of 10% annually [3]. Consequently, in 2017 and 2018, the average annual growth of sales was 20% for industrial robots, and 58.5% for professional services robots [4]. The theory of artificial intelligence job replacement [5] predicts that the robot infusion will be particularly intense in

Á. Rocha et al. (eds.), *Marketing and Smart Technologies*, Smart Innovation, Systems and Technologies 167, https://doi.org/10.1007/978-981-15-1564-4_42

the short term for jobs involving mechanical and analytical task, and would reach a broader scope by affecting intuitive and even empathetic skills in the next decades [5].

According to the Spanish Association of Hotel Managers [6], 96% of hotel receptionist will be replaced by autonomous devices by 2029, and 42% of food and drinks delivery in hotels will be performed by robots in 2023. Humanoid robots are already carrying out basic tasks such as providing help in banks or transporting drugs in hospitals [3]. As well, more than 10,000 humanoid robots such as "Pepper" are already performing waiter tasks in many restaurants around the world [1].

Consequently, within the field of frontline service technology infusion, the introduction of humanoid automated social agents is becoming particularly a matter of interest [3, 7]. Indeed, the emergence of humanoid robots is becoming one of the most dramatic evolution in the service realm [1].

However, research on this disruptive innovation is still scarce. Despite the emerging body of knowledge specifically focusing on service robots [3, 5, 7], previous research only present theoretical predictions about how service robot innovation may affect customers' experiences and organizations' service management, but empirical evidence to support these insights is still very exceptional [1].

To shed some light on this underexplored emerging field of research, we investigate how customers' perceptions toward humanoid robots may affect their behavioral intentions to use such service innovation. In particular, the following previous research on human–robot interaction [8], we hypothesize that customer's perceptions of human-likeness may contribute to increase in humanoid use intention. We also propose that customer's personal characteristics in terms of risk aversion may play a relevant role in deciding whether to interact with a humanoid service robot. We predict that risk aversion has a direct negative influence on use intention. In addition, our research hypothesizes a moderate effect of risk aversion [9]; that is the level of robot human-likeness may be relevant depending on user's risk aversion or risk seeking.

We test the proposed hypotheses by means of an empirical study presenting a human waiter scenario to a well-balanced sample of US customer in terms of age and gender. This research contributes to previous literature by extending previous knowledge on human–robot interaction and risk aversion to the marketing field. Specifically, our approach empirically assesses individuals' reactions toward robots from a marketing focus; that is, investigating how customers' perceptions and personal features may determine a successful introduction of humanoid robots in the service domain. The results and findings of the study are discussed together with the managerial implications derived from the research. The limitations and further research avenues are also commented at the end in order to inspire scholars to contribute to advances in this emerging field.

42.2 Hypotheses Development

Like in human–human interaction, social perceptions of robots depend to a large extent on its appearance [10]. The Computers Are Social Actors (CASA) paradigm proposes that human–computer interaction follows the same rules of human–human interaction [11], especially when machines have anthropomorphic cues that individuals recognize as social rules. The anthropomorphization of technical devices is a common phenomenon considered a "default schema" to be applied to non-social objects [12]. Using robots' physical appearance to infer human cues is a heuristic attribution difficult to avoid [10]. Thus, humanoid entities are perceived as having human identity to some extent.

In this line, the uncanny valley theory [13] proposes that people's affinity toward robots enhances as a consequence of human-likeness, that is, of robot similarity with human features; however, robots with very developed human features (i.e. droids with skin, hair, teeth) may be perceived as unpleasant, eerie or uncanny. Nevertheless, empirical research carried out to prove the uncanny valley existence (i.e. perceptions of eeriness) has been inconclusive or against support [8]. Indeed, the uncanny valley effect is widely debated and considered controversial [10].

From a complementary approach, studies in other research fields also found that human decision making related to technology is influenced by the technology amount of anthropomorphism [12]; a finding that may be applied to humanoid service robots. Indeed, anthropomorphism has been already found to be the most relevant factor for robot acceptance by consumers in hotels [14]. Some evidence support that people systematically prefer robotic employees whose human-likeness matched the sociability required to carry out those jobs [15], suggesting that humans judge humanoids favorably in terms of appearance and human similarity [8].

Therefore, our first hypothesis proposes:

H1: Customer's perception of robot's human-likeness increases customer's intention to use a humanoid service robot.

Risk aversion is a key variable in marketing and many other individual's economic decisions (e.g. finance, new product purchase) [16]. Hofstede and Bond ([17], p. 419), define risk aversion as "the extent to which people feel threatened by ambiguous situations, and have created beliefs and institutions that try to avoid these". Risk aversion belongs to the self (either individual or group) and has been described as a cultural value, a personality trait or a consumer decision-making style [18].

In general, it is assumed that humans are risk averse but this aversion may vary between people [19]. Risk aversion has been found critical in decisions involving uncertainty and it is closely linked to the concept of uncertainty avoidance [20]. Consumers with low risk aversion feel less threatened by uncertainty, whereas those with high risk aversion feel more threatened [19].

In comparison to products, risk aversion is more salient for services due to its greater intangibility [21]. There is a long tradition suggesting that risk-averse individuals are particularly reluctant to rely on technology-driven services [22]. In this regard, previous research found that risk-averse consumers avoid engaging in online shopping [23] or online banking [22], and are hesitant about using energy-saving systems [24].

As past research has not explicitly analyzed the role of risk aversion on robot use, assuming that humanoid introduction represents a disruptive technology innovation, our second hypothesis is proposed:

H2: Customer's risk aversion reduces the intention to use a humanoid service robot.

Previous literature suggests that service managers should carefully deal with customer's risk perceptions. In particular, customers are often risk averse toward technology-driven services offered by a company, which should offer employees-based alternatives to those customers particularly fearful to technology [25]. In this regard, previous research indicates that risk-averse individuals avoid adopting self-managed technological systems in banking service encounters, but prefer personal treatment [26].

As a personal feature of customers, risk aversion is often proposed as a moderator variable in decision making [9]. Considering that humanoid robot systems may be perceived as more human-like or more mechanical-like, and based on the previous argumentations, we propose that risk aversion moderates the influence of human-likeness on customer's intention to use the humanoid service robot [8]. Specifically, we propose that as far as human-likeness tend to arise customer's perceptions of humanness and affinity [8, 13], high risk-averse customers would avoid using less human-like robots, whereas low risk-averse customers (i.e. risk seekers) would be more willing to use less human-like robots. Consequently, the last hypothesis is proposed:

H3: Customer's risk aversion moderates the influence of human-likeness perceptions on intention to use a humanoid service robot.

42.3 Method

42.3.1 Procedure

A study was designed to test the proposed influence of human-likeness and risk aversion on intentions to use services provided by a humanoid. Participants in the study were 168 US customers recruited by means of an online market research company in exchange for 1$ per participant. This procedure allowed us to obtain a balanced sample in terms of age (6.6% aged 18–24, 35.7% 25–34, 29.7% 35–44, 17.9% 45–54, 10.1% more than 54 years old) and gender (53.6% men, 46.4% women).

The study was presented as a research about innovation in services. The scenario presented the robot Casey (fictitious name specifically chosen to avoid gender or brand biases [27]) as a new humanoid waiter recently introduced in some restaurants and with equivalent performance skills to regular human waiters. Figure 42.1 depicts the photographs that were presented together with the scenario description.

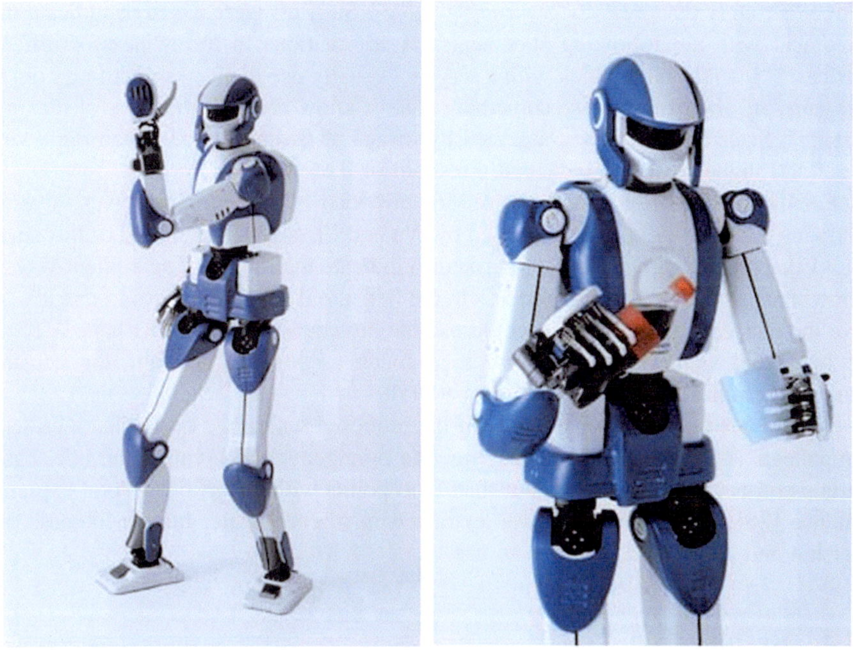

Fig. 42.1. Photographs included in the scenario description about the humanoid waiter

The humanoid robot was actually HRP-4, which is a prototype robot developed by the National Institute of Advanced Industrial Science and Technology in Tokyo. This prototype was selected among five humanoids (including also RMC, Atlas, Robothesphian, HRP-4C) because of scoring an intermediate level of human-likeness in a pretest previously conducted with a different sample of 91 participants.

After reading the scenario description and the pictures, participants of the main study were invited to answer to some questions related to the variables of our study. That is, they had to indicate their perceptions of the humanoid level of human-likeness, their intention to use this humanoid provided service, their level or risk aversion and some socio-demographic information.

42.3.2 Measures

Human-likeness was measured by two items obtained from Rosenthal-von der Püthen and Kramer [8]. Because of being a two-item scale, reliability was analyzed by means of the Spearman correlation (0.86). More precisely, customers assessed to what extent the humanoid was human-like and mechanical-like (reversed question) in a seven-point scale ranging from 1 "not at all" to 5 "very much".

Risk aversion was measured by three items borrowed from Bao et al. [18]. Specifically, participants responded to the following seven-point Likert scale ranging from 1 “strongly disagree” to 7 “strongly agree”. Subjects were asked to indicate their agreement with the following statements: “I am cautious in trying new or different services”, “I would rather stick with a service I usually use than try something I am not very sure of” and “I never use something I don’t know about at the risks of making a mistake”. Scale reliability was assessed by means of Cronbach’s α, obtaining a value ($\alpha = 0.87$) higher than the indicated threshold of 0.65 [28].

Finally, the intention to use the service provided by the humanoid was measured following the three-item scale proposed by Wu and Chen [29]. By means of this seven-point Likert scale, participants were asked to indicate their level of agreement with the following statements: “If I have access to the humanoid waiter, I intend to use it”, “If I have the chance, I predict that I would use the humanoid waiter”, “If I have access to the humanoid waiter, I want to use it as much as possible”. Again, the construct presented a high level of reliability ($\alpha = 0.95$).

To evaluate the unidimensionality of the proposed scales, we conducted a principal components analysis, with a factor extraction on eigenvalues greater than one. Factor loadings on each scale were greater than 0.50 points with a significant total explained variance [28]. Only one factor was extracted from each scale: human-likeness, risk aversion and behavioral intention to use.

42.4 Results

42.4.1 Test of the Hypotheses

A regression model was used to test the proposed hypotheses. In every case, the variables introduced were calculated as the average of their respective measures. To evaluate the moderating effect, the interaction term of human-likeness and risk aversion was calculated (before conducting this analysis, constructs were mean-centered to avoid multicollinearity in the presence of interaction terms [30]).

Results indicate that perceptions of human-likeness increases the customer intention to use the humanoid service robot ($\beta = 0.56$, $p < 0.01$), in support of the direct effect suggested in H1. As proposed in H2, customer’s level of risk aversion reduces the intention to use the service robot, but this direct effect is only marginally significant ($\beta = -0.16$, $p < 0.10$). Interestingly, as hypothesized, risk aversion moderates the effect of human-likeness on intention to use ($\beta = 0.27$, $p < 0.01$), in support of the interaction effect proposed in H3.

Finally, the proposed influence of the two independent variables, as well as the interaction term, allows us to explain to a great extent the variance of the dependent variable. In particular, the intention to use explained variance ($R^2 = 0.47$) reached higher levels than those usually reported in behavioral intention literature [31].

42.4.2 Post Hoc Analysis

In order to better understand the interaction effect of human-likeness and risk aversion on customer’s intention to use humanoid service robots, we conducted an additional post hoc ANOVA analysis.

Following the standardized procedures to compare between groups [32], participants were split up taking the average of the independent variable plus or minus one half of the standard deviation. Despite the reduction of the sample size due to subsequent dichotomizations, this method allowed us to compare between users with higher and lower levels of risk aversion and higher and lower levels of human-likeness perceptions. Results of the ANOVA analysis reinforce the previous findings. As hypothesized, human-likeness exerts a positive influence on customers' intention to use the service robot ($F = 30.86$, $p < 0.01$), in support of H1. Nevertheless, the effect of risk aversion proposed in H2 turns non-significant when comparing high versus low risk-averse customers ($F = 2.37$, $p > 0.10$). Again, as proposed in our moderating hypothesis, there is a significant interaction effect between human-likeness and risk aversion on humanoid use intention ($F = 4.99$, $p < 0.01$), in support of H3.

Figure 42.2 presents the level of intention to use the humanoid robot depending on the levels of human-likeness and risk aversion. More precisely, the post hoc analysis reveals that customers level of risk aversion is not relevant when facing a robot perceived as highly human-like ($t = 0.47$, $p > 0.10$). However, risk aversion is significantly influencing customers' intention to use the humanoid when it is perceived as lowly human-like ($t = -2.82$, $p < 0.01$). This means that risk-averse customers will be more reluctant to use mechanical-like humanoids, whereas risk seekers will be more prone to use a humanoid service robot, even when they perceive the robot as highly mechanical-like.

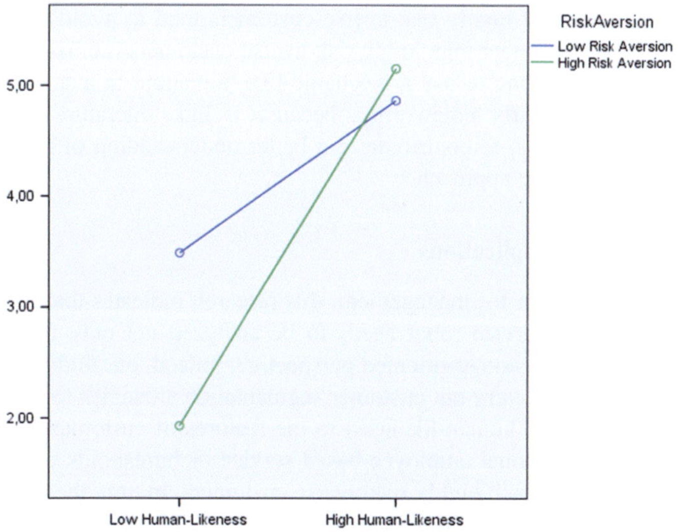

Fig. 42.2. Interaction effect of perceptions of robot human-likeness and customer's risk aversion on the intention to use the humanoid service robot

42.5 Discussion

42.5.1 Conclusions

Robotics is a technology innovation that increases productivity and profitability in manufacture and also in service provision. Thus, scholars, practitioners, international organizations and the overall society identifies this phenomenon as having a relevant impact on economy and employment; and as an emerging field in frontline operations affecting customer choices that are crucial for the future of service encounters [1, 5, 7].

Assuming the emergence of humanoid service infusion in the frontline [1], our research analyzes to what extent perceptions of human-likeness and customer's risk aversion may determine a successful introduction of such a disruptive technology.

The results of our empirical study found that robot's human-likeness increases customer's intention to use the humanoid agent. This result agrees with previous findings in literature about human–robot interaction [8], and contributes to evaluate this influence from a marketing approach and in a service context. Our research also found that risk-averse customers tend to be more reluctant to use humanoid service robots; however, this effect was only marginally significant, suggesting that the risk aversion may not play a direct influence but a moderating one.

The test of the moderation effect, further extended in a post hoc analysis, revealed an interaction effect between human-likeness and risk aversion. In particular, risk aversion is found relevant when robots are less human-like (i.e., more mechanical-like), whereas it is found irrelevant when dealing with highly human-like humanoid robots. Our findings suggest that highly risk-averse customers tend to avoid mechanical-like humanoids to a greater extent than lowly risk-averse customers (i.e. risk seekers), who will be in turn more willing to use a mechanic-like humanoid in a service encounter. This finding is particularly noteworthy, because it links literature on both robot adoption and risk aversion to contribute to a better understanding of humanoid introduction from a marketing approach.

42.5.2 Managerial Implications

As a principal implication for management, this research indicates that the decision to introduce a humanoid service robot needs to be analyzed not only from an internal perspective but from a customer-oriented perspective. Indeed, our findings suggest that service managers should carry out customer segmentation strategies to better adapt the features of the robot (i.e. human-likeness) to the features of customers (i.e. risk aversion). Offering the traditional employee-based service or humanoids with human features would help to approach highly risk-averse customers. In turn, the identification of lowly risk-averse customers could help to address them as a group of pioneers for using prototypes when launching humanoid service robots for the first time in the company.

42.5.3 Limitations and Further Research

The limitations of this work open avenues for further research. We rely on a single service scenario of a robot waiter. To confirm the generalizability of our findings, the empirical study should be replicated in different contexts and using different kinds of humanoid robots. The study should also be replicated in other countries, as far as cultural values may greatly impact the level of uncertainty avoidance (and risk aversion). Our research focuses on humanoid service robots because they are the most commonly used in service provision; however, some companies are designing droids (very human-like robots like Sophia, with human features such as skin, face, etc.). Further research is thus needed to explore whether the findings of this research could be applied not only to humanoids but also to droids that could be potentially employed in service encounters.

References

1. Mende, M., Scott, M.L., van Doorn, J., Grewal, D., Shanks, I.: Service robots rising: how humanoid robots influence service experiences and elicit compensatory consumer responses. J. Mark. Res. **56**(4), 1–22 (2019)
2. Negahban, A., Smith, J.S.: Simulation for manufacturing system design and operation: literature review and analysis. J. Manuf. Syst. **33**, 241–261 (2014)
3. Wirtz, J., Patterson, P.G., Kunz, W.H., Gruber, T., Lu, V.N., Paluch, S., Martins, A.: Brave new world: service robots in the frontline. J. Serv. Manag. **29**(5), 907–931 (2018)
4. International Federation of Robotics, Robot density rises globally, https://ifr.org/ifr-press-releases/news/robot-density-rises-globally (2018). Accessed 03 June 2019
5. Huang, M.H., Rust, R.T.: Artificial intelligence in service. J. Serv. Res. **21**(2), 155–172 (2018)
6. Asociación Española de Directores de Hoteles, Annual Report, https://www.abc.es/espana/comunidad-valenciana/abci-como-seran-hoteles-futuro-sin-recepcionista-robots-camarero-o-cocinero-y-comida-sana-201906112007_noticia.html, https://www.elmundo.es/viajes/el-baul/2019/07/27/5d3ac4defdddff83558b45b7.html. Accessed 03 June 2019
7. Van Doorn, J., Mende, M., Noble, S.M., Hulland, J., Ostrom, A.L., Grewal, D., Petersen, J. A.: Domo arigato Mr. Roboto: Emergence of automated social presence in organizational frontlines and customers' service experiences. J. Serv. Res. **20**(1), 43–58 (2017)
8. Rosenthal-von der Püthen, A.M., Kramer, N.C.: How design characteristics of robots determine evaluation and uncanny valley related responses. Computers in Human Behavior 36, 422–439 (2014)
9. Allen, D.G., Weeks, K.P., Moffitt, K.R.: Turnover intentions and voluntary turnover: the moderating roles of self-monitoring, locus of control, proactive personality, and risk aversion. J. Appl. Psychol. **90**(5), 980–996 (2005)
10. Walters, M.L., Syrdal, D.S., Dautenhahn, K., Te Boekhorst, R., Koay, K.L.: Avoiding the uncanny valley: robot appearance, personality and consistency of behavior in an attention-seeking home scenario for a robot companion. Auton. Robots **24**(2), 159–178 (2008)
11. Nass, C., Moon, Y., Fogg, B.J., Reeves, B., Dryer, D.C.: Can computer personalities be human personalities? Int. J. Hum Comput Stud. **43**, 223–239 (1995)
12. Luczak, H., Roetting, M., Schmidt, L.: Let's talk: anthropomorphization as means to cope with stress of interacting with technical devices. Ergonomics **46**(13–14), 1361–1374 (2003)
13. Mori, M.: The uncanny valley. Energy **7**(4), 33–35 (1970)
14. Tussyadiah, I.P., Park, S.: Consumer evaluation of hotel service robots. In: Stangl, B., Pesonen, J. (eds.) Information and Communication Technologies in Tourism 2018, pp. 308–320. Springer (2018)

15. Gray, K., Wegner, D.M.: Feeling robots and human zombies: mind perception and the uncanny valley. Cognition **125**(1), 125–130 (2012)
16. Pratt, J.W.: Risk aversion in the small and in the large. In: Uncertainty in Economics, pp. 59–79. Academic Press (1978)
17. Hofstede, G., Bond, M.H.: Hofstede's culture dimensions: an independent validation using Rokeach's value survey. J. Cross Cult. Psychol. **15**, 417–433 (1984)
18. Bao, Y., Zhou, K.Z., Su, C.: Face consciousness and risk aversion: do they affect consumer decision-making? Psychol. Market. **20**(8), 733–755 (2003)
19. Sun, J.: How risky are services? An empirical investigation on the antecedents and consequences of perceived risk for hotel service. Int. J. Hosp. Manag. **37**, 171–179 (2014)
20. Meroño-Cerdán, A.L., López-Nicolás, C., Molina-Castillo, F.J.: Risk aversion, innovation and performance in family firms. Econ. Innov. New Technol. **27**(2), 189–203 (2017)
21. Wu, C.H.J., Liao, H.C., Hung, K.P., Ho, Y.H.: Service guarantees in the hotel industry: their effects on consumer risk and service quality perceptions. Int. J. Hosp. Manag. **31**(3), 757–763 (2012)
22. Bauer, K., Hein, S.E.: The effect of heterogeneous risk on the early adoption of Internet banking technologies. J. Bank. Finance **30**(6), 1713–1725 (2006)
23. Tan, S.J.: Strategies for reducing consumers' risk aversion in Internet shopping. J. Consum. Market. **16**(2), 163–180 (1999)
24. Schleich, J., Gassmann, X., Meissner, T., Faure, C.: A large-scale test of the effects of time discounting, risk aversion, loss aversion, and present bias on household adoption of energy-efficient technologies. Energy Econ. **80**, 377–393 (2019)
25. Dedeke, A.: Service quality: a fulfilment-oriented and interactions-centered approach. Manag. Serv. Qual. Int. J. **13**(4), 276–289 (2003)
26. De Jong, A., De Ruyter, K., Lemmink, J.: The adoption of information technology by self-managing service teams. J. Serv. Res. **6**(2), 162–179 (2003)
27. Cauberghe, V., De Pelsmacker, P.: The impact of banners on digital television: the role of program interactivity and product involvement. CyberPsychol. Behav. **11**(1), 91–94 (2008)
28. Steenkamp, J.B., Geyskens, I.: How country characteristics affect the perceived value of a Website. J. Market. **70**(3), 136–150 (2006)
29. Wu, L., Chen, J.L.: An extension of trust and TAM model with TPB in the initial adoption of on-line tax: an empirical study. Int. J. Hum Comput Stud. **62**(6), 784–808 (2005)
30. Jaccard, J., Turrisi, R.: Interaction effects in multiple regression, 72. Sage (2003)
31. Venkatesh, V., Morris, M.G., Davis, G.B., Davis, F.D.: User acceptance of information technology: toward a unified view. MIS Q. **27**(3), 425–478 (2003)
32. Jaccard, J., Wan, C.: LISREL approaches to interaction effects in multiple regressions. Sage Publications, Thousand Oaks, CA (1996)

MIX
Papier aus verantwortungsvollen Quellen
Paper from responsible sources
FSC® C105338

If you have any concerns about our products,
you can contact us on
ProductSafety@springernature.com

In case Publisher is established outside the EU,
the EU authorized representative is:
Springer Nature Customer Service Center GmbH
Europaplatz 3, 69115 Heidelberg, Germany

Printed by Libri Plureos GmbH
in Hamburg, Germany